VOLUME ONE HUNDRED AND FIFTY

PROGRESS IN MOLECULAR BIOLOGY AND TRANSLATIONAL SCIENCE

Prion Protein

VOLUME ONE HUNDRED AND FIFTY

PROGRESS IN MOLECULAR BIOLOGY AND TRANSLATIONAL SCIENCE

Prion Protein

Edited by

GIUSEPPE LEGNAME

Scuola Internazionale Superiore di Studi Avanzati (SISSA), Trieste, Italy

SILVIA VANNI

Scuola Internazionale Superiore di Studi Avanzati (SISSA), Trieste, Italy

ELSEVIER

ACADEMIC PRESS

An imprint of Elsevier

Academic Press is an imprint of Elsevier
50 Hampshire Street, 5th Floor, Cambridge, MA 02139, United States
525 B Street, Suite 1800, San Diego, CA 92101-4495, United States
The Boulevard, Langford Lane, Kidlington, Oxford OX5 1GB, United Kingdom
125 London Wall, London EC2Y 5AS, United Kingdom

First edition 2017

ISBN: 978-0-12-811226-7
ISSN: 1877-1173

For information on all Academic Press publications
visit our website at https://www.elsevier.com/books-and-journals

Publisher: Zoe Kruze
Acquisition Editor: Ashlie M. Jackman
Editorial Project Manager: Helene Kabes
Production Project Manager: James Selvam
Cover Designer: Miles Hitchen

Typeset by SPi Global, India

CONTENTS

CONTRIBUTORS

James D. Alibhai
The National CJD Research and Surveillance Unit, Centre for Clinical Brain Sciences, University of Edinburgh, Edinburgh, United Kingdom

Olivier Andréoletti
UMR INRA-ENVT 1225, Interactions Hôte Agent Pathogène, Ecole Nationale Vétérinaire de Toulouse, Toulouse, France

Ivana Biljan
Faculty of Science, University of Zagreb, Zagreb, Croatia

Edoardo Bistaffa
IRCCS Foundation Carlo Besta Neurological Institute, Milan; Scuola Internazionale Superiore di Studi Avanzati (SISSA), Trieste, Italy

Maria Laura Bolognesi
Alma Mater Studiorum-University of Bologna, Bologna, Italy

Matilde Bongianni
University of Verona, Verona, Italy

Elena Bozzetta
Italian Reference Centre for Animal Encephalopathies, Istituto Zooprofilattico Sperimentale del Piemonte, Liguria e Valle d'Aosta, Turin, Italy

Maria Caramelli
Italian Reference Centre for Animal Encephalopathies, Istituto Zooprofilattico Sperimentale del Piemonte, Liguria e Valle d'Aosta, Turin, Italy

Paolo Carloni
Institute of Neuroscience and Medicine (INM-9) and Institute for Advanced Simulation (IAS-5), Jülich; RTWH-Aachen, Aachen, Germany

George A. Carlson
McLaughlin Research Institute, Great Falls, MT, United States

Cristina Casalone
Italian Reference Centre for Animal Encephalopathies, Istituto Zooprofilattico Sperimentale del Piemonte, Liguria e Valle d'Aosta, Turin, Italy

Byron Caughey
Laboratory of Persistent Viral Diseases, Rocky Mountain Laboratories, National Institute for Allergy and Infectious Diseases, National Institutes of Health, Hamilton, MT, United States

Cristiano Corona
Italian Reference Centre for Animal Encephalopathies, Istituto Zooprofilattico Sperimentale del Piemonte, Liguria e Valle d'Aosta, Turin, Italy

Elena Vallino Costassa
Italian Reference Centre for Animal Encephalopathies, Istituto Zooprofilattico Sperimentale del Piemonte, Liguria e Valle d'Aosta, Turin, Italy

Chiara M.G. De Luca
IRCCS Foundation Carlo Besta Neurological Institute, Milan; Università degli Studi di Pavia, Pavia, Italy

Rosanna Desiato
Italian Reference Centre for Animal Encephalopathies, Istituto Zooprofilattico Sperimentale del Piemonte, Liguria e Valle d'Aosta, Turin, Italy

Abigail B. Diack
The Roslin Institute & R(D)SVS, University of Edinburgh, Easter Bush, United Kingdom

Juan C. Espinosa
Centro de Investigación en Sanidad Animal, CISA-INIA, Madrid, Spain

Eric G.B. Evans
University of Washington, Seattle, WA, United States

Natalia Fernández-Borges
Centro de Investigación en Sanidad Animal, CISA-INIA, Madrid, Spain

Michele Fiorini
University of Verona, Verona, Italy

Annachiara Gandini
Alma Mater Studiorum-University of Bologna, Bologna; Scuola Internazionale Superiore di Studi Avanzati (SISSA), Trieste, Italy

Bradley R. Groveman
Laboratory of Persistent Viral Diseases, Rocky Mountain Laboratories, National Institute for Allergy and Infectious Diseases, National Institutes of Health, Hamilton, MT, United States

Théo Z. Hirsch
Université Paris Descartes, Sorbonne Paris Cité; INSERM UMR 1124, Paris, France

Andrew G. Hughson
Laboratory of Persistent Viral Diseases, Rocky Mountain Laboratories, National Institute for Allergy and Infectious Diseases, National Institutes of Health, Hamilton, MT, United States

Gregor Ilc
Slovenian NMR Centre, National Institute of Chemistry; EN-FIST Centre of Excellence, Ljubljana, Slovenia

James W. Ironside
National CJD Research & Surveillance Unit, Centre for Clinical Brain Sciences, School of Clinical Sciences, University of Edinburgh, Western General Hospital, Edinburgh, United Kingdom

Barbara Iulini
Italian Reference Centre for Animal Encephalopathies, Istituto Zooprofilattico Sperimentale del Piemonte, Liguria e Valle d'Aosta, Turin, Italy

Richard Knight
National CJD Research & Surveillance Unit, Centre for Clinical Brain Sciences, University of Edinburgh, Edinburgh, United Kingdom

Allison Kraus
Laboratory of Persistent Viral Diseases, Rocky Mountain Laboratories, National Institute for Allergy and Infectious Diseases, National Institutes of Health, Hamilton, MT, United States

Giuseppe Legname
Scuola Internazionale Superiore di Studi Avanzati (SISSA), Trieste, Italy

Neil A. Mabbott
The Roslin Institute & Royal (Dick) School of Veterinary Sciences, University of Edinburgh, Midlothian, United Kingdom

Matteo Manca
Laboratory of Persistent Viral Diseases, Rocky Mountain Laboratories, National Institute for Allergy and Infectious Diseases, National Institutes of Health, Hamilton, MT, United States

Jean C. Manson
The Roslin Institute & R(D)SVS, University of Edinburgh, Easter Bush, United Kingdom

Alba Marín-Moreno
Centro de Investigación en Sanidad Animal, CISA-INIA, Madrid, Spain

Séverine Martin-Lannerée
Université Paris Descartes, Sorbonne Paris Cité; INSERM UMR 1124, Paris, France

Candace K. Mathiason
Colorado State University, Fort Collins, CO, United States

Maria Mazza
Italian Reference Centre for Animal Encephalopathies, Istituto Zooprofilattico Sperimentale del Piemonte, Liguria e Valle d'Aosta, Turin, Italy

Glenn L. Millhauser
University of California, Santa Cruz, CA, United States

Fabio Moda
IRCCS Foundation Carlo Besta Neurological Institute, Milan, Italy

Salvatore Monaco
University of Verona, Verona, Italy

Sophie Mouillet-Richard
Université Paris Descartes, Sorbonne Paris Cité; INSERM UMR 1124, Paris, France

Christina D. Orru
Laboratory of Persistent Viral Diseases, Rocky Mountain Laboratories, National Institute for Allergy and Infectious Diseases, National Institutes of Health, Hamilton, MT, United States

Anna Pepe
University of Naples "Federico II", Naples, Italy; Unité de Trafic Membranaire et Pathogenese, Institut Pasteur, Paris, France

Janez Plavec
Slovenian NMR Centre, National Institute of Chemistry; EN-FIST Centre of Excellence; Faculty of Chemistry and Chemical Technology, University of Ljubljana, Ljubljana, Slovenia

Brent Race
Laboratory of Persistent Viral Diseases, Rocky Mountain Laboratories, National Institute for Allergy and Infectious Diseases, National Institutes of Health, Hamilton, MT, United States

Gregory J. Raymond
Laboratory of Persistent Viral Diseases, Rocky Mountain Laboratories, National Institute for Allergy and Infectious Diseases, National Institutes of Health, Hamilton, MT, United States

Lynne D. Raymond
Laboratory of Persistent Viral Diseases, Rocky Mountain Laboratories, National Institute for Allergy and Infectious Diseases, National Institutes of Health, Hamilton, MT, United States

Jesús R. Requena
CIMUS Biomedical Research Institute, University of Santiago de Compostela-IDIS, Santiago de Compostela, Spain

Diane L. Ritchie
National CJD Research & Surveillance Unit, Centre for Clinical Brain Sciences, School of Clinical Sciences, University of Edinburgh, Western General Hospital, Edinburgh, United Kingdom

Giulia Rossetti
Institute of Neuroscience and Medicine (INM-9) and Institute for Advanced Simulation (IAS-5); Jülich Supercomputing Centre, Jülich; University Hospital Aachen, RWTH-Aachen, Aachen, Germany

Martina Rossi
Scuola Internazionale Superiore di Studi Avanzati (SISSA), Trieste, Italy

Giuseppe Ru
Italian Reference Centre for Animal Encephalopathies, Istituto Zooprofilattico Sperimentale del Piemonte, Liguria e Valle d'Aosta, Turin, Italy

Eri Saijo
Laboratory of Persistent Viral Diseases, Rocky Mountain Laboratories, National Institute for Allergy and Infectious Diseases, National Institutes of Health, Hamilton, MT, United States

Daniela Sarnataro
University of Naples "Federico II"; Ceinge-Biotecnologie avanzate, s.c.a r.l., Naples, Italy

Juan M. Torres
Centro de Investigación en Sanidad Animal, CISA-INIA, Madrid, Spain

Silvia Vanni
Scuola Internazionale Superiore di Studi Avanzati (SISSA), Trieste, Italy

Holger Wille
Centre for Prions and Protein Folding Diseases, University of Alberta, Edmonton, Alberta, Canada

Gianluigi Zanusso
University of Verona, Verona, Italy

Chiara Zurzolo
University of Naples "Federico II", Naples, Italy; Unité de Trafic Membranaire et Pathogenese, Institut Pasteur, Paris, France

PREFACE

The idea of this book came to us following the requests of many students who lamented the lack of a textbook for preparing their exam on the molecular aspects of neuropathology and neurodegeneration of prion diseases. Over the years, several books have been compiled on the subject, but very few met our expectation on covering every single aspect of this complex subject. To date over 17,000 citations are present in PubMed when entering the key words "prion" or "prions." For students this number is insurmountable. Thus, we decided to embark in this journey, to call on world experts to provide insights in the different facets of this theme.

The book comprises different chapters covering key aspects of prion biology and disease.

The contributions begin with a thorough review of what it is known about the physiological functions of the prion protein. Sophie Mouillet-Richard and colleagues examine the various biological processes in which the prion protein is involved. This is an important overview on a much-debated issue on what is the precise function of the protein.

Glenn Millhauser and colleagues author the second chapter dealing with the interaction of the prion protein with metals and its relevance in neurotoxicity and function. Prion protein metal interaction has been much of the focus of recent research aimed at understanding its involvement in physiology and pathology.

Chiara Zurzolo and coauthors cover relevant cell biological aspects of the protein. Maturation and trafficking of the prion protein are dealt in great details. Most importantly, these processes are of relevance also for prion conversion and replication.

Next, Janez Plavec and collaborators provide an up-to-date survey of many different molecular structures of wild-type prion protein and several pathogenic mutant variants. By reviewing their own work as well as the one of others, they provide a detailed atomistic evidence of the structural changes in the overall folding of the protein in physiology and disease.

From experiments to simulations, the chapter of Paolo Carloni and Giulia Rossetti presents a survey of most pathogenic mutants of the human prion protein using molecular dynamics. The striking significance of this approach is well highlighted in the manuscript.

The next block of chapters deals with the different characteristics of prion disease. It starts with a chapter authored by George Carlson on the genetic susceptibility to prion disease. It is a most welcome recount on the genes, and in particular the prion protein gene, that influence the duration of the disease.

The following chapter is coauthored by one of us and Fabio Moda on the fundamental prion concept and synthetic prions. Validation of the prion concept through the development of synthetic prions has been a milestone in the field of prion biology and disease.

Jean Manson and colleagues describe all the mouse models employed in prion research. This is a timely account on transgenic models developed so far.

Juan Maria Torres and colleagues report the complexity of prion transmission and replication in different hosts harboring diverse prion proteins. The so-called transmission barrier due to the different sequences of the prion protein in different species is illustrated in much depth.

The immunological aspects in which the prion protein is involved are discussed in the chapter compiled by Neil Mabbott. This section is of particular relevance if considering the implication for acquired prion disorders and the role of immunity in modulating the disease.

Cristiano Corona and colleagues have contributed the chapter that deals with the various forms of bovine spongiform encephalopathy (BSE). Besides the epidemic of classical BSE that affected the cattle population in Europe and worldwide, this chapter deals with the discovery of the amyloidogenic variant of BSE by the same group.

An emerging concern is posed by chronic wasting disease (CWD), a disorder that affects the cervids population in North America. Candace Mathiason, a renowned world expert in CWD, discusses the past, present, and future of the disease together with other two animal pathologies such as scrapie and transmissible mink encephalopathy.

After dealing with prion disorders in the animal realm, we then find two chapters that discuss in great details about these devastating diseases in humans.

Richard Knight has contributed a chapter on infectious and sporadic prion diseases, while James Ironside and Diane Ritchie discuss in detail the neuropathological aspects of human prion diseases.

Holger Wille and Jesús Requena provide an update of what we know about the structure of infectious prions. This subject is of paramount importance for understanding the molecular details of prion strain diversity. Only an atomistic knowledge of infectious prions can offer new tools/resources/ means for researchers for understanding prions.

In recent years, new technical tools have been developed for an early diagnosis of these disorders. Fabio Moda gives an overview of one of these emerging techniques: protein misfolding cyclic amplification, commonly known as PMCA. In another chapter, Byron Caughey and colleagues discuss the fascinating implication of yet another useful technique denoted as RT-QuIC. The advent of these two assays promises unprecedented advancement in terms of developing robust diagnostic tests, and identifying novel therapeutics and disinfection strategies.

Gianluigi Zanusso and collaborators present a description of complete biochemical signatures of the various forms of prion diseases.

In the next chapter, one of us discusses in detail the diverse *omics* approaches that have been exploited for better understanding the disease in animals and humans. From GWAS to transcriptomics, to genomics and proteomics, all these methodologies can lead to unbiased identification of pathways involved in prion diseases.

The book closes with two important contributions. Maria Laura Bolognesi gives an overview of the different attempts reported in the scientific literature on the development of effective therapeutic strategies for these still incurable disorders.

Fabio Moda closes the book with the important/relevant subject of biosafety of prions. When dealing with these pathogens, researchers are confronted with disposal and containment of prions. The chapter offers a detailed list of effective methodologies for limiting the biological contamination of laboratories and workers alike.

Giuseppe Legname and Silvia Vanni
Trieste, July 13, 2017

CHAPTER ONE

Functions of the Prion Protein

Théo Z. Hirsch*[,†], Séverine Martin-Lannerée*[,†], Sophie Mouillet-Richard*[,†,1]

*INSERM UMR 1124, Paris, France
†Université Paris Descartes, Sorbonne Paris Cité, UMR 1124, Paris, France
[1]Corresponding author: e-mail address: sophie.mouillet-richard@parisdescartes.fr

Contents

Progress in Molecular Biology and Translational Science, Volume 150
ISSN 1877-1173
http://dx.doi.org/10.1016/bs.pmbts.2017.06.001

Abstract

Although initially disregarded compared to prion pathogenesis, the functions exerted by the cellular prion protein PrP^C have gained much interest over the past two decades. Research aiming at unraveling PrP^C functions started to intensify when it became appreciated that it would give clues as to how it is subverted in the context of prion infection and, more recently, in the context of Alzheimer's disease. It must now be admitted that PrP^C is implicated in an incredible variety of biological processes, including neuronal homeostasis, stem cell fate, protection against stress, or cell adhesion. It appears that these diverse roles can all be fulfilled through the involvement of PrP^C in cell signaling events. Our aim here is to provide an overview of our current understanding of PrP^C functions from the animal to the molecular scale and to highlight some of the remaining gaps that should be addressed in future research.

1. INTRODUCTION

Historically, the cellular prion protein PrP^C was discovered as the normal host counterpart of the pathogenic scrapie prion protein (PrP^{Sc}), itself the main component of the infectious agent responsible for transmissible spongiform encephalopathies.[1] In the early years of prion research, unraveling PrP^C normal function seemed not as much a priority as grasping the properties of PrP^{Sc} and understanding TSE-associated pathogenesis. This view was reversed when it came to light that unraveling PrP^C function would illuminate how it is corrupted by pathogenic prions to promote neurodegeneration. The first generation of PrP-knockout mice in the early 1990s [2] set the stage for studying PrP^C physiological role. Although somehow disappointing at first glance since these mice develop with no overt phenotype, they have actually allowed bringing to light subtle defects and remain a major tool to dissect PrP^C function. At the molecular scale, much insight has been gained through the identification of PrP^C-interacting molecules, as well as PrP^C-dependent signaling events. In addition, in more recent years, our view of PrP^C has evolved from a protein majorly involved in neuronal function to a broad player in cellular homeostasis, in various tissues and at various stages of life. Another equally important concept that has emerged is that PrP^C exerts a broad protective role against various types of stresses, which has notably been exemplified upon challenging PrP-null animals. In this chapter, we will address these different points with the aim to provide a comprehensive view of our current knowledge of PrP^C functions.

2. THE BASICS OF PrP^C

2.1 PrP^C From Gene to Function

PrP^C is encoded by a single gene denoted *Prnp*, which is located on chromosome 20 in man and 2 in mouse. It possesses two paralogues, *Prnd* and *Sprn*, encoding the Doppel and Shadoo proteins, respectively, whose properties have been reviewed recently[3] and will not be discussed further in this chapter. According to species, the *Prnp* gene contains 2 or 3 exons, the entire coding region being contained in the last exon, excluding the possibility of alternative splicing.[4] The resulting native protein of 253 amino acids (human numbering) is processed to generate a mature form of 208 residues lacking its N-terminal-most signal peptide and modified in its C-terminus with the attachment of a glycosyl-phosphatidyl (GPI) moiety allowing for its anchoring to the extracellular plasma membrane.[5] Two topological transmembrane variants with either their N-terminus (CtmPrP) or C-terminus (NtmPrP) portions in the cytosol have been described[6] and may account for the interaction of PrP^C with several cytosolic partners. In terms of secondary/tertiary structure, it can be globally divided into an N-terminally intrinsically disordered domain—often referred to as "flexible tail"; a central hydrophobic domain; and a C-terminal globular domain (Fig. 1). The latter region contains two asparagine residues that can be N-glycosylated, flanked by two cysteine residues forming a disulfide bridge. In addition to the full-length isoforms, PrP^C can undergo proteolytic processing at various positions.[7] The best characterized and most conserved cleavage site lies in the central domain (alpha-cleavage site) and leads to the production of the N1 soluble

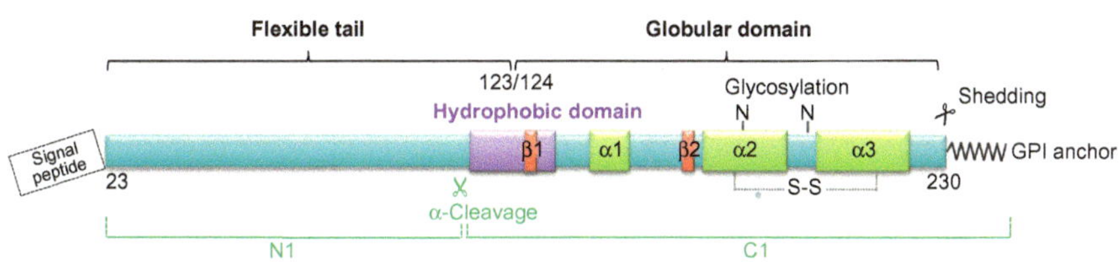

Fig. 1 Secondary structure of mouse PrP^C emphasizing the N-terminal flexible tail and the C-terminal globular domain. N1 and C1 fragments are generated following alpha-cleavage. N-glycosylation sites and cysteine disulfide bridge formation are indicated. Beta-sheets and alpha-helices are highlighted in *red* and *green*, respectively, while the hydrophobic domain is highlighted in *purple*.

fragment and the GPI-anchored C1 fragment (Fig. 1). Finally, PrP^C may exist under soluble full-length isoforms, resulting either from a phospholipase-dependent GPI cleavage or from proteolytic cleavage at residues 228/229.[7,8] Thus, the various posttranslational modifications that can impact on PrP^C give rise to a plethora of isoforms, which add to the complexity when studying PrP^C function.

From an evolutionary perspective, it is noteworthy that the structural features of PrP^C are conserved in vertebrates,[9] most notably in mammals,[10] underscoring their functional importance. Interestingly, bioinformatics analyses suggest that the *Prnp* gene family may originate from a common ancestor of the ZIP family of metal ion transporter.[11] In this scheme, the metal binding capacity of PrP^C (see below) would represent one of its ancestral functions.

2.2 PrP^C Expression and Localization

Delineating the functions of PrP^C necessitates taking into account when and where it is expressed. During embryonic development, *Prnp* mRNA expression was reported to start around embryonic day 6.5 (E6.5) in extraembryonic tissues,[12] followed by the cardiac mesoderm at E7.5[13] and the developing CNS at E8.[14] Around E9, PrP^C is detected in the developing nervous and cardiovascular systems as well as in extraembryonic regions,[15] in line with its role in stem cell self-renewal (see below). The observation that PrP^C expression in extraembryonic annexes is maintained at early differentiation stages fits in with an important—albeit little explored—role for this protein in the physiology of the placenta (reviewed in Ref. 16) and in angiogenesis.[17,18]

During neurodevelopment, PrP^C remains abundant in sites of active neurogenesis such as the ventricular zone, while it is also highly expressed in regions where maturing neurons migrate such as the hippocampus and the thalamic and hypothalamic regions, consistent with its involvement in neuronal differentiation.[19] Analysis of PrP^C expression during forebrain development in human also indicates that synapses exhibit increasing amounts of PrP^C throughout synaptogenesis.[20] In the adult, while PrP^C still is enriched in the brain not only in neurons but also in glia and endothelial cells of the blood–brain barrier,[21] it is also widely expressed in all tissues.

At the subcellular scale, most PrP^C is found at the membrane, and is notably enriched in lipid rafts, or at the synapse in neurons.[22,23] This plasma membrane location led to the assumption that it may act as a cell surface receptor or coreceptor and instruct signal transduction events. PrP^C-dependent signaling now represents one of the most fertile areas of research related to PrP^C function and will be addressed in detail in Section 8.

Beyond the membrane, a less studied spatial location is the cytosol, where PrP^C seems to accumulate in conditions of stress.[24] Other locations include the base of the primary cilium,[15] the nucleus,[25] or extracellular vesicles such as exosomes.[26] At present, the mechanisms sustaining the trafficking of PrP^C to the cilia or its shuttling to the nucleus remain enigmatic.

3. LESSONS FROM KNOCKOUT MODELS

In the early 1990s, several groups reported on the generation of PrP-null mice. Beyond a drastic neurodegenerating phenotype due to the accidental upregulation of the PrP paralogue Doppel, whose gene is located downstream from *Prnp*,[27] and a mis-attributed regulation of phagocytosis due to confounding effects of the flanking gene *Sirpa*,[28] these various models have revealed that PrP^C deficiency is associated with a variety of neurological defects. These include alterations in memory and locomotor activity,[29] circadian rythms,[30,31] olfaction,[32] pain sensitivity,[31,33] and a depressive-like behavior.[34] Interestingly, some defects appear to worsen with age,[35] suggesting that the lack of PrP^C may be more detrimental at defined periods of life.[36] *Prnp*-null mice have also been instrumental in revealing that PrP^C is necessary for the long-term maintenance of hematopoietic stem cells (HSCs)[37] and for the proliferation of neural progenitors,[38] opening the question of the role exerted by this protein in stem cell biology (see below and Ref. 39). Another set of observations has exemplified an increased vulnerability of PrP-null mice to various types of stresses: ischemia,[40] behavioral stresses,[41] or inflammatory conditions (reviewed in Ref. 42). Perhaps the best understood phenotype at present is the demyelinating disease not only observed in first-generation PrP-knockout mice[43] but also confirmed in *Prnp*-ablated mice generated through genome editing.[44] Indeed, molecular studies have shown that the PrP^C N-terminal acts as an agonistic ligand of the Adgrg6 receptor, whose activation is necessary for myelin maintenance.[45] Finally, in contrast to the mild phenotypes observed in mice invalidated for PrP^C, a drastic impact of the knockdown of the fish *Prnp* orthologue *PrP-1* was reported in zebrafish, with a lethal phenotype due to cell adhesion and gastrulation defects.[46] Interestingly, the phenotype of *PrP-1* morphants could be rescued using mouse PrP^C, highlighting an evolutionary conserved function.[46] Readers are referred to the recent review by

Halliez *et al.* for a comprehensive view of the consequences of PrP^C invalidation.[47]

4. MISCELLANEOUS ROLES IN CNS FUNCTION

4.1 Cognitive and Behavioral Functions

Because TSEs are neurodegenerative diseases and because PrP^C is most abundantly expressed in the brain, PrP^C function has been the most studied in the context of neurons. The notion that pathogenic prions exert their toxicity by deviating the normal function of PrP^C in neurons[48] has further boosted research on the physiological role exerted by PrP^C in these cells. As mentioned above, synaptic deficits were among the first defects to be described in PrP-null mice,[49] although initially questioned.[50] Further studies have consistently reported on alterations of synaptic activity in the absence of PrP^C,[51,52] including in the context where PrP^C is ablated at a postnatal stage.[53] Incidentally, the notion that PrP^C may contribute to memory is also indirectly supported by the identification of SNPs in the human *PRNP* gene associated with enhanced long-term memory[54] and the demonstration that *PRNP* is a target gene of FOXP2, a key transcription factor involved in language development.[55] The regulation of *Prnp* gene expression by the E4BP4 transcription factor,[56] which is involved in the regulation of the circadian clock,[57] may on another hand accommodate the link between PrP^C and circadian rhythms[30,31] and contribute to the induction of PrP^C mRNAs in response to melatonin.[58]

4.2 Neuritogenesis and Neuronal Differentiation

From a mechanistic point of view, the contribution of PrP^C to synaptic function may operate at different levels. A first axis relates to neuronal differentiation and brain wiring. The contribution of PrP^C to neuronal differentiation has been mostly investigated from a cell architecture point of view, although the implementation of neuronal polarity and that of neurotransmitter-associated functions are intimately connected[59] and are both altered in the context of prion infection.[60] Several studies indeed have exemplified a participation of PrP^C to neurite elongation and synapse development.[61–66] This positive action of PrP^C on neuritogenesis may depend on the functional interaction of PrP^C with various partners: the neural cell adhesion molecule NCAM,[61,66] laminin,[63,67] STI1,[68,69] integrins,[65,70,71] vitronectin,[70] or itself[61] and the recruitment of signal transduction cascades, as will be described below. On another hand, PrP^C was shown to potentiate the inhibitory effect of the contactin-associated protein (Caspr) on neurite outgrowth in a Reelin-dependent manner.[72] In mature neurons, PrP^C may further contribute to the structural maintenance of the synapse by regulating cytoskeletal components such as cofilin or local protein translation, two processes that are impaired under prion infection.[60,73]

4.3 Balancing Neuroprotection and Neurotoxicity

Given the beneficial functions exerted by PrP^C in neuronal physiology (see also Section 6 for the protective role of PrP^C against stress), it came as a surprise that antibody-mediated cross-linking of PrP^C caused extensive neuronal apoptosis in vivo,[74] hinting at a proapoptotic role of this protein. Several observations obtained in mice expressing mutant forms of PrP^C have additionally brought insight into the dichotomous role of PrP^C as a trigger of prosurvival or proapoptotic cascades in neurons. For instance, mice expressing an N-terminally truncated form of PrP^C (Δ32–121 or Δ32–134) were shown to suffer from ataxia and cerebellar degeneration,[75] pointing to a protective role of the N-terminal flexible tail of PrP^C. Deletions of the central hydrophobic region (Δ94–134 or Δ105–125) also caused drastic spontaneous cerebellar disorders.[76,77] Deleting the central domain in PrP^C notably obliterates the alpha-cleavage site and precludes the generation of the soluble N1 fragment, raising the possibility that these severe phenotypes are caused—at least partly—by the lack of N1.

More recently, several elegant studies lead by the Aguzzi's group and exploiting a panel of antibodies targeting various epitopes of PrP^C have shed further light on the respective roles of the N-terminal and C-terminal regions of PrP^C. One outstanding conclusion of these works is that antibodies targeting the globular domain exert a neurotoxic effect in mice or cerebellar organotypic slices and seem to mimic the effect of prion infection,[78] while those directed against the flexible tail are innocuous.[79] Of note, the latter protect against the deleterious effects of antibodies to the globular domain,[78] as well as against the toxicity of PrP^C mutants bearing a deletion of the central domain (PrP Δ94–134).[79] These observations hence corroborate the notion that the flexible tail of PrP^C and its globular domain exert distinct action. Since the flexible tail of PrP^C was recently identified as an agonistic ligand of the Adgrg6 receptor to sustain myelin maintenance,[45] we may envision that it exerts broader agonistic effects on yet-to-be-identified receptor(s) to maintain brain homeostasis. Other members of the adhesion GPCR family beyond Adgrg6 itself clearly represent candidates for relaying the protective action of PrP^C N-terminal tail in view of their conspicuous roles in nervous system development and function.[80]

4.4 Beyond Neurons: PrP^C in Glia

It has recently emerged that neural circuits are also shaped by complex neuron–glia interactions.[81,82] In this scheme, the contribution of PrP^C to proper brain function may involve dialogues between neurons and other cell types, as described for the PrP^C-Adgrg6 duo at the interface between neurons and Schwann cells.[45] For instance, PrP^C from astrocytes was shown to promote neuritogenesis and neuronal survival.[83] More generally, PrP^C from astrocytes appears to play a protective role under stress conditions, both on astrocytes themselves[84] and on neighboring neurons.[85] In the latter study, PrP^C was shown to be transferred from astrocytes to neurons via exosomes.[85] Adding yet another layer of complexity to the picture, PrP^C appears to regulate the secretion of exosomes itself.[86] We may hence suspect that the absence of PrP^C may have broader impact on brain extracellular vesicles, which represents an important means of intercellular communication.[87] Other consequences of PrP deficiency in astrocytes include defects in lactate transport,[88] as well as altered excitatory amino acid transporter activity.[89,90]

4.5 The Metal Connection

The involvement of PrP^C in neuronal functions may further arise from its ability to bind divalent cations in its N-terminal region. Following the first report that PrP^C binds copper in vivo,[91] several studies have suggested that PrP^C may be involved in the regulation of copper concentration at the synapse (see Ref. 92 for review). Interestingly, copper appears to influence some PrP^C-mediated cell signaling events.[93,94] More recently, Gasperini *et al.* found that PrP^C cooperates with copper to induce S-nitrosylation of the NMDA receptor.[95] Beyond copper, PrP^C also binds zinc ions and facilitates their uptake in neuronal cells via AMPA receptors.[96] Altogether, several reports converge onto a protective role against glutamate or

AMPA/kainate-induced excitotoxicity.[95,97–99] It is noteworthy that this protective role with respect to neuron excitability is conserved in fish.[100]

4.6 From Neurotransmitter Receptors to Neurotransmitter-Associated Functions

The above studies further point to cross talks between PrP^C and neurotransmitter receptors. These functional interactions, which have been reviewed recently,[101] extend beyond NMDA and AMPA receptors and also include serotonergic[94,102,103] or cholinergic[104,105] receptors. An interaction between PrP^C and the purinergic P2X4R has also been recently described.[106] A related emerging theme is the broader implication of PrP^C in the fine-tuning of neurotransmitter-associated functions. Indeed, at the animal scale, the absence of PrP^C appears to induce defects in the dopaminergic,[107,108] cholinergic,[109] or serotonergic[31,107] systems. As for the latter, PrP^C contributes to the regulation of monoamine synthesis,[110] catabolism,[111] or release[94] through various cell signaling-dependent mechanisms. Interestingly, a modulatory role of PrP^C on synaptic release was recently exemplified at the Drosophila neuromuscular junction,[112] further adding support to a potentiating role of PrP^C in synaptic transmission[113] (for review).

5. CELL FATE AND DIFFERENTIATION

5.1 Stem Cell Self-renewal and Expansion

Interest for a potential contribution of PrP^C to the biology of stem cells arose with the discovery that it is abundantly expressed in HSCs, including the most immature $Lin^-Sca^+Endoglin^+$ subpopulation, and that it is necessary for the long-term maintenance of these cells.[37] PrP^C was subsequently shown to be highly expressed in neural,[38] cardiac,[13] mammary,[114] skeletal muscle,[115] or mesenchymal[116] progenitors. Thus, high PrP^C expression appears as a generic feature of various types of stem cells. In line with this, PrP^C was shown to be upregulated following the induction of multipotential embryonic stem (ES) or carcinoma (EC) cells toward a more restricted neural[117,118] or cardiac[13] fate, or during their spontaneous differentiation.[119,120] Furthermore, expressing PrP^C in ES cells grown under self-renewal conditions was reported to promote their induction toward lineages of the three germ layers.[121] Altogether, these observations posit PrP^C as an important player in stem cell fate determination and raise the question as to the cellular processes to which it may participate in a stem cell context. One recurring theme is its contribution to proliferation. For instance, PrP^C levels correlate with the proliferation rate of neural progenitors in two neurogenic regions in adult mice: the subventricular zone and the dentate gyrus.[38] Ex vivo, neurospheres from fetal[122] or adult[123] brain further proliferate less efficiently when derived from *Prnp*-knockout vs wild-type mice. These defects may notably be explained by a lack of response to the PrP^C ligand STI1.[122] The contribution of PrP^C to stem cell proliferation also holds true for mesenchymal progenitors.[116] Interestingly, PrP^C expression was recently shown to be induced after hypoxia preconditioning and to enhance the survival and proliferation of transplanted mesenchymal stem cells (MSCs).[124] That PrP^C may strengthen tissue repair was also exemplified in the context of HSC transplantation[37] or muscle regeneration.[115] Another notable example is the participation of PrP^C to recovery from ischemia,[18,40,125,126] possibly after mobilization by its ligand STI1.[127] Finally, because PrP^C remains highly expressed in adult neurogenic zones, it is likely to take part in brain repair following stress or injury, a process that appears to be defective under prion infection.[128]

5.2 A Hub in Developmental Signaling Pathways

At a mechanistic level, we reported on altered cell cycling and reduced expression of cyclin D1 after knockdown of PrP^C in a neuronal progenitor cell line.[15] These changes may notably be accounted for by defects in the Hedgehog pathway that we identified in vitro and further confirmed in vivo within a narrow time window during embryonic development.[15] Another important clue to the PrP^C-Hedgehog connection is the location of PrP^C at the base of the primary cilium, where Hedgehog signaling operates,[129] highlighting a potential role for PrP^C in primary cilium turnover.[15] It is of note that PrP^C was found in the proteome of the primary cilium in three independent studies,[130–132] calling for a better knowledge of the function exerted by PrP^C in this specific organelle. Beyond the Hedgehog pathway, we recently added another piece to the PrP-stem cell puzzle by uncovering a connection between PrP^C and the Notch pathway.[133] We found that the absence of PrP^C in neural progenitors is associated with a marked reduction in the expression of various effectors of the Notch pathway, which plays an overriding role in neural stem cell maintenance.[134] This correlates with subtle differences in the developing nervous system of *Prnp*-knockout mice, including transient decreases in the expression of the neural stem cell marker Nestin, and altered distribution of N-cadherin, which is important for the maintenance of neural progenitors in the ventricular zone.[135] The latter observation also fits in with the proposed role of PrP^C in cell adhesion, notably during embryonic development (see Section 7). Our study further allowed delineating a gene regulatory network involving PrP^C, Notch, and the EGF receptor (EGFR), where PrP^C positively controls the expression of EGFR via Notch while EGFR exerts a negative feedback on both PrP^C and Notch.[133] Adding another layer to this network, we documented a positive feedback regulation of the Notch pathway on PrP^C. Incidentally, the Notch target gene Hes1 was reported to bind to the *Prnp* promoter[136] and thus represent a good candidate for relaying this process. Finally, because PrP^C, Notch, and EGFR all regulate HSC self-renewal in the bone marrow niche,[137,138] we may speculate that this network also operates in HSCs.

5.3 Choice of Cell Fate and Differentiation

A corollary to the reduced self-renewal of neural progenitors in the absence of PrP^C would be a precocious engagement into a more mature state. This notion is supported by the increase in doublecortin-positive neuroblasts in *Prnp*-null mice.[123] The absence of PrP^C further promotes an increase in $NG2^+Olig2^+$ oligodendrocyte precursor cells, both during embryonic development and in the adult brain.[139] In line with the notion that Notch signaling negatively regulates Olig2 expression during brain patterning,[140] we proposed that Notch pathway alterations may account for the increased Olig2 expression in the neural tube of PrP-null embryos.[133] Curiously, while PrP^C is necessary for the maintenance of neural progenitor cells in an immature state, it is on another hand required for the early steps of neurite elongation[65] and subsequent neuronal differentiation (see Section 4.2). One explanation for this apparent discrepancy would be the multiplicity of PrP^C isoforms, likely endowed with specific—or even opposite—functions. We may also suspect the involvement of defined partners, which would engage preferentially with a subset of PrP^C isoforms. Although much speculative at present, such a view is nevertheless supported by the reported competitive action of the two PrP^C ligands STI1 and Aβ for the recruitment of protective and toxic pathways, respectively.[141] Apart from neurons, high levels of PrP^C were shown to enhance the development of astrocytes.[142] Finally, PrP^C is differentially regulated according to cell fate along the hematopoietic lineage as well (reviewed in Refs. 39,42). For instance, PrP^C expression is decreased along erythroid differentiation.[143] In this context, it is noteworthy that mutant goats naturally devoid of PrP^C exhibit elevated red blood cell number.[144]

6. A BROAD PROTECTIVE ROLE AGAINST STRESS

6.1 PrP^C Is Induced Under Stress

Dissecting the regulation of a given gene can teach us about the function of the corresponding protein. In the case of PrP^C, several analyses of the *Prnp* promoter have notably revealed binding sites for various transcription factors, including Sp1 and AP-1,[145] p53,[146] the heat-shock factor HSF1,[147] sXBP1,[148] or FOXO3A.[149] Interestingly, Qin *et al.*[150] reported on the induction of PrP^C expression by Sp1 and p53 following the activation of the ATM kinase under oxidative stress conditions. In line with this, the PrP^C protein content was shown to correlate with altered redox states in prostate spheroids.[151] The Sp1 transcription factor, which is positively regulated by the hypoxia-inducible factor HIFα, was also proposed to account for the induction of PrP^C under hypoxic conditions.[152] The upregulation of PrP^C in response to hypoxia has been further corroborated by diverse studies.[18,153] The regulation of the *Prnp* gene by sXBP1 was on another hand shown to promote PrP^C expression under ER stress conditions.[148,154] As for the control of *Prnp* gene expression by FOXO3A, it may notably have relevance as to contribution of PrP^C to the long-term maintenance of HSCs since FOXO3A is essential for this process.[155] A FOXO3A-dependent induction of PrP^C could also possibly account for the protection afforded by FOXO3A in HSCs against oxidative DNA damage.[156] In addition to these findings, an interesting observation is the upregulation of PrP^C in neutrophils after systemic exposure to LPS, which was shown to involve TGFβ and glucocorticoids.[157] The latter study hints at a potential role of PrP^C in the response to neuroimmunoendocrine stress. In agreement, PrP^C was recently shown to mitigate the inflammatory response to LPS.[158]

6.2 Protection Against Oxidative Stress

That PrP^C protects against oxidative stress is a recurring theme in PrP^C function and has been addressed in several reviews.[9,92,159,160] A global notion is that PrP^C may take part in antioxidant defense either by directly buffering ROS,[161] undergoing processing within its N-terminal region (so-called beta-cleavage),[162,163] or by promoting the induction of antioxidant enzyme activities and increasing glutathione levels.[89,124,164] These actions are directly connected to the copper-binding capacity of PrP^C and depend on its N-terminal domain.[160] The protective action of PrP^C against oxidative stress is not restricted to neurons, but was also reported in various cell contexts including astrocytes,[84] MSCs,[124] or the placenta.[17] This function underlies, at least partly, the beneficial role exerted by PrP^C under hypoxic conditions.[18,124]

6.3 Protection Against Genotoxic Stress

The beneficial action of PrP^C against oxidative stress extends to the protection against reactive oxygen species-induced DNA damage[165,166] and also involves the N-terminal domain of the protein.[166] Recently, Bravard *et al.* brought further insight into the relationship between PrP^C and the DNA damage response by demonstrating that *PRNP* gene transcription is upregulated in neurons after exposure to a genotoxic stress.[167] The resulting increased PrP^C was shown to accumulate in the nucleus, where it interacts with the endonuclease APE1 and stimulates its base excision repair activity. The involvement of PrP^C in protection against genotoxic stress is also supported by its positive regulation by the ATM kinase,[150] a major guardian of genome integrity.[168]

6.4 Protection Against Apoptosis

The protective role exerted by PrP^C against other types of deleterious inputs has been extensively exemplified in various settings (see Refs. 159,169 for review). The prosurvival action of PrP^C has notably been documented in the context of serum deprivation,[170] caspase activation by staurosporine,[171–173] protein synthesis inhibition,[174] calcium overload,[172] ischemia,[18,40,125,126] or excitotoxicity.[95,97–99] From a mechanistic point of view, the cytoprotective role of PrP^C is imparted by its N-terminal domain[99,125,171,175] (see also Section 4.3). It may relate to its mobilization by STI1[68,127,176,177] and the downstream activation of PI3K/Akt[126,175] or PKA.[68] An alternative mode of action relies on the shedding of the soluble N1 fragment and the downregulation of p53.[171]

6.5 An Immune Privilege Molecule?

These well-established cytoprotective actions of PrP^C together with its high and regulated expression in the immune system led to the proposal that it may contribute to immune quiescence.[42] Immune quiescence, also termed immune privilege, encompasses a combination of anatomical, physical, and immunoregulatory processes that restrain immune-mediated inflammation in sites where it may cause irreversible damage.[178] The PrP^C-immune privilege connection is supported by the abundant expression of PrP^C in tissues with immune privilege, such as the brain and the placenta, together with a set of observations highlighting that the absence of PrP^C aggravates pathology in various experimental inflammatory settings.[42] The concept that PrP^C may be considered as an immune privilege molecule provides a new angle for investigating PrP^C function and will certainly constitute an interesting area for future research.

6.6 Involvement in Proteostasis

Proteostasis refers to the global network allowing for cellular proteins to maintain a proper biological activity according to the cell environment.[179] A strong indicator that PrP^C may contribute to the proteostatic network is the regulation of *Prnp* gene transcription by various factors of this network,[179] such as HSF1,[147] sXBP1,[148] or FOXO3A.[149] In addition, in terms of protein translation, one interesting observation is that in the course of TSEs, while activation of the PERK pathway halts the translation of most proteins, including synaptic constituents, *Prnp* transcripts escape this negative regulation.[73] From a functional point of view, PrP^C may take part in the proteostatic network at several levels: by promoting protein synthesis via mTOR,[180] activating autophagy by recruiting Beclin1[181] or mediating the action of the cholinergic alpha7nAchR,[105] organizing the interplay between exosome secretion and autophagosome formation,[86] or protecting against ER stress.[182] Finally, the location of PrP^C at the base of the primary cilium and its potential impact on ciliary turnover[15] may also have relevance as to this issue in view of the emerging relationship between cellular proteostasis and ciliogenesis.[183]

7. CELL ADHESION AND ECM CONTACTS

7.1 Multiple Adhesion Partners

The involvement of PrP^C in cell adhesion was first supported by the demonstration that it interacts with several proteins relevant to this process such as the ECM molecules laminin[63,184] or vitronectin,[70] or several transmembrane proteins such as the laminin receptor precursor[185] or NCAM.[186]

PrP^C has further been shown to interact with β1 integrins[71,187] and to modulate their activity.[65,188] This interplay with integrins may endow PrP^C with the capacity to regulate the turnover of focal adhesions.[65,189] The contribution of PrP^C to cell adhesion processes may further involve homophilic interactions, as suggested for instance in the case of brain endothelial cells.[21]

7.2 Special Links With Cadherins

Cadherins represent another class of adhesion molecules sustaining cell–cell contacts whose functional interactions with PrP^C have received much interest. In intestinal cells, which form a highly polarized epithelium, PrP^C was found to colocalize with E-cadherin at cell–cell junctions.[190] While E-cadherin did not coprecipitate with PrP^C in that study, it is noteworthy that PrP^C was however found in the interactome of E-cadherin.[191] Subsequent reports suggested that, according to cell density, PrP^C may shuttle between the membrane, where it interacts with desmosomal proteins, and the nucleus, where it regulates the transcriptional activity of β-catenin.[192,193] These observations are consistent with a role for PrP^C in the balance between proliferation and differentiation of intestinal epithelial cells. They are further in line with the reported alteration in the intestinal epithelial barrier in PrP-null mice and the decreased levels of PrP^C in the intestine of patients with inflammatory bowel diseases.[194] Besides, in the context of embryonic development, the seminal study by Malaga-Trillo *et al.* demonstrated that the lethal phenotype obtained upon knockdown of *PrP-1* in zebrafish correlated with impaired delivery of E-cadherin at the cell surface.[46] PrP^C actually appears to broadly affect the cycling of E-cadherin since it also negatively controls its endocytosis via src kinases.[195] A second member of the cadherin family, namely, N-cadherin, may also be influenced by PrP^C, in terms of trafficking to the neuronal cell surface[196] or distribution in the developing neuroepithelium.[133]

7.3 Modulating the ECM Composition

Finally, PrP^C may impact on the composition of the ECM itself, as described in astrocytes[83] or neuronal progenitor cells,[65] or on its remodeling, for instance by controlling the activity of the metalloprotease MMP9.[197] Another context where we may suspect PrP^C to act on the ECM composition is in dental tissue in view of the defects in dentin structure reported in *Prnp*-knockout mice.[198]

8. FROM PARTNERS TO CELL SIGNALING

8.1 Cell Signaling: The Unifying Function of PrP^C?

The involvement of PrP^C in such a broad range of biological processes raises the question as to how these heterogeneous roles can be unified. A consensual view is to consider PrP^C from a cell communication point of view: its interaction with a given set of partners within a given cell environment will mobilize signal transduction cascades. In this paradigm, PrP^C will serve as a "hub," connecting an input to cell adaptation. This function may operate according to different modes of action: in some instances PrP^C (or its cleavage products) would serve as a ligand, as described for the activation for Adgrg6,[45] while in others it may act as a cell surface receptor or coreceptor, as reported in the case of the Aβ-PrP^C-mGluR5 complex.[199] Finally, PrP^C may also indirectly regulate the activity of various signaling molecules such as β1 integrins,[65] E-cadherin,[46] or EGFR[200] by impinging on their trafficking.

8.2 A Great Variety of Partners

A plethora of PrP^C partners have been identified through various methods (see Refs. 9,201 for review), including using proteomics-based strategies.[187,202,203] Among those, a subset whose interaction with PrP^C was shown to have functional relevance is listed in Table 1. Two of the most studied PrP^C-interacting molecules until now are the extracellular chaperone STI1 and the amyloid peptide Aβ, involved in Alzheimer's disease. STI1 was initially identified as a PrP^C ligand through complementary hydropathy.[222] Subsequent studies have highlighted an involvement of the PrP^C-STI1 duo in multiple biological processes including neural stem cell self-renewal,[122] neuronal differentiation,[68,69] neuronal survival,[68,177] neuronal protein synthesis,[180] memory consolidation,[204] potentiation of neurotransmitter release,[94] or protection against ischemia.[127,176] As mentioned above, STI1 may further protect against the toxicity of the PrP^C alternative ligand Aβ.[141] Aβ has entered the PrP^C field more recently, with the seminal report by Lauren *et al.* that PrP^C serves as a receptor for Aβ-derived diffusible ligands and partially relays their toxicity.[205] The Aβ-PrP^C interplay now represents an area of intense research and has been the topic of several recent reviews.[223–225] Of note, some studies anterior to the Lauren paper had identified the Aβ precursor protein APP or its paralogues APLP1 and APLP2 as

Table 1 PrP^C Ligands and Their Related Biological Processes

Protein Name	**Biological Outcome**
Extracellular ligands	
STI1	Neural stem cell self-renewal[122] Neuronal differentiation[68,69] Neuronal survival[68,177] Neuronal protein synthesis[180] Memory consolidation[204] Potentiation of neurotransmitter release[94] Protection against ischemia[127,176]
Aβ	Neurotoxicity[205] NMDA turnover[206] Tau phosphorylation[207]
ECM proteins	
Laminin	Neurite outgrowth[63,67,69] Memory[208]
Vitronectin	Neurite outgrowth[70]
Scaffold proteins	
Caveolin-1	Activation of src kinases downstream from PrP^C[71,94,209] Balance between exosomes and autophagosomes[86]
Flotillin	Trafficking of N-cadherin[196] Trafficking of EGFR[200]
Cell adhesion receptors	
APP	Cell adhesion and CNS development[210]
Integrin β1	Neuritogenesis[65,70,71]
37 kDa/67 kDa Laminin receptor	Cell survival[211]
NCAM	Neurite outgrowth[66] Neuronal differentiation[123]
Growth factor receptors	
EGFR	Cell proliferation[212]
Focal adhesion components	
Desmosomal proteins	Maintenance of epithelial polarity[193]

Continued

Table 1 PrP^C Ligands and Their Related Biological Processes—cont'd

Protein Name	Biological Outcome
Neurotransmitter receptors	
mGluR1	Neurite outgrowth[67]
mGluR2	Astrocytic lactate transport[88]
mGluR5	Neurite outgrowth[67] Long-term depression[213] Aβ-dependent neurotoxicity[199]
mGluR6/7	Protection against excitotoxicity[214]
GluA1, GluA2	Neuronal zinc transport[96]
GluN2D	Protection against excitotoxicity[215]
α7nAchR	Regulation of autophagy[105] Calcium signaling[104]
Miscellaneous cell surface proteins	
LRP1	Endocytosis[216] Potentiation of neurotransmitter release[94]
Adgrg6	Peripheral myelin maintenance[45]
Na^+/K^+-ATPase	Astrocytic lactate transport[88]
Glypican-1	Raft association[217]
TNAP	Laminin dephosphorylation[110]
BACE1	APP cleavage[218]
Miscellaneous intracellular proteins	
Members of the Rab family of small GTPases	PrP^C trafficking[219]
Argonaute	MicroRNA processing[220]
Lactate dehydrogenase A (LDHA)	Activation of LDHA[221]
β-Catenin	Transcriptional activity of the β-catenin/TCF7L2 complex[192]

PrP^C partners.[203,226] From a functional point of view, the interaction of APP with PrP^C was shown to be conserved from zebrafish to human and to contribute to cell adhesion and CNS development.[210] Another central PrP^C partner that deserves special attention is the scaffold protein caveolin-1.

Caveolin-1 not only serves as a relay for the activation of src kinases downstream from PrP^C[71,94,209] but is also involved in regulating the balance between exosomes and autophagosomes.[86] Caveolin may actually act as a bridging molecule between PrP^C and several of its binding partners: cell signaling effectors such as src kinases[71,94,209] or β-catenin,[192,227] integrins,[71] and cytoskeletal components such as actin.[193,228] The reader is referred to Table 1 for other relevant PrP^C-interacting molecules.

8.3 Overview of PrP^C-Dependent Signal Transduction

The involvement of PrP^C in cell signaling events has been addressed in several recent reviews, with special focus on neuronal[101,229] or immune[42] contexts. We further refer the reader to these reviews for a list of PrP^C downstream targets. Mechanistically, it is now generally accepted that the mobilization of signaling cascades by GPI-anchored PrP^C relies on the formation of multimolecular complexes in specific subdomains of the plasma membrane known as lipid rafts. The participation of raft constituents such as caveolin[71,94,209] or flotillin[230] has been documented in some studies and is suspected in some others, where raft integrity is an important parameter.[231,232] In this line, it is noteworthy that the removal of sialic acid within the GPI anchor interferes with the targeting of PrP^C to the synapse,[233] directs PrP^C to a subset of rafts distinct from those where it normally segregates, and hinders its activation of the cytosolic form of phospholipase A2.[234] Thus, compartmentalization represents an important parameter with respect to PrP^C signaling, which may provide spatial proximity or, to the opposite, insulation between PrP^C and its effectors. This notion is all the more important since prion infection may alter the spatial organization and composition of rafts.[235,236] Endocytosis represents a connected issue, with relevance to signal duration and formation of signaling endosomes.[237] Although little explored until now, PrP^C endocytosis was shown to be important for the recruitment of ERK1/2 in response to STI1.[238] The cell surface residency of PrP^C was further reported to be increased in cells exposed to Aβ,[239] while decreased in prion-infected cells.[240] Another interesting point is the notion that PrP^C-dependent signaling may necessitate the formation of PrP^C dimers[99,241] and in turn favor proteolytic processing.[93,242] Metal and ligand binding, internalization, and cleavage may actually be intertwined processes that clearly deserve further investigation.

Finally, despite tremendous progress in the PrP^C signaling field, several gray areas remain. One unresolved question relates to the subset of PrP^C

isoforms that specifically engages into a given signaling platform, according to the cell type or even according to the subcellular location, as in the case of neuronal cell bodies vs neurites.[94,209] Delineating the mechanisms imparting spatial and temporal control on PrP^C signaling also features among the priority axes for future investigations, since their deregulation is likely to contribute to the deleterious action of pathogenic prions[243] or other amyloid peptides such as Aβ (see Ref. 244 for review).

9. CONCLUDING REMARKS

Because it fulfills myriads of functions and at the same time appears to be dispensable for normal development and most physiologic processes, PrP^C is both a tricky and fascinating molecule. Recent years have witnessed important advance in our understanding of the cellular and molecular pathways to which PrP^C contributes and we may expect exciting new discoveries in the coming years. Achieving a comprehensive picture of PrP^C biological role goes well beyond curiosity as it may have implications for the development of drugs aimed at restoring PrP^C functions in TSEs and other pathological settings.

REFERENCES

1. Oesch B, Westaway D, Walchli M, et al. A cellular gene encodes scrapie PrP 27-30 protein. *Cell.* 1985;40(4):735–746.
2. Bueler H, Fischer M, Lang Y, et al. Normal development and behaviour of mice lacking the neuronal cell-surface PrP protein. *Nature.* 1992;356(6370):577–582.
3. Ciric D, Rezaei H. Biochemical insight into the prion protein family. *Front Cell Dev Biol.* 2015;3:5. http://dx.doi.org/10.3389/fcell.2015.00005.
4. Mastrangelo P, Westaway D. The prion gene complex encoding PrPC and Doppel: insights from mutational analysis. *Gene.* 2001;275(1):1–18. http://dx.doi.org/10.1016/S0378-1119(01)00627-8.
5. Stahl N, Baldwin MA, Hecker R, Pan KM, Burlingame AL, Prusiner SB. Glycosylinositol phospholipid anchors of the scrapie and cellular prion proteins contain sialic acid. *Biochemistry (Mosc).* 1992;31(21):5043–5053.
6. Harris DA. Trafficking, turnover and membrane topology of PrP. *Br Med Bull.* 2003;66:71–85.
7. McDonald AJ, Dibble JP, Evans EG, Millhauser GL. A new paradigm for enzymatic control of alpha-cleavage and beta-cleavage of the prion protein. *J Biol Chem.* 2014;289(2):803–813. http://dx.doi.org/10.1074/jbc.M113.502351.
8. Parkin ET, Watt NT, Turner AJ, Hooper NM. Dual mechanisms for shedding of the cellular prion protein. *J Biol Chem.* 2004;279(12):11170–11178. http://dx.doi.org/10.1074/jbc.M312105200.
9. Aguzzi A, Baumann F, Bremer J. The prion's elusive reason for being. *Annu Rev Neurosci.* 2008;31:439–477. http://dx.doi.org/10.1146/annurev.neuro.31.060407.125620.

10. van Rheede T, Smolenaars MMW, Madsen O, de Jong WW. Molecular evolution of the mammalian prion protein. *Mol Biol Evol*. 2003;20(1):111–121.
11. Schmitt-Ulms G, Ehsani S, Watts JC, Westaway D, Wille H. Evolutionary descent of prion genes from the ZIP family of metal ion transporters. *PLoS One*. 2009;4(9):e7208.
12. Manson J, West JD, Thomson V, McBride P, Kaufman MH, Hope J. The prion protein gene: a role in mouse embryogenesis? *Development*. 1992;115(1):117–122.
13. Hidaka K, Shirai M, Lee JK, et al. The cellular prion protein identifies bipotential cardiomyogenic progenitors. *Circ Res*. 2010;106(1):111–119. http://dx.doi.org/10.1161/CIRCRESAHA.109.209478.
14. Tremblay P, Bouzamondo-Bernstein E, Heinrich C, Prusiner SB, DeArmond SJ. Developmental expression of PrP in the post-implantation embryo. *Brain Res*. 2007;1139:60–67. http://dx.doi.org/10.1016/j.brainres.2006.12.055.
15. Halliez S, Martin-Lanneree S, Passet B, et al. Prion protein localizes at the ciliary base during neural and cardiovascular development and its depletion affects α-tubulin post-translational modifications. *Sci Rep*. 2015;5:17146.
16. Makzhami S, Passet B, Halliez S, et al. The prion protein family: a view from the placenta. *Front Cell Dev Biol*. 2014;2:35.
17. Alfaidy N, Chauvet S, Andrei S, et al. Prion protein expression and functional importance in developmental angiogenesis: role in oxidative stress and copper homeostasis. *Antioxid Redox Signal*. 2012;18:400–411. http://dx.doi.org/10.1089/ars.2012.4637.
18. Doeppner TR, Kaltwasser B, Schlechter J, et al. Cellular prion protein promotes post-ischemic neuronal survival, angioneurogenesis and enhances neural progenitor cell homing via proteasome inhibition. *Cell Death Dis*. 2015;6:e2024. http://dx.doi.org/10.1038/cddis.2015.365.
19. Benvegnù S, Poggiolini I, Legname G. Neurodevelopmental expression and localization of the cellular prion protein in the central nervous system of the mouse. *J Comp Neurol*. 2010;518(11):1879–1891. http://dx.doi.org/10.1002/cne.22357.
20. Adle-Biassette H, Verney C, Peoc'h K, et al. Immunohistochemical expression of prion protein (PrPC) in the human forebrain during development. *J Neuropathol Exp Neurol*. 2006;65(7):698–706. http://dx.doi.org/10.1097/01.jnen.0000228137.10531.72.
21. Viegas P, Chaverot N, Enslen H, Perrière N, Couraud P-O, Cazaubon S. Junctional expression of the prion protein PrPC by brain endothelial cells: a role in trans-endothelial migration of human monocytes. *J Cell Sci*. 2006;119(pt 22):4634–4643. http://dx.doi.org/10.1242/jcs.03222.
22. Collins MO, Husi H, Yu L, et al. Molecular characterization and comparison of the components and multiprotein complexes in the postsynaptic proteome. *J Neurochem*. 2006;97(suppl 1):16–23. http://dx.doi.org/10.1111/j.1471-4159.2005.03507.x.
23. Moya KL, Salès N, Hässig R, Créminon C, Grassi J, Di Giamberardino L. Immunolocalization of the cellular prion protein in normal brain. *Microsc Res Tech*. 2000;50(1):58–65. http://dx.doi.org/10.1002/1097-0029(20000701)50:1<58::AID-JEMT9>3.0.CO;2-5.
24. Dubnikov T, Ben-Gedalya T, Reiner R, et al. PrP-containing aggresomes are cytosolic components of an ER quality control mechanism. *J Cell Sci*. 2016;129(19):3635–3647. http://dx.doi.org/10.1242/jcs.186981.
25. Strom A, Wang G-S, Picketts DJ, Reimer R, Stuke AW, Scott FW. Cellular prion protein localizes to the nucleus of endocrine and neuronal cells and interacts with structural chromatin components. *Eur J Cell Biol*. 2011;90(5):414–419. http://dx.doi.org/10.1016/j.ejcb.2010.11.015.
26. Vella LJ, Sharples RA, Lawson VA, Masters CL, Cappai R, Hill AF. Packaging of prions into exosomes is associated with a novel pathway of PrP processing. *J Pathol*. 2007;211(5):582–590. http://dx.doi.org/10.1002/path.2145.

27. Moore RC, Lee IY, Silverman GL, et al. Ataxia in prion protein (PrP)-deficient mice is associated with upregulation of the novel PrP-like protein doppel. *J Mol Biol*. 1999;292(4):797–817.
28. Nuvolone M, Kana V, Hutter G, et al. SIRPα polymorphisms, but not the prion protein, control phagocytosis of apoptotic cells. *J Exp Med*. 2013;210(12):2539–2552. http://dx.doi.org/10.1084/jem.20131274.
29. Coitinho AS, Roesler R, Martins VR, Brentani RR, Izquierdo I. Cellular prion protein ablation impairs behavior as a function of age. *Neuroreport*. 2003;14(10):1375–1379. http://dx.doi.org/10.1097/01.wnr.0000078541.07662.90.
30. Tobler I, Gaus SE, Deboer T, et al. Altered circadian activity rhythms and sleep in mice devoid of prion protein. *Nature*. 1996;380(6575):639–642.
31. Vidal C, Herzog C, Haeberle AM, et al. Early dysfunction of central 5-HT system in a murine model of bovine spongiform encephalopathy. *Neuroscience*. 2009;160(4): 731–743.
32. Le Pichon CE, Valley MT, Polymenidou M, et al. Olfactory behavior and physiology are disrupted in prion protein knockout mice. *Nat Neurosci*. 2009;12(1):60–69. http://dx.doi.org/10.1038/nn.2238.
33. Gadotti VM, Zamponi GW. Cellular prion protein protects from inflammatory and neuropathic pain. *Mol Pain*. 2011;7:59. http://dx.doi.org/10.1186/1744-8069-7-59.
34. Gadotti VM, Bonfield SP, Zamponi GW. Depressive-like behaviour of mice lacking cellular prion protein. *Behav Brain Res*. 2012;227(2):319–323. http://dx.doi.org/10.1016/j.bbr.2011.03.012.
35. Schmitz M, Greis C, Ottis P, et al. Loss of prion protein leads to age-dependent behavioral abnormalities and changes in cytoskeletal protein expression. *Mol Neurobiol*. 2014;50:923–936. http://dx.doi.org/10.1007/s12035-014-8655-3.
36. Gasperini L, Legname G. Prion protein and aging. *Front Cell Dev Biol*. 2014;2:44. http://dx.doi.org/10.3389/fcell.2014.00044.
37. Zhang CC, Steele AD, Lindquist S, Lodish HF. Prion protein is expressed on long-term repopulating hematopoietic stem cells and is important for their self-renewal. *Proc Natl Acad Sci USA*. 2006;103(7):2184–2189.
38. Steele AD, Emsley JG, Ozdinler PH, Lindquist S, Macklis JD. Prion protein (PrPc) positively regulates neural precursor proliferation during developmental and adult mammalian neurogenesis. *Proc Natl Acad Sci USA*. 2006;103(9):3416–3421.
39. Martin-Lannerée S, Hirsch TZ, Hernandez-Rapp J, et al. PrPC from stem cells to cancer. *Front Cell Dev Biol*. 2014;2:55.
40. McLennan NF, Brennan PM, McNeill A, et al. Prion protein accumulation and neuroprotection in hypoxic brain damage. *Am J Pathol*. 2004;165(1):227–235.
41. Nico PBC, De-Paris F, Vinadé ER, et al. Altered behavioural response to acute stress in mice lacking cellular prion protein. *Behav Brain Res*. 2005;162(2):173–181. http://dx.doi.org/10.1016/j.bbr.2005.02.003.
42. Bakkebø MK, Mouillet-Richard S, Espenes A, Goldmann W, Tatzelt J, Tranulis MA. The cellular prion protein: a player in immunological quiescence. *Front Immunol*. 2015;6:450. http://dx.doi.org/10.3389/fimmu.2015.00450.
43. Bremer J, Baumann F, Tiberi C, et al. Axonal prion protein is required for peripheral myelin maintenance. *Nat Neurosci*. 2010;13(3):310–318.
44. Nuvolone M, Hermann M, Sorce S, et al. Strictly co-isogenic C57BL/6J-Prnp-/- mice: a rigorous resource for prion science. *J Exp Med*. 2016;213(3):313–327. http://dx.doi.org/10.1084/jem.20151610.
45. Küffer A, Lakkaraju AKK, Mogha A, et al. The prion protein is an agonistic ligand of the G protein-coupled receptor Adgrg6. *Nature*. 2016;536(7617):464–468. http://dx.doi.org/10.1038/nature19312.

46. Malaga-Trillo E, Solis GP, Schrock Y, et al. Regulation of embryonic cell adhesion by the prion protein. *PLoS Biol.* 2009;7(3):e55.
47. Halliez S, Passet B, Martin-Lanneree S, et al. To develop with or without the prion protein. *Front Cell Dev Biol.* 2014;2:58.
48. Westergard L, Christensen HM, Harris DA. The cellular prion protein (PrP(C)): its physiological function and role in disease. *Biochim Biophys Acta.* 2007;1772(6): 629–644. http://dx.doi.org/10.1016/j.bbadis.2007.02.011.
49. Collinge J, Whittington MA, Sidle KC, et al. Prion protein is necessary for normal synaptic function. *Nature.* 1994;370(6487):295–297.
50. Lledo PM, Tremblay P, DeArmond SJ, Prusiner SB, Nicoll RA. Mice deficient for prion protein exhibit normal neuronal excitability and synaptic transmission in the hippocampus. *Proc Natl Acad Sci USA.* 1996;93(6):2403–2407.
51. Brown DR. Prion and prejudice: normal protein and the synapse. *Trends Neurosci.* 2001;24(2):85–90.
52. Herms J, Tings T, Gall S, et al. Evidence of presynaptic location and function of the prion protein. *J Neurosci.* 1999;19(20):8866–8875.
53. Mallucci GR, Ratté S, Asante EA, et al. Post-natal knockout of prion protein alters hippocampal CA1 properties, but does not result in neurodegeneration. *EMBO J.* 2002;21(3):202–210. http://dx.doi.org/10.1093/emboj/21.3.202.
54. Papassotiropoulos A, Wollmer MA, Aguzzi A, Hock C, Nitsch RM, de Quervain DJ-F. The prion gene is associated with human long-term memory. *Hum Mol Genet.* 2005;14(15):2241–2246. http://dx.doi.org/10.1093/hmg/ddi228.
55. Konopka G, Bomar JM, Winden K, et al. Human-specific transcriptional regulation of CNS development genes by FOXP2. *Nature.* 2009;462(7270):213–217. http://dx.doi.org/10.1038/nature08549.
56. Burgess STG, Shen C, Ferguson LA, et al. Identification of adjacent binding sites for the YY1 and E4BP4 transcription factors in the ovine PrP (prion) gene promoter. *J Biol Chem.* 2009;284(11):6716–6724. http://dx.doi.org/10.1074/jbc.M807065200.
57. Ben-Moshe Z, Alon S, Mracek P, et al. The light-induced transcriptome of the zebrafish pineal gland reveals complex regulation of the circadian clockwork by light. *Nucleic Acids Res.* 2014;42(6):3750–3767. http://dx.doi.org/10.1093/nar/gkt1359.
58. Dupré SM, Burt DW, Talbot R, et al. Identification of melatonin-regulated genes in the ovine pituitary pars tuberalis, a target site for seasonal hormone control. *Endocrinology.* 2008;149(11):5527–5539. http://dx.doi.org/10.1210/en.2008-0834.
59. Jung H, Yoon BC, Holt CE. Axonal mRNA localization and local protein synthesis in nervous system assembly, maintenance and repair. *Nat Rev Neurosci.* 2012;13(5):308–324. http://dx.doi.org/10.1038/nrn3210.
60. Alleaume-Butaux A, Nicot S, Pietri M, et al. Double-edge sword of sustained ROCK activation in prion diseases through neuritogenesis defects and prion accumulation. *PLoS Pathog.* 2015;11(8):e1005073. http://dx.doi.org/10.1371/journal.ppat.1005073.
61. Amin L, Nguyen XTA, Rolle IG, et al. Characterization of prion protein function by focal neurite stimulation. *J Cell Sci.* 2016;129(20):3878–3891. http://dx.doi.org/10.1242/jcs.183137.
62. Chen S, Mange A, Dong L, Lehmann S, Schachner M. Prion protein as trans-interacting partner for neurons is involved in neurite outgrowth and neuronal survival. *Mol Cell Neurosci.* 2003;22(2):227–233.
63. Graner E, Mercadante AF, Zanata SM, et al. Cellular prion protein binds laminin and mediates neuritogenesis. *Brain Res Mol Brain Res.* 2000;76(1):85–92.
64. Kanaani J, Prusiner SB, Diacovo J, Baekkeskov S, Legname G. Recombinant prion protein induces rapid polarization and development of synapses in embryonic rat hippocampal neurons in vitro. *J Neurochem.* 2005;95(5):1373–1386.

65. Loubet D, Dakowski C, Pietri M, et al. Neuritogenesis: the prion protein controls beta1 integrin signaling activity. *FASEB J*. 2012;26:678–690. http://dx.doi.org/10.1096/fj.11-185579.
66. Santuccione A, Sytnyk V, Leshchyns'ka I, Schachner M. Prion protein recruits its neuronal receptor NCAM to lipid rafts to activate p59fyn and to enhance neurite outgrowth. *J Cell Biol*. 2005;169(2):341–354.
67. Beraldo FH, Arantes CP, Santos TG, et al. Metabotropic glutamate receptors transduce signals for neurite outgrowth after binding of the prion protein to laminin gamma1 chain. *FASEB J*. 2011;25(1):265–279. http://dx.doi.org/10.1096/fj.10-161653.
68. Lopes MH, Hajj GN, Muras AG, et al. Interaction of cellular prion and stress-inducible protein 1 promotes neuritogenesis and neuroprotection by distinct signaling pathways. *J Neurosci*. 2005;25(49):11330–11339.
69. Santos TG, Beraldo FH, Hajj GN, et al. Laminin-gamma1 chain and stress inducible protein 1 synergistically mediate PrPC-dependent axonal growth via Ca2+ mobilization in dorsal root ganglia neurons. *J Neurochem*. 2013;124(2):210–223. http://dx.doi.org/10.1111/jnc.12091.
70. Hajj GN, Lopes MH, Mercadante AF, et al. Cellular prion protein interaction with vitronectin supports axonal growth and is compensated by integrins. *J Cell Sci*. 2007;120(pt 11):1915–1926.
71. Pantera B, Bini C, Cirri P, et al. PrPc activation induces neurite outgrowth and differentiation in PC12 cells: role for caveolin-1 in the signal transduction pathway. *J Neurochem*. 2009;110(1):194–207. http://dx.doi.org/10.1111/j.1471-4159.2009.06123.x.
72. Devanathan V, Jakovcevski I, Santuccione A, et al. Cellular form of prion protein inhibits Reelin-mediated shedding of Caspr from the neuronal cell surface to potentiate Caspr-mediated inhibition of neurite outgrowth. *J Neurosci*. 2010;30(27):9292–9305. http://dx.doi.org/10.1523/JNEUROSCI.5657-09.2010.
73. Moreno JA, Radford H, Peretti D, et al. Sustained translational repression by eIF2alpha-P mediates prion neurodegeneration. *Nature*. 2012;485(7399):507–511. http://dx.doi.org/10.1038/nature11058.
74. Solforosi L, Criado JR, McGavern DB, et al. Cross-linking cellular prion protein triggers neuronal apoptosis in vivo. *Science*. 2004;303(5663):1514–1516.
75. Shmerling D, Hegyi I, Fischer M, et al. Expression of amino-terminally truncated PrP in the mouse leading to ataxia and specific cerebellar lesions. *Cell*. 1998;93(2):203–214.
76. Baumann F, Tolnay M, Brabeck C, et al. Lethal recessive myelin toxicity of prion protein lacking its central domain. *EMBO J*. 2007;26(2):538–547.
77. Li A, Christensen HM, Stewart LR, Roth KA, Chiesa R, Harris DA. Neonatal lethality in transgenic mice expressing prion protein with a deletion of residues 105-125. *EMBO J*. 2007;26(2):548–558. http://dx.doi.org/10.1038/sj.emboj.7601507.
78. Herrmann US, Sonati T, Falsig J, et al. Prion infections and anti-PrP antibodies trigger converging neurotoxic pathways. *PLoS Pathog*. 2015;11(2):e1004662. http://dx.doi.org/10.1371/journal.ppat.1004662.
79. Sonati T, Reimann RR, Falsig J, et al. The toxicity of antiprion antibodies is mediated by the flexible tail of the prion protein. *Nature*. 2013;501(7465):102–106. http://dx.doi.org/10.1038/nature12402.
80. Langenhan T, Piao X, Monk KR. Adhesion G protein-coupled receptors in nervous system development and disease. *Nat Rev Neurosci*. 2016;17(9):550–561. http://dx.doi.org/10.1038/nrn.2016.86.
81. Clarke LE, Barres BA. Emerging roles of astrocytes in neural circuit development. *Nat Rev Neurosci*. 2013;14(5):311–321. http://dx.doi.org/10.1038/nrn3484.
82. Wake H, Moorhouse AJ, Miyamoto A, Nabekura J. Microglia: actively surveying and shaping neuronal circuit structure and function. *Trends Neurosci*. 2013;36(4):209–217. http://dx.doi.org/10.1016/j.tins.2012.11.007.

83. Lima FR, Arantes CP, Muras AG, Nomizo R, Brentani RR, Martins VR. Cellular prion protein expression in astrocytes modulates neuronal survival and differentiation. *J Neurochem.* 2007;103(6):2164–2176.
84. Bertuchi FR, Bourgeon DMG, Landemberger MC, Martins VR, Cerchiaro G. PrPC displays an essential protective role from oxidative stress in an astrocyte cell line derived from PrPC knockout mice. *Biochem Biophys Res Commun.* 2012;418(1):27–32. http://dx.doi.org/10.1016/j.bbrc.2011.12.098.
85. Guitart K, Loers G, Buck F, Bork U, Schachner M, Kleene R. Improvement of neuronal cell survival by astrocyte-derived exosomes under hypoxic and ischemic conditions depends on prion protein. *Glia.* 2016;64(6):896–910. http://dx.doi.org/10.1002/glia.22963.
86. Dias MVS, Teixeira BL, Rodrigues BR, et al. PRNP/prion protein regulates the secretion of exosomes modulating CAV1/caveolin-1-suppressed autophagy. *Autophagy.* 2016;12(11):2113–2128. http://dx.doi.org/10.1080/15548627.2016.1226735.
87. Budnik V, Ruiz-Cañada C, Wendler F. Extracellular vesicles round off communication in the nervous system. *Nat Rev Neurosci.* 2016;17(3):160–172. http://dx.doi.org/10.1038/nrn.2015.29.
88. Kleene R, Loers G, Langer J, Frobert Y, Buck F, Schachner M. Prion protein regulates glutamate-dependent lactate transport of astrocytes. *J Neurosci.* 2007;27(45):12331–12340. http://dx.doi.org/10.1523/JNEUROSCI.1358-07.2007.
89. Guitart K, Loers G, Schachner M, Kleene R. Prion protein regulates glutathione metabolism and neural glutamate and cysteine uptake via excitatory amino acid transporter 3. *J Neurochem.* 2015;133(4):558–571. http://dx.doi.org/10.1111/jnc.13071.
90. Pathmajeyan MS, Patel SA, Carroll JA, et al. Increased excitatory amino acid transport into murine prion protein knockout astrocytes cultured in vitro. *Glia.* 2011;59(11):1684–1694. http://dx.doi.org/10.1002/glia.21215.
91. Brown DR, Qin K, Herms JW, et al. The cellular prion protein binds copper in vivo. *Nature.* 1997;390(6661):684–687.
92. Vassallo N, Herms J. Cellular prion protein function in copper homeostasis and redox signalling at the synapse. *J Neurochem.* 2003;86(3):538–544.
93. Haigh CL, Marom SY, Collins SJ. Copper, constitutive endoproteolytic processing of the prion protein and cell signalling. *Front Biosci.* 2010;15:1086–1104.
94. Hernandez-Rapp J, Martin-Lanneree S, Hirsch TZ, et al. A PrP(C)-caveolin-Lyn complex negatively controls neuronal GSK3beta and serotonin 1B receptor. *Sci Rep.* 2014;4:4881. http://dx.doi.org/10.1038/srep04881.
95. Gasperini L, Meneghetti E, Pastore B, Benetti F, Legname G. Prion protein and copper cooperatively protect neurons by modulating NMDA receptor through S-nitrosylation. *Antioxid Redox Signal.* 2015;22(9):772–784. http://dx.doi.org/10.1089/ars.2014.6032.
96. Watt NT, Taylor DR, Kerrigan TL, et al. Prion protein facilitates uptake of zinc into neuronal cells. *Nat Commun.* 2012;3:1134. http://dx.doi.org/10.1038/ncomms2135.
97. Black SA, Stys PK, Zamponi GW, Tsutsui S. Cellular prion protein and NMDA receptor modulation: protecting against excitotoxicity. *Front Cell Dev Biol.* 2014;2:45. http://dx.doi.org/10.3389/fcell.2014.00045.
98. Carulla P, Llorens F, Matamoros-Angles A, et al. Involvement of PrP(C) in kainate-induced excitotoxicity in several mouse strains. *Sci Rep.* 2015;5:11971. http://dx.doi.org/10.1038/srep11971.
99. Rambold AS, Muller V, Ron U, Ben-Tal N, Winklhofer KF, Tatzelt J. Stress-protective signalling of prion protein is corrupted by scrapie prions. *EMBO J.* 2008;27(14):1974–1984.
100. Fleisch VC, Leighton PLA, Wang H, et al. Targeted mutation of the gene encoding prion protein in zebrafish reveals a conserved role in neuron excitability. *Neurobiol Dis.* 2013;55:11–25. http://dx.doi.org/10.1016/j.nbd.2013.03.007.

101. Hirsch TZ, Hernandez-Rapp J, Martin-Lanneree S, Launay JM, Mouillet-Richard S. PrPC signalling in neurons: from basics to clinical challenges. *Biochimie*. 2014;104:2–11.
102. Mouillet-Richard S, Pietri M, Schneider B, et al. Modulation of serotonergic receptor signaling and cross-talk by prion protein. *J Biol Chem*. 2005;280(6):4592–4601.
103. Mouillet-Richard S, Schneider B, Pradines E, et al. Cellular prion protein signaling in serotonergic neuronal cells. *Ann NY Acad Sci*. 2007;1096:106–119.
104. Beraldo FH, Arantes CP, Santos TG, et al. Role of alpha7 nicotinic acetylcholine receptor in calcium signaling induced by prion protein interaction with stress-inducible protein 1. *J Biol Chem*. 2010;285(47):36542–36550. http://dx.doi.org/10.1074/jbc.M110.157263.
105. Jeong J-K, Park S-Y. Neuroprotective effect of cellular prion protein (PrPC) is related with activation of alpha7 nicotinic acetylcholine receptor (α7nAchR)-mediated autophagy flux. *Oncotarget*. 2015;6(28):24660–24674. http://dx.doi.org/10.18632/oncotarget.4953.
106. Carneiro MV, Americo TA, Guimarães MZP, Linden R. The prion protein selectively binds to and modulates the content of purinergic receptor P2X4R. *Biochem Biophys Res Commun*. 2016;472(2):293–298. http://dx.doi.org/10.1016/j.bbrc.2016.02.122.
107. Beckman D, Santos LE, Americo TA, Ledo JH, de Mello FG, Linden R. Prion protein modulates monoaminergic systems and depressive-like behavior in mice. *J Biol Chem*. 2015;290(33):20488–20498. http://dx.doi.org/10.1074/jbc.M115.666156.
108. Rial D, Pamplona FA, Moreira ELG, et al. Cellular prion protein is present in dopaminergic neurons and modulates the dopaminergic system. *Eur J Neurosci*. 2014;40(3):2479–2486. http://dx.doi.org/10.1111/ejn.12600.
109. Rial D, Duarte FS, Xikota JC, et al. Cellular prion protein modulates age-related behavioral and neurochemical alterations in mice. *Neuroscience*. 2009;164(3):896–907. http://dx.doi.org/10.1016/j.neuroscience.2009.09.005.
110. Ermonval M, Baudry A, Baychelier F, et al. The cellular prion protein interacts with the tissue non-specific alkaline phosphatase in membrane microdomains of bioaminergic neuronal cells. *PLoS One*. 2009;4(8):e6497.
111. Pradines E, Loubet D, Mouillet-Richard S, et al. Cellular prion protein coupling to TACE-dependent TNF-alpha shedding controls neurotransmitter catabolism in neuronal cells. *J Neurochem*. 2009;110(3):912–923.
112. Robinson SW, Nugent ML, Dinsdale D, Steinert JR. Prion protein facilitates synaptic vesicle release by enhancing release probability. *Hum Mol Genet*. 2014;23(17):4581–4596. http://dx.doi.org/10.1093/hmg/ddu171.
113. Steinert JR. Prion protein as a mediator of synaptic transmission. *Commun Integr Biol*. 2015;8(4):e1063753. http://dx.doi.org/10.1080/19420889.2015.1063753.
114. Liao MJ, Zhang CC, Zhou B, et al. Enrichment of a population of mammary gland cells that form mammospheres and have in vivo repopulating activity. *Cancer Res*. 2007;67(17):8131–8138.
115. Stella R, Massimino ML, Sandri M, Sorgato MC, Bertoli A. Cellular prion protein promotes regeneration of adult muscle tissue. *Mol Cell Biol*. 2010;30(20):4864–4876. http://dx.doi.org/10.1128/MCB.01040-09.
116. Mohanty ST, Cairney CJ, Chantry AD, et al. A small molecule modulator of prion protein increases human mesenchymal stem cell lifespan, ex vivo expansion, and engraftment to bone marrow in NOD/SCID mice. *Stem Cells*. 2012;30(6):1134–1143. http://dx.doi.org/10.1002/stem.1065.
117. Mouillet-Richard S, Laurendeau I, Vidaud M, Kellermann O, Laplanche JL. Prion protein and neuronal differentiation: quantitative analysis of prnp gene expression in a murine inducible neuroectodermal progenitor. *Microbes Infect*. 1999;1(12):969–976.
118. Peralta OA, Huckle WR, Eyestone WH. Expression and knockdown of cellular prion protein (PrPC) in differentiating mouse embryonic stem cells. *Differentiation*. 2011;81(1):68–77. http://dx.doi.org/10.1016/j.diff.2010.09.181.

119. Lee YJ, Baskakov IV. Treatment with normal prion protein delays differentiation and helps to maintain high proliferation activity in human embryonic stem cells. *J Neurochem*. 2010;114:362–373.
120. Miranda A, Pericuesta E, Ramirez MA, Gutierrez-Adan A. Prion protein expression regulates embryonic stem cell pluripotency and differentiation. *PLoS One*. 2011;6(4):e18422. http://dx.doi.org/10.1371/journal.pone.0018422.
121. Lee YJ, Baskakov IV. The cellular form of the prion protein is involved in controlling cell cycle dynamics, self-renewal, and the fate of human embryonic stem cell differentiation. *J Neurochem*. 2013;124(3):310–322. http://dx.doi.org/10.1111/j.1471-4159.2012.07913.x.
122. Santos TG, Silva IR, Costa-Silva B, Lepique AP, Martins VR, Lopes MH. Enhanced neural progenitor/stem cells self-renewal via the interaction of stress-inducible protein 1 with the prion protein. *Stem Cells*. 2011;29(7):1126–1136. http://dx.doi.org/10.1002/stem.664.
123. Prodromidou K, Papastefanaki F, Sklaviadis T, Matsas R. Functional cross-talk between the cellular prion protein and the neural cell adhesion molecule is critical for neuronal differentiation of neural stem/precursor cells. *Stem Cells*. 2014;32(6):1674–1687. http://dx.doi.org/10.1002/stem.1663.
124. Han Y-S, Lee JH, Yoon YM, Yun CW, Noh H, Lee SH. Hypoxia-induced expression of cellular prion protein improves the therapeutic potential of mesenchymal stem cells. *Cell Death Dis*. 2016;7(10):e2395. http://dx.doi.org/10.1038/cddis.2016.310.
125. Mitteregger G, Vosko M, Krebs B, et al. The role of the octarepeat region in neuroprotective function of the cellular prion protein. *Brain Pathol*. 2007;17(2):174–183.
126. Weise J, Sandau R, Schwarting S, et al. Deletion of cellular prion protein results in reduced Akt activation, enhanced postischemic caspase-3 activation, and exacerbation of ischemic brain injury. *Stroke*. 2006;37(5):1296–1300.
127. Lee S-D, Lai TW, Lin S-Z, et al. Role of stress-inducible protein-1 in recruitment of bone marrow derived cells into the ischemic brains. *EMBO Mol Med*. 2013;5(8):1227–1246. http://dx.doi.org/10.1002/emmm.201202258.
128. Relano-Gines A, Gabelle A, Hamela C, et al. Prion replication occurs in endogenous adult neural stem cells and alters their neuronal fate: involvement of endogenous neural stem cells in prion diseases. *PLoS Pathog*. 2013;9(8):e1003485. http://dx.doi.org/10.1371/journal.ppat.1003485.
129. Corbit KC, Aanstad P, Singla V, Norman AR, Stainier DYR, Reiter JF. Vertebrate smoothened functions at the primary cilium. *Nature*. 2005;437(7061):1018–1021. http://dx.doi.org/10.1038/nature04117.
130. Liu Q, Tan G, Levenkova N, et al. The proteome of the mouse photoreceptor sensory cilium complex. *Mol Cell Proteomics*. 2007;6(8):1299–1317. http://dx.doi.org/10.1074/mcp.M700054-MCP200.
131. Mick DU, Rodrigues RB, Leib RD, et al. Proteomics of primary cilia by proximity labeling. *Dev Cell*. 2015;35(4):497–512. http://dx.doi.org/10.1016/j.devcel.2015.10.015.
132. Narita K, Kozuka-Hata H, Nonami Y, et al. Proteomic analysis of multiple primary cilia reveals a novel mode of ciliary development in mammals. *Biol Open*. 2012;1(8):815–825. http://dx.doi.org/10.1242/bio.20121081.
133. Martin-Lannerée S, Halliez S, Hirsch TZ, et al. The cellular prion protein controls notch signalling in neural stem/progenitor cells. *Stem Cells*. 2017;35:754–765.
134. Louvi A, Artavanis-Tsakonas S. Notch signalling in vertebrate neural development. *Nat Rev Neurosci*. 2006;7(2):93–102.
135. Rousso DL, Pearson CA, Gaber ZB, et al. Foxp-mediated suppression of N-cadherin regulates neuroepithelial character and progenitor maintenance in the CNS. *Neuron*. 2012;74(2):314–330. http://dx.doi.org/10.1016/j.neuron.2012.02.024.

136. Wright JA, McHugh PC, Stockbridge M, Lane S, Kralovicova S, Brown DR. Activation and repression of prion protein expression by key regions of intron 1. *Cell Mol Life Sci*. 2009;66(23):3809–3820. http://dx.doi.org/10.1007/s00018-009-0154-8.
137. Doan PL, Himburg HA, Helms K, et al. Epidermal growth factor regulates hematopoietic regeneration after radiation injury. *Nat Med*. 2013;19(3):295–304. http://dx.doi.org/10.1038/nm.3070.
138. Poulos MG, Guo P, Kofler NM, et al. Endothelial jagged-1 is necessary for homeostatic and regenerative hematopoiesis. *Cell Rep*. 2013;4(5):1022–1034. http://dx.doi.org/10.1016/j.celrep.2013.07.048.
139. Bribian A, Fontana X, Llorens F, et al. Role of the cellular prion protein in oligodendrocyte precursor cell proliferation and differentiation in the developing and adult mouse CNS. *PLoS One*. 2012;7(4):e33872. http://dx.doi.org/10.1371/journal.pone.0033872.
140. Kong JH, Yang L, Dessaud E, et al. Notch activity modulates the responsiveness of neural progenitors to sonic hedgehog signaling. *Dev Cell*. 2015;33(4):373–387. http://dx.doi.org/10.1016/j.devcel.2015.03.005.
141. Ostapchenko VG, Beraldo FH, Mohammad AH, et al. The prion protein ligand, stress-inducible phosphoprotein 1, regulates amyloid-beta oligomer toxicity. *J Neurosci*. 2013;33(42):16552–16564. http://dx.doi.org/10.1523/JNEUROSCI.3214-13.2013.
142. Hartmann CA, Martins VR, Lima FRS. High levels of cellular prion protein improve astrocyte development. *FEBS Lett*. 2013;587(2):238–244. http://dx.doi.org/10.1016/j.febslet.2012.11.032.
143. Panigaj M, Glier H, Wildova M, Holada K. Expression of prion protein in mouse erythroid progenitors and differentiating murine erythroleukemia cells. *PLoS One*. 2011;6(9):e24599. http://dx.doi.org/10.1371/journal.pone.0024599.
144. Reiten MR, Bakkebø MK, Brun-Hansen H, et al. Hematological shift in goat kids naturally devoid of prion protein. *Front Cell Dev Biol*. 2015;3:44. http://dx.doi.org/10.3389/fcell.2015.00044.
145. Westaway D, Cooper C, Turner S, Da Costa M, Carlson GA, Prusiner SB. Structure and polymorphism of the mouse prion protein gene. *Proc Natl Acad Sci USA*. 1994;91(14):6418–6422.
146. Vincent B, Sunyach C, Orzechowski HD, St George-Hyslop P, Checler F. p53-dependent transcriptional control of cellular prion by presenilins. *J Neurosci*. 2009;29(20):6752–6760. http://dx.doi.org/10.1523/JNEUROSCI.0789-09.2009.
147. Shyu W-C, Harn H-J, Saeki K, et al. Molecular modulation of expression of prion protein by heat shock. *Mol Neurobiol*. 2002;26(1):1–12. http://dx.doi.org/10.1385/MN:26:1:001.
148. Misiewicz M, Déry M-A, Foveau B, Jodoin J, Ruths D, LeBlanc AC. Identification of a novel endoplasmic reticulum stress response element regulated by XBP1. *J Biol Chem*. 2013;288(28):20378–20391. http://dx.doi.org/10.1074/jbc.M113.457242.
149. Liu T, Yi W, Feng B, Zhou Z, Xiao G. IGF-1-induced enhancement of PRNP expression depends on the negative regulation of transcription factor FOXO3a. *PLoS One*. 2013;8(8):e71896. http://dx.doi.org/10.1371/journal.pone.0071896.
150. Qin K, Zhao L, Ash RD, McDonough WF, Zhao RY. ATM-mediated transcriptional elevation of prion in response to copper-induced oxidative stress. *J Biol Chem*. 2009;284(7):4582–4593. http://dx.doi.org/10.1074/jbc.M808410200.
151. Sauer H, Dagdanova A, Hescheler J, Wartenberg M. Redox-regulation of intrinsic prion expression in multicellular prostate tumor spheroids. *Free Radic Biol Med*. 1999;27(11 – 12):1276–1283.
152. Jeong J-K, Park S-Y. Transcriptional regulation of specific protein 1 (SP1) by hypoxia-inducible factor 1 alpha (HIF-1α) leads to PRNP expression and neuroprotection from

toxic prion peptide. *Biochem Biophys Res Commun.* 2012;429(1–2):93–98. http://dx.doi.org/10.1016/j.bbrc.2012.10.086.
153. Park J-Y, Jeong J-K, Lee J-H, et al. Induction of cellular prion protein (PrPc) under hypoxia inhibits apoptosis caused by TRAIL treatment. *Oncotarget.* 2015;6(7):5342–5353. http://dx.doi.org/10.18632/oncotarget.3028.
154. Dery MA, Jodoin J, Ursini-Siegel J, et al. Endoplasmic reticulum stress induces PRNP prion protein gene expression in breast cancer. *Breast Cancer Res.* 2013;15(2):R22. http://dx.doi.org/10.1186/bcr3398.
155. Miyamoto K, Araki KY, Naka K, et al. Foxo3a is essential for maintenance of the hematopoietic stem cell pool. *Cell Stem Cell.* 2007;1(1):101–112. http://dx.doi.org/10.1016/j.stem.2007.02.001.
156. Bigarella CL, Li J, Rimmelé P, Liang R, Sobol RW, Ghaffari S. FOXO3 is essential for protecting hematopoietic stem and progenitor cells from oxidative DNA damage. *J Biol Chem.* 2017;292(7):3005–3015. http://dx.doi.org/10.1074/jbc.M116.769455.
157. Mariante RM, Nobrega A, Martins RA, Areal RB, Bellio M, Linden R. Neuroimmunoendocrine regulation of the prion protein in neutrophils. *J Biol Chem.* 2012;287(42):35506–35515. http://dx.doi.org/10.1074/jbc.M112.394924.
158. Liu J, Zhao D, Liu C, et al. Prion protein participates in the protection of mice from lipopolysaccharide infection by regulating the inflammatory process. *J Mol Neurosci.* 2015;55(1):279–287. http://dx.doi.org/10.1007/s12031-014-0319-2.
159. Roucou X, Leblanc AC. Cellular prion protein neuroprotective function: implications in prion diseases. *J Mol Med.* 2005;83(1):3–11.
160. Singh N, Singh A, Das D, Mohan ML. Redox control of prion and disease pathogenesis. *Antioxid Redox Signal.* 2010;12(11):1271–1294. http://dx.doi.org/10.1089/ars.2009.2628.
161. Nadal RC, Abdelraheim SR, Brazier MW, Rigby SEJ, Brown DR, Viles JH. Prion protein does not redox-silence Cu2+, but is a sacrificial quencher of hydroxyl radicals. *Free Radic Biol Med.* 2007;42(1):79–89. http://dx.doi.org/10.1016/j.freeradbiomed.2006.09.019.
162. McMahon HE, Mange A, Nishida N, Creminon C, Casanova D, Lehmann S. Cleavage of the amino-terminus of the prion protein by reactive oxygen species. *J Biol Chem.* 2001;276(3):2286–2291.
163. Watt NT, Taylor DR, Gillott A, Thomas DA, Perera WSS, Hooper NM. Reactive oxygen species-mediated beta-cleavage of the prion protein in the cellular response to oxidative stress. *J Biol Chem.* 2005;280(43):35914–35921. http://dx.doi.org/10.1074/jbc.M507327200.
164. Rachidi W, Vilette D, Guiraud P, et al. Expression of prion protein increases cellular copper binding and antioxidant enzyme activities but not copper delivery. *J Biol Chem.* 2003;278(11):9064–9072. http://dx.doi.org/10.1074/jbc.M211830200.
165. Senator A, Rachidi W, Lehmann S, Favier A, Benboubetra M. Prion protein protects against DNA damage induced by paraquat in cultured cells. *Free Radic Biol Med.* 2004;37(8):1224–1230. http://dx.doi.org/10.1016/j.freeradbiomed.2004.07.006.
166. Watt NT, Routledge MN, Wild CP, Hooper NM. Cellular prion protein protects against reactive-oxygen-species-induced DNA damage. *Free Radic Biol Med.* 2007;43(6):959–967. http://dx.doi.org/10.1016/j.freeradbiomed.2007.06.004.
167. Bravard A, Auvré F, Fantini D, et al. The prion protein is critical for DNA repair and cell survival after genotoxic stress. *Nucleic Acids Res.* 2015;43(2):904–916. http://dx.doi.org/10.1093/nar/gku1342.
168. Shiloh Y. ATM: expanding roles as a chief guardian of genome stability. *Exp Cell Res.* 2014;329(1):154–161. http://dx.doi.org/10.1016/j.yexcr.2014.09.002.
169. Resenberger UK, Winklhofer KF, Tatzelt J. Neuroprotective and neurotoxic signaling by the prion protein. *Top Curr Chem.* 2011;305:101–119. http://dx.doi.org/10.1007/128_2011_160.

170. Kuwahara C, Takeuchi AM, Nishimura T, et al. Prions prevent neuronal cell-line death. *Nature*. 1999;400(6741):225–226.
171. Guillot-Sestier MV, Sunyach C, Druon C, Scarzello S, Checler F. The alpha-secretase-derived N-terminal product of cellular prion, N1, displays neuroprotective function in vitro and in vivo. *J Biol Chem*. 2009;284(51):35973–35986.
172. Wang KKW, Yang Z, Chiu A, Lin F, Rubenstein R. Examining the neural and astroglial protective effects of cellular prion protein expression and cell death protease inhibition in mouse cerebrocortical mixed cultures. *Mol Neurobiol*. 2016;53(7): 4821–4832. http://dx.doi.org/10.1007/s12035-015-9407-8.
173. Zafar S, Behrens C, Dihazi H, et al. Cellular prion protein mediates early apoptotic proteome alternation and phospho-modification in human neuroblastoma cells. *Cell Death Dis*. 2017;8(1):e2557. http://dx.doi.org/10.1038/cddis.2016.384.
174. Chiarini LB, Freitas AR, Zanata SM, Brentani RR, Martins VR, Linden R. Cellular prion protein transduces neuroprotective signals. *EMBO J*. 2002;21(13):3317–3326.
175. Vassallo N, Herms J, Behrens C, et al. Activation of phosphatidylinositol 3-kinase by cellular prion protein and its role in cell survival. *Biochem Biophys Res Commun*. 2005;332(1):75–82.
176. Beraldo FH, Soares IN, Goncalves DF, et al. Stress-inducible phosphoprotein 1 has unique cochaperone activity during development and regulates cellular response to ischemia via the prion protein. *FASEB J*. 2013;27(9):3594–3607. http://dx.doi.org/10.1096/fj.13-232280.
177. Zanata SM, Lopes MH, Mercadante AF, et al. Stress-inducible protein 1 is a cell surface ligand for cellular prion that triggers neuroprotection. *EMBO J*. 2002;21(13): 3307–3316. http://dx.doi.org/10.1093/emboj/cdf325.
178. Niederkorn JY. See no evil, hear no evil, do no evil: the lessons of immune privilege. *Nat Immunol*. 2006;7(4):354–359. http://dx.doi.org/10.1038/ni1328.
179. Roth DM, Balch WE. Modeling general proteostasis: proteome balance in health and disease. *Curr Opin Cell Biol*. 2011;23(2):126–134. http://dx.doi.org/10.1016/j.ceb.2010.11.001.
180. Roffe M, Beraldo FH, Bester R, et al. Prion protein interaction with stress-inducible protein 1 enhances neuronal protein synthesis via mTOR. *Proc Natl Acad Sci USA*. 2010;107(29):13147–13152. http://dx.doi.org/10.1073/pnas.1000784107.
181. Nah J, Pyo JO, Jung S, et al. BECN1/Beclin 1 is recruited into lipid rafts by prion to activate autophagy in response to amyloid beta 42. *Autophagy*. 2013;9(12):2009–2021.
182. Yoon YM, Lee JH, Yun SP, et al. Tauroursodeoxycholic acid reduces ER stress by regulating of Akt-dependent cellular prion protein. *Sci Rep*. 2016;6:39838. http://dx.doi.org/10.1038/srep39838.
183. Malicki JJ, Johnson CA. The cilium: cellular antenna and central processing unit. *Trends Cell Biol*. 2017;27(2):126–140. http://dx.doi.org/10.1016/j.tcb.2016.08.002.
184. Graner E, Mercadante AF, Zanata SM, Martins VR, Jay DG, Brentani RR. Laminin-induced PC-12 cell differentiation is inhibited following laser inactivation of cellular prion protein. *FEBS Lett*. 2000;482(3):257–260 [in process citation].
185. Rieger R, Edenhofer F, Lasmezas CI, Weiss S. The human 37-kDa laminin receptor precursor interacts with the prion protein in eukaryotic cells. *Nat Med*. 1997;3(12):1383–1388 [see comments].
186. Schmitt-Ulms G, Legname G, Baldwin MA, et al. Binding of neural cell adhesion molecules (N-CAMs) to the cellular prion protein. *J Mol Biol*. 2001;314(5):1209–1225.
187. Watts JC, Huo H, Bai Y, et al. Interactome analyses identify ties of PrP and its mammalian paralogs to oligomannosidic N-glycans and endoplasmic reticulum-derived chaperones. *PLoS Pathog*. 2009;5(10).e1000608:
188. Richardson DD, Tol S, Valle-Encinas E, et al. The prion protein inhibits monocytic cell migration by stimulating β1 integrin adhesion and uropod formation. *J Cell Sci*. 2015;128(16):3018–3029. http://dx.doi.org/10.1242/jcs.165365.

189. Schrock Y, Solis GP, Stuermer CAO. Regulation of focal adhesion formation and filopodia extension by the cellular prion protein (PrPC). *FEBS Lett.* 2009;583(2): 389–393. http://dx.doi.org/10.1016/j.febslet.2008.12.038.
190. Morel E, Fouquet S, Chateau D, et al. The cellular prion protein PrPc is expressed in human enterocytes in cell-cell junctional domains. *J Biol Chem.* 2004;279(2): 1499–1505. http://dx.doi.org/10.1074/jbc.M308578200.
191. Guo Z, Neilson LJ, Zhong H, Murray PS, Zanivan S, Zaidel-Bar R. E-cadherin interactome complexity and robustness resolved by quantitative proteomics. *Sci Signal.* 2014;7(354):rs7. http://dx.doi.org/10.1126/scisignal.2005473.
192. Besnier LS, Cardot P, Da Rocha B, et al. The cellular prion protein PrPc is a partner of the Wnt pathway in intestinal epithelial cells. *Mol Biol Cell.* 2015;26(18):3313–3328. http://dx.doi.org/10.1091/mbc.E14-11-1534.
193. Morel E, Fouquet S, Strup-Perrot C, et al. The cellular prion protein PrP(c) is involved in the proliferation of epithelial cells and in the distribution of junction-associated proteins. *PLoS One.* 2008;3(8):e3000. http://dx.doi.org/10.1371/journal.pone.0003000.
194. Petit CSV, Barreau F, Besnier L, et al. Requirement of cellular prion protein for intestinal barrier function and mislocalization in patients with inflammatory bowel disease. *Gastroenterology.* 2012;143(1) 122–132.e15. http://doi.org.10.1053/j.gastro.2012.03.029.
195. Sempou E, Biasini E, Pinzón-Olejua A, Harris DA, Málaga-Trillo E. Activation of zebrafish Src family kinases by the prion protein is an amyloid-β-sensitive signal that prevents the endocytosis and degradation of E-cadherin/β-catenin complexes in vivo. *Mol Neurodegener.* 2016;11:18. http://dx.doi.org/10.1186/s13024-016-0076-5.
196. Bodrikov V, Solis GP, Stuermer CA. Prion protein promotes growth cone development through reggie/flotillin-dependent N-cadherin trafficking. *J Neurosci.* 2011;31(49):18013–18025. http://dx.doi.org/10.1523/JNEUROSCI.4729-11.2011.
197. Pradines E, Loubet D, Schneider B, Launay JM, Kellermann O, Mouillet-Richard S. CREB-dependent gene regulation by prion protein: impact on MMP-9 and beta-dystroglycan. *Cell Signal.* 2008;20(11):2050–2058.
198. Schneider K, Korkmaz Y, Addicks K, Lang H, Raab WH-M. Prion protein (PrP) in human teeth: an unprecedented pointer to PrP's function. *J Endod.* 2007;33(2):110–113. http://dx.doi.org/10.1016/j.joen.2006.11.010.
199. Um JW, Kaufman AC, Kostylev M, et al. Metabotropic glutamate receptor 5 is a coreceptor for Alzheimer abeta oligomer bound to cellular prion protein. *Neuron.* 2013;79(5):887–902. http://dx.doi.org/10.1016/j.neuron.2013.06.036.
200. Solis GP, Schrock Y, Hulsbusch N, Wiechers M, Plattner H, Stuermer CA. Reggies/flotillins regulate E-cadherin-mediated cell contact formation by affecting EGFR trafficking. *Mol Biol Cell.* 2012;23(10):1812–1825. http://dx.doi.org/10.1091/mbc.E11-12-1006.
201. Linden R, Martins VR, Prado MA, Cammarota M, Izquierdo I, Brentani RR. Physiology of the prion protein. *Physiol Rev.* 2008;88(2):673–728.
202. Rutishauser D, Mertz KD, Moos R, et al. The comprehensive native interactome of a fully functional tagged prion protein. *PLoS One.* 2009;4(2):e4446.
203. Schmitt-Ulms G, Hansen K, Liu J, et al. Time-controlled transcardiac perfusion cross-linking for the study of protein interactions in complex tissues. *Nat Biotechnol.* 2004;22(6):724–731.
204. Coitinho AS, Lopes MH, Hajj GN, et al. Short-term memory formation and long-term memory consolidation are enhanced by cellular prion association to stress-inducible protein 1. *Neurobiol Dis.* 2007;26(1):282–290.
205. Lauren J, Gimbel DA, Nygaard HB, Gilbert JW, Strittmatter SM. Cellular prion protein mediates impairment of synaptic plasticity by amyloid-beta oligomers. *Nature.* 2009;457(7233):1128–1132.

206. Um JW, Nygaard HB, Heiss JK, et al. Alzheimer amyloid-beta oligomer bound to postsynaptic prion protein activates Fyn to impair neurons. *Nat Neurosci.* 2012;15(9):1227–1235. http://dx.doi.org/10.1038/nn.3178.
207. Larson M, Sherman MA, Amar F, et al. The complex PrP(c)-Fyn couples human oligomeric Abeta with pathological tau changes in Alzheimer's disease. *J Neurosci.* 2012;32(47) 16857–71a. http://doi.org.10.1523/JNEUROSCI.1858-12.2012.
208. Coitinho AS, Freitas AR, Lopes MH, et al. The interaction between prion protein and laminin modulates memory consolidation. *Eur J Neurosci.* 2006;24(11):3255–3264.
209. Mouillet-Richard S, Ermonval M, Chebassier C, et al. Signal transduction through prion protein. *Science.* 2000;289(5486):1925–1928.
210. Kaiser DM, Acharya M, Leighton PL, et al. Amyloid beta precursor protein and prion protein have a conserved interaction affecting cell adhesion and CNS development. *PLoS One.* 2012;7(12):e51305. http://dx.doi.org/10.1371/journal.pone.0051305.
211. Mbazima V, Da Costa Dias B, Omar A, Jovanovic K, Weiss S. Interactions between PrPc and other ligands with the 37-kDa/67-kDa laminin receptor. *Front Biosci.* 2010;15:1150–1163.
212. Llorens F, Carulla P, Villa A, et al. PrP(C) regulates epidermal growth factor receptor function and cell shape dynamics in Neuro2a cells. *J Neurochem.* 2013;127(1):124–138. http://dx.doi.org/10.1111/jnc.12283.
213. Hu NW, Nicoll AJ, Zhang D, et al. mGlu5 receptors and cellular prion protein mediate amyloid-beta-facilitated synaptic long-term depression in vivo. *Nat Commun.* 2014;5:3374. http://dx.doi.org/10.1038/ncomms4374.
214. Carulla P, Bribián A, Rangel A, et al. Neuroprotective role of PrPC against kainate-induced epileptic seizures and cell death depends on the modulation of JNK3 activation by GluR6/7-PSD-95 binding. *Mol Biol Cell.* 2011;22(17):3041–3054. http://dx.doi.org/10.1091/mbc.E11-04-0321.
215. Khosravani H, Zhang Y, Tsutsui S, et al. Prion protein attenuates excitotoxicity by inhibiting NMDA receptors. *J Cell Biol.* 2008;181(3):551–565. http://dx.doi.org/10.1083/jcb.200711002.
216. Parkyn CJ, Vermeulen EG, Mootoosamy RC, et al. LRP1 controls biosynthetic and endocytic trafficking of neuronal prion protein. *J Cell Sci.* 2008;121(pt 6):773–783.
217. Taylor DR, Whitehouse IJ, Hooper NM. Glypican-1 mediates both prion protein lipid raft association and disease isoform formation. *PLoS Pathog.* 2009;5(11):e1000666.
218. Griffiths HH, Whitehouse IJ, Baybutt H, et al. Prion protein interacts with BACE1 protein and differentially regulates its activity toward wild type and Swedish mutant amyloid precursor protein. *J Biol Chem.* 2011;286(38):33489–33500. http://dx.doi.org/10.1074/jbc.M111.278556.
219. Zafar S, von Ahsen N, Oellerich M, et al. Proteomics approach to identify the interacting partners of cellular prion protein and characterization of Rab7a interaction in neuronal cells. *J Proteome Res.* 2011;10(7):3123–3135. http://dx.doi.org/10.1021/pr2001989.
220. Gibbings D, Leblanc P, Jay F, et al. Human prion protein binds Argonaute and promotes accumulation of microRNA effector complexes. *Nat Struct Mol Biol.* 2012;19(5):517–524. S1. http://doi.org.10.1038/nsmb.2273.
221. Ramljak S, Schmitz M, Zafar S, et al. Cellular prion protein directly interacts with and enhances lactate dehydrogenase expression under hypoxic conditions. *Exp Neurol.* 2015;271:155–167. http://dx.doi.org/10.1016/j.expneurol.2015.04.025.
222. Martins VR, Graner E, Garcia-Abreu J, et al. Complementary hydropathy identifies a cellular prion protein receptor. *Nat Med.* 1997;3(12):1376–1382.

223. Hernandez-Rapp J, Martin-Lanneree S, Hirsch TZ, Launay JM, Mouillet-Richard S. Hijacking PrP(c)-dependent signal transduction: when prions impair Abeta clearance. *Front Aging Neurosci.* 2014;6:25. http://dx.doi.org/10.3389/fnagi.2014.00025.
224. Jarosz-Griffiths HH, Noble E, Rushworth JV, Hooper NM. Amyloid-β receptors: the good, the bad, and the prion protein. *J Biol Chem.* 2016;291(7):3174–3183. http://dx.doi.org/10.1074/jbc.R115.702704.
225. Lauren J. Cellular prion protein as a therapeutic target in Alzheimer's disease. *J Alzheimers Dis.* 2014;38(2):227–244. http://dx.doi.org/10.3233/JAD-130950.
226. Yehiely F, Bamborough P, Da Costa M, et al. Identification of candidate proteins binding to prion protein. *Neurobiol Dis.* 1997;3(4):339–355. http://dx.doi.org/10.1006/nbdi.1997.0130.
227. Mo S, Wang L, Li Q, et al. Caveolin-1 regulates dorsoventral patterning through direct interaction with beta-catenin in zebrafish. *Dev Biol.* 2010;344(1):210–223. http://dx.doi.org/10.1016/j.ydbio.2010.04.033.
228. Head BP, Patel HH, Insel PA. Interaction of membrane/lipid rafts with the cytoskeleton: impact on signaling and function: membrane/lipid rafts, mediators of cytoskeletal arrangement and cell signaling. *Biochim Biophys Acta.* 2014;1838(2):532–545. http://dx.doi.org/10.1016/j.bbamem.2013.07.018.
229. Schneider B, Pietri M, Pradines E, et al. Understanding the neurospecificity of prion protein signaling. *Front Biosci.* 2011;16:169–186.
230. Stuermer CA, Langhorst MF, Wiechers MF, et al. PrPc capping in T cells promotes its association with the lipid raft proteins reggie-1 and reggie-2 and leads to signal transduction. *FASEB J.* 2004;18(14):1731–1733.
231. Haigh CL, Lewis VA, Vella LJ, et al. PrPC-related signal transduction is influenced by copper, membrane integrity and the alpha cleavage site. *Cell Res.* 2009;19(9):1062–1078. http://dx.doi.org/10.1038/cr.2009.86.
232. Hugel B, Martinez MC, Kunzelmann C, Blattler T, Aguzzi A, Freyssinet JM. Modulation of signal transduction through the cellular prion protein is linked to its incorporation in lipid rafts. *Cell Mol Life Sci.* 2004;61(23):2998–3007.
233. Bate C, Nolan W, McHale-Owen H, Williams A. Sialic acid within the glycosylphosphatidylinositol anchor targets the cellular prion protein to synapses. *J Biol Chem.* 2016;291(33):17093–17101. http://dx.doi.org/10.1074/jbc.M116.731117.
234. Bate C, Nolan W, Williams A. Sialic acid on the glycosylphosphatidylinositol anchor regulates PrP-mediated cell signaling and prion formation. *J Biol Chem.* 2016;291(1):160–170. http://dx.doi.org/10.1074/jbc.M115.672394.
235. Pietri M, Dakowski C, Hannaoui S, et al. PDK1 decreases TACE-mediated alpha-secretase activity and promotes disease progression in prion and Alzheimer's diseases. *Nat Med.* 2013;19(9):1124–1131. http://dx.doi.org/10.1038/nm.3302.
236. Russelakis-Carneiro M, Hetz C, Maundrell K, Soto C. Prion replication alters the distribution of synaptophysin and caveolin 1 in neuronal lipid rafts. *Am J Pathol.* 2004;165(5):1839–1848.
237. Sorkin A, von Zastrow M. Endocytosis and signalling: intertwining molecular networks. *Nat Rev Mol Cell Biol.* 2009;10(9):609–622. http://dx.doi.org/10.1038/nrm2748.
238. Caetano FA, Lopes MH, Hajj GN, et al. Endocytosis of prion protein is required for ERK1/2 signaling induced by stress-inducible protein 1. *J Neurosci.* 2008;28(26):6691–6702.
239. Caetano FA, Beraldo FH, Hajj GN, et al. Amyloid-beta oligomers increase the localization of prion protein at the cell surface. *J Neurochem.* 2011;117(3):538–553. http://dx.doi.org/10.1111/j.1471-4159.2011.07225.x.

240. Uchiyama K, Muramatsu N, Yano M, Usui T, Miyata H, Sakaguchi S. Prions disturb post-Golgi trafficking of membrane proteins. *Nat Commun*. 2013;4:1846. http://dx.doi.org/10.1038/ncomms2873.
241. Roucou X. Regulation of PrP(C) signaling and processing by dimerization. *Front Cell Dev Biol*. 2014;2:57. http://dx.doi.org/10.3389/fcell.2014.00057.
242. Beland M, Motard J, Barbarin A, Roucou X. PrP(C) homodimerization stimulates the production of PrPC cleaved fragments PrPN1 and PrPC1. *J Neurosci*. 2012;32(38):13255–13263. http://dx.doi.org/10.1523/JNEUROSCI.2236-12.2012.
243. Pradines E, Hernandez-Rapp J, Villa-Diaz A, et al. Pathogenic prions deviate PrP(C) signaling in neuronal cells and impair A-beta clearance. *Cell Death Dis*. 2013;4:e456. http://dx.doi.org/10.1038/cddis.2012.195.
244. Salazar SV, Strittmatter SM. Cellular prion protein as a receptor for amyloid-β oligomers in Alzheimer's disease. *Biochem Biophys Res Commun*. 2016;483(4):1143–1147. http://dx.doi.org/10.1016/j.bbrc.2016.09.062.

CHAPTER TWO

Copper- and Zinc-Promoted Interdomain Structure in the Prion Protein: A Mechanism for Autoinhibition of the Neurotoxic N-Terminus

Eric G.B. Evans[*,1], **Glenn L. Millhauser**[†,1]
[*]University of Washington, Seattle, WA, United States
[†]University of California, Santa Cruz, CA, United States
[1]Corresponding authors: e-mail address: egevans@uw.edu; glennm@ucsc.edu

Contents

Abstract

The function of the cellular prion protein (PrP^C), while still poorly understood, is increasingly linked to its ability to bind physiological metal ions at the cell surface. PrP^C binds divalent forms of both copper and zinc through its unstructured N-terminal domain, modulating interactions between PrP^C and various receptors at the cell surface and ultimately tuning downstream cellular processes. In this chapter, we briefly discuss the molecular features of copper and zinc uptake by PrP^C and summarize evidence implicating these metal ions in PrP-mediated physiology. We then focus our review on recent biophysical evidence revealing a physical interaction between the flexible N-terminal and globular C-terminal domains of PrP^C. This interdomain *cis* interaction is electrostatic in nature and is promoted by the binding of Cu^{2+} and Zn^{2+} to the N-terminal octarepeat domain. These findings, along with recent cellular studies,

Progress in Molecular Biology and Translational Science, Volume 150
ISSN 1877-1173
http://dx.doi.org/10.1016/bs.pmbts.2017.06.005

suggest a mechanism whereby N—C interactions serve to regulate the activity and/or toxicity of the PrP^C N-terminus. We discuss this potential mechanism in relation to familial prion disease mutations, lethal deletions of the PrP^C central region, and neurotoxicity induced by certain globular domain ligands, including *bona fide* prions and toxic amyloid-β oligomers.

1. INTRODUCTION

Prion diseases are fatal neurodegenerative diseases of mammals that involve the templated conversion of the cellular prion protein (PrP^C) into a protease-resistant and β-sheet rich isoform known as PrP-scrapie (PrP^{Sc}).[1–3] PrP^C is present in various cell types throughout the body, but it is most abundantly expressed in the central nervous system where it localizes to the pre- and postsynaptic terminals of neurons.[4,5] Because PrP^C serves as the substrate for PrP^{Sc} propagation, host expression of PrP^C is required for prion disease progression.[6,7] However, there is now strong evidence that in addition to its requirement as a substrate for PrP^{Sc}, cell-surface PrP^C is directly responsible for transmitting the neurotoxic signal in prion pathogenesis.[8–11] In other words, it appears that PrP^C serves as the receptor for PrP^{Sc} and, by an unknown mechanism, triggers a cascade of downstream pathways that result in cellular dysfunction and neurodegeneration. Interestingly, PrP^C has also been implicated as the receptor responsible for transmitting the synaptotoxic effects of amyloid-β oligomers (Aβo) in Alzheimer's disease (AD), suggesting that stimulation of toxic PrP^C signaling may be an early event common to multiple neurodegenerative diseases.[5,12–16]

Owing to PrP^C's central involvement in the earliest molecular events of at least two neurodegenerative diseases in humans, elucidating the mechanism of PrP^C-mediated toxicity is clearly of great importance. Progress has been complicated, however, by the fact that the physiological function of PrP^C in healthy tissue is still under debate. Transgenic mice lacking the gene that encodes cellular PrP are viable and display no obvious developmental or reproductive deficits.[17] Likewise, cattle devoid of PrP^C live seemingly normal lives.[18] These observations suggest that PrP^C is nonessential. However, more detailed studies involving PrP^C-ablated animal models have revealed subtle behavioral and physiological derangements, including altered circadian rhythms, impaired synaptic function, deficits in peripheral myelin maintenance, and increased susceptibility to Cu^{2+}-mediated oxidative damage.[19–23]

Consistent with its localization to cholesterol-rich lipid rafts of pre- and postsynaptic membranes, PrP^C is increasingly being recognized as a promiscuous regulator of neuronal signaling.[24] PrP^C has been shown to interact with, and modulate, a surprisingly large number of membrane receptors, including ionotropic and metabotropic glutamate receptors, ion channels, and amino acid transporters. Together these examples highlight a fairly ubiquitous role for PrP^C in neuronal maintenance; however, recent experiments point toward a particularly conserved role in modulating the excitability of neurons. Functional interactions with PrP^C have recently been reported for the α-amino-3-hydroxy-5-methyl-4-isoxazolepropionic acid (AMPA) and *N*-methyl-D-aspartate (NMDA) glutamate receptors, as well as the type 1 metabotropic glutamate receptors mGluR1 and mGluR5.[25–30] In addition, PrP^C was shown to regulate glutamate uptake by the excitatory amino acid transporter EAAT3, lending further support to a role in neurotransmission and protection from excitotoxicity.[31] Nevertheless, PrP^C-null mice, cattle, and sheep develop normally, arguing that the regulatory pathways disrupted by PrP^C deficiency are either subtle or can be at least partially compensated for by alternative mechanisms.

What causes the dramatic transformation from a promiscuous, yet subtle, modulator of cell surface receptors into a strong effector of neurotoxicity? Cellular pathways and molecular mechanisms are beginning to emerge that may help answer this important question. Here, we review a recently discovered molecular interaction within the PrP^C molecule that may have important effects on PrP-mediated neurodegeneration.

1.1 PrP^C Structure

Fig. 1A summarizes the main structural elements of the PrP^C molecule. After C-terminal attachment to the outer leaflet of the plasma membrane by way of a glycophosphatidylinositol (GPI) anchor and removal of the 22-amino acid signal peptide, mature PrP^C (23 – 231) consists of 209 amino acids and can possess up to two N-linked glycans at asparagines 181 and 197 (human sequence). The N-terminal half of the protein (residues 23–120) is intrinsically disordered and is sometimes referred to as the "flexible tail."[32,33] This domain contains the metal-binding octarepeat (OR) domain as well as two polybasic charged clusters and a hydrophobic linker domain. The C-terminal half of PrP^C (residues 121–231) is globular, with a structure composed of three α-helices and one short antiparallel β-sheet.[34] A single

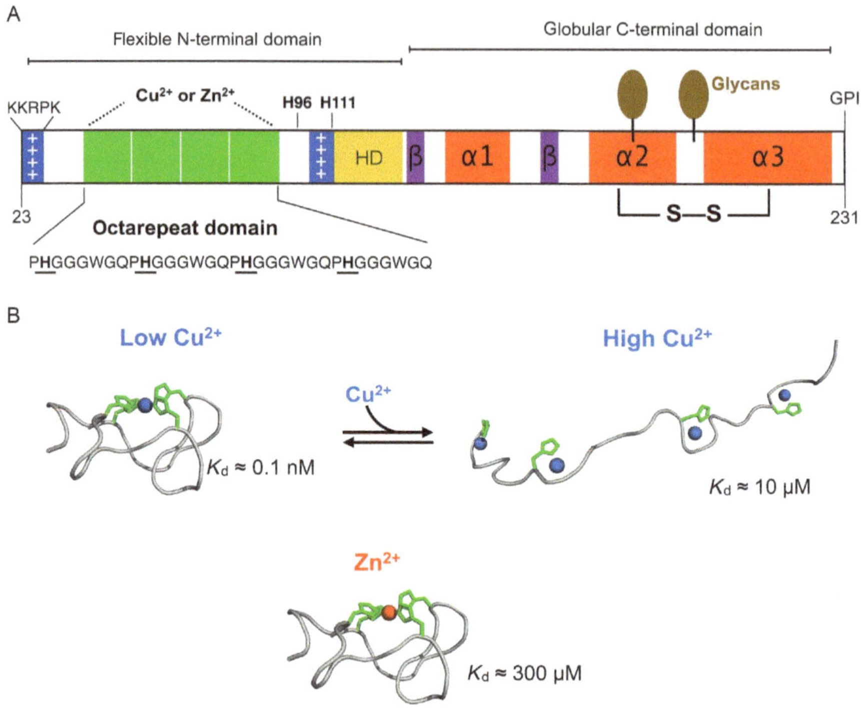

Fig. 1 The cellular prion protein and its interactions with Cu^{2+} and Zn^{2+}. (A) Mature PrP^{C} contains 209 amino acids and is C-terminally tethered to the membrane with a glycosylphosphatidylinositol (GPI) anchor. The globular C-terminal domain consists of three alpha helices (*red boxes*) and two short beta strands (*purple boxes*) and contains a single disulfide bond (S–S) between helices 2 and 3. Two sites of potential N-linked glycosylation are also shown (*brown ovals*). The unstructured N-terminal domain features the Cu^{2+}- and Zn^{2+}-binding octarepeat domain (*green boxes*) along with two positively charged clusters (*blue boxes*) and a hydrophobic domain (*yellow box*). Two additional sites of Cu^{2+} binding at histidines 96 and 111 are indicated. (B) Cu^{2+} and Zn^{2+} coordination by the PrP^{C} octarepeat domain (OR). Histidine side chains are colored *green*, and approximate binding affinities for each coordination mode are indicated.

disulfide bond connects helices 2 and 3 and is essential in forming and maintaining the tertiary fold.[35,36]

Although early nuclear magnetic resonance (NMR) experiments hinted that transient interactions between the N- and C-terminal regions of PrP^{C} might exist, these two domains have by-in-large been thought of as mutually independent.[33] Only recently has it become apparent that these domains can physically associate and that the interdomain *cis* interaction is promoted by metal binding. This association and its implications in PrP-mediated toxicity are discussed in Sections 3 and 4.

1.2 Molecular Features of Cu^{2+} and Zn^{2+} Binding

PrP^C takes up the physiological transition metal ions Cu^{2+} and Zn^{2+} through its flexible N-terminal domain.[37,38] Binding occurs primarily through the OR region, which is comprised of four tandem repeats of the eight-residue sequence PHGGGWGQ. Zinc uptake by PrP^C arises through coordination of a single Zn^{2+} ion by imidazole side chains from all four OR histidine residues and occurs with a dissociation constant of approximately 200 μM.[38,39] Copper coordination, on the other hand, is more complex. Cu^{2+} binds to the PrP^C OR in a multicomponent fashion that critically depends on the relative concentration of Cu^{2+} and the pH of the surrounding environment.[40–42] Measured dissociation constants for OR-bound Cu^{2+} range from ~0.1 nM in the low-occupancy mode to 10 μM in the fully loaded high-occupancy mode (Fig. 1B). Hence, Cu^{2+} binding to the PrP^C OR displays negative cooperativity.[43] In addition, two sites extrinsic to the OR, defined by histidines 96 and 111, are also capable of coordinating Cu^{2+} with high affinity.[44–47]

The ability of PrP^C to respond dynamically to fluctuating concentrations of Cu^{2+} and Zn^{2+}, with dissociation constants that closely mirror reported Cu^{2+} and Zn^{2+} levels at the synapse, is highly suggestive of a role in neuronal metal ion homeostasis.[48,49] From its GPI-anchored position at pre- and postsynaptic terminals, PrP^C would be poised to quickly respond to changing local concentrations of metals created, for example, by neuronal action potentials or neurotransmitter release. These properties have led to proposed roles for PrP^C in metal-dependent signaling and neuroprotection.[27,37] However, precisely how these dynamic interactions with Cu^{2+} and Zn^{2+}, observed in vitro, contribute to the physiological function of PrP^C has yet to be fully understood.

1.3 Metal Ions and PrP^C Function

Since the first observation of PrP^C as a copper-binding protein nearly two decades ago, substantial effort has been made to understand the role of metal ions in both normal PrP^C physiology as well as in prion disease pathogenesis.[50] As a result, the molecular details of PrP^C metal binding are now reasonably well understood, as summarized in Sections 1.1 and 1.2. In contrast, the effects of metal binding on the in vivo function of PrP^C are largely unknown, in part due to a lack of consensus regarding the true function of PrP^C in healthy tissue. Nonetheless, a growing body of evidence now supports the idea that the physiological function of PrP^C is related to its metal-binding properties.[51,52] PrP^C binds copper in vivo, and cultured cells and neurons chronically exposed to elevated Cu^{2+} levels show increased expression of the gene encoding PrP^C.[50,53,54] Both Cu^{2+} and Zn^{2+} stimulate the rapid endocytosis and trafficking of PrP^C, an action that requires the metal-binding OR.[55–58] In addition, PrP^C has been phylogenetically linked to the ZIP (Zrt-, Irt-like protein) family of transmembrane zinc transporters, supporting the idea that its evolved function is related to metal ion regulation.[59] In support of this conclusion, X-ray fluorescence data demonstrate that concentrations and distributions of copper, zinc, and iron in the mouse brain correlate with overall PrP^C expression levels.[60]

Recent reports demonstrating PrP^C-mediated regulation of cell surface receptors have provided additional evidence of the essential roles of PrP Cu^{2+}- and Zn^{2+} binding. Watt *et al.* demonstrated that PrP^C stimulates the influx of Zn^{2+} ions into neuronal cells through AMPA receptors in a manner that requires direct interaction with the receptor and is independent of endocytosis.[25] It was additionally shown that this activity required the presence of the Zn^{2+}-binding OR domain of PrP^C. It has also been established that PrP^C modulates desensitization of NMDA receptors, protecting neurons from excitotoxicity and excessive calcium influx caused by prolonged glutamate exposure.[26,61] Interestingly, proper activation of the NMDA desensitization mechanism required interaction of the receptor with Cu^{2+}-bound PrP^C. Finally, a recent study found that PrP^C interaction with the glutamate transporter EAAT3 in neurons was essential for proper glutamate uptake and protection from oxidative stress.[31] Again, this regulation required the PrP OR. Together, these studies suggest that the modulation of receptor function by PrP^C depends critically on the ability of PrP^C to bind Cu^{2+} and Zn^{2+} through it flexible N-terminal domain.

2. THE N-TERMINAL TOXICITY HYPOTHESIS

Intriguing clues as to how PrP^C might transduce neurotoxic signals have come from two converging lines of experimental evidence in cellular and animal models. The first stems from experiments involving small internal deletions of the PrP^C N-terminal domain that result in profound neurotoxicity in transgenic mouse models.[62–64] These toxic phenotypes depend on both the location and the extent of the deletion and, importantly, can be ameliorated with coexpression of wild-type PrP^C. The second line of evidence involves the observed toxicity induced by ligands (monoclonal antibodies and fragments thereof) targeting the globular C-terminal domain of PrP^C.[65–67] This globular domain (GD) ligand toxicity is epitope dependent and is rescued by deletion or neutralization of certain N-terminal domain segments. These experiments, outlined later, argue for a model wherein the PrP^C GD serves to regulate the activity of the N-terminal tail. Disruptions to this regulation, either through central region deletions or GD ligands, allow toxic signaling carried out through the PrP^C N-terminus. We refer to this model as the "N-terminal toxicity hypothesis."

2.1 Internal Deletions of PrP^C

As mentioned earlier, deletion of the gene encoding PrP^C is well tolerated in cells and transgenic animals. Likewise, deletion of the entire N-terminal domain (residues 23–124) is benign.[68] However, certain internal deletions of the PrP^C N-terminal domain produce varying degrees of spontaneous neurotoxicity in transgenic mice.[62–64] These mice display axonal demyelination, cerebellar neurodegeneration, and neonatal lethality without accumulation of PrP^{Sc}. Importantly, this neurotoxic phenotype can be suppressed in a dose-dependent manner by coexpression of wild-type PrP^C, suggesting that the normal and mutant PrP molecules either interact with each other directly or compete for binding to a specific molecule—or set of molecules—at the cell surface. The severity of the neurotoxic phenotype additionally depends on the length of the amino acid deletion, with mutants that leave larger portions of the N-terminal domain intact being progressively more toxic.[63] The most toxic mutant studied thus far, as judged by the relative amount of wild-type expression required to rescue the toxic phenotype, involves deletion of a mere 21-amino-acid segment in the PrP^C central region (mouse PrP residues 105–125) and is referred to as PrP ΔCR.[64] It was subsequently discovered that the polybasic region at the extreme N-terminus of the protein, consisting of residues 23–32, is essential both for mutant PrP toxicity as well as for the rescuing properties of wild-type PrP^C.[68,69] Interestingly, this polybasic-charged cluster is also important for the interaction of PrP^C with AMPA receptors and toxic Aβ oligomers in glutaminergic neurons, as well as the G-protein-coupled receptor GPR126 in Schwann cells.[25,70,71]

Neuronal cells expressing PrP ΔCR and cultured brain slices from PrP ΔCR transgenic mice display spontaneous ionic currents that can be observed electrophysiologically.[72,73] As with the histopathological markers of ΔCR-mediated neurodegeneration, these currents are inhibited dose dependently by coexpression of wild-type PrP^C and suggest that altered membrane electrical activity is central to the toxic phenotype observed in PrP ΔCR mice. Two main theories have arisen from these observations. The first posits that the observed currents result from a direct disruption of the plasma membrane by the mutant PrP molecules and that wild-type PrP^C blocks the toxic membrane interactions through homophilic interactions with the mutant molecule.[74] While this possibility cannot be ruled out, compelling evidence supports a second mechanism whereby aberrant interactions between the mutant PrP molecules and ionotropic membrane receptors (i.e., NMDA and AMPA receptors) induce spontaneous currents through channel hyperactivation, resulting in subsequent downstream pathology.[72] In this scenario, wild-type PrP^C reverses the toxic signaling by outcompeting the mutant PrP for receptor binding, thus restoring normal PrP^C-mediated regulation. This second mechanism is supported by experiments showing enhanced sensitivity to glutamate-induced excitotoxicity in PrP ΔCR neurons that is reversible with an NMDA receptor antagonist.[72]

2.2 N-Terminal Toxicity Induced by GD Ligands

Scrapie-induced neurodegeneration and neurotoxicity of Aβ oligomers require expression of PrP^C at the neuronal surface and seem to involve toxic, gain-of-function mechanisms mediated by PrP^C. This assessment, combined with observations that PrP^C ablation in animals seems to be tolerated, has promoted the development of monoclonal antibodies (mAbs) targeted to PrP^C as attractive therapeutics for both prion and Alzheimer's diseases.[75,76] However, Sonati *et al.* observed that certain mAbs and mAb fragments with epitopes in the structured C-terminal region of PrP^C produce severe neurotoxicity in mice and cultured cerebellar brain slices.[65] Interestingly, the toxicity of these GD ligands could be reversed by coincubation with mAbs targeted to the PrP^C OR in the N-terminal domain. Moreover, mice lacking the OR were resistant to the toxicity elicited by GD ligands. In order to explain these observations, it was proposed that the GD of PrP^C acts as a regulator of the flexible tail. In this model, mAb binding at specific GD epitopes disrupts normal regulation and releases the N-terminal domain to carry out toxic, gain-of-function effects. Toxic signaling of the dysregulated flexible tail can be blocked either by neutralization with N-terminal mAbs or by deletion of specific N-terminal segments. In support of this hypothesis, transgenic mice expressing a GPI-anchored PrP construct lacking the GD display spontaneous neurodegeneration of cerebellar granule neurons, similar to that seen in with toxic GD ligands.[77] Interestingly, injection of the protective N-terminal mAb POM2 into mice harboring the neonatal lethal central domain deletion PrP Δ94–134 resulted in prolonged survival, supporting the idea that deleterious interactions of the N-terminal tail are also responsible for the neurotoxic phenotype resulting from internal deletions of PrP^{C}.[65]

Sonati *et al.* suggest the possibility that toxic GD ligands mimic the binding of PrP^{Sc} in *bona fide* prion disease. This hypothesis was bolstered by subsequent experiments showing that the neurotoxic pathways induced by toxic GD ligands—namely, calpain activation, increased production of reactive oxidative species, and induction of the unfolded protein response (UPR)—were also induced in scrapie-infected cerebellar brain slice preparations.[66] The most toxic GD-specific mAb studied thus far, POM1, recognizes a noncontiguous epitope consisting of PrP^{C} residues at the exposed surface of helix 1, helix 3, and the C-terminal end of the β1–α1 loop.[65] Indeed, there is evidence from cell-free conversion assays, suggesting that PrP^{Sc} may also bind at this exposed surface of PrP^{C}.[78,79] A more recent study found that another PrP^{C} mAb, ICSM-18, which recognizes an epitope overlapping with that of POM1, also stimulates rapid, on-target neurotoxicity, suggesting that engagement of this specific region of the PrP^{C} GD is required for ligand-induced toxicity.[67] In addition to raising questions regarding the safety of potential PrP^{C}-targeted therapeutic mAbs, these studies provide valuable clues toward uncovering the molecular basis of PrP-mediated toxicity in neurodegenerative disease.

3. CU^{2+} AND ZN^{2+} PROMOTE INTERDOMAIN INTERACTION IN *CIS*

The cellular data presented earlier as evidence of the N-terminal toxicity hypothesis clearly support a paradigm in which the N- and C-terminal domains of PrP^{C} are functionally coupled, with disruption of the coupling leading to protein dysregulation and neurotoxicity carried out by the N-terminal domain. But exactly how this coupling takes place at the molecular level is unclear. Perhaps the simplest mechanism by which the GD could regulate the activity of the N-terminal domain is by a direct interaction. In such a scenario, a domain–domain *cis* interaction would allow for the sequestration of the N-terminal domain, thereby regulating its interactions with cell surface receptors through an autoinhibitory-like mechanism. Suggestive evidence for such N—C interactions of PrP^{C} came initially from NMR experiments demonstrating increased helical order of the GD in full-length

PrP^C compared to a PrP construct lacking the N-terminal domain.[80] Furthermore, extended X-ray absorption fine structure studies revealed that the disease-associated mutation Q212P, occurring on helix 3 of the GD, alters the coordination geometry of the N-terminal, non-OR Cu^{2+}-binding site at His111.[81] The direct involvement of metal ions in *cis* interactions of PrP^C was initially proposed based on Cu^{2+}-induced changes to the NMR spectrum of full-length—but not N-terminally truncated—PrP constructs and was supported by small-angle X-ray scattering data indicating a marginal compaction of the PrP^C structure in the presence of Cu^{2+}.[82]

Given the emerging evidence of functional N—C interactions in PrP^C and their potential relevance to PrP-mediated neurodegeneration, we sought to understand how the physiologically relevant metal ions Cu^{2+} and Zn^{2+} contribute to the global structure of PrP^C. Using NMR and electron paramagnetic resonance (EPR) spectroscopy, we found that both Cu^{2+} and Zn^{2+} promote *cis* interaction between the GD and the flexible N-terminal domain of PrP^C.[83,84] As shown in Fig. 2, metal binding to the PrP^C OR results in the selective broadening of cross peaks in the ^1H–^{15}N heteronuclear single-quantum coherence (HSQC) NMR spectrum of full-length PrP^C. These peak broadenings are manifested as intensity reductions in the 2D HSQC spectrum that can be quantified on a residue-by-residue basis. Those residues significantly broadened by the addition of Cu^{2+} or Zn^{2+}, when mapped onto the structure of the mouse PrP GD, reveal a localized interaction surface defined by helix 2, helix 3, and the distal section of the β1–α1 loop leading into helix 1 (Fig. 2B). The selective broadening of these GD residues is highly dependent on the presence of the N-terminal domain, indicating that the effect is not caused by direct binding of Cu^{2+} or Zn^{2+} to the GD, but instead by long-range tertiary interactions with the N-terminal domain.

While NMR definitively identified the residues in the PrP^C GD that participate in *cis* interaction, the amino acids of the flexible N-terminal domain are generally not resolvable by NMR, and therefore, it was unclear which region—or regions—of the N-terminal domain were involved in the tertiary interaction. In an attempt to answer this question, we employed double electron–electron resonance (DEER) EPR to directly probe distances between nitroxide (NO) spin labels incorporated at various residues in the N- and C-terminal domains of PrP^C. As shown in Fig. 2, DEER reveals a broad and relatively featureless distance distribution when spin labels are placed simultaneously in the OR (residue 66, mouse numbering) and in the GD (residue 200). However, upon the addition of Zn^{2+}, the

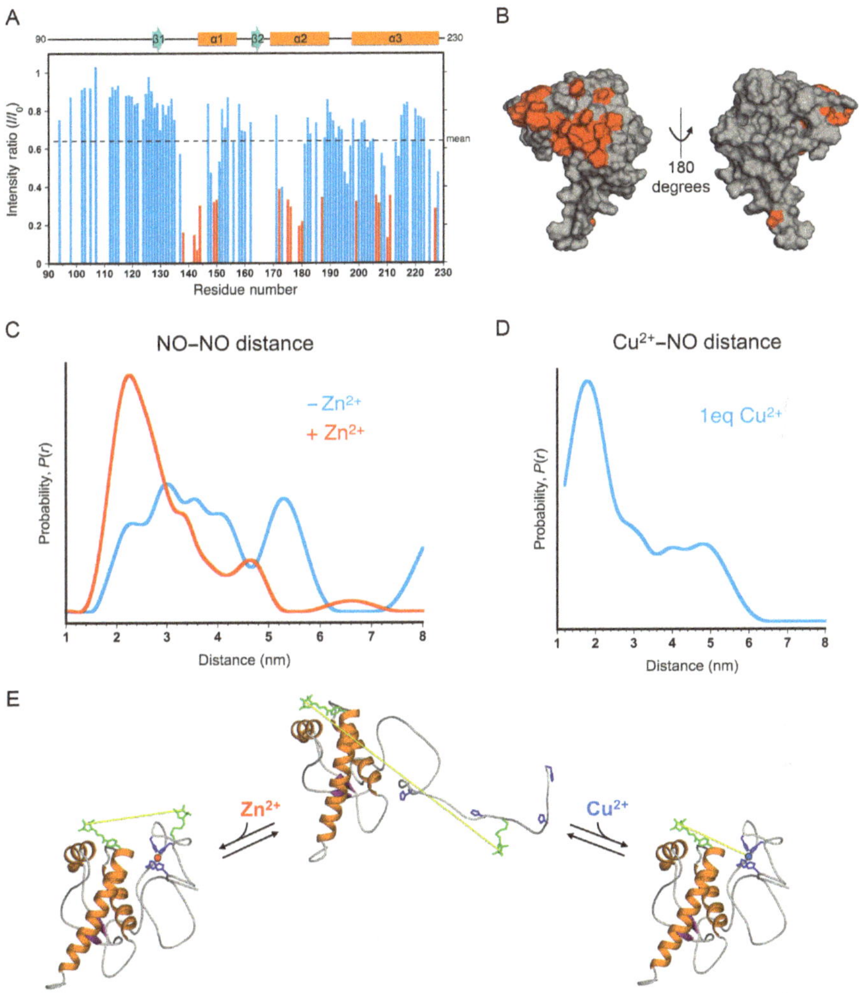

Fig. 2 Octarepeat-bound Cu^{2+} and Zn^{2+} promote *cis* association of the PrP^C N- and C-terminal domains. (A) ^1H–^{15}N HSQC NMR intensity ratios (I/I_0) as a function of primary sequence for recombinant mouse PrP H95Y/H110Y. Intensity ratios are defined as the cross-peak intensity in the presence of 1 equivalent of Cu^{2+} divided by the peak intensity in the absence of metal. Significantly perturbed residues (*Z*-score < -1) are colored in *red*. (B) Surface plot of mouse PrP(121 – 230) (PDB 1XYX) showing residues significantly affected by OR Cu^{2+} binding as determined from ^1H–^{15}N HSQC. (C) DEER EPR distance distributions between nitroxide (NO) spin labels placed at residue 200 of the PrP^C GD and within the OR at residue 66, both in the absence of metal (*blue*) and in the presence of 1 mM Zn^{2+} (*red*). (D) Distance distribution between OR-bound Cu^{2+} and an NO spin label introduced at residue 200 of the PrP^C GD. (E) Structural models of interdomain association from DEER distance measurements. Spin-labeled side chains are indicated in *green*, and OR histidines are colored *blue*.

distribution sharpens significantly and a dominant interspin distance emerges at approximately 2.2 nm, indicating that the Zn^{2+}-bound OR occupies a relatively stable position proximate to helix 3 of the GD. Cumulative analysis of distributions from spin labels placed at several positions in the GD and at an additional site near the charged cluster at residue 102 indicates that the Zn^{2+}-driven *cis* interaction is likely to involve the OR and residues close to the N-terminal half of helix 3, consistent with the NMR results.

Cu^{2+} is a d^9 transition metal ion with a single unpaired electron in its valence shell. Hence, it is paramagnetic and can be used as an EPR spin probe in much the same way as NO spin labels. However, as discussed in Section 1.2, Cu^{2+} binding to the N-terminal domain of PrP^C is complex, with at least six Cu^{2+} ions bound to the protein at saturation. In order to mitigate the challenges this complexity creates in both NMR and DEER experiments, we used a combination of mutagenesis, pH adjustment, and careful control of Cu^{2+} stoichiometry to limit Cu^{2+} coordination to the high-affinity, low-occupancy OR-binding mode (see Fig. 1B). We then used DEER to measure distances between the endogenous Cu^{2+} ion and NO spin labels placed in the GD. Fig. 2D shows the distance distribution obtained by DEER between the OR-bound Cu^{2+} and a spin label introduced at residue 200 of the GD. A dominant peak at approximately 1.8 nm is observed, in remarkable agreement with the NO–NO distance obtained by DEER on Zn^{2+}-bound PrP^C with labels at residues 66 of the OR and 200 of the GD, suggesting that Cu^{2+}—at least in the low-occupancy-binding mode—induces a similar interaction between the N- and C-terminal domains of PrP^C as does Zn^{2+}.

The GD interaction surface revealed by our NMR experiments shows a striking overlap with a well-known, highly conserved electronegative pocket of PrP^C. Fig. 3A shows the calculated electrostatic surface potential from several mammalian PrP^C structures, illustrating the conservation of this negatively charged pocket across species. The glaring coincidence between the *cis* interaction surface and the conserved electronegative patch, combined with our DEER results indicating the proximity of the metal-bound OR, strongly suggests that the observed interdomain association is driven by electrostatic attraction. In fact, trilateration based on the combined NO–NO and Cu^{2+}–NO distances from DEER localize the Cu^{2+} ion to the exposed surface of PrP^C near the negative pocket, further supporting a direct involvement of OR-bound divalent ions in stabilizing domain–domain tertiary structure.[84] However, others have shown that *cis* interaction in PrP^C can exist even in the absence of Cu^{2+} or Zn^{2+} and is critically dependent

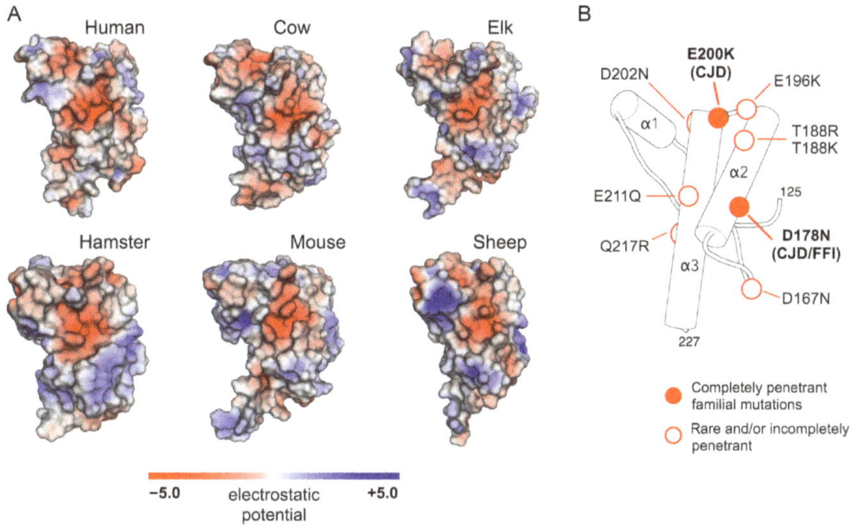

Fig. 3 PrP^C surface electrostatics and familial prion disease mutations. (A) Surface electrostatic calculations showing a negatively charged pocket of the globular domain that is conserved among species. Plots were generated using the adaptive Poisson–Boltzmann solver (http://www.poissonboltzmann.org) in PyMOL (http://www.pymol.org) with structural coordinates deposited for human (PDB 1HJM), cow (PDB 1DWY), elk (PDB 1XYW), hamster (PDB 1B10), mouse (PDB 1XYX), and sheep (PDB 1UW3) PrP^C. (B) All reportedly pathogenic human mutations of PrP^C involving a change of charge are displayed relative to the globular domain structure. Notably, all mutations involving a charge change result in a decrease of negative charge, an increase of positive charge, or both. Assessment of pathogenicity taken from Refs. 85 and 86.

on charge complementarity between the positively charged residues of the N-terminal polybasic region[23–32] and negatively charged residues of helix 3.[87] In any case, it is evident that *cis* association between the N- and C-terminal domains of PrP^C relies on electrostatic attraction and is stabilized by the binding of Cu^{2+} and Zn^{2+} to the N-terminal OR.

4. DISRUPTION OF *CIS* INTERACTION AS A MECHANISM OF NEUROTOXICITY

Recent cellular and transgenic animal studies point to a highly interdependent relationship between the N- and C-terminal regions of PrP^C and implicate a dysregulated N-terminal domain in multiple types of PrP-mediated toxicity. At the same time, biophysical studies with recombinant PrP^C show an interdomain *cis* interaction that is stabilized by coordination of Cu^{2+} and Zn^{2+}. Might this physical, intramolecular interaction be important for PrP^{C}'s functional regulation of cell surface receptors, and might the disruption of normal N—C interactions in PrP^{C} be a general mechanism of PrP-mediated neurotoxicity? While this question is still far from being answered, several lines of evidence offer initial support for this hypothesis and provide an underpinning for future study.

Approximately 15% of human prion diseases are caused by inherited mutations in *PRNP*, the gene that encodes PrP^{C}. A majority of these pathogenic mutations, which include Creutzfeldt–Jakob disease (CJD), fatal familial insomnia (FFI), and Gerstmann–Sträusller–Scheinker syndrome (GSS), result in amino acid substitutions in either the PrP^{C} GD, or in the linker region that separates the GD from the flexible N-terminal domain.[85,86] These mutations were initially hypothesized to cause disease by destabilizing the PrP^{C} structure, thereby increasing the likelihood of misfolding to a PrP^{Sc}-like state and ultimately triggering neurodegeneration.[88] However, subsequent studies revealed no systematic reductions in the overall stability of PrP^{C} harboring pathogenic point mutations.[89,90] Interestingly, many of the pathogenic mutations of the GD have the effect of decreasing the overall negative charge of the domain.[91] In fact, as illustrated in Fig. 3B, every reportedly pathogenic mutation of the GD that involves a charge change results in either the gain of a positive charge, the loss of a negative charge, or both.

Given the biophysical evidence suggesting the importance of the electronegative GD pocket in promoting *cis* interaction with the N-terminal domain, we performed HSQC NMR experiments on several of the most common pathogenic PrP^{C} mutants to test for Zn^{2+}-promoted interdomain contacts.[83] The mouse PrP^{C} mutations E199K and D177N, corresponding to the human E200K and D178N mutations responsible for familial CJD and either CJD or FFI (depending on the codon at residue 129), respectively, showed weakened *cis* interactions, as evidenced by an overall decrease in Zn^{2+}-induced peak broadening compared to wild-type PrP^{C}. Despite the resulting increase in positive charge, the protective PrP^{C} polymorphism Q219K did not display reduced N—C interactions by NMR, indicating that altered interdomain structure may be a feature restricted to pathogenic GD mutants of PrP^{C}. Surprisingly, the GSS-associated mutation P101L (P102L in humans) also produced a minor attenuation of the apparent strength of *cis* interaction, suggesting that the conformational rigidity of this proline may play a role in orienting the two PrP domains for proper *cis* interaction.

There is also reason to believe that Cu^{2+}- and Zn^{2+}-promoted N—C interactions of PrP^{C} may also be compromised by the binding of toxic ligands to the PrP^C GD.[84] Two mAbs that have demonstrated severe neurotoxicity in vivo, POM1 and ICSM18, recognize a discontinuous epitope on PrP^C consisting of residues from helix 1, helix 3, and part of the β1–α1 loop.[65,67] This epitope lies entirely within the *cis* interaction surface identified by NMR and DEER EPR experiments. Binding of these mAbs would therefore preclude binding of the metal-bound N-terminal domain to its target electronegative surface of the GD and disrupt the tertiary fold. If it is indeed true that neurotoxic ligands, including PrP^{Sc}, compete with the metal-bound OR for binding to the exposed GD surface identified in Fig. 2B, it could explain why both Cu^{2+} and Zn^{2+} strongly inhibit the in vitro amplification of infectious and protease-resistant PrP aggregates.[92] Assessment of this theory will have to await validation by future experiments. However, the occlusion of a physiological intramolecular structure in PrP^C, namely, metal-mediated *cis* interaction, by various neurotoxic species may represent the initial molecular event in prion-mediated neurodegeneration. As such, understanding the molecular details of this process may lead to alternative strategies for treating neurodegenerative disease.

As described in previous sections, GPI-anchored PrP^C regulates the activity of several cell surface receptors—including the ionotropic AMPA and NMDA glutamate receptors—through interactions involving PrP's metal-binding N-terminal domain.[25,26,28] Fig. 4 presents a schematic model depicting potential mechanisms of PrP-mediated neurotoxicity, using the functional regulation of NMDAR signaling as an example. In this model, *cis* interaction promoted by Cu^{2+}- or Zn^{2+} binding to the N-terminal OR results in a compact tertiary structure essential for proper regulation of the receptor by PrP^C. Disruption of *cis* interaction in PrP^C by antibody-based ligands that bind a specific GD epitope encompassing portions of helices 1 and 3, as well as the β1–α1 loop, releases the N-terminal domain to engage in aberrant or uncontrolled interactions with the receptor, stimulating toxic signaling and downstream neurodegenerative pathways. The toxicity induced by these GD ligands, which may include PrP^{Sc} and other β-sheet-rich oligomers, is blocked by the administration of certain N-terminal mAbs such as POM2 and polyanionic compounds such as glycosaminoglycans (GAGs).[65,68] In a similar manner, we propose that internal deletions in the central region linking the GD and the OR, as exemplified by the lethal ΔCR phenotype, may also disrupt normal *cis* interaction, resulting in toxic signaling mediated by the PrP N-terminus. This model would suggest that binding of N-terminal ligands (POM2, GAGs) might also disrupt *cis*

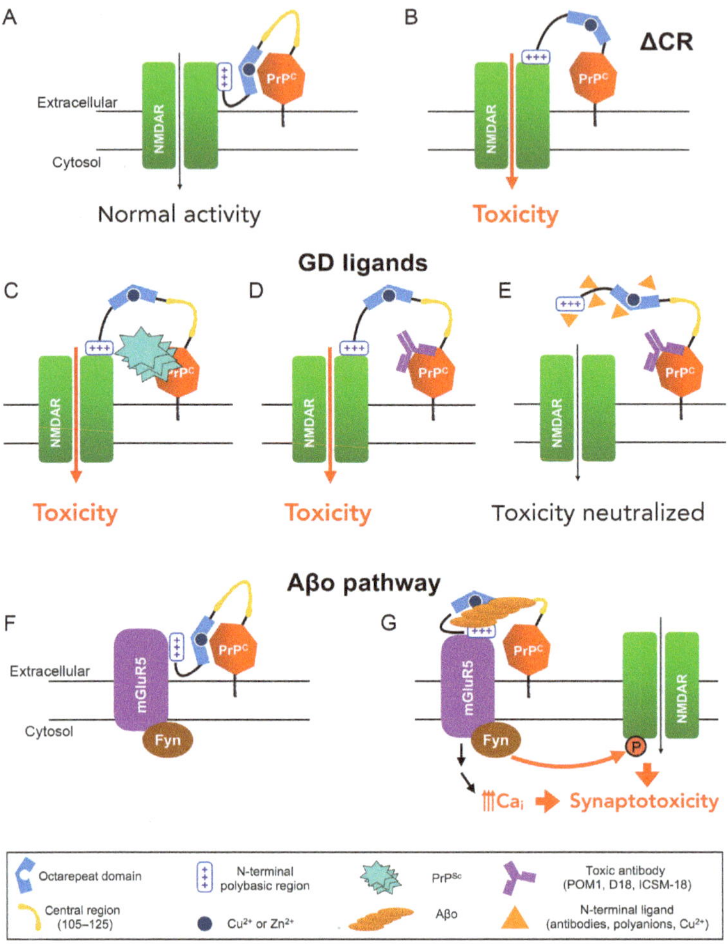

Fig. 4 A theoretical model of PrP^C-mediated neurotoxicity involving dysregulation of the N-terminal domain. (A) Normal modulation of cell surface receptors such as the *N*-methyl D-aspartate receptor (NMDAR) by PrP^C requires proper regulation of the N-terminal domain via Cu^{2+}/Zn^{2+}-promoted N—C interaction. (B) Internal deletions of the PrP^C N-terminal domain that preserve the N-terminal polybasic region, such as ΔCR, may disrupt normal N—C interactions, resulting in dysregulation of the N-terminal domain and uncontrolled activation of NMDA receptors. In a similar manner, toxic ligands of the globular domain, including PrP^{Sc} (C) and certain antibodies (D), may also disrupt *cis* interactions of PrP^C. (E) The toxic effects of the dysregulated N-terminal domain are neutralized by N-terminal ligands such as antibodies, polyanionic compounds, and perhaps Cu^{2+} ions. (F) and (G) Synaptotoxic effects of amyloid-β oligomers (Aβo) are mediated by PrP^C. Aβo bind directly to PrP^C at the cell surface, activating intracellular Fyn kinase through the metabotropic glutamate receptor mGluR5. Activation of this pathway causes phosphorylation of NMDAR subunits and an increase in intracellular calcium (Ca_i), resulting in impaired long-term potentiation (LTP) and a retraction of dendritic spines.

interaction in PrP^C; however, because the N-terminal domain is responsible for inducing the toxic signal, the effect of these ligands would be akin to functional deletion of PrP^C which, as we know, is generally nontoxic. As a consequence, compounds aimed at neutralizing the PrP^C N-terminal domain might be reasonable targets for therapeutic intervention in prion disease.

Increasing evidence now points to a central role for PrP^C in transducing early neurotoxic signals associated with AD.[12,15] As shown in Fig. 4, PrP^C interacts with the metabotropic glutamate receptor 5 (mGluR5) at the post-synaptic terminals of neurons, where it is coupled to various intracellular signaling cascades.[13,29] Toxic species of oligomeric amyloid-beta peptide (Aβo), implicated in AD pathogenesis, bind specifically to PrP^C at two regions in the flexible N-terminal domain and trigger mGluR5 activation of Fyn kinase, phosphorylation of NMDA receptor subunits, and a rise in the intracellular calcium concentration, ultimately leading to impairment of synaptic long-term potentiation (LTP) and retraction of dendritic spines.[5,12,13,93,94] Understanding how normal PrP^C regulation is subverted upon binding of Aβo is, therefore, of great interest, as it may lead to new and desperately needed treatments for AD.

5. CONCLUDING REMARKS

Studies conducted over the past decade have revealed a host of cellular and molecular pathways that are now believed to be regulated by PrP^C. In particular, PrP^C is increasingly recognized as a promiscuous mediator of neuronal excitability, with an important role in protecting neurons against excitotoxicity. Interestingly, functional modulation of cellular receptors by PrP^C seems to be related to its ability to bind Cu^{2+} and Zn^{2+} through its unstructured N-terminal domain. In addition to clarifying the physiological role of PrP^C in healthy cells, these studies have also shed new light on the neurotoxic signals that occur when these regulatory properties are altered, with important implications for prion and other neurodegenerative diseases. At the same time, recent structural and biophysical studies have revealed previously unseen physical interactions between the N- and C-terminal regions of PrP^C that are stabilized by Cu^{2+}- and Zn^{2+} binding to the N-terminal OR. Here, we contend that this interdomain *cis* interaction in PrP^C may represent an important molecular mechanism for regulating the activity of the N-terminal domain. Moreover, we believe that disruption of physiological N—C interactions may be a general mechanism, common to multiple types of PrP-mediated neurotoxicity, that results in toxic signaling induced by the PrP^C N-terminal domain. Our results on metal-mediated *cis* interaction have thus far been limited to situations in which a single Zn^{2+} or Cu^{2+} ion is bound to the PrP^C OR. However, as discussed earlier, PrP^C also binds Cu^{2+} at two sites extrinsic to the OR and can adopt additional OR coordination environments when exposed to higher concentrations of Cu^{2+}. Future studies exploring *cis* interaction in PrP^C with varying concentrations of Cu^{2+} and Zn^{2+} are sure to provide important insights into the physiological regulation of receptor function by PrP^C and may also shed light on how these regulatory interactions are subverted in PrP-mediated neurodegeneration.

ACKNOWLEDGMENT

The authors acknowledge generous support from the National Institutes of Health to GLM (R01GM 065790).

REFERENCES

1. Prusiner SB. Novel proteinaceous infectious particles cause scrapie. *Science.* 1982;216(4542):136–144.
2. Prusiner SB. Prions. *Proc Natl Acad Sci USA.* 1998;95(23):13363–13383.
3. Collinge J. Mammalian prions and their wider relevance in neurodegenerative diseases. *Nature.* 2016;539(7628):217–226.
4. Herms J, Tings T, Gall S, et al. Evidence of presynaptic location and function of the prion protein. *J Neurosci.* 1999;19(20):8866–8875.
5. Um JW, Nygaard HB, Heiss JK, et al. Alzheimer amyloid-β oligomer bound to postsynaptic prion protein activates Fyn to impair neurons. *Nat Neurosci.* 2012;15(9):1227–1235.
6. Büeler H, Aguzzi A, Sailer A, et al. Mice devoid of PrP are resistant to scrapie. *Cell.* 1993;73(7):1339–1347.
7. Sailer A, Büeler H, Fischer M, Aguzzi A, Weissmann C. No propagation of prions in mice devoid of PrP. *Cell.* 1994;77(7):967–968.
8. Brandner S, Isenmann S, Raeber A, et al. Normal host prion protein necessary for scrapie-induced neurotoxicity. *Nature.* 1996;379(6563):339–343.
9. Mallucci G, Dickinson A, Linehan J, Klöhn P-C, Brandner S, Collinge J. Depleting neuronal PrP in prion infection prevents disease and reverses spongiosis. *Science.* 2003;302(5646):871–874.
10. Chesebro B, Trifilo M, Race R, et al. Anchorless prion protein results in infectious amyloid disease without clinical scrapie. *Science.* 2005;308(5727):1435–1439.
11. Saá P, Harris DA, Cervenakova L. Mechanisms of prion-induced neurodegeneration. *Expert Rev Mol Med.* 2016;18: e5.
12. Laurén J, Gimbel DA, Nygaard HB, Gilbert JW, Strittmatter SM. Cellular prion protein mediates impairment of synaptic plasticity by amyloid-beta oligomers. *Nature.* 2009;457(7233):1128–1132.
13. Um JW, Kaufman AC, Kostylev M, et al. Metabotropic glutamate receptor 5 is a coreceptor for Alzheimer aβ oligomer bound to cellular prion protein. *Neuron.* 2013;79(5):887–902.

14. Kostylev MA, Kaufman AC, Nygaard HB, et al. Prion-protein-interacting amyloid-β oligomers of high molecular weight are tightly correlated with memory impairment in multiple Alzheimer mouse models. *J Biol Chem*. 2015;290(28):17415–17438.
15. Resenberger UK, Harmeier A, Woerner AC, et al. The cellular prion protein mediates neurotoxic signalling of β-sheet-rich conformers independent of prion replication. *EMBO J*. 2011;30(10):2057–2070.
16. Rushworth JV, Griffiths HH, Watt NT, Hooper NM. Prion protein-mediated toxicity of amyloid-β oligomers requires lipid rafts and the transmembrane LRP1. *J Biol Chem*. 2013;288(13):8935–8951.
17. Büeler H, Fischer M, Lang Y, et al. Normal development and behaviour of mice lacking the neuronal cell-surface PrP protein. *Nature*. 1992;356(6370):577–582.
18. Richt JA, Kasinathan P, Hamir AN, et al. Production of cattle lacking prion protein. *Nat Biotechnol*. 2007;25(1):132–138.
19. Steele AD, Lindquist S, Aguzzi A. The prion protein knockout mouse: a phenotype under challenge. *Prion*. 2007;1(2):83–93.
20. Tobler I, Gaus SE, Deboer T, et al. Altered circadian activity rhythms and sleep in mice devoid of prion protein. *Nature*. 1996;380(6575):639–642.
21. Collinge J, Whittington MA, Sidle KC, et al. Prion protein is necessary for normal synaptic function. *Nature*. 1994;370(6487):295–297.
22. Canello T, Friedman-Levi Y, Mizrahi M, et al. Copper is toxic to PrP-ablated mice and exacerbates disease in a mouse model of E200K genetic prion disease. *Neurobiol Dis*. 2012;45(3):1010–1017.
23. Bremer J, Baumann F, Tiberi C, et al. Axonal prion protein is required for peripheral myelin maintenance. *Nat Neurosci*. 2010;13(3):310–318.
24. Mouillet-Richard S, Vilotte J-L. Promiscuous functions of the prion protein family. *Front Cell Dev Biol*. 2015;3:7.
25. Watt NT, Taylor DR, Kerrigan TL, et al. Prion protein facilitates uptake of zinc into neuronal cells. *Nat Commun*. 2012;3:1134.
26. You H, Tsutsui S, Hameed S, et al. Aβ neurotoxicity depends on interactions between copper ions, prion protein, and N-methyl-D-aspartate receptors. *Proc Natl Acad Sci USA*. 2012;109(5):1737–1742.
27. Watt NT, Hooper NM. The prion protein and neuronal zinc homeostasis. *Trends Biochem Sci*. 2003;28(8):406–410.
28. Black SAG, Stys PK, Zamponi GW, Tsutsui S. Cellular prion protein and NMDA receptor modulation: protecting against excitotoxicity. *Front Cell Dev Biol*. 2014;2:45.
29. Haas LT, Salazar SV, Kostylev MA, Um JW, Kaufman AC, Strittmatter SM. Metabotropic glutamate. Receptor 5 couples cellular prion protein to intracellular signalling in Alzheimer's disease. *Brain*. 2016;139(pt 2):526–546.
30. Beraldo FH, Arantes CP, Santos TG, et al. Metabotropic glutamate receptors transduce signals for neurite outgrowth after binding of the prion protein to laminin γ1 chain. *FASEB J*. 2011;25(1):265–279.
31. Guitart K, Loers G, Schachner M, Kleene R. Prion protein regulates glutathione metabolism and neural glutamate and cysteine uptake via excitatory amino acid transporter 3. *J Neurochem*. 2015;133(4):558–571.
32. Riek R, Hornemann S, Wider G, Glockshuber R, Wüthrich K. NMR characterization of the full-length recombinant murine prion protein, mPrP (23-231). *FEBS Lett*. 1997;413(2):282–288.
33. Donne DG, Viles JH, Groth D, et al. Structure of the recombinant full-length hamster prion protein PrP(29-231): the N terminus is highly flexible. *Proc Natl Acad Sci USA*. 1997;94(25):13452–13457.
34. Riek R, Hornemann S, Wider G, Billeter M, Glockshuber R, Wüthrich KNMR. Structure of the mouse prion protein domain PrP(121-231). *Nature*. 1996;382(6587): 180–182.

35. Maiti NR, Surewicz WK. The role of disulfide bridge in the folding and stability of the recombinant human prion protein. *J Biol Chem*. 2001;276(4):2427–2431.
36. Benetti F, Biarnés X, Attanasio F, Giachin G, Rizzarelli E, Legname G. Structural determinants in prion protein folding and stability. *J Mol Biol*. 2014;426(22):3796–3810.
37. Millhauser GL. Copper and the prion protein: methods, structures, function, and disease. *Annu Rev Phys Chem*. 2007;58:299–320.
38. Walter ED, Stevens DJ, Visconte MP, Millhauser GL. The prion protein is a combined zinc and copper binding protein: Zn2+ alters the distribution of Cu2+ coordination modes. *J Am Chem Soc*. 2007;129(50):15440–15441.
39. Pushie MJ, Nienaber KH, Mcdonald A, Millhauser GL, George GN. Combined EXAFS and DFT structure calculations provide structural insights into the 1:1 multi-histidine complexes of Cu(II), Cu(I), and Zn(II) with the tandem octarepeats of the mammalian prion protein. *Chemistry*. 2014;20(31):9770–9783.
40. Aronoff-Spencer E, Burns CS, Avdievich NI, et al. Identification of the Cu2+ binding sites in the N-terminal domain of the prion protein by EPR and CD spectroscopy. *Biochemistry*. 2000;39(45):13760–13771.
41. Chattopadhyay M, Walter ED, Newell DJ, et al. The octarepeat domain of the prion protein binds Cu(II) with three distinct coordination modes at pH 7.4. *J Am Chem Soc*. 2005;127(36):12647–12656.
42. Whittal RM, Ball HL, Cohen FE, Burlingame AL, Prusiner SB, Baldwin MA. Copper binding to octarepeat peptides of the prion protein monitored by mass spectrometry. *Protein Sci*. 2008;9(2):332–343.
43. Walter ED, Chattopadhyay M, Millhauser GL. The affinity of copper binding to the prion protein octarepeat domain: evidence for negative cooperativity. *Biochemistry*. 2006;45(43):13083–13092.
44. Burns CS, Aronoff-Spencer E, Legname G, et al. Copper coordination in the full-length, recombinant prion protein. *Biochemistry*. 2003;42(22):6794–6803.
45. Jones CE, Abdelraheim SR, Brown DR, Viles JH. Preferential Cu2+ coordination by His96 and His111 induces beta-sheet formation in the unstructured amyloidogenic region of the prion protein. *J Biol Chem*. 2004;279(31):32018–32027.
46. Jones CE, Klewpatinond M, Abdelraheim SR, Brown DR, Viles JH. Probing copper2+ binding to the prion protein using diamagnetic nickel2+ and 1H NMR: the unstructured N terminus facilitates the coordination of six copper2+ ions at physiological concentrations. *J Mol Biol*. 2005;346(5):1393–1407.
47. Walter ED, Stevens DJ, Spevacek AR, Visconte MP, Dei Rossi A, Millhauser GL. Copper binding extrinsic to the octarepeat region in the prion protein. *Curr Protein Pept Sci*. 2009;10(5):529–535.
48. Kramer ML, Kratzin HD, Schmidt B, et al. Prion protein binds copper within the physiological concentration range. *J Biol Chem*. 2001;276(20):16711–16719.
49. Hopt A, Korte S, Fink H, et al. Methods for studying synaptosomal copper release. *J Neurosci Methods*. 2003;128(1–2):159–172.
50. Brown DR, Qin K, Herms JW, et al. The cellular prion protein binds copper in vivo. *Nature*. 1997;390(6661):684–687.
51. Millhauser GL. Copper and prion protein function: a brief review of emerging theories of neuroprotection. In: Milardi D, Rizzarelli E, eds. *Neurodegeneration: Metallostasis and Proteostasis*. Cambridge, UK: Royal Society of Chemistry; 2011:249–258. RSC Drug Discovery, vol 7.
52. Watt NT, Griffiths HH, Hooper NM. Neuronal zinc regulation and the prion protein. *Prion*. 2013;7(3):203–208.
53. Armendariz AD, Gonzalez M, Loguinov AV, Vulpe CD. Gene expression profiling in chronic copper overload reveals upregulation of Prnp and App. *Physiol Genomics*. 2004;20(1):45–54.

54. Varela-Nallar L, Toledo EM, Larrondo LF, Cabral ALB, Martins VR, Inestrosa NC. Induction of cellular prion protein gene expression by copper in neurons. *Am J Physiol Cell Physiol.* 2006;290(1):C271–C281.
55. Pauly PC, Harris DA. Copper stimulates endocytosis of the prion protein. *J Biol Chem.* 1998;273(50):33107–33110.
56. Perera WS, Hooper NM. Ablation of the metal ion-induced endocytosis of the prion protein by disease-associated mutation of the octarepeat region. *Curr Biol.* 2001;11(7): 519–523.
57. Brown LR, Harris DA. Copper and zinc cause delivery of the prion protein from the plasma membrane to a subset of early endosomes and the Golgi. *J Neurochem.* 2003;87(2):353–363.
58. Hooper NM, Taylor DR, Watt NT. Mechanism of the metal-mediated endocytosis of the prion protein. *Biochem Soc Trans.* 2008;36(pt 6):1272–1276.
59. Schmitt-Ulms G, Ehsani S, Watts JC, Westaway D, Wille H. Evolutionary descent of prion genes from the ZIP family of metal ion transporters. *PLoS One.* 2009;4(9): e7208.
60. Pushie MJ, Pickering IJ, Martin GR, Tsutsui S, Jirik FR, George GN. Prion protein expression level alters regional copper, iron and zinc content in the mouse brain. *Metallomics.* 2011;3(2):206–214.
61. Stys PK, You H, Zamponi GW. Copper-dependent regulation of NMDA receptors by cellular prion protein: implications for neurodegenerative disorders. *J Physiol.* 2012;590(6):1357–1368.
62. Shmerling D, Hegyi I, Fischer M, et al. Expression of amino-terminally truncated PrP in the mouse leading to ataxia and specific cerebellar lesions. *Cell.* 1998;93(2):203–214.
63. Baumann F, Tolnay M, Brabeck C, et al. Lethal recessive myelin toxicity of prion protein lacking its central domain. *EMBO J.* 2007;26(2):538–547.
64. Li A, Christensen HM, Stewart LR, Roth KA, Chiesa R, Harris DA. Neonatal lethality in transgenic mice expressing prion protein with a deletion of residues 105-125. *EMBO J.* 2007;26(2):548–558.
65. Sonati T, Reimann RR, Falsig J, et al. The toxicity of antiprion antibodies is mediated by the flexible tail of the prion protein. *Nature.* 2013;501(7465):102–106.
66. Herrmann US, Sonati T, Falsig J, et al. Prion infections and anti-PrP antibodies trigger converging neurotoxic pathways. *PLoS Pathog.* 2015;11(2): e1004662.
67. Reimann RR, Sonati T, Hornemann S, et al. Differential toxicity of antibodies to the prion protein. *PLoS Pathog.* 2016;12(1): e1005401.
68. Westergard L, Turnbaugh JA, Harris DA. A nine amino acid domain is essential for mutant prion protein toxicity. *J Neurosci.* 2011;31(39):14005–14017.
69. Turnbaugh JA, Westergard L, Unterberger U, Biasini E, Harris DA. The N-terminal, polybasic region is critical for prion protein neuroprotective activity. *PLoS One.* 2011;6(9): e25675.
70. Chen S, Yadav SP, Surewicz WK. Interaction between human prion protein and amyloid-beta (Abeta) oligomers: role of N-terminal residues. *J Biol Chem.* 2010;285(34):26377–26383.
71. Küffer A, Lakkaraju AKK, Mogha A, et al. The prion protein is an agonistic ligand of the G protein-coupled receptor Adgrg 6. *Nature.* 2016;536(7617):464–468.
72. Biasini E, Unterberger U, Solomon IH, et al. A mutant prion protein sensitizes neurons to glutamate-induced excitotoxicity. *J Neurosci.* 2013;33(6):2408–2418.
73. Solomon IH, Huettner JE, Harris DA. Neurotoxic mutants of the prion protein induce spontaneous ionic currents in cultured cells. *J Biol Chem.* 2010;285(34): 26719–26726.
74. Solomon IH, Biasini E, Harris DA. Ion channels induced by the prion protein: mediators of neurotoxicity. *Prion.* 2012;6(1):40–45.
75. White AR, Enever P, Tayebi M, et al. Monoclonal antibodies inhibit prion replication and delay the development of prion disease. *Nature.* 2003;422(6927):80–83.

76. Klyubin I, Nicoll AJ, Khalili-Shirazi A, et al. Peripheral administration of a humanized anti-PrP antibody blocks Alzheimer's disease Aβ synaptotoxicity. *J Neurosci.* 2014;34(18):6140–6145.
77. Dametto P, Lakkaraju AKK, Bridel C, et al. Neurodegeneration and unfolded-protein response in mice expressing a membrane-tethered flexible tail of PrP. *PLoS One.* 2015;10(2): e0117412.
78. Horiuchi M, Caughey B. Specific binding of normal prion protein to the scrapie form via a localized domain initiates its conversion to the protease-resistant state. *EMBO J.* 1999;18(12):3193–3203.
79. Li L, Guest W, Huang A, Plotkin SS, Cashman NR. Immunological mimicry of PrPC-PrPSc interactions: antibody-induced PrP misfolding. *Protein Eng Des Sel.* 2009;22(8):523–529.
80. Zahn R, Liu A, Lührs T, et al. NMR solution structure of the human prion protein. *Proc Natl Acad Sci USA.* 2000;97(1):145–150.
81. D'Angelo P, Longa Della S, Arcovito A, et al. Effects of the pathological Q212P mutation on human prion protein non-octarepeat copper-binding site. *Biochemistry.* 2012;51(31):6068–6079.
82. Thakur AK, Srivastava AK, Srinivas V, Chary KVR, Rao CM. Copper alters aggregation behavior of prion protein and induces novel interactions between its N- and C-terminal regions. *J Biol Chem.* 2011;286(44):38533–38545.
83. Spevacek AR, Evans EGB, Miller JL, Meyer HC, Pelton JG, Millhauser GL. Zinc drives a tertiary fold in the prion protein with familial disease mutation sites at the interface. *Structure.* 2013;21(2):236–246.
84. Evans EGB, Pushie MJ, Markham KA, Lee H-W, Millhauser GL. Interaction between prion protein's copper-bound octarepeat domain and a charged C-terminal pocket suggests a mechanism for N-terminal regulation. *Structure.* 2016;24(7):1057–1067.
85. Beck JA, Poulter M, Campbell TA, et al. PRNP allelic series from 19 years of prion protein gene sequencing at the MRC prion unit. *Hum Mutat.* 2010;31(7):E1551–E1563.
86. Minikel EV, Vallabh SM, Lek M, et al. Quantifying prion disease penetrance using large population control cohorts. *Sci Transl Med.* 2016;8(322):322ra9.
87. Martínez J, Sánchez R, Castellanos M, et al. PrP charge structure encodes interdomain interactions. *Sci Rep.* 2015;5: 13623.
88. Huang Z, Gabriel JM, Baldwin MA, Fletterick RJ, Prusiner SB, Cohen FE. Proposed three-dimensional structure for the cellular prion protein. *Proc Natl Acad Sci USA.* 1994;91(15):7139–7143.
89. Swietnicki W, Petersen RB, Gambetti P, Surewicz WK. Familial mutations and the thermodynamic stability of the recombinant human prion protein. *J Biol Chem.* 1998;273(47):31048–31052.
90. Liemann S, Glockshuber R. Influence of amino acid substitutions related to inherited human prion diseases on the thermodynamic stability of the cellular prion protein. *Biochemistry.* 1999;38(11):3258–3267.
91. Shen L, Ji H-F. Mutation directional selection sheds light on prion pathogenesis. *Biochem Biophys Res Commun.* 2011;410(2):159–163.
92. Orem NR, Geoghegan JC, Deleault NR, Kascsak R, Supattapone S. Copper (II) ions potently inhibit purified PrPres amplification. *J Neurochem.* 2006;96(5):1409–1415.
93. Nicoll AJ, Panico S, Freir DB, et al. Amyloid-β nanotubes are associated with prion protein-dependent synaptotoxicity. *Nat Commun.* 2013;4:2416.
94. Nieznanski K, Surewicz K, Chen S, Nieznanska H, Surewicz WK. Interaction between prion protein and Aβ amyloid fibrils revisited. *ACS Chem Nerosci.* 2014;5:340–345.

CHAPTER THREE

Cell Biology of Prion Protein

Daniela Sarnataro*,†,1, Anna Pepe*,‡, Chiara Zurzolo*,‡
*University of Naples "Federico II", Naples, Italy
†Ceinge-Biotecnologie avanzate, s.c.a r.l., Naples, Italy
‡Unité de Trafic Membranaire et Pathogenese, Institut Pasteur, Paris, France
[1]Corresponding author: e-mail address: sarnatar@unina.it

Contents

Abstract

Cellular prion protein (PrP^C) is a mammalian glycoprotein which is usually found anchored to the plasma membrane via a glycosylphosphatidylinositol (GPI) anchor. The precise function of PrP^C remains elusive but may depend upon its cellular localization. PrP^C misfolds to a pathogenic isoform PrP^{Sc}, the causative agent of neurodegenerative prion diseases. Nonetheless some forms of prion disease develop in the apparent absence of infectious PrP^{Sc}, suggesting that molecular species of PrP distinct from PrP^{Sc} may represent the primary neurotoxic culprits. Indeed, in some inherited cases of human prion disease, the predominant form of PrP detectable in the brain is not PrP^{Sc} but rather ^{Ctm}PrP, a transmembrane form of the protein. The relationship between the neurodegeneration occurring in prion diseases involving PrP^{Sc} and that associated with ^{Ctm}PrP remains unclear. However, the different membrane topology of the PrP mutants, as well as the presence of the GPI anchor, could influence both the function and the intracellular localization and trafficking of the protein, all being potentially very important in the pathophysiological mechanism that ultimately causes the disease.

Here, we review the latest findings on the fundamental aspects of prions biology, from the PrP^C biosynthesis, function, and structure up to its intracellular traffic and analyze the possible roles of the different topological isoforms of the protein, as well as the GPI anchor, in the pathogenesis of the disease.

Progress in Molecular Biology and Translational Science, Volume 150
ISSN 1877-1173
http://dx.doi.org/10.1016/bs.pmbts.2017.06.018

1. EXPRESSION AND FUNCTION OF PRION PROTEIN

The cellular prion protein (PrP^C) is a ubiquitous glycosylphosphatidylinositol (GPI)-linked glycoprotein, highly conserved among mammalian species.[1,2] Early in embryogenesis PrP mRNA is expressed in the neural tube of mice and its level increases in the brain as development proceeds. In the adult central nervous system, PrP^C and its mRNA are widely distributed in neocortical and hippocampal neurons, cerebellar Purkinje cells, and spinal motor neurons. However, suggested PrP^C function in synaptic transmission and neuronal excitability comes from finding of its localization to synaptic membranes, as well as presynaptic domain.[3] Substantial amounts of PrP^C have also been found in heart,[4] skeletal muscle,[5,6] intestinal tissues,[7] uterus, and testis.[8] PrP^C is also expressed in the immune system, in hematopoietic stem cells and mature lymphoid and myeloid compartments.[9] The normal function of PrP^C remains unknown, although many biological functions have been attributed. Cell surface localization of PrP^C would reflect its proposed role in cell adhesion, ligand uptake, and transmembrane signaling.[10]

Studies on PrP^C knockout mice have not been very helpful in understanding PrP^C function as it might be redundant. Several lines of PrP^C knockout mice are viable, fertile, and display no gross developmental or anatomical defects; however in the hippocampus they are reported to have electrophysiological and structural abnormalities, loss of cerebellar Purkinje cells, changes in both learning and memory, and alterations in circadian rhythm and sleep pattern.[11]

Different studies reported a functional role of PrP^C in the copper metabolism, an essential metal required for the catalytic activity of diverse enzymes, which play a fundamental role in the human nervous system (for review see Ref. 12). PrP^C is able to bind selectively copper (II)ions, as reported by electrospray ionization mass spectroscopy studies, through four histidine-containing five or six octapeptide tandem repeats present in its N-terminal half.[13] This binding is pH-dependent and in the presence of copper, the highly flexible amino terminus of PrP^C becomes more structured contributing to PrP^C conformation. PrP^C is endocytosed from the cell surface via clathrin-coated pits after binding copper, that is then transferred to other copper-carrier cytosolic proteins. Studies demonstrating 50% decrement of copper concentration in synaptosomal preparations of Prnp 0/0 mice with respect to wild-type mice,[14] indicate that PrP^C might regulate the copper concentration in the synaptic region, and that may play a role in the reuptake of copper into presynapse.[15] Thus, PrP^C copper binding sites support the reduction of Cu^{2+} to Cu^+ and prevention of reactive oxygen species (ROS) formation.[16] Indeed, cells derived from Prnp knockout mice show reduced resistance to oxidative stress, probably due to a decreased activity of the enzymes Cu/Zn SOD-1, because PrP^C is able to regulate the incorporation of copper into these enzymes consequently regulating their activity.[13] Overall, it has been suggested that PrP^C maintains synaptic integrity and neurophysiological function thanks to copper binding, becoming able to activate calcium, an intracellular messenger. Regulation of calcium is critical leading to structural and functional changes in neurons and the description of a role for PrP^C in the modulation of calcium is related to *N*-methyl-D-aspartate (NMDA) receptor activity.[17] In support of this finding, prion-deficient neuronal cells showed an alteration in neuronal calcium homeostasis.

Later studies subsequently provided evidence for a role of PrP^C in the regulation of cell–cell adhesion through processing and/or control of the trafficking of E-cadherin,[18] whose downregulation and consequent destabilization of adherent junctions[19] was induced by knockdown *PrP-1* gene in zebrafish, which led to a lethal phenotype with loss of embryonic cell adhesion and arrest of gastrulation.[19]

A biological role of PrP^C in cell–cell adhesion was demonstrated not only with the identification of its interacting partners, such as the junctional proteins desmoglein-2, plakoglobin, desmoplakin, but also with PrP^C downregulation in the enterocytes that led to an alteration of the three types of junctional proteins (desmosomes, adherens, and tight junction) and to a disorganization of the monolayer.[20] The intestinal epithelium of PrP^C knockdown mice showed an increase in paracellular permeability and impairment of intestinal barrier function, suggesting that PrP^C is a key factor in the organization and integrity of tissue barriers. In addition, the role of PrP^C in cell adhesion was strongly supported by the demonstration of PrP^C binding with proteins of the extracellular matrix such as sulphated glycosaminoglycans, N-CAM, and 37/67 kDa nonintegrin laminin receptor.[21–24]

Mouillet-Richard and coworkers have been the first to identify PrP^C as signaling molecule by using antibodies against prion protein to mimic a putative ligand. Indeed, they observed a marked decrease in the phosphorylation level of tyrosine kinase Fyn in the 1C11 murine cell line that can differentiate in both serotoninergic and noradrenergic neurons.[10] The coupling of PrP^C and kinase Fyn is closely related to the differentiation of 1C11 cells. Thus, PrP^C is defined as a modulator of serotoninergic receptor coupling to G proteins and thereby as a protagonist for the homeostasis of serotoninergic neurons.[25]

Another PrP^C function may include a neuroprotective activity, which has been described to derive from its dimerization.[26] Also, the neuroprotective PrP^C role is clearly due to its antiapoptotic action.[27] Indeed, it is involved in survival or cell death. In yeast two-hybrid system[28] PrP^C has been found to selectively bind Bcl-2, the suppressor of the programmed cell death, and to exhibit antiapoptotic activity on human breast carcinoma cell lines, MCF-7, in response to cytotoxic action of TNF.[29] In addition to these effects, it has been reported that the cytosolic form of PrP^C, which derives from ER retrotranslocation, is able to protect human primary neurons from Bax-mediated apoptosis.

As described above, PrP^C can have a great variety of binding partners and for this, it appears to control different biological processes; moreover, the activity of PrP^C appears to be critical in situation of stress as its downregulation makes the cells more vulnerable to different types of stress.

2. PrP^C: THE PRINCIPAL ACTOR IN PRION REPLICATION

PrP^C plays a role in the Transmissible Spongiform Encephalopathies (TSEs), fatal neurodegenerative disorders that affect humans, and a wide variety of animals. In these disorders, PrP^C can misfold into a pathogenic scrapie isoform (PrP^{Sc}), an insoluble and partially protease-resistant isoform that is able to propagate itself by interacting with PrP^C, its normal counterpart (Fig. 1).

PrP^{Sc} has a strong tendency to polymerize, forming amyloid aggregates which accumulate in the brain. The accretion of this abnormal protein takes place mainly in the brain and in the lymphoreticular system, accompanied by neuronal vacuolation (spongiosis) and neuronal death. Affected individuals, after long incubation time, show progressive neurological symptoms all terminating with death.[30] PrP^{Sc} has a great resistance to radiation and nucleases and it is insoluble and resistant to proteases. Indeed, after PrP^{Sc} digestion with Proteinase K, a 27- to 30-kDa fragment, termed PrP^{res}, generated.[31] PrP^C and PrP^{Sc} share the same primary sequence and their principal difference seemed to involve posttranslational modifications, such as their conformation: PrP^C is monomeric and digested by proteinase K, whereas PrP^{Sc} forms highly insoluble aggregates and shows higher resistance to proteolytic digestion. Spectroscopic studies demonstrated a conformational difference

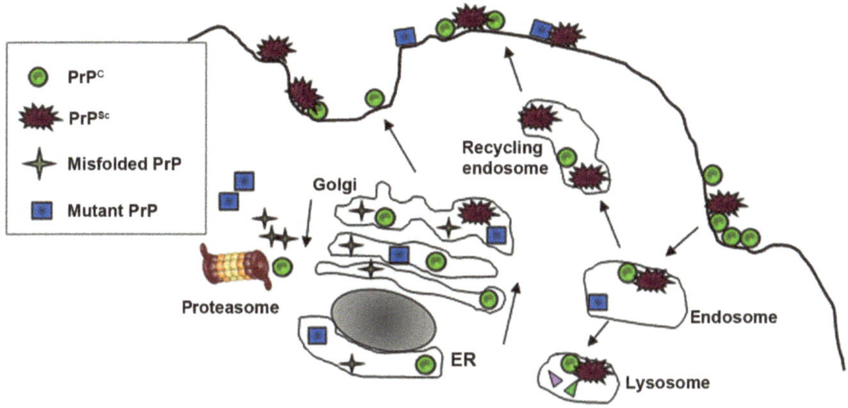

Fig. 1 Possible pathways of PrP^{Sc} formation. PrP^C is synthesized in the ER, in which it acquires posttranslational modifications and in which misfolded PrP and hereditary pathological mutants are partially degraded by the proteasome after retrotranslocation through the cytosol. After ER quality control, PrP^C is transported through the Golgi apparatus to the cell surface, at which it is internalized by clathrin- and/or caveolin-dependent mechanisms. A reduction in PrP^C internalization decreases PrP^{Sc} formation. Both PrP isoforms are found in early endosomes, pass through late endosomes, and are degraded in acidic lysosomes. Finally, the ER has been postulated to have a role in prion conversion by amplifying PrP^{Sc} formation after the Rab6a-dependent retrograde transport of PrP^C.

between PrP^C and PrP^{Sc}; PrP^C has a high α-helical content of approx. 42%, with little or no β-sheets (approx. 3%), while PrP^{Sc} contains approx. 30% α-helices and approx. 45% β-sheets.[32]

Although the possible mechanism of the prion conversion reaction into the abnormal PrP^{Sc} isoform is still unknown, the most accredited model of prion formation and replication proposes that physical interaction between the endogenous PrP^C and the pathogenic PrP^{Sc} may be required for the formation of nascent prions.[33] However, as shown by studies in transgenic mice, it is possible that a host factor, a protein X, could be involved in the conversion reaction.[34] Moreover, formation of PrP^{Sc} might take place at the plasma membrane, where the endogenous PrP^C and exogenous PrP^{Sc} interact, or immediately after its internalization in the endolysosomal or recycling compartment (Fig. 1).[35,36] Interestingly, Rouvinski *et al.*[37] have observed by live imaging the membrane anchored PrP^{Sc} amyloids as thin and micrometer-long strings that reside for hours on cell surface, sheltered from endocytosis. However, prion conversion may also occur in endosomal recycling compartment where, at steady state, high amounts of PrP^{Sc} reside (Fig. 1).[36] Although the mechanism that leads PrP^C misfolding is not still known, it may be considered the main actor in prion replication. PrP^{C} knockout mice are resistant to the disease as PrP^{Sc} cannot be formed; in addition PrP^{C} appears to be required to induce neurodegeneration in host neurons, as the presence of PrP^{Sc} only is not sufficient to cause disease.[38] This resulted in the postulate that neurodegeneration results form a subversion of the normal signaling function of PrP^{C} in the presence of PrP^{Sc}.[39] In the case of infection from the outside, interestingly, PrP^{C} seems to be not required for the primary binding and uptake of PrP^{Sc} in the cells,[40] while other possible cell surface receptors could be involved in PrP^{Sc} uptake such as laminin receptor, glycosaminoglycans (GAGs), and the neuronal receptor LRP1.[41] This has led to the proposal that the targeting of PrP receptor might be promising strategy against prion diseases.[23]

3. STRUCTURE AND BIOSYNTHESIS OF PrP^{C}

The mammalian PrP^{C} is the product of a single gene which directs the synthesis of a protein containing approximately 250 amino acids, with an N-terminal signal peptide near a series of five proline- and glycine-rich octapeptide repeats that bind copper. Moreover, PrP^{C} carries a highly conserved central hydrophobic segment (HD), two glycosylation sites, one disulfide bridge, and a C-terminal hydrophobic region which contains a signal for GPI anchor attachment (Fig. 2). The primary signal targeting PrP^{C} to the ER is the N-terminal signal peptide (aa 1–22), which is recognized and bound by SRP (the signal recognizing particles) that after binding to the receptor on ER, protein synthesis continues and PrP polypeptide translocates into ER, proceeding via cotranslational manner. Integrity of N-terminal signal peptide is critical for the efficient import and the formation of topological isoforms of PrP^{C}. The abnormal accumulation of PrP^{C} in the cytosol is a consequence of N-signal peptide inefficient import of PrP to ER, causing interference with cell viability.[42]

The observation that the HD region comprising aa 106–126 may have amyloidogenic properties and was able to form fibrils in vitro and to induce cell death in cultured cells,[43] together with findings that transgenic mice that expressed PrP lacking HD region, did not develop a transmissible prion infection,[44] led to the conclusion that the HD region is linked to the toxic effect of cytosolic PrP^{C}.[45]

PrP^{C} has also a GPI signal peptide (GPI-SP), where a GPI anchor mediating its anchoring to the membrane is attached to the omega site, the acceptor amino acids (Fig. 2).[46] Specifically, GPI anchor addition occurs through

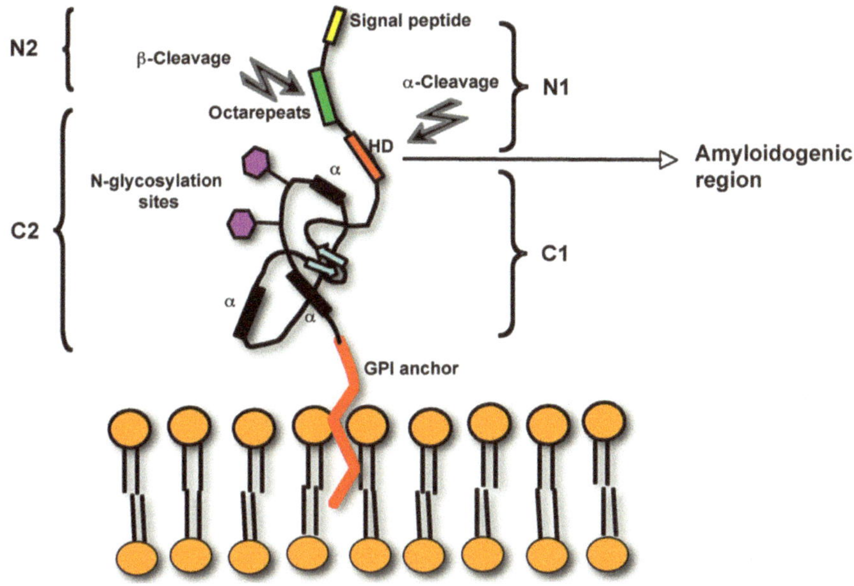

Fig. 2 An outline of the structure of the cellular prion protein including posttranslational modifications and representation of PrP cleavages. A secretory signal peptide resides at the extreme N-terminus. HD defines the hydrophobic region; α indicates alpha-helix and the arrows β-sheets; hexagons define N-glycosylation; the GPI anchors PrP to the cell membranes. α-Cleavage acts in the HD generating N1 and C1 fragments, while β-cleavage acts in the octarepeat region generating N2 and C2.

the action of multimeric GPI transamidase in the lumen of ER, where the preassembled GPI anchor is added. The GPI-SP is specifically degraded by the proteasome, but is spared from degradation by proteasome when it carries two autosomal dominant pathogenic mutations responsible for genetic forms of prion diseases. Under these conditions, GPI-SP is targeted to the mitochondria causing loss of membrane potential (MMP) and their fragmentation.[47] Interestingly, also the C-terminal GPI anchor signal peptide could mediate the targeting of PrP^C to the ER, promoting posttranslational PrP import as shown both in vitro studies and in cultured cells.[48,49] Furthermore, information in the C-terminal domain, and in particular its alpha helical motif, promotes the PrP import in the ER, leading to the conclusion that the secondary structure of the nascent polypeptide chain could modulate translocation efficiency.[50] Additionally, essential for the stability of the protein, in the ER the two Cys residues (179 and 214) are engaged in the formation of a disulfide bond and the N-linked oligosaccharide chains, highly mannose type and sensitive to Endoglycosidase-H digestion, are added to two asparagines (N) (residues 181 and 197 in humans) in the C-terminal part of PrP^C. In the Golgi they are subsequently modified to more complex oligosaccharides that contain sialic acid and are Endoglycosidase-H resistant.[51]

Overall, PrP^C has a flexible unstructured N-terminal region and a well ordered C-terminal globular domain that contains three α-helices and two antiparallel β-sheets, as described by nuclear magnetic resonance; the secondary structure of PrP^C is mainly composed of α-helices, while PrP^{Sc} contains mainly β-sheets.[52]

A feature of the PrP^C is its endoproteolitic processing; two internal cleavage processes, α-cleavage and β-cleavage, are responsible for the main C- and N-terminal fragments (Fig. 2). Under healthy conditions, α-cleavage of PrP^C occurring in the hydrophobic domain, amino acids 110/111 (a region that plays an important role in the pathological conversion of PrP^C), produces a 17-kDa C-terminal fragment, named "C1" that is tethered to the plasma membrane, and a 9-kDa N-terminal fragment, N1. Together they are the products of the normal prion protein metabolism increasing under PrP^C intracellular dimerization.[53] Unwanted dimerization may initiate prion conversion and result in neuronal toxicity.[54]

During disease, α-cleavage occurs less frequently than β-cleavage which preserves the amyloidogenic region at amino acids 110/111, indeed occurring within or adjacent to the octarepeat region, between residues 90/91 leading to generation of 21-kDa C2 and N2 fragments. Because the predominant C-terminal fragment switches from C1 to C2 fragment during disease progression, endoproteolysis of prion protein probably plays a fundamental role in PrP^{Sc} propagation. The α-cleavage gives rise to C1/N1 products, considered positive fragments because after the cleavage they lose the amyloidogenic region and to the C1 fragment, which unlike C2, has a p53-dependent pro-apoptotic function that may be balanced by N1, which in turn by modulating the p53 pathway, protected against pressure-induced ischaemia in models of rat retina.[55] Furthermore, mice expressing a transgene encoding C1 in the absence of wild-type PrP remained healthy and were resistant to prion disease.[56]

Experiments conducted in cultured cells suggest that β-cleavage appears to have a dual role; it has been described as a protective response to oxidative insult within the cells and it may move prion aggregation as the amyloidogenic core is left intact. Moreover, in prion-infected brains, in which the levels of β-cleavage were elevated was found increment of ROS levels, correlating with increase in the prion conversion process. Thus, more studies suggest that β-cleavage provides an essential process to reduce oxidative stress; probably in the infective system the protective β-cleavage could play as a double-edged sword.[57]

4. TRAFFICKING OF CELLULAR PRION PROTEIN

In normal tissues, PrP^C is localized predominantly on the plasma membranes, as illustrated by cryo-immunogold electron microscopy of brain.[58] Like other secretory proteins, after its synthesis in the rough endoplasmic reticulum (ER), PrP^C traffics through the Golgi apparatus experiencing many posttranslational modifications to both the proteic part and the anchor.[59] In hippocampal neurons, PrP^C transport along the secretory pathway, specifically along the axons, is assisted by both the molecular motors kinesin and dynein.[60] Interestingly, a delayed post-Golgi trafficking of PrP^C, and also of other raft-associated proteins, such as Thy-1 and insulin receptor, has been described for RML-prion-infected mouse brain, suggesting that prion infection could cause impairment of normal post-Golgi trafficking.[61]

The majority of GPI-anchored proteins (GPI-APs) have an apical localization in polarized cells and PrP^C is the first native GPI-AP to be found enriched on the basolateral surface of epithelial cells,[62] from where it has been described to be transcytosed to the apical surface.[63] Moreover, its localization may have a role in PrP^C function or in spreading of the infection. Thus, on one hand, PrP^C intracellular localization could be of fundamental importance for its functions and, on the other hand alteration in its intracellular traffic therefore might be involved in its normal function and in the conversion process, that could be prevented by aberrant sorting.

PrP^C can be endocytosed from the plasma membrane.[64] As discussed before this pathway could have a relevant role in prion conversion.

However, the route and mechanism of internalization of PrP^C, as well as the mechanisms regulating internalization, are controversial because both clathrin-dependent and -independent pathways could be involved (Fig. 3). At the plasma membrane, PrP^C and its pathological isoform PrP^{Sc} have been found to be associated to lipid rafts,[62,64–66] dynamic assemblies enriched in cholesterol and sphingolipids, which play a key role in the endocytosis of PrP^C in epithelial cells.[67] Moreover, an early association of PrP^C during its biosynthesis, with lipid rafts in the early secretory pathway, stabilizes its own conformation.[68] An impairment of rafts-association by cholesterol depletion during the early stage of PrP^C biosynthesis leads to protein

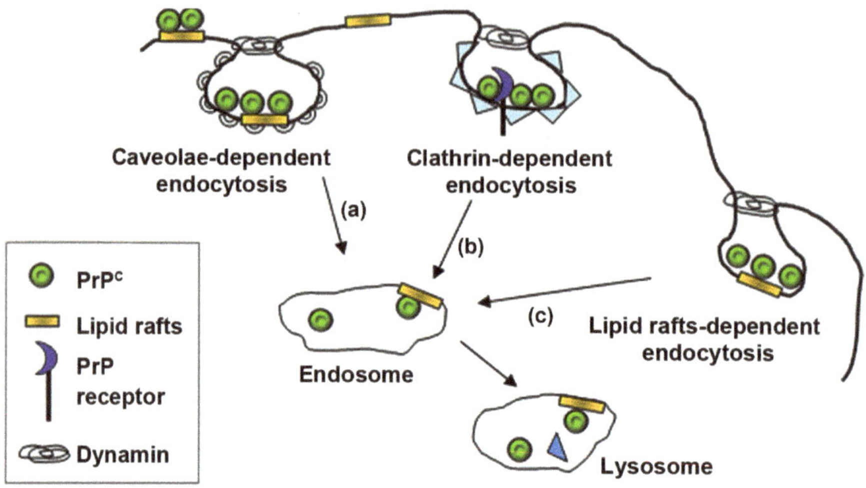

Fig. 3 Pathways of PrP^C internalization. Caveolin-related endocytosis and trafficking have been implicated in PrP^C transport in Chinese hamster ovary and glial cells (a). At the plasma membrane, PrP^C can be constitutively internalized and its endocytosis can be increased by extracellular copper ions (not shown). A chief pathway of PrP^C internalization in neuronal cells seems to depend on clathrin-mediated endocytosis (b). Finally, nonclathrin and noncaveolin but raft-dependent endocytosis has been proposed to participate in the internalization and conversion of prion protein (c). The GTPase dynamin is localized on the neck of invaginations.

misfolding in ER, suggesting a functional role for rafts in the protection of PrP^C folding[68,69] (Fig. 1).

Some studies showed that chicken PrP^C, transfected in murine neuroblastoma cells, was internalized via clathrin-coated pits.[70] However, at cell surface PrP^C associates with lipid rafts and prior to endocytosis, it leaves the lipid rafts to cluster in nonraft membranes such as coated pits, suggesting that the trafficking of PrP^C differs from that of some other GPI proteins. Interestingly, the N-terminal half of PrP^C is essential for an efficient clathrin-mediated endocytosis.[71] Furthermore, ultrastructural investigation has shown that PrP^C, after its internalization via coated pits, could be associated with the transferrin receptor, indicating that it can recycle back to the cell surface by recycling endosomes.[71] PrP^C endocytosis through clathrin-coated pits is surprising, since PrP^C as a GPI-AP lacks intracellularly oriented amino acid sequences capable of interacting with adaptor proteins necessary for clathrin-mediated endocytosis. Thus, co-internalization of PrP^C with other "clathrin adaptable" proteins has been suggested, such as the laminin receptor precursor LRP/LR and the low-density lipoprotein receptor-related protein 1 (LRP1) that belongs to the low-density lipoprotein (LDL) receptor family, a group of cell-surface transmembrane proteins binding a variety of ligands which internalize via clathrin-coated pits. LRP1 is expressed in neurons and hepatocytes and is required for the Cu^{2+}-mediated endocytosis of PrP in SH-SY5Y neuronal cells.[41] Interestingly, the binding between PrP^C and LRP1 could occur within lipid rafts, after the octapeptide repeats become fully loaded with Cu^{2+}. Subsequently, under these conditions, PrP^C endocytosis has been described to be allowed by flotillins which are membrane association proteins.[72] Moreover, the polybasic region of its N-terminus half is required to engage PrP^C with a transmembrane adaptor protein, which in turn links to the clathrin endocytic machinery and the conformational change in the N-terminus is also critical for its clathrin-dependent endocytosis for which HSPG (heparin sulfate proteoglycans) has been shown to be required.

Following the yeast two-hybrid screen technology in *Saccharomyces cerevisiae*, the 37 kDa laminin receptor precursor and the 67 kDa laminin receptor (mature form of 37 kDa) have been discovered as cell surface binding partners of PrP^C[73]; moreover, in vitro studies in neuronal and non-neuronal cells proved that both the 37 kDa LRP and the 67 kDa receptor act as the receptor for the PrP^C.[23,74]

Internalization by caveolae may be one of several mechanisms involved in nonclathrin-mediated endocytosis (Fig. 3). It has been shown that in CHO (Chinese Hamster Ovary) cells, stably transfected with either Syrian hamster or chimeric mouse–hamster (MH2M), PrP^C is endocytosed via caveolae. However, caveolae containing endocytosed PrP^C do not intersect with the classical early endocytic recycling compartment but are delivered to late endosomal/lysosomal compartments via nonclassical endosomes.[75] Moreover, the internalization of PrP^C via caveolae raises new questions, such as whether endocytosis via caveolae could be a necessary step in the conversion of PrP^C into PrP^{Sc} and whether the signaling activity of PrP^C could occur at the plasma membrane or after caveolae-mediated endocytosis.[10,34]

Galvan *et al.*[76] have shown that disruption of rafts by sphingolipid or cholesterol depletion leads to the nonpolarized distribution and impaired endocytosis of PrP^C, suggesting that they are essential for proper trafficking and distribution of PrP^C in hippocampal primary neurons. In agreement with this, we have demonstrated that the internalization of mouse PrP (moPrP), transfected in FRT cells (Fisher Rat Thyroid), which lack caveolin-1 (cav-1) and caveolae, is dependent on cooperation between lipid rafts and clathrin.[67]

This is also supported by the finding that PrP^C endocytosis is completely blocked only when both lipid rafts and clathrin coat formation are affected. Furthermore, PrP^C remains associated to lipid rafts through all endocytic processes, suggesting that clathrin assembly occurs within these domains.[67]

Overall, it is not clear whether PrP^C can undertake multiple internalization pathways in the same cell and whether the apparent contrasting data on PrP^C endocytosis might be due to differences in cell types, as well as to differences in trafficking of endogenous PrP^C and stably expressed PrP^C. Understanding the endocytic itinerary of PrP^C is fundamental, since this could be very relevant for both understanding the prion pathogenesis and to reveal the basis of the different prion strains ability to infect and replicate in different cells.[77]

By using GFP-linked proteins, it has been found that GFP-PrP^C internalization is dynamin-regulated and mediated by Rab5-positive endocytic vesicles. PrP^C may be degraded via the endosomal/lysosomal pathway after the transit in these endosomes.[78] Indeed, PrP^C has been also found to be associated to Rab7 positive vesicles, suggesting that an amount of protein undergoes lysosomal degradation. However, in prion-infected cells an impairment of this degradation pathway has been shown; this could induce accumulation of PrP^{Sc} thus supporting the persistence of prions replication.[79] In contrast to GFP-PrP^C, GFP-GPI is not found in the same endocytic organelles, indicating that PrP^C trafficking differs from that of other standard GPI-APs and may depend on additional internalization signals present in the protein.

5. SEVERAL TOPOLOGICAL FORMS OF PrP^C

PrP^C is anchored to the outer leaflet of the plasma membrane by the C-terminal GPI and can adopt two transmembrane forms which are generated in a small amount as part of its normal biosynthesis. These species are designated as ${}^{Ntm}PrP$ and ${}^{Ctm}PrP$ and both of them have the same hydrophobic region, TM1 (aa 111–134), but span the membrane with an opposite orientation (Fig. 4). Their generation is due to an unusual mode of stop transfer of the polypeptide chain across the ER membrane and is regulated by an inefficient N-terminal signal sequence that prevents cotranslational translocation of PrP^C, conversely, allowing some PrP^C molecules to be first fully synthesized, and then posttranslationally translocated through the C-terminal hydrophobic domain, which acts as signal sequence.[49]

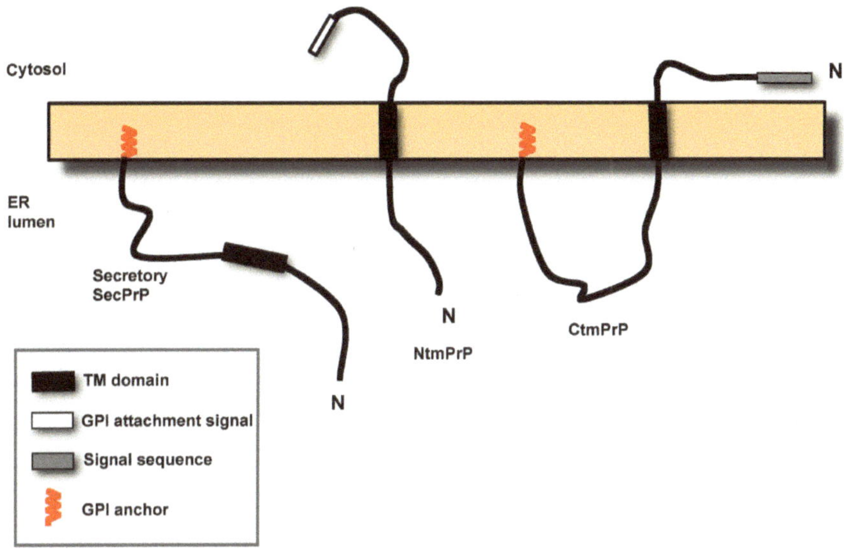

Fig. 4 Depiction of different topologic forms of prion protein. The topologic forms of PrP differ in two important ways. One is the N-terminal location: either in the cytosol (^{Ctm}PrP) or in the lumen of the endoplasmic reticulum (^{Sec}PrP and ^{Ntm}PrP). The other is whether the central hydrophobic domain becomes membrane integrated (^{Ntm}PrP and ^{Ctm}PrP) or not (^{sec}PrP). ^{sec}PrP, Secretory prion protein; ^{Ctm}PrP, transmembrane form of prion protein that has the N terminus in the cytosol and C terminus in the endoplasmic reticulum lumen; ^{Ntm}PrP, N-transmembrane form of prion protein that has the N terminus in the endoplasmic reticulum lumen and C terminus in the cytosol; ER, endoplasmic reticulum; GPI, glycosylphosphatidyl inositol; TM, transmembrane.

^{Ntm}PrP, whose translocation is initiated by the N-terminal signal peptide subsequently stopped by the transmembrane segment into bilayer,[80] has the N-terminal in the ER lumen, while ^{Ctm}PrP is integrated in the opposite orientation exposing the N-terminus on the cytoplasmic side of the ER. Furthermore, ^{Ctm}PrP crosses the membrane with the hydrophobic region because during biosynthesis, the C-terminal region is inserted into the membrane and then cleaved and replaced by a GPI anchor.[81] It has been observed that an alteration of the delicate balance of the different forms of PrP that increases the expression of ^{Ctm}PrP leads to neurodegenerative prion disease.[82,83] Indeed, increased levels of ^{Ctm}PrP were found in the brain of transgenic mice, which developed a scrapie-like neurological disorder because of expressing PrP carrying a mutation (A117V) in the transmembrane domain. Further evidence highlighting involvement of ^{Ctm}PrP in the pathogenesis of prion diseases comes from experiments conducted in PrP transgenic mice inoculated with mouse prions, which showed an increase in the amount of ^{Ctm}PrP and PrP^{Sc} in the brain, indicating that PrP^{Sc} may induce its formation.[83,84] However, mutations found in/or around the central transmembrane domain lead to an increase of ^{Ctm}PrP isoform which, thus, can be considered a toxic intermediate involved in the neurodegeneration.

6. QUALITY CONTROL OF CELLULAR PRION PROTEIN

Several degradation mechanisms have been described to be employed by the cell in the clearance of altered proteins in the folding: the endosomal–lysosomal, the autophagic–lysosomal pathway, and proteasomal degradation, that includes clearance of cytosolic proteins and endoplasmic reticulum-associated degradation (ERAD). Before reaching the membrane, PrP^C, being a secretory protein, is subjected to a quality control process to ensure its correct folding but when misfolded, could be retrotranslocated out of the ER in the cytosol, where it is degraded by proteasome: an overloaded PrP^C into cytosol was shown to be toxic.[85,86] Conversely, it has been discovered that a small fraction of PrP molecules are degraded via proteasome (Fig. 1), but by a pathway that does not involve retrotranslocation from the ER lumen, suggesting that PrP has an abortive translocation remaining closely associated with the cytoplasmic face of the ER membrane, where it is rapidly degraded. However, inhibition of proteasome could cause the cytosolic accumulation of both mutant and wild-type PrP molecules and could induce an increase of protein synthesis, explaining the cytosolic PrP aggregation into cytosol.[87] Thus, a small amount of PrP^C may be mis-translocated from ER and not retrotranslocated; however, also a PrP glycosylation mutant, associated with a genetic form of prion disease, was not degraded by the proteasome, indicating that the ERAD pathway is not the primary pathway involved in the degradation of PrP^C.[88]

Observations from Ashok and Hegde[89] hinted at the possibility that GPI-anchored misfolded PrP was degraded by an undefined non-ERAD route especially important during ER stress. Indeed, when ER folding activity is saturated during stress, misfolded GPI proteins dissociate from ER chaperones and leave the ER for secretory transport to the Golgi, accessing the cell surface transiently before lysosome degradation.[90] As reviewed by Senatore *et al.*,[91] in mouse models of genetic prion disease the ER retention of misfolded PrP, but not ER stress, affects synaptic function proposing that impairment of secretory protein trafficking due to intracellular accumulation of misfolded proteins, could be a major cause of neuronal dysfunction in prion diseases.

Another cytosolic clearance mechanism that the cells use to catabolize and recycle damaged constituents is the autophagy degradation pathway, whose defects are implicated in different neurodegenerative disorders. Although the role of autophagy in prion disease remains controversial, it has been proposed that it may be involved in the clearance of misfolded prion protein. Indeed, the appearance of autophagic vacuoles in both prion-infected neuronal cells and biopsies from prion-infected patients, suggests a role of autophagy in the disease.[92,93] Conversely, recent work demonstrated that in prion-infected cells, PrP^{Sc} was degraded by lysosomes in an autophagic-independent manner. Indeed, the tamoxifen and hydroxytamoxifen, the classical inducers of autophagy, are able to exert their antiprion effects only after perturbation of cholesterol metabolism, inducing prion scrapie degradation into lysosomes.[94,95] Thus, lysosomal degradation could be considered an important event for clearance of prion aggregates and lysosomal dysfunction could contribute to prion propagation.

7. GPI-ANCHOR: STRUCTURE AND BIOSYNTHESIS

Glycosylphosphatidylinositol anchoring of proteins is a conserved posttranslational modification in eukaryotes. The GPI anchor has been described in more than 150 proteins in mammalian cells and mediates the attachment of these proteins to the cell surface, conferring them specific properties. Mature GPI anchor is synthesized in the ER and is transferred to a protein by a GPI-transamidase complex that simultaneously removes GPI attachment signal peptide and attaches the anchor to the C-terminal of proteins termed the ω-site. Some studies have showed that the GPI backbone structure was conserved among mammals, yeast, parasites, protozoan, and also plants. A phosphoethanolamine (EtN-P) linker, that is covalently linked by an amide bond to the C-terminal ω-site of proteins and a glycan core that consists of a mannose(α1, 2)mannose(α1, 6)mannose(α1, 4)-glucosamine(α1, 6) and inositol-phospholipid tail, render the GPI anchor a complex structure.[96,97] Depending on both cell type and protein type, the three core mannose can be modified by side chains, such as phosphoethanolamine group, mannose, galactose, sialic acid, or other sugars (Fig. 5).

Many GPI anchors can also contain an additional fatty acid, as well as palmitic acid on the 2-hydroxyl of the inositol ring[98] and thus can result resistant to the cleavage by phospholipase C enzyme. However, there is great variability in the side chains and lipid moieties of GPI anchors

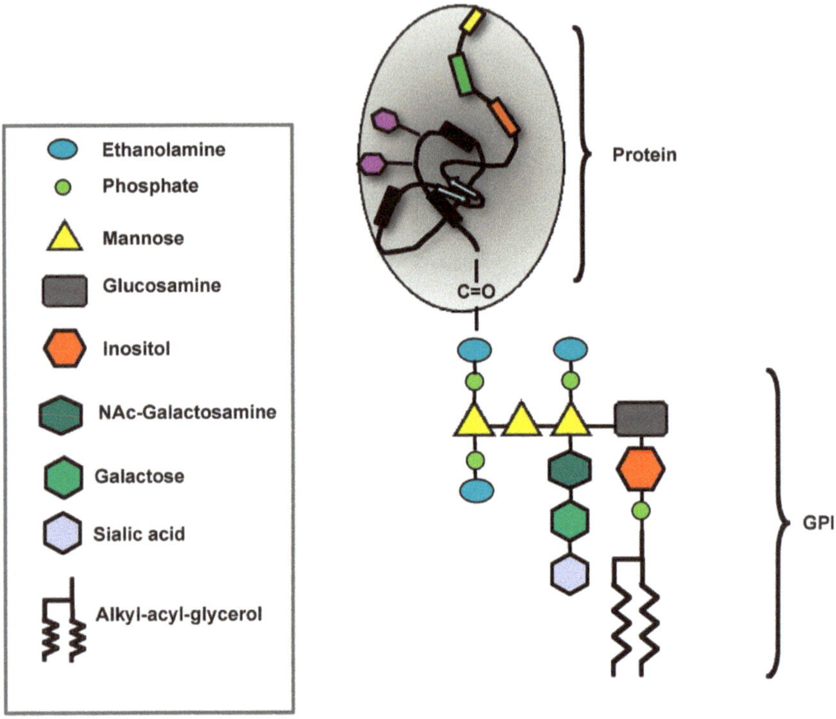

Fig. 5 Schematic representation of the GPI-anchor. The polypeptide chain (protein) is covalently linked through its C-terminus to the core GPI structure ethanolammine-phosphate-(mannose)3-glucosamine-phosphatidylinositol. The lipid alkyl-acyl-glycerol in the example shown interacts with the outer leaflet of the lipid bilayer. Additional ethanolamine phosphate groups and sialic acid are shown on the first and third mannose residues.

dependent upon proteins and species and this suggests that GPI structures are remodeled during biosynthesis and along the secretory pathway, where they undergo a series of structural modifications, including the acquisition of a saturated lipid moiety, which could lead the GPI-APs to associate with saturated lipids into specialized domains.[99]

8. FUNCTIONS OF GPI-ANCHORED PROTEINS

GPI-APs are functionally diverse and could be involved in several biological pathways, as well as in signal transduction, immune response, and prion disease pathogenesis.[97,100] However, they are critical for the maintenance of normal cellular morphology and are essential for viability; indeed, abrogation of GPI biosynthesis results in embryonic lethality. GPI-APs functions range from enzymatic activity, such as the alkaline phosphatase, that catalyzes the removal of phosphates groups from biomolecules, to involvement in cell–cell interaction and adhesion, described for NCAM and LFA3; while certain GPI proteins, such as CD55 and CD59, which are complement regulators, exert organisms protection activity from pathogens. The GPI-AP, Thy-1, is the major cell surface protein of thymocytes and neurons of rat and it is able to mediate lymphokine release and cell division.[101] The *Trypanosoma brucei* VSG reported the presence of GPI-AP and is able to form a protective coat around the trypanosome parasite.[102] However, many GPI-APs do not have known functions yet.

The GPI-APs, through their anchor, are attached to the outer leaflet of the membrane where they are preferentially partitioned into the ordered, tightly packed lipid rafts domains[103] and result insoluble in the nonionic detergent Triton X-100 at 4°C, more probably due to their association with lipid rafts.[104] However, in lipid rafts, GPI-APs are organized as clusters, which are stabilized by galectins through a joint recognition of the GPI glycan core and of the surrounding glycosphingolipids for apical delivery.[99,105]

Involvement of these proteins in the signal transduction has been suggested, indeed, in the lipid platforms, have been found proteins that play a role in cell signaling.[106]

9. GPI-ANCHOR FUNCTION

The known role of GPI-anchor is to provide the anchorage of a wide variety of extracellular proteins to the membrane. However, compared to other lipid anchoring domain, its structure is fairly complicated, suggesting that the GPI-anchor may have also biological function. Numerous studies have shown the ability of GPI to participate in cell signaling mechanisms or to act directly as second messenger of biological receptors. The phospholipase C mediates the hydrolysis of GPI, producing diacylglycerol and inositol phosphoglycans, that could be intracellular signaling mediators. However, the GPI-anchor may be involved in the signal transduction; indeed, it has a similarity with the free GPI phospholipids that participate in cell signaling. Other evidence shows that the antibody cross-linking of GPI-APs, such as Thy-1, CD48, CD59, can induce the cellular activation, resulting in an intracellular flux of calcium, protein tyrosine phosphorylation, and elaboration of growth-promoting factors. Confirmation of GPI role in intracellular signaling comes from a loss of these effects after the replacement of GPI-anchor with a transmembrane domain.[107] However, the exact function of GPI anchor is unknown; some studies suggest that GPI-anchor may serve as an intracellular signal for transport of the proteins to the apical surface in polarized cells.[108] The lack in GPI-APs of cytosolic part, which could be recognized by the machinery employed in the generation and transport of vesicles, differently from transmembrane proteins, suggests that the sorting determinant of these proteins could be ensured by association with lipid rafts, which are destined for the apical membrane, in MDCK cells.[108] However, in epithelial FRT cells, most of GPI-APs are sorted to the basolateral surface, and some apical protein in MDCK cells are not found in lipid rafts,[109] implicating that the only association with rafts did not ensure the apical targeting. Thus, other mechanisms, such as N-glycosylation and oligomerization, could be responsible for their sorting.[110]

GPI-anchor is able to bind bacterial toxin, such as aerolysin, the toxin required for the virulence of *Aeromonas* hydrophila, which binds specific GPI-anchored receptors on sensitive cells, such as Thy-1 and the EAR (erythrocyte aerolysin receptor). Also other channel-forming protein as alpha-toxin produced by *Clostridium septicum,* used GPI-anchor proteins to bind sensitive cells; indeed, as demonstrated by Gordon *et al.*, the mutant CHO cells that lack the ability to synthesize GPI-anchors are resistant to alpha toxin.[111] GPI-anchor may also influence the structure of proteins to which it is attached and may hold the proteins close to the cell surface.

10. GPI-ANCHOR AND PRION

Recently, the GPI anchor has been described to play a potential role in prion disease pathogenesis. Both PrP^{C} and PrP^{Sc} are bound to the cells via GPI anchor. However, the role of the anchor in the folding of PrP^{C} remains unclear.

Extensive evidence accumulated in recent years shows that PrP^{C} is the main actor in prion replication and that it is absolutely necessary for prion-induced neurotoxicity and prion scrapie production.[38,112] Implication of the GPI anchor in prion conversion comes from use of transgenic mice expressing only a secreted form of PrP^{C} lacking GPI-anchor: in neurons of these mice, PrP^{C} was not detected on the plasma membrane but was partially retained in the ER and Golgi apparatus and secreted in the medium of cultured cells. However, the mice did not develop clinical features of prion disease but were healthy and viable when inoculated with PrP^{Sc}. The resistance of these mice to develop prion diseases was observed also after the infection with various mouse strains such as RML, ME7, and 22L.[100,113] Interestingly, the brain of these mice presented widespread PrP^{Sc} amyloid plaques; thus, it is possible that the lack of GPI anchor on PrP^{C} did not impair prion replication and PrP^{Sc} deposition, but likely prevented the delivery of neurotoxic or clinical disease, reminiscent of neuronal knockout PrP mice (reviewed in Ref. 114).

Furthermore, after replacing the GPI anchor of PrP^{C} with transmembrane segment fused to the C-terminal, PrP^{C} did not localize in the lipid rafts and was not a substrate for PrP^{Sc} formation, inhibiting prion infection.[115] Thus, the data discussed here provide support for a crucial role of GPI-anchor in prion neurotoxicity. Based on the evidence that PrP^{Sc}, contrary to other misfolded proteins, are infectious, many studies have investigated the mechanisms of intercellular spread of scrapie isoform and whether its membrane anchoring could modulate this feature.[116] Evidence suggests that cell-to-cell spread of GPI-anchored prion protein from dendritic cells to neurons is allowed through mother–daughter cell division, cell-to-cell contact, exosomes and tunneling nanotubes (TNTs) that may contribute to neuroinvasion.[117,118]

TNTs have been recently identified as an efficient route for PrP^{Sc} spreading in neuronal cells, since they allow transfer of exogenous and endogenous PrP^{Sc} between infected and naive neuronal cells.[119] Any of these processes could involve the cell-to-cell transfer of PrP^{Sc} in membrane containing particles.[120,121] Specifically the transfer of PrP^{Sc} in TNTs has been shown to occur inside organelles of the endolysosomal pathway, consistently with its subcellular accumulation.[122] Other studies show that GPI-anchor on PrP^{Sc} allows the formation of smaller oligomeric assemblies that have an enhanced seeding activity, neurotoxicity, and infectivity while limiting the formation of large and stable aggregates, as amyloid plaques.[123,124] More recently, Marshall and coworkers have shown that GPI anchoring and the localization of PrP^{C} to lipid rafts are crucial to the ability of PrP^{C} to propagate as a prion.[125] However, when PrP^{C} is targeted to lipid rafts, its GPI anchor is able to gain information about its subcellular trafficking, indeed, directs it to lipid rafts where it misfolds and an efficient formation of scrapie prion occurs.[126,127] Sialic acid composition of GPI is important to target PrP^{C} in specific subset of lipid rafts that are heterogeneous and only in a specific subset of them PrP^{C} conversion may take place. Indeed, Bate *et al.*[128] observed that in neuronal cells, where PrP^{C} was desialylated, the conversion of PrP^{C} to PrP^{Sc} was significantly reduced, proposing that the binding of desialylated PrP^C to scrapie prion is not in the correct membrane environment to support the conversion process.[129,130]

Furthermore, current data suggest the possibility that the PrP^C signal peptide attachment for GPI anchor could play a role in neurodegenerative prion disease. This peptide is released after each GPI-anchor attachment and is specifically degraded by proteasome but a point mutation (P238S), within this signal peptide, was responsible for genetic form of prion diseases and able to target mitochondria, inducing their fragmentation and vacuolization.[47] Thus, GPI signal peptide might be responsible for mitochondrial dysfunction and potentially could be one of the initial causes of synaptic pathology observed in prion diseases.

REFERENCES

1. Schätzl HM, Da Costa M, Taylor L, Cohen FE, Prusiner SB. Prion protein gene variation among primates. *J Mol Biol.* 1995;245:362–374.
2. Wopfner F, Weidenhöfer G, Schneider R, et al. Analysis of 27 mammalian and 9 avian PrPs reveals high conservation of flexible regions of the prion protein. *J Mol Biol.* 1999;289:1163–1178.
3. Fournier JG, Escaig-Haye F, Grigoriev V. Ultrastructural localization of prion proteins: physiological and pathological implications. *Microsc Res Tech.* 2000;50:76–88.
4. Brown HR, Goller NL, Rudelli RD, et al. The mRNA encoding the scrapie agent protein is present in a variety of non-neuronal cells. *Acta Neuropathol.* 1990;80:1–6.
5. Bosque PJ, Ryou C, Telling G, et al. Prions in skeletal muscle. *Proc Natl Acad Sci USA.* 2002;99:3812–3817.
6. Brown P. Donor pool size and the risk of blood-borne Creutzfeldt-Jakob disease. *Transfusion.* 1998;38:312–315.
7. Morel E, Fouquet S, Chateau D, et al. The cellular prion protein PrPc is expressed in human enterocytes in cell-cell junctional domains. *J Biol Chem.* 2004;279:1499–1505.
8. Tanji K, Saeki K, Matsumoto Y, et al. Analysis of PrPc mRNA by in situ hybridization in brain, placenta, uterus and testis of rats. *Intervirology.* 1995;38:309–315.
9. Isaacs JD, Jackson GS, Altmann DM. The role of the cellular prion protein in the immune system. *Clin Exp Immunol.* 2006;146:1–8.
10. Mouillet-Richard S, Ermonval M, Chebassier C, et al. Signal transduction through prion protein. *Science.* 2000;289:1925–1928.
11. Büeler H, Fischer M, Lang Y, et al. Normal development and behaviour of mice lacking the neuronal cell-surface PrP protein. *Nature.* 1992;356:577–582.
12. Walter ED, Stevens DJ, Spevacek AR, Visconte MP, Dei Rossi A, Millhauser GL. Copper binding extrinsic to the octarepeat region in the prion protein. *Curr Protein Pept Sci.* 2009;10:529–535.
13. Brown DR, Qin KF, Herms JW, et al. The cellular prion protein binds copper in vivo. *Nature.* 1997;390:684–687.
14. Herms J, Tings T, Gall S, et al. Evidence of presynaptic location and function of the prion protein. *J Neurosci.* 1999;19:8866–8875.
15. Kretzschmar HA, Tings T, Madlung A, Giese A, Herms J. Function of PrP(C) as a copper-binding protein at the synapse. *Arch Virol Suppl.* 2000;16:239–249.

16. Liu L, Jiang D, McDonald A, Hao Y, Millhauser GL, Zhou F. Copper redox cycling in the prion protein depends critically on binding mode. *J Am Chem Soc.* 2011;133:12229–12237.
17. Khosravani H, Zhang Y, Tsutsui S, et al. Prion protein attenuates excitotoxicity by inhibiting NMDA receptors. *J Gen Physiol.* 2008;131:5.
18. Sempou E, Biasini E, Pinzón-Olejua A, Harris DA, Málaga-Trillo E. Activation of zebrafish Src family kinases by the prion protein is an amyloid-β-sensitive signal that prevents the endocytosis and degradation of E-cadherin/β-catenin complexes in vivo. *Mol Neurodegener.* 2016;11:18.
19. Málaga-Trillo E, Solis GP, Schrock Y, et al. Regulation of embryonic cell adhesion by the prion protein. *PLoS Biol.* 2009;7:e55.
20. Petit CS, Barreau F, Besnier L, et al. Requirement of cellular prion protein for intestinal barrier function and mislocalization in patients with inflammatory bowel disease. *Gastroenterology.* 2012;143:122–132.
21. Caughey B, Brown K, Raymond GJ, Katzenstein GE, Thresher W. Binding of the protease-sensitive form of PrP (prion protein) to sulfated glycosaminoglycan and congo red. *J Virol.* 1994;68:4107.
22. Li C, Wong P, Pan T, et al. Normal cellular prion protein is a ligand of selectins: binding requires Le(X) but is inhibited by sLe(X). *Biochem J.* 2007;406:333–341.
23. Sarnataro D, Pepe A, Altamura G, et al. The 37/67 kDa laminin receptor (LR) inhibitor, NSC47924, affects 37/67 kDa LR cell surface localization and interaction with the cellular prion protein. *Sci Rep.* 2016;6:24457.
24. Schmitt-Ulms G, Legname G, Baldwin MA, et al. Binding of neural cell adhesion molecules (N-CAMs) to the cellular prion protein. *J Mol Biol.* 2001;314:1209–1225.
25. Mouillet-Richard S, Pietri M, Schneider B, et al. Modulation of serotonergic receptor signaling and cross-talk by prion protein. *J Biol Chem.* 2005;280:4592–4601.
26. Rambold AS, Muller V, Ron U, Ben-Tal N, Winklhofer KF, Tatzelt J. Stress-protective signaling of prion protein is corrupted by scrapie prions. *EMBO J.* 2008;27:1974–1984.
27. Bounhar Y, Zhang Y, Goodyer CG, LeBlanc A. Prion protein protects human neurons against Bax-mediated apoptosis. *J Biol Chem.* 2001;276:39145–39149.
28. Kurschner C, Morgan JI. The cellular prion protein (PrP) selectively binds to Bcl-2 in the yeast two-hybrid system. *Brain Res Mol Brain Res.* 1995;30:165–168.
29. Diarra-Mehrpour M, Arrabal S, Jalil A, et al. Prion protein prevents human breast carcinoma cell line from tumor necrosis factor alpha-induced cell death. *Cancer Res.* 2004;64:719–727.
30. Prusiner SB, DeArmond SJ. Prion diseases and neurodegeneration. *Annu Rev Neurosci.* 1994;17:311–339.
31. Bolton DC, McKinley MP, Prusiner SB. Identification of a protein that purifies with the scrapie prion. *Science.* 1982;218:1309–1311.
32. Pan KM, Baldwin M, Nguyen J, et al. Conversion of alpha-helices into beta-sheets features in the formation of the scrapie prion proteins. *Proc Natl Acad Sci USA.* 1993;90:10962–10966.
33. Prusiner SB. Prions. *Proc Natl Acad Sci USA.* 1998;95:13363–13383.
34. Telling GC, Scott M, Mastrianni J, et al. Prion propagation in mice expressing human and chimeric PrP transgenes implicates the interaction of cellular PrP with another protein. *Cell.* 1995;83:79–90.
35. Goold R, Rabbanian S, Sutton L, et al. Rapid cell-surface prion protein conversion revealed using a novel cell system. *Nat Commun.* 2011;2:281.
36. Marijanovic Z, Caputo A, Campana V, Zurzolo C. Identification of an intracellular site of prion conversion. *PLoS Pathog.* 2009;5:e1000426.

37. Rouvinski A, Karniely S, Kounin M, et al. Live imaging of prions reveals nascent PrPSc in cell-surface, raft-associated amyloid strings and webs. *J Cell Biol*. 2014;204: 423–441.
38. Büeler H, Aguzzi A, Sailer A, et al. Mice devoid of PrP are resistant to scrapie. *Cell*. 1993;73:1339–1347.
39. Pradines E, Hernandez-Rapp J, Villa-Diaz A, et al. Pathogenic prions deviate PrP(C) signaling in neuronal cells and impair A-beta clearance. *Cell Death Dis*. 2013;4:e456.
40. Hijazi N, Kariv-Inbal Z, Gasset M, Gabizon R. PrPSc incorporation to cells requires endogenous glycosaminoglycan expression. *J Biol Chem*. 2005;280:17057–17061.
41. Taylor DR, Hooper NM. The low-density lipoprotein receptor-related protein 1 (LRP1) mediates the endocytosis of the cellular prion protein. *Biochem J*. 2007; 402:17–23.
42. Heller U, Winklhofer KF, Heske J, Reintjes A, Tatzelt J. Post-translational import of the prion protein into the endoplasmic reticulum interferes with cell viability: a critical role for the putative transmebrane domain. *J Biol Chem*. 2003;278:36139–36147.
43. Tagliavini F, Prelli F, Verga L, et al. Synthetic peptides homologous to prion protein residues 106-147 form amyloid-like fibrils in vitro. *Proc Natl Acad Sci USA*. 1993;90:9678–9682.
44. Baumann F, Tolnay M, Brabeck C, et al. Lethal recessive myelin toxicity of prion protein lacking its central domain. *EMBO J*. 2007;26:538–547.
45. Rambold AS, Miesbauer M, Rapaport D, et al. Association of Bcl-2 with misfolded prion protein is linked to the toxic potential of cytosolic PrP. *Mol Biol Cell*. 2006;17:3356–3368.
46. Stahl N, Borchelt DR, Hsiao K, Prusiner SB. Scrapie prion protein contains a phosphatidylinositol glycolipid. *Cell*. 1987;51:229–240.
47. Guizzunti G, Zurzolo C. The fate of PrP GPI-anchor signal peptide is modulated by P238S pathogenic mutation. *Traffic*. 2014;15:78–93.
48. Gu Y, Singh A, Bose S, Singh N. Pathogenic mutations in the glycosylphosphatidylinositol signal peptide of PrP modulate its topology in neuroblastoma cells. *Mol Cell Neurosci*. 2008;37:647–656.
49. Hölscher C, Bach UC, Dobberstein B. Prion protein contains a second endoplasmic reticulum targeting signal sequence located at its C terminus. *J Biol Chem*. 2001;276:13388–13394.
50. Miesbauer M, Rambold AS, Winklhofer KF, Tatzelt J. Targeting of the prion protein to the cytosol: mechanisms and consequences. *Curr Issues Mol Biol*. 2010;12:109–118.
51. Caughey B, Race RE, Ernst D, Buchmeier MJ, Chesebro B. Prion protein biosynthesis in scrapie-infected and uninfected neuroblastoma cells. *J Virol*. 1989;63:175–181.
52. Cohen FE, Pan KM, Huang Z, Baldwin M, Fletterick RJ, Prusiner SB. Structural clues to prion replication. *Science*. 1994;264:530–531.
53. Beland M, Motard J, Barbarin A, Roucou X. PrP(C) homodimerization stimulates the production of PrPC cleaved fragments PrPN1 and PrPC1. *J Neurosci*. 2012;32: 13255–13263.
54. Tompa P, Tusnady GE, Friedrich P, Simon I. The role of dimerization in prion replication. *Biophys J*. 2002;82:1711–1718.
55. Sunyach C, Cisse MA, da Costa CA, Vincent B, Checler F. The C-terminal products of cellular prion protein processing, C1 and C2, exert distinct influence on p53-dependent staurosporine-induced caspase-3 activation. *J Biol Chem*. 2007;282: 1956–1963.
56. Westergard L, Turnbaugh JA, Harris DA. A naturally occurring C-terminal fragment of the prion protein (PrP) delays disease and acts as a dominant-negative inhibitor of PrPSc formation. *J Biol Chem*. 2011;286:44234–44242.

57. Watt NT, Hooper NM. Reactive oxygen species (ROS)-mediated beta-cleavage of the prion protein in the mechanism of the cellular response to oxidative stress. *Biochem Soc Trans*. 2005;33:1123–1125.
58. Godsave SF, Wille H, Kujala P, et al. Cryo-immunogold electron microscopy for prions: toward identification of a conversion site. *J Neurosci*. 2008;28:12489–12499.
59. Orlean P, Menon AK. Thematic review series: lipid posttranslational modifications. GPI anchoring of protein in yeast and mammalian cells, or: how we learned to stop worrying and love glycophospholipids. *J Lipid Res*. 2007;48:993–1011.
60. Encalada SE, Szpankowski L, Xia CH, Goldstein LS. Stable kinesin anddynein assemblies drive the axonal transport of mammalian prion protein vesicles. *Cell*. 2011;144:551–565.
61. Uchiyama K, Muramatsu N, Yano M, Usui T, Miyata H, Sakaguchi S. Prions disturb post-Golgi trafficking of membrane proteins. *Nat Commun*. 2013;4:1846.
62. Sarnataro D, Paladino S, Campana V, Grassi J, Nitsch L, Zurzolo C. PrPC is sorted to the basolateral membrane of epithelial cells independently of its association with rafts. *Traffic*. 2002;3:810–821.
63. Arkhipenko A, Syan S, Victoria GS, Lebreton S, Zurzolo C. PrPC undergoes basal to apical transcytosis in polarized epithelial MDCK cells. *PLoS One*. 2016;11. e0157991.
64. Harris DA. Cellular biology of prion diseases. *Clin Microbiol Rev*. 1999;12:429–444.
65. Naslavsky N, Stein R, Yanai A, Friedlander G, Taraboulos A. Characterization of detergent-insoluble complexes containing the cellular prion protein and its scrapie isoform. *J Biol Chem*. 1997;272:6324–6331.
66. Taraboulos A, Raeber AJ, Borchelt DR, Serban D, Prusiner SB. Synthesis and trafficking of prion proteins in cultured cells. *Mol Biol Cell*. 1992;3:851–863.
67. Sarnataro D, Caputo A, Casanova P, et al. Lipid rafts and clathrin cooperate in the internalization of PrP in epithelial FRT cells. *PLoS One*. 2009;4. e5829.
68. Sarnataro D, Campana V, Paladino S, Stornaiuolo M, Nitsch L, Zurzolo C. PrP(C) association with lipid rafts in the early secretory pathway stabilizes its cellular conformation. *Mol Biol Cell*. 2004;15:4031–4042.
69. Campana V, Sarnataro D, Zurzolo C. The highways and byways of prion protein trafficking. *Trends Cell Biol*. 2005;15:102–111.
70. Shyng SL, Heuser JE, Harris DA. A glycolipid-anchored prion protein is endocytosed via clathrin-coated pits. *J Cell Biol*. 1994;125:1239–1250.
71. Sunyach C, Jen A, Deng J, et al. The mechanism of internalization of glycosylphosphatidylinositol-anchored prion protein. *EMBO J*. 2003;22:3591–3601.
72. Ren K, Gao C, Zhang J, et al. Flotillin-1 mediates PrPc endocytosis in the cultured cells during Cu^{2+} stimulation through molecular interaction. *Mol Neurobiol*. 2013; 48:631–646.
73. Rieger R, Edenhofer F, Lasmézas CI, Weiss S. The human 37-kDa laminin receptor precursor interacts with the prion protein in eukaryotic cells. *Nat Med*. 1997;3:1383–1388.
74. Gauczynski S, Peyrin JM, Haïk S, et al. The 37-kDa/67-kDa laminin receptor acts as the cell-surface receptor for the cellular prion protein. *EMBO J*. 2001;20:5863–5875.
75. Peters PJ, Mironov Jr A, Peretz D, et al. Trafficking of prion proteins through a caveolae-mediated endosomal pathway. *J Cell Biol*. 2003;162:703–717.
76. Galvan C, Camoletto PG, Dotti CG, Aguzzi A, Ledesma MD. Proper axonal distribution of PrP(C) depends on cholesterol-sphingomyelin-enriched membrane domains and is developmentally regulated in hippocampal neurons. *Mol Cell Neurosci*. 2005;30:304–315.
77. Mahal SP, Baker CA, Demczyk CA, Smith EW, Julius C, Weissmann C. Prion strain discrimination in cell culture: the cell panel assay. *Proc Natl Acad Sci USA*. 2007;104:20908–20913.

78. Magalhães AC, Silva JA, Lee KS, et al. Endocytic intermediates involved with the intracellular trafficking of a fluorescent cellular prion protein. *J Biol Chem*. 2002;277: 33311–33318.
79. Shim SY, Karri S, Law S, Schatzl HM, Gilch S. Prion infection impairs lysosomal degradation capacity by interfering with rab7 membrane attachment in neuronal cells. *Sci Rep*. 2016;6:21658.
80. Yost CS, Lopez CD, Prusiner SB, Myers RM, Lingappa VR. Non-hydrophobic extracytoplasmic determinant of stop transfer in the prion protein. *Nature*. 1990;343:669–672.
81. Udenfriend S, Kodukula K. How glycosylphosphatidylinositol-anchored membrane proteins are made. *Annu Rev Biochem*. 1995;64:563–591.
82. Hegde RS, Mastrianni JA, Scott MR, et al. A transmembrane form of the prion protein in neurodegenerative disease. *Science*. 1998;279:827–834.
83. Hegde RS, Tremblay P, Groth D, DeArmond SJ, Prusiner SB, Lingappa VR. Transmissible and genetic prion diseases share a common pathway of neurodegeneration. *Nature*. 1999;402:822–826.
84. Harris DA. Trafficking, turnover and membrane topology of PrP. *Br Med Bull*. 2003;66:71–85.
85. Ma J, Lindquist S. Wild-type PrP and a mutant associated with prion disease are subject to retrograde transport and proteasome degradation. *Proc Natl Acad Sci USA*. 2001;98:14955–14960.
86. Yedidia Y, Horonchik L, Tzaban S, Yanai A, Taraboulos A. Proteasomes and ubiquitin are involved in the turnover of the wild-type prion protein. *EMBO J*. 2001;20: 5383–5391.
87. Drisaldi B, Stewart RS, Adles C, et al. Mutant PrP is delayed in its exit from the endoplasmic reticulum, but neither wild-type nor mutant PrP undergoes retrotranslocation prior to proteasomal degradation. *J Biol Chem*. 2003;278:21732–21743.
88. Campana V, Sarnataro D, Fasano C, Casanova P, Paladino S, Zurzolo C. Detergent-resistant membrane domains but not the proteasome are involved in the misfolding of a PrP mutant retained in the endoplasmic reticulum. *J Cell Sci*. 2006;119:433–442.
89. Ashok A, Hegde RS. Retrotranslocation of prion proteins from the endoplasmic reticulum by preventing GPI signal transamidation. *Mol Biol Cell*. 2008;19:3463–3476.
90. Satpute-Krishnan P, Ajinkya M, Bhat S, Itakura E, Hegde RS, Lippincott-Schwartz J. ER stress-induced clearance of misfolded GPI-anchored proteins via the secretory pathway. *Cell*. 2014;158:522–533.
91. Senatore A, Restelli E, Chiesa R. Synaptic dysfunction in prion diseases: a trafficking problem? *Int J Cell Biol*. 2013;2013:543803.
92. Liberski PP, Brown DR, Sikorska B, Caughey B, Brown P. Cell death and autophagy in prion diseases (transmissible spongiform encephalopathies). *Folia Neuropathol*. 2008;46:1–25.
93. Sikorska B, Liberski PP, Giraud P, Kopp N, Brown P. Autophagy is a part of ultrastructural synaptic pathology in Creutzfeldt-Jakob disease: a brain biopsy study. *Int J Biochem Cell Biol*. 2004;36:2563–2573.
94. Browman D, Zurzolo C. Not on the menu: autophagy-independent clearance of prions. *Prion*. 2013;7:286–290.
95. Marzo L, Marijanovic Z, Browman D, Chamoun Z, Caputo A, Zurzolo C. 4-hydroxytamoxifen leads to PrPSc clearance by conveying both PrPC and PrPSc to lysosomes independently of autophagy. *J Cell Sci*. 2013;126:1345–1354.
96. Low MG, Stütz P. Inhibition of the plasma glycosylphosphatidylinositol-specific phospholipase D by synthetic analogs of lipid A and phosphatidic acid. *Arch Biochem Biophys*. 1999;371:332–339.

97. Nosjean O, Briolay A, Roux B. Mammalian GPI proteins: sorting, membrane residence and functions. *Biochim Biophys Acta*. 1997;133:153–186.
98. McConville MJ, Collidge TA, Ferguson MA, Schneider P. The glycoinositol phospholipids of Leishmania mexicana promastigotes. Evidence for the presence of three distinct pathways of glycolipid biosynthesis. *J Biol Chem*. 1993;268:15595–15604.
99. Muñiz M, Zurzolo C. Sorting of GPI-anchored proteins from yeast to mammals—common pathways at different sites? *J Cell Sci*. 2014;127:2793–2801.
100. Chesebro B, Trifilo M, Race R, et al. Anchorless prion protein results in infectious amyloid disease without clinical scrapie. *Science*. 2005;308:1435–1439.
101. Williams AF, Tse AG. A glycophospholipid covalently attached to the C-terminus of the Thy-1 glycoprotein. *Biosci Rep*. 1985;5:999–1005.
102. Ferguson MA, Low MG, Cross GA. Glycosylsn-1,2-dimyristylphosphatidylinositol is covalently linked to Trypanosoma brucei variant surface glycoprotein. *J Biol Chem*. 1985;260:14547–14555.
103. Chatterjee S, Mayor S. The GPI-anchor and protein sorting. *Cell Mol Life Sci*. 2001;58:1969–1987.
104. Rajendran L, Simons K. Lipid rafts and membrane dynamics. *J Cell Sci*. 2005;118:1099–1102.
105. Surma MA, Klose C, Simons K. Lipid-dependent protein sorting at the trans-Golgi network. *Biochim Biophys Acta*. 2012;1821:1059–1067.
106. Simons K, Toomre D. Lipid rafts and signal transduction. *Nat Rev Mol Cell Biol*. 2000;1:31–39.
107. Jones DR, Varela-Nieto I. The role of glycosyl-phosphatidylinositol in signal transduction. *Int J Biochem Cell Biol*. 1998;30:313–326.
108. Schuck S, Simons K. Controversy fuels trafficking of GPI-anchored proteins. *J Cell Biol*. 2006;172:963–965.
109. Zurzolo C, Lisanti MP, Caras IW, Nitsch L, Rodriguez-Boulan E. Glycosylphosphatidylinositol-anchored proteins are preferentially targeted to the basolateral surface in Fischer rat thyroid epithelial cells. *J Cell Biol*. 1993;121:1031–1039.
110. Paladino S, Sarnataro D, Pillich R, Tivodar S, Nitsch L, Zurzolo C. Protein oligomerization modulates raft partitioning and apical sorting of GPI-anchored proteins. *J Cell Biol*. 2004;167:699–709.
111. Gordon VM, Nelson KL, Buckley JT, et al. Clostridium septicum alpha toxin uses glycosylphosphatidylinositol-anchored protein receptors. *J Biol Chem*. 1999;274: 27274–27280.
112. Radford HE, Mallucci GR. The role of GPI-anchored PrP C in mediating the neurotoxic effect of scrapie prions in neurons. *Curr Issues Mol Biol*. 2010;12:119–127.
113. Aguzzi A. Cell biology. Prion toxicity: all sail and no anchor. *Science*. 2005;3(308): 1420–1421.
114. Priola SA, McNally KL. The role of the prion protein membrane anchor in prion infection. *Prion*. 2009;3:134–138.
115. Kaneko K, Vey M, Scott M, Pilkuhn S, Cohen FE, Prusiner SB. COOH-terminal sequence of the cellular prion protein directs subcellular trafficking and controls conversion into the scrapie isoform. *Proc Natl Acad Sci USA*. 1997;94:2333–2338.
116. Caughey B, Baron GS, Chesebro B, Jeffrey M. Getting a grip on prions: oligomers, amyloids, and pathological membrane interactions. *Annu Rev Biochem*. 2009;78: 177–204.
117. Gousset K, Zurzolo C. Tunnelling nanotubes: a highway for prion spreading? *Prion*. 2009;3:94–98.
118. Porto-Carreiro I, Février B, Paquet S, Vilette D, Raposo G. Prions and exosomes: from PrPc trafficking to PrPsc propagation. *Blood Cells Mol Dis*. 2005;35:143–148.

119. Gousset K, Schiff E, Langevin C, et al. Prions hijack tunnelling nanotubes for intercellular spread. *Nat Cell Biol.* 2009;11:328–336.
120. Baron GS, Wehrly K, Dorward DW, Chesebro B, Caughey B. Conversion of raft associated prion protein to the protease-resistant state requires insertion of PrP-res (PrP(Sc)) into contiguous membranes. *EMBO J.* 2002;21:1031–1040.
121. Mishra RS, Basu S, Gu Y, et al. Protease-resistant human prion protein and ferritin are cotransported across Caco-2 epithelial cells: implications for species barrier in prion uptake from the intestine. *J Neurosci.* 2004;24:11280–11290.
122. Zhu S, Victoria GS, Marzo L, Ghosh R, Zurzolo C. Prion aggregates transfer through tunneling nanotubes in endocytic vesicels. *Prion.* 2015;9:125–135.
123. Caughey B, Raymond GJ, Callahan MA, Wong C, Baron GS, Xiong LW. Interactions and conversions of prion protein isoforms. *Adv Protein Chem.* 2001;57:139–169.
124. Silveira JR, Raymond GJ, Hughson AG, et al. The most infectious prion protein particles. *Nature.* 2005;437:257–261.
125. Marshall KE, Hughson A, Vascellari S, et al. PrP knockout cells expressing transmembrane PrP resist prion infection. *J Virol.* 2017;91:e01686-16.
126. Taylor DR, Hooper NM. The prion protein and lipid rafts. *Mol Membr Biol.* 2006;23:89–99.
127. Taraboulos A, Scott M, Semenov A, Avraham D, Laszlo L, Prusiner SB. Cholesterol depletion and modification of COOH-terminal targeting sequence of the prion protein inhibit formation of the scrapie isoform. *J Cell Biol.* 1995;129:121–132.
128. Bate C, Tayebi M, Williams A. Cholesterol esterification reduces the neurotoxicity of prions. *Neuropharmacology.* 2008;54:1247–1253.
129. Bate C, Williams A. Neurodegeneration induced by clustering of sialylated glycosylphosphatidylinositols of prion proteins. *J Biol Chem.* 2012;287:7935–7944.
130. Bate C, Nolan W, Williams A. Glycosylphosphatidylinositols: more than just an anchor? *Commun Integr Biol.* 2016;9:e1149671.

CHAPTER FOUR

Understanding the Effect of Disease-Related Mutations on Human Prion Protein Structure: Insights From NMR Spectroscopy

Ivana Biljan*, Gregor Ilc†,‡, Janez Plavec†,‡,§,1
*Faculty of Science, University of Zagreb, Zagreb, Croatia
†Slovenian NMR Centre, National Institute of Chemistry, Ljubljana, Slovenia
‡EN-FIST Centre of Excellence, Ljubljana, Slovenia
§Faculty of Chemistry and Chemical Technology, University of Ljubljana, Ljubljana, Slovenia
[1]Corresponding author: e-mail address: janez.plavec@ki.si

Contents

Abstract

Prion diseases or transmissible spongiform encephalopathies constitute a group of fatal neurodegenerative diseases that can be of sporadic, genetic, or acquired origin. The central molecular event of prion diseases is the conformational conversion of the physiological cellular prion protein, PrP^C, into a disease-associated form known as prion or PrP^{Sc}. Spontaneous generation of prions in genetic prion diseases is caused by mutations in the human prion protein gene (*PRNP*). Understanding of the earliest conformational changes during misfolding of PrP^C in genetic forms of prion diseases may benefit from detailed structural characterization of various human (Hu) PrP variants. Nuclear magnetic resonance (NMR) spectroscopy offers unique opportunities to obtain detailed atomic-level structure information. In this chapter we present an overview of high-resolution NMR studies on several HuPrPs with disease-associated mutations at mildly acidic and physiological pH conditions that provided valuable insights into possible key structural determinants underlying the formation of prions.

Progress in Molecular Biology and Translational Science, Volume 150
ISSN 1877-1173
http://dx.doi.org/10.1016/bs.pmbts.2017.06.006

1. INTRODUCTION

Prion diseases, also known as transmissible spongiform encephalopathies, are a family of invariably fatal neurodegenerative diseases affecting humans and animals. The term "prion," coined by Stanley B. Prusiner in 1982, describes a small proteinaceous infectious particle considered to be the sole causative agent of prion diseases.[1] Within the context of this generally accepted "protein-only hypothesis," the central molecular event of prion diseases is the conformational conversion of the physiological cellular prion protein, PrP^{C}, into a disease-associated form, prion or PrP^{Sc} (scrapie PrP).[2,3] It has been proposed that PrP^{Sc} propagates in an autocatalytic process by binding to PrP^{C} and templating its conformational conversion into the malicious PrP^{Sc} form.[4]

Prion diseases can be of sporadic, genetic, or acquired origin.[5] Sporadic prion diseases are the most common of all human prion diseases accounting for approximately 85% of all cases. They include sporadic Creutzfeldt–Jakob disease (sCJD), which is the most frequent form, sporadic fatal insomnia, and the variably protease-sensitive prionopathies. Genetic forms of prion diseases are associated with mutations in the human prion protein gene (*PRNP*) and include familial CJD (fCJD), Gerstmann–Sträussler–Scheinker (GSS) syndrome, fatal familial insomnia (FFI), and prion protein cerebral amyloid angiopathy. Acquired prion diseases are very rare occurring in less than 1% of the cases. They are transmitted from human to human, as iatrogenic CJD and Kuru, or from animals to humans, as variant CJD (vCJD).[6] Prion diseases affecting animals are scrapie in sheep and goats, bovine spongiform encephalopathy in cattle, and chronic wasting disease in deer, elk, and moose.[7] The common neuropathological features of prion diseases include spongiform degeneration, neuronal loss, and gliosis.[8] In addition, similar to some other neurodegenerative disorders such as Alzheimer's and Parkinson's diseases, prion diseases are often characterized by formation of amyloid plaques. It is interesting to note that the recent studies support a unifying role for prions in neurodegenerative diseases, suggesting that proteins causing many diseases including Alzheimer's and Parkinson's diseases also show prion-like behavior.[9]

A major focus in prion biology is unraveling the molecular mechanism leading to conformational change of PrP^{C} into pathological PrP^{Sc} form. One of the strongest arguments in favor of the "protein-only hypothesis" is the link between genetic prion diseases and mutations in the *PRNP* gene.

Numerous pathological mutations responsible for genetic forms of prion diseases in humans have been identified in the *PRNP* gene, but the mechanism by which these mutations facilitate spontaneous generation of prions remains largely unresolved.[10] Many of the mutations that cause genetic prion diseases are clustered within the globular domain of human PrP (HuPrP).[11] It has been suggested that mutations within the globular domain may increase the likelihood of misfolding of PrP^{C} by its thermodynamic destabilization.[12–14] In addition, these mutations may increase the population of partially folded intermediate of the PrP^{C} proposed as a direct monomeric precursor of the pathological PrP^{Sc} form.[15] Mutations may also induce an aberrant trafficking and accumulation inside the cells or alter surface properties of PrP^{C} triggering abnormal interactions with other not yet identified cofactors.[16–18] Understanding of the earliest structural and dynamical changes in PrP^{C} during its pathological conversion in genetic forms of prion diseases may benefit from detailed structural characterization of various HuPrP variants. Nuclear magnetic resonance (NMR) spectroscopy has proven to be a powerful tool for revealing atomic-level structural features. The focus of this chapter is to provide an overview of high-resolution NMR structural studies of HuPrPs with different disease-associated mutations that provided valuable insights into possible key structural determinants underlying misfolding of PrP^{C}.[19–22]

2. STRUCTURAL FEATURES OF PrP^{C} AND PrP^{Sc}

In order to fully understand a PrP^{C} to PrP^{Sc} conversion it is essential to reveal the atomic-level structural features of both PrP forms. The mature HuPrP is found in vivo as a glycosylphosphatidylinositol-anchored glycoprotein composed of 209 residues.[23] The majority of high-resolution structural studies were carried out with bacterially expressed full-length or truncated recombinant PrPs lacking posttranslational modifications. Most of the three-dimensional (3D) structures of PrPs were determined by using high-resolution solution-state NMR spectroscopy,[24–32] while the X-ray crystallography was employed less frequently.[33–37] The first 3D structure of the wild-type (WT) HuPrP, determined in year 2000 by NMR spectroscopy in solution, revealed that the protein contains two parts with different structural and dynamical features.[32] The first part is a highly flexible N-terminal segment encompassing residues 23–124, whereas the second part is a folded C-terminal globular domain comprising residues 125–231. The globular domain contains three α-helices, spanning residues 144–154 (α1), 173–194 (α2), and 200–228 (α3), and two short antiparallel β-strands comprising residues 128–131 (β1) and 161–164 (β2). Helices α2 and α3 are covalently linked by a disulfide bond between Cys179 and Cys214. Structures of PrP expressed by other mammalian species show similar overall fold. Local sequence and structure variations are most prominently localized at the interface of the β2–α2 loop and in the C-terminal part of the helix α3. The first X-ray structure of the WT HuPrP reported in year 2001 revealed an unexpected 3D domain-swapped dimer.[36] The major differences between the crystal and solution structures of the WT protein can be found in helix α3 which packs against helix α2 from the other monomer through inter-subunit Cys179-Cys214 disulfide bond in the crystal structure rather than against its own polypeptide chain as in solution NMR structure. The X-ray-derived structure of the WT protein suggested that oligomerization through domain swapping may provide a possible mechanism underlying conversion of PrP^{C} into pathological PrP^{Sc}. Interestingly, domain swapping has been also proposed to be implicated in oligomerization of other amyloidogenic proteins.[38]

In contrast to the wealth of structural information available for PrP^{C}, there is a lack of high-resolution 3D structural data for PrP^{Sc} which is mostly due to its insolubility and propensity for aggregation.[39] Although PrP^{C} and PrP^{Sc} display no variations in the primary structure, they exhibit several different properties. Spectroscopy techniques such as Fourier transform infrared and circular dichroism spectroscopy revealed that on contrary to mainly α-helical PrP^{C}, PrP^{Sc} contains a significant content of β-sheets.[40,41] Several models of PrP^{Sc} have been proposed[42] of which the most prominent are the β-helix model,[43] the β-spiral model,[44] and the extended in-register β-sheet model.[45,46] The β-helix model, based on electron microscopy studies, proposes a major refolding of the N-terminal region of PrP which forms left-handed β-helices that associate into trimers, whereas the C-terminal part maintains its α-helical structure. According to the β-spiral model developed by molecular dynamics (MD) simulations, PrP^{Sc} consists of a spiraling core of extended sheets comprising short β-strands localized at the N-terminal part of the protein, whereas the α-helical structure of the C-terminus is preserved. The extended in-register β-sheet model is substantially different when compared to the β-helix and β-spiral models. This model was suggested on the basis of the site-directed spin labeling in combination with electron paramagnetic resonance and hydrogen/deuterium exchange coupled with mass spectrometry. In the extended in-register β-sheet model of PrP^{Sc}, C-terminal residues form single-molecule layers stacking on top of one another with parallel, in-register β-strands. This model suggests that conversion of PrP^{C} into PrP^{Sc} involves refolding of the entire protein into the mainly β-sheet structure, with no α-helices present.

3. STRATEGY FOR RESONANCE ASSIGNMENT AND NMR STRUCTURE DETERMINATION OF HUMAN PRION PROTEIN VARIANTS

The first NMR experiment to be recorded on isotopically labeled protein is typically two-dimensional ^{1}H–^{15}N heteronuclear single-quantum coherence (HSQC) spectrum which serves as a protein's fingerprint. Inspection of ^{1}H–^{15}N HSQC spectrum provides a quick insight into suitability of a sample for structure determination. A good dispersion of signals in the ^{1}H–^{15}N HSQC spectra of HuPrP variants (e.g., HuPrP(Q212P), HuPrP(V210I), and HuPrP(E219K)) indicates that the proteins are well folded (Fig. 1).

In order to determine the 3D structure of a protein by NMR spectroscopy, the backbone and side-chain ^{1}H, ^{15}N, and ^{13}C resonances must be assigned. For proteins such as HuPrP with molecular masses close to 19 kDa, assignments are usually accomplished by recording a series of 3D

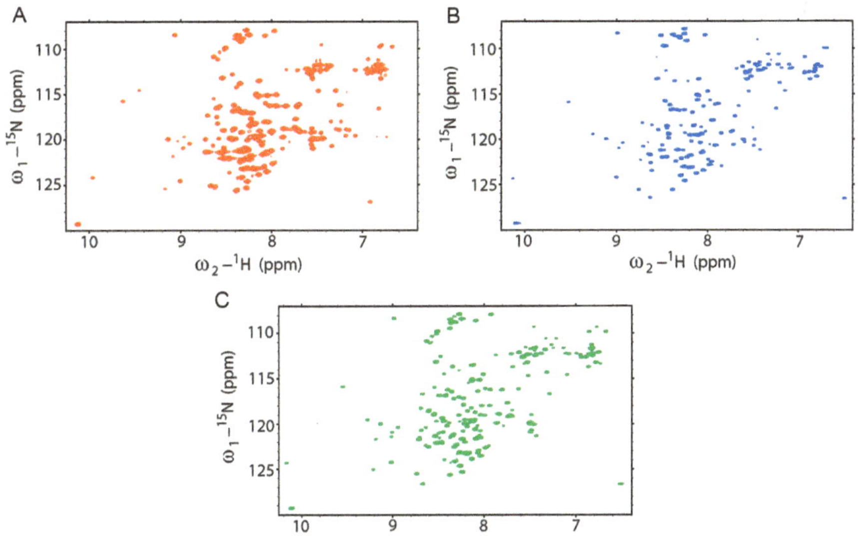

Fig. 1 ^{1}H–^{15}N HSQC spectra of HuPrP variants at pH 5.5. (A) HuPrP(90–231, M129V, Q212P), (B) HuPrP(90–231, M129V, V210I), and (C) HuPrP(90–231, M129V, E219K).

NMR experiments on ^{13}C and ^{15}N uniformly labeled protein samples.[47] The sequence-specific assignments of backbone resonances of HuPrP variants can be achieved by combining HNCO, CA(CO)NH, HNCA, CBCA(CO)NH, and HNCACB 3D NMR experiments.[47] ^{1}H and ^{13}C resonances in aliphatic side chains of HuPrP variants can be assigned by using 3D HAHB(CO)NH, (H)CCH-total correlation spectroscopy (TOCSY), and ^{13}C-edited nuclear Overhauser enhancement spectroscopy (NOESY)-HSQC experiments.[47] The assignment of the aromatic ^{1}H and ^{13}C resonances is carried out by analyzing the aromatic region of the 3D ^{13}C-edited NOESY-HSQC spectrum. For assignment of the side-chain amide groups of Asn and Gln residues 3D ^{15}N- and ^{13}C-edited NOESY-HSQC experiments are typically used.

Values of C_{β} chemical shifts of Cys residues can be employed as sensitive indicators of disulfide bond formations.[48] For example, C_{β} chemical shifts of Cys179 and Cys214 in HuPrP(Q212P), HuPrP(V210I), and HuPrP(E219K) are 40.3 and 41.5 ppm, 40.6 and 41.8 ppm, and 40.8 and 41.8 ppm, respectively, confirming the presence of Cys179-Cys214 disulfide bonds in all three proteins.

The most commonly used parameters in structure determination from NMR data are NOE distance restraints extracted from NOESY spectra. The intensity of NOE is proportional to the inverse sixth power of the interproton distance, and thus only the cross-peaks between protons separated by less than 5–6 Å will be observable in NOESY spectra. Crucial for defining the protein fold are long-range NOE contacts between protons that are close in space but distant along the protein sequence. The large number of NOE distance restraints (over 2200) derived from 3D ^{15}N- and ^{13}C-edited NOESY-HSQC experiments supplemented by backbone torsion angle restraints obtained from chemical shifts[49] allowed high-resolution structure determination of HuPrP variants. By using program CYANA, automatic NOE assignment and structure calculations of HuPrP variants can be performed.[50,51] The 20 conformers of each protein with the lowest residual target function values can be energy-minimized in a water shell with the YASARA program suite.[52] PROCHECK-NMR[53] and WhatIF[54] programs can be used for validation of NMR structures of HuPrP variants. This procedure demonstrated that the final family of 3D structures of HuPrP(Q212P), HuPrP(V210I), and HuPrP(E219K) is in agreement with the distance restraints and offers good geometry and side-chain packing. As evidenced by average root mean square deviation (r.m.s.d.) values lying in the range from 0.66 to 1.01 Å for backbone atoms and from 1.06 to 1.45 Å for all heavy atoms (residues 125–228), NMR structures of HuPrP(Q212P), HuPrP(V210I), and HuPrP(E219K) are well defined.[19,21,22]

4. STRUCTURAL FEATURES OF HUMAN PRION PROTEINS WITH PATHOLOGICAL MUTATIONS

4.1 Mildly Acidic pH Conditions

NMR spectroscopy in solution was used to determine high-resolution structures of HuPrPs with pathological Q212P and V210I mutations at mildly acidic conditions of pH 5.5.[21,22] The Q212P is a very rare point mutation associated with GSS syndrome. Residue Gln212 is situated in the middle of the helix α3 close to the Cys179-Cys214 disulfide bridge holding the core of the protein together. Substitution of Gln by Pro at codon 212 did not affect an overall fold of HuPrP which highly resembles that of the WT protein consisting of an unstructured N-terminal tail (residues 90–124) and a globular C-terminal domain (residues 125–231). The globular domain of HuPrP(Q212P) comprises four α-helices and a short antiparallel β-sheet (Fig. 2A). Interestingly, Q212P mutation introduces several local structural differences at the C-terminal part of the HuPrP(Q212P) in comparison to the structure of the WT protein determined under the same experimental conditions.[19] In particular, Q212P mutation induces a break of helix α3 between Glu200 and Tyr226 into two helices, denoted as α3 and α4. In HuPrP(Q212P), helix α3 spans residues from Glu200 to Arg220, while a new, short helix α4 is composed of residues 223–226. Moreover, the mutual orientation of helices α2 and α3 is altered in the HuPrP(Q212P). The α2–α3 interhelical angle is 46° in the WT protein, whereas the corresponding angle is 33° in the HuPrP(Q212P). Notable local structural changes in HuPrP(Q212P) when compared to the WT protein can be also found at the interface of the β2–α2 loop and the C-terminal end of helix α3 (Fig. 2B). In both, WT and HuPrP(Q212P) proteins the β2–α2 loop region is not structurally well defined. However, several long-range hydrophobic interactions between residues from the β2–α2 loop (e.g., Met166) and residues from the C-terminal part of helix α3 (e.g., Tyr225) that are present in the WT protein are absent in the HuPrP(Q212P). The loss of tertiary contacts between the β2–α2 loop and the C-terminus of helix α3 results in exposure of hydrophobic surface in HuPrP(Q212P) to solvent (Fig. 2C). The NMR structure of the HuPrP(Q212P) represents the first known example of HuPrP structure in which single amino acid substitution induced significant local structural perturbations. Previously determined NMR

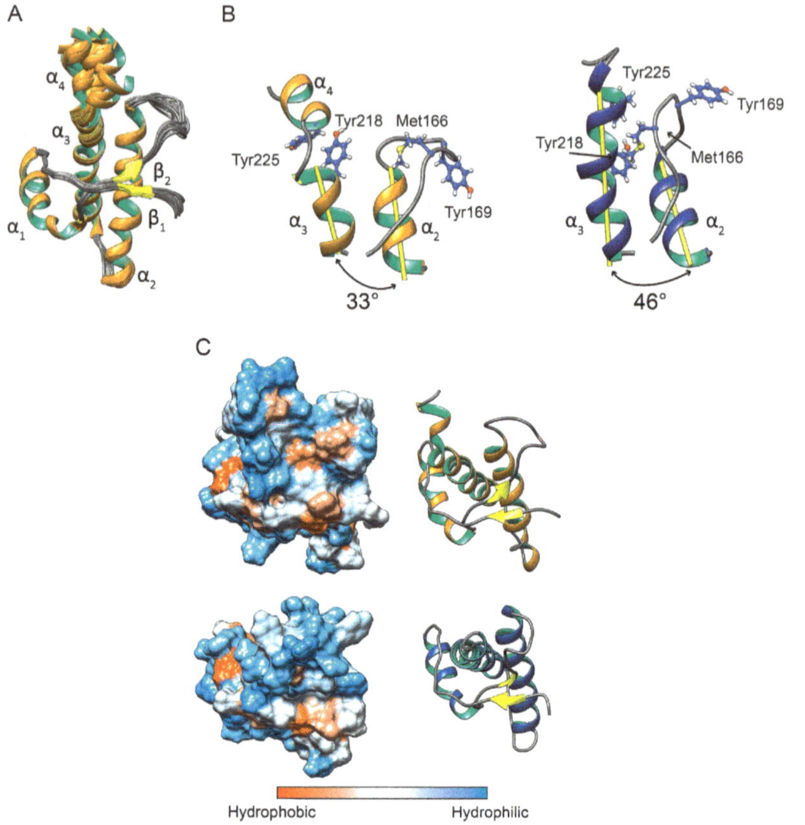

Fig. 2 Structural features of HuPrP(Q212P) (pdb id 2KUN) and comparison with the WT protein (pdb id 2LSB). (A) NMR structure of HuPrP(Q212P). (B) Tertiary contacts at the interface of the β2–α2 loop and the C-terminus of helix α3 in HuPrP(Q212P) and WT protein. (C) Presentation of surface hydrophobicity in HuPrP(Q212P) and WT protein.

structure of fCJD-linked E200K HuPrP mutant is practically identical to the structure of the WT HuPrP apart from minor differences in the flexible regions.[55] The only major consequence of the E200K mutation is the alteration of electrostatic potential on the protein surface.

The V210I mutation is linked with fCJD[56] and presents one of the most common mutations among genetic prion diseases in the European population.[57] Residue Val210 is located in the middle of the helix α3 being a part of a tightly packed hydrophobic core that is important for the overall stability of the globular domain of HuPrP.[32,58–60] NMR spectroscopy demonstrated that HuPrP(V210I) exhibits global structural features similar to the WT protein.[21] The 3D structure of HuPrP(V210I) contains an N-terminal flexible tail (residues 90–124) and a C-terminal globular region (residues 125–231) consisting of three α-helices and a short antiparallel β-sheet (Fig. 3A). However, substitution of Val by bulkier Ile at position 210 introduces steric hindrance at the α2–α3 interhelical interface. As a result, side chains of several residues involved in the α2–α3 hydrophobic interactions change their orientations when compared to the WT protein (Fig. 3B). Rearrangements are especially pronounced for Val180, located in the helix α2 opposite to Ile210, and for Val176 and Ile184. The aforementioned rearrangements induce reorganization of several hydrophobic contacts between residues in helices α2 and α3 with respect to the WT protein. The most pronounced changes upon Val-to-Ile mutation are the loss of hydrophobic contacts of Val180 with Val176 and Ile184 and, on the other hand, formation of long-range interactions of Val176 with Tyr218 that are not present in the WT protein. Moreover, mutual orientation of helices α2 and α3 is altered as evidenced by

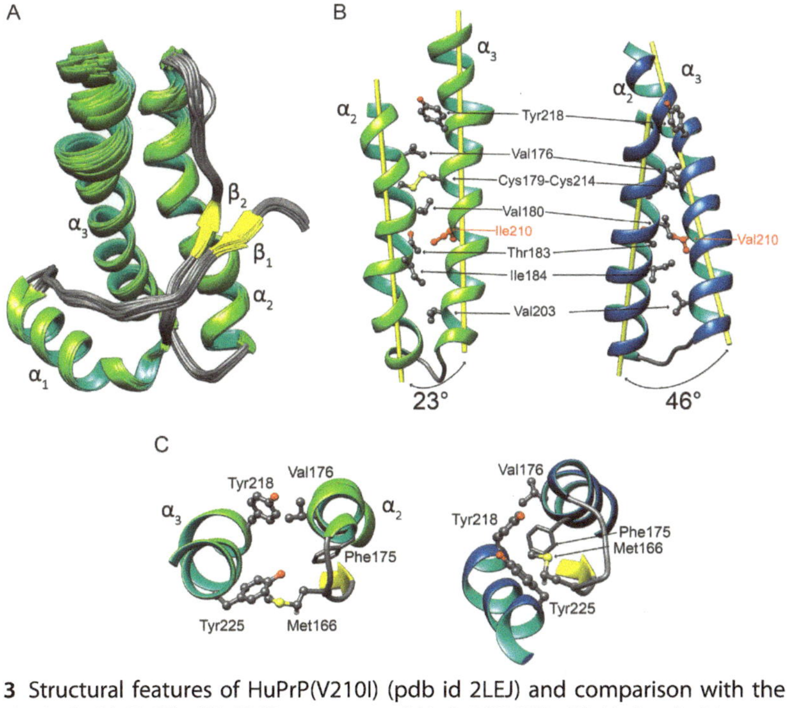

Fig. 3 Structural features of HuPrP(V210I) (pdb id 2LEJ) and comparison with the WT protein (pdb id 2LSB). (A) NMR structure of HuPrP(V210I). (B) Hydrophobic contacts at the α2–α3 interhelical interface in HuPrP(V210I) and WT protein. (C) Aromatic contacts at the interface of the β2–α2 loop and the C-terminus of helix α3 in HuPrP(V210I) and WT protein.

the decrease in the interhelical angle from 46° in the WT protein to 23° in the V210I mutant. Another region showing structural variations with respect to the WT protein is the interface of the β2–α2 loop and the C-terminal part of helix α3. Several aromatic interactions between residues from the structurally poorly defined β2–α2 loop region (e.g., Tyr163-Phe175) and between residues from the loop (e.g., Phe175) and the C-terminal part of helix α3 (e.g., Tyr218) that are normally present in the WT protein are lost in the HuPrP(V210I) (Fig. 3C). Consequently, hydrophobic cluster in HuPrP(V210I) shows increased exposure to solvent.

Earlier studies suggested that the partially unfolded monomer is an intermediate in formation of PrP^{Sc}.[61] The experimental data obtained by NMR spectroscopy[62,63] and kinetic stopped-flow studies[64] supported this hypothesis. Urea-induced unfolding experiments suggested that V210I mutation increases the population of intermediate states with respect to the WT protein[15] but has no significant effect on the global stability of the protein.[12] In fact, similar results were obtained for the majority of HuPrP variants linked to genetic prions diseases. These findings suggested that partially structured intermediate could be a direct precursor of the misfolded PrP^{Sc} in genetic prion diseases acting as a seed for the aggregation process.[65,66] Interestingly, partially folded intermediates seem to be involved in formation of amyloid fibrils in several other proteins and their population is increased by disease-associated mutations.[67–69] Protein folding intermediates usually feature an increased exposure of hydrophobic surface to solvent, enabling intermolecular interactions.[70–72] Several spectroscopic and cellular studies revealed that the increased exposure of hydrophobic residues is one of the main determinants of PrP neurotoxicity.[73–75] HuPrPs with pathological Q212P and V210I mutations both show increased exposure of hydrophobic residues to solvent resulting from loosening of contacts between the β2–α2 loop and helix α3. Similar structural perturbations were also reported in the NMR structure of E200K HuPrP mutant,[55] crystal structures of D178N and F198S HuPrP mutants,[37] and in several MD studies.[76–78] Different studies suggested that the conformation of the β2–α2 loop and its long-range interactions with residues in the C-terminal part of helix α3 play a role in susceptibility and transmission of prion diseases.[17,18,30,31,79] NMR structural studies demonstrated that species that display resistance to prion infection and do not develop spontaneous prion diseases (e.g., horse, rabbit, and tammar wallaby) are characterized by a rigid β2–α2 loop and pronounced tertiary contacts between the loop and helix α3. On the other hand, the increased exposure of hydrophobic surface to solvent in pathological mutants may facilitate intermolecular interactions involved in spontaneous generation of prions in genetic prion diseases.

4.2 Physiological pH Conditions

Another unresolved issue in prion biology is the identification of the subcellular site where conversion of PrP^C into PrP^{Sc} occurs. It has been suggested that PrP^C misfolding takes place either in the endosomal compartments where pH can vary from 4.7 to 6.5 or on the cell surface where the pH is around 7.[80,81] Several studies proposed that PrP^C misfolding and accumulation most likely occurs during the endocytosis and internalization pathway.[82,83] Reports using cell culture models described accumulation of PrP^{Sc} in both late[83] and recycling endosomes[84] of which the latter are the probable sites of prion conversion. Several biophysical and MD studies suggested that the decrease in pH provokes significant structural rearrangements in the PrP^C what may trigger its conversion to PrP^{Sc}.[85–89] Moreover, it seems that conformational rearrangements of HuPrP already start under mildly acidic conditions in the pH range from 4.4 to 6.[90] The NMR structure of the WT HuPrP at neutral pH revealed only minor conformational changes with respect to the structure determined at pH 4.5.[91] At pH 7.0 in comparison to pH 4.5, C-terminal ends of helices α1 and α2 in the WT HuPrP display conformation that is more regular. Authors suggested that protonation of His155 in helix α1 and His187 in helix α2 at pH 4.5 might contribute to destabilization of the two helices in the WT HuPrP. Thermal unfolding and hydrogen exchange experiments conducted in the same study indicated that the WT HuPrP is more stable at neutral pH conditions.

More recently, NMR spectroscopy was used to investigate the effect of pH on the structure of HuPrP carrying V210I mutation. The NMR structure of HuPrP(V210I) was determined at pH 7.2 and compared with the previously solved NMR structure of the same mutant at pH 5.5.[20] The comparison revealed that the 3D structures of HuPrP(V210I) at pH 5.5 and 7.2 display global similarities and several local structural variations. Differences between the two structures are mostly localized at the α2–α3 interhelical region, at the interface of the β1–α1 loop, in helices α1 and α3, and in the β2–α2 loop region.

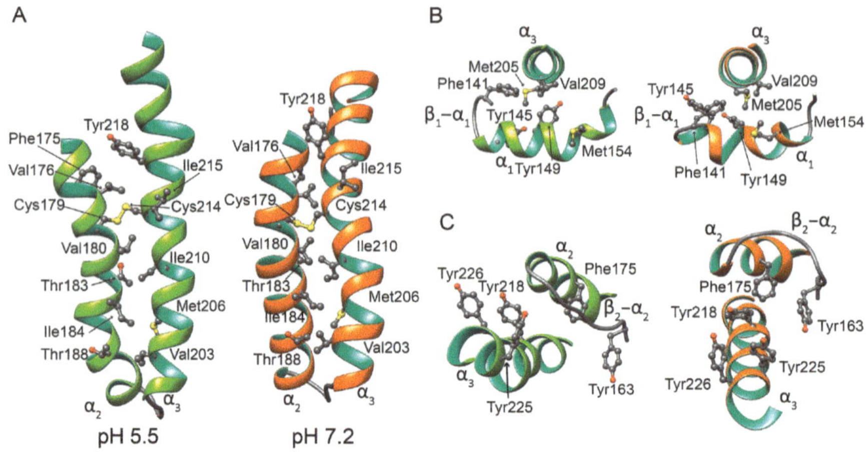

Fig. 4 Structural variations in HuPrP(V210I) at pH 5.5 (pdb id 2LEJ) and 7.2 (pdb id 2LV1). (A) Hydrophobic contacts at the α2–α3 interhelical interface. (B) Hydrophobic contacts at the interface of β1–α1 loop and helices α1 and α3. (C) Aromatic contacts at the interface of the β2–α2 loop and the C-terminus of helix α3.

The Val-to-Ile hydrophobic mutation at position 210 provokes rearrangements at the α2–α3 interhelical region at both mildly acidic and neutral pH conditions. Whereas at pH 5.5 several hydrophobic interactions between residues in helices α2 and α3 are lost, at pH 7.2 the majority of hydrophobic contacts are preserved (Fig. 4A). In addition, at pH 7.2 in comparison to pH 5.5, HuPrP(V210I) exhibits stronger hydrophobic interactions between residues at the interface of the β1–α1 loop (e.g., Phe141) and residues in helix α1 (e.g., Tyr145, Tyr149, and Met154) and helix α3 (e.g., Met205 and Val209) (Fig. 4B). Interestingly, several studies suggested that helix α1 promotes aggregation of PrP^C and that the separation of the β1–α1 loop, helix α1, and the α1–β2 loop from the α2–α3 region marks the early stages of conversion of PrP^C into PrP^{Sc}.[92–95] Results of MD simulations also indicated that lowering of pH leads to a loosening of contacts between helix α1 and other secondary structure elements in the globular domain of PrP^C.[85,86,89] Local structural differences in HuPrP(V210I) at mildly acidic and neutral pH conditions are also present at the interface of the β_2–α_2 loop and helix α_3. Aromatic interactions between residues from the loop (e.g., Tyr163 and Phe175) and helix α_3 (e.g., Tyr218) seem to be more pronounced at pH 7.2 (Fig. 4C).

Comparison of the ^{13}Cα chemical shifts of HuPrP(V210I) at pH 5.5 and 7.2 suggested an increase in helical content for the C-terminus of helix α2 spanning residues 187–193. The NMR structure of the WT HuPrP at pH 7.0 also shows an increase in population of helical structure in the same region.[91] The reduced stability of the C-terminal end of helix α_2 under acidic pH conditions is probably due to the positive charge on His187 leading to structural perturbations.[88,89,91,95,96] Notably, several pathological mutations associated with genetic forms of prion diseases destabilize helix α_2 by introducing the positive charge in the C-terminus of helix α_2 (e.g., H187R, T188R/K, E200K, and D202N). The comparison of distribution of electrostatic surface potential in HuPrP(V210I) at two pH values shows that the areas around the four His residues (His140, His155, His177, and His187) are predominantly positively charged at pH 5.5 and neutral at pH 7.2.

NMR studies of pH-related structural changes in HuPrP(V210I) demonstrated that the decrease in pH leads to a reduction in interactions between secondary structure elements. These findings indicate that HuPrP(V210I) exhibits higher structural stability under neutral pH, suggesting that spontaneous formation of prions in genetic prion diseases is favored under acidic pH conditions present in the endosomal compartments.

5. STRUCTURAL FEATURES OF HUMAN PRION PROTEIN WITH PROTECTIVE E219K POLYMORPHISM

Polymorphisms in PRNP affect etiology and neuropathology of prion diseases. The common M129V polymorphism at position 129 in the HuPrP influences the susceptibility to sporadic and acquired cases of the prion diseases.[97–100] The majority of individuals who developed sCJD are homozygous with respect to polymorphism at position 129 with M129 homozygosity being linked with higher predisposition for propagation of prion diseases. To date, all known cases of vCJD are homozygous for Met at position 129. On the other hand, the M129V heterozygosity seems to confer resistance to prion diseases.[101,102] M129V polymorphism also affects the phenotype of genetic prion diseases. For example, D178N/M129 variant is linked with FFI, whereas D178N/V129 is associated with fCJD.[103] NMR spectroscopy was used to study the effect of M129V polymorphism on the structure of HuPrP.[104] It was found that polymorphism does not have significant effect on the folding, dynamics, and stability of HuPrP. Another naturally occurring HuPrP polymorphism, E219K, shows a high incidence among the East Asian population and is considered to act protectively against sCJD.[105,106] The dominant-negative effect of E219K polymorphism has been supported by several experimental studies both in vivo in transgenic mouse[107] and in vitro in scrapie-infected neuroblastoma cells.[23,108] The authors of these studies suggested that the dominant-negative PrP^C variants inhibit conversion to PrP^{Sc} by binding to "protein X" and sequestering it from the replication process. The experimental evidence that formation of PrP^{Sc} is inhibited in knock-in (KI) mice expressing HuPrP heterozygous for Lys at codon 219 whereas KI mice homozygous for Glu or Lys at codon 219 showed susceptibility to prion infection indicates that only heterozygosity at codon 219 is protective against prion diseases.[109] Consequently, it was proposed that individuals heterozygous for Lys at codon 219 are protected against sCJD because the PrPs derived from two allelic variants interfere with each other in the conversion process due to their incompatible structures.

The structural basis for the protective effect of E219K polymorphism was investigated by solving an NMR structure of the recombinant HuPrP with E219K substitution.[19] It was found that HuPrP(E219K) has the same tertiary fold as the WT protein, comprising three α-helices and a short antiparallel β-sheet (Fig. 5A). Comparison of the NMR structure of HuPrP(E219K) with the structures of the WT and HuPrPs carrying pathological Q212P and V210I mutations revealed several local structural variations. NOE data identified close contacts between the side chains of several residues located at the interface of the β1–α1 loop, helix α1, α1–β2 loop, and helix α3 in both HuPrP(E219K) and WT protein (Fig. 5B). However, some tertiary contacts that are present in the HuPrP(E219K) (e.g., Phe141-Val209) do not exist in the WT protein. Moreover, HuPrP(E219K) exhibits strong aromatic and hydrophobic interactions between residues situated at the interface of the β2–α2 loop (e.g., Met166 and Phe175) and the C-terminus of the helix α3 (e.g., Tyr218 and Tyr225) (Fig. 5C). This is in contrast to the HuPrPs with pathological Q212P and V210I mutations where tertiary contacts within this region are interrupted leading to increased exposure of hydrophobic surface to solvent.[21,22] In addition, the β2–α2 loop in HuPrP(E219K) shows a tendency toward 3_{10}-helical conformation. These structural features suggested that better definition of the β2–α2 loop along with stronger tertiary contacts at the interface of the loop and helix α3 in HuPrP(E219K) may account for the protective influence of E219K polymorphism against sCJD.

In addition, substitution induced a marked change in the distribution of electrostatic surface potential from negatively charged or neutral in the WT protein to mostly positively charged in the HuPrP(E219K) (Fig. 5D).

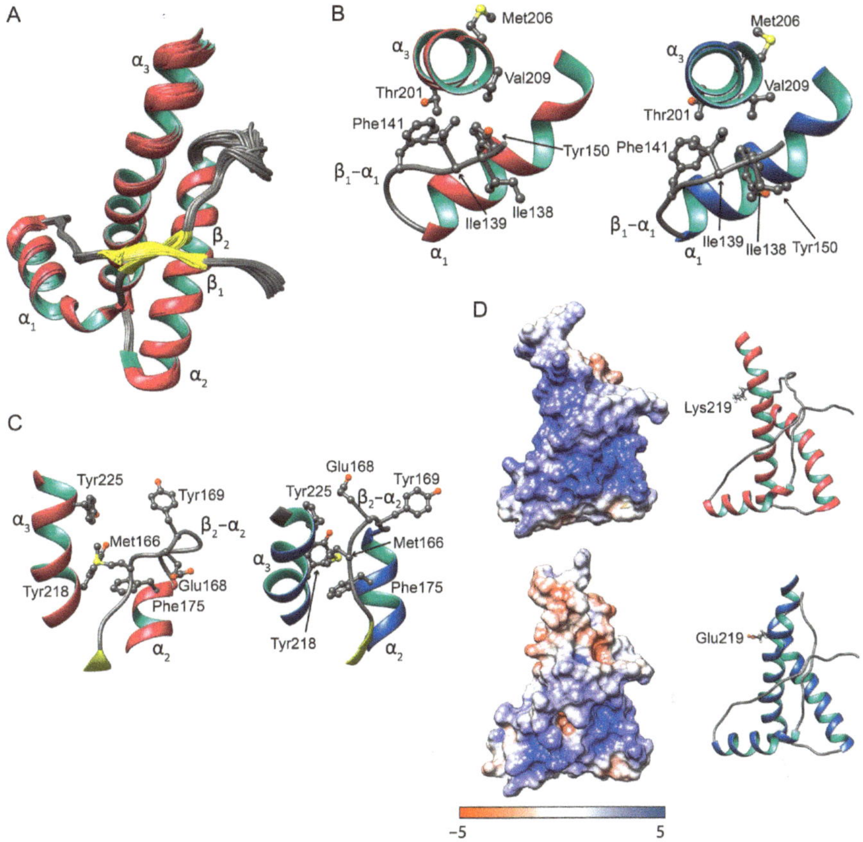

Fig. 5 Structural features of HuPrP(E219K) (pdb id 2LFT) and comparison with the WT protein (pdb id 2LSB). (A) NMR structure of HuPrP(E219K). (B) Hydrophobic contacts at the interface of the β1–α1 loop, helix α1, α1–β2 loop, and helix α3 in HuPrP(E219K) and WT protein. (C) Aromatic and hydrophobic contacts at the interface of the β2–α2 loop and the C-terminus of helix α3 in HuPrP(E219K). (D) Electrostatic surface potential of HuPrP(E219K) and WT protein.

Variations in surface charge distribution are mostly clustered around the site of substitution, at the interface of the β2–α2 loop and the C-terminus of the helix α3, and at the N-terminal part of the helix α3. These findings suggested that alterations in distribution of charges in the WT and HuPrP(E219K) proteins may facilitate intermolecular interactions between the two allelic variants, thus sequestering them from the early stages of the fibrillization process or inhibiting interactions with an as yet unidentified facilitator of PrP^{C} to PrP^{Sc} conversion.

6. CONCLUDING REMARKS

Understanding the molecular mechanism by which PrP^{C} misfolds into PrP^{Sc} is a crucial step toward preventing and curing prion diseases. NMR structural studies of HuPrP variants have provided valuable insights into the detailed atomic-level structural effects of mutations on HuPrP structure, thus contributing to our knowledge of the earliest stages of the pathological conversion process. NMR structures of HuPrPs with pathological mutations highlighted several common structural perturbations. These include an altered conformation of the β2–α2 loop together with the loosening of tertiary interactions between the loop and the C-terminal part of helix α3, increased exposure of hydrophobic surface to solvent, and disruption of hydrophobic contacts at the α2–α3 interhelical interface. In contrast to pathological mutants, hydrophobic interactions between residues from the β2–α2 loop and helix α3 are not interrupted in HuPrP with protective E219K polymorphism. Although our knowledge about the molecular mechanism of prion diseases is still far from complete, the findings presented in this chapter offer new insights into the early conformational changes leading to the formation of PrP^{Sc} and might suggest new approaches for development of antiprion drugs.

ACKNOWLEDGMENT

Authors acknowledge financial support of the Slovenian Research Agency (ARRS, grants P1-0242 and J1-6733).

REFERENCES

1. Prusiner SB. Novel proteinaceous infectious particles cause scrapie. *Science*. 1982;216:136–144.
2. Prusiner SB. Prions. *Proc Natl Acad Sci USA*. 1998;95(23):13363–13383.
3. Colby DW, Prusiner SB. Prions. Vol 3, Plainview, NY: Cold Spring Harbor Laboratory Press; 2011.
4. Kocisko DA, Come JH, Priola SA, et al. Cell-free formation of protease-resistant prion protein. *Nature*. 1994;370:471–474.
5. Collinge J. Prion diseases of humans and animals: their causes and molecular basis. *Annu Rev Neurosci*. 2001;24:519–550.
6. Head MW, Ironside JW. Review: Creutzfeldt-Jakob disease: prion protein type, disease phenotype and agent strain. *Neuropathol Appl Neurobiol*. 2012;38:296–310.
7. Imran M, Mahmood S. An overview of animal prion diseases. *Virol J*. 2011;8:493.
8. Budka H. Neuropathology of prion diseases. *Br Med Bull*. 2003;66:121–130.
9. Prusiner SB. A unifying role for prions in neurodegenerative disorders. *Science*. 2012;336:1511–1513.

10. Van der Kamp MW, Daggett V. The consequences of pathogenic mutations to the human prion protein. *Protein Eng Des Sel.* 2009;22(8):461–468.
11. Mead S. Prion disease genetics. *Eur J Hum Genet.* 2006;14(3):273–281.
12. Liemann S, Glockshuber R. Influence of amino acid substitutions related to inherited human prion diseases on the thermodynamic stability of the cellular prion protein. *Biochemistry.* 1999;38(11):3258–3267.
13. Swietnicki W, Petersen RB, Gambetti P, Surewicz WK. Familial mutations and the thermodynamic stability of the recombinant human prion protein. *J Biol Chem.* 1998;273(47):31048–31052.
14. Vanik DL, Surewicz WK. Disease-associated F198S mutation increases the propensity of the recombinant prion protein for conformational conversion to scrapie-like form. *J Biol Chem.* 2002;277:49065–49070.
15. Apetri AC, Surewicz K, Surewicz WK. The effect of disease-associated mutations on the folding pathway of human prion protein. *J Biol Chem.* 2004;279(17):18008–18014.
16. Ashok A, Hegde RS. Selective processing and metabolism of disease-causing mutant prion proteins. *PLoS Pathog.* 2009;5(6): e1000479.
17. Kaneko K, Zulianello L, Scott M, et al. Evidence for protein X binding to a discontinuous epitope on the cellular prion protein during scrapie prion propagation. *Proc Natl Acad Sci USA.* 1997;94(19):10069–10074.
18. Telling GC, Scott M, Mastrianni J, et al. Prion propagation in mice expressing human and chimeric PrP transgenes implicates the interaction of cellular PrP with another protein. *Cell.* 1995;83(1):79–90.
19. Biljan I, Giachin G, Ilc G, Zhukov I, Plavec J, Legname G. Structural basis for the protective effect of the human prion protein carrying the dominant-negative E219K polymorphism. *Biochem J.* 2012;446:243–251.
20. Biljan I, Ilc G, Giachin G, Plavec J, Legname G. Structural rearrangements at physiological pH: nuclear magnetic resonance insights from the V210I human prion protein mutant. *Biochemistry.* 2012;51:7465–7474.
21. Biljan I, Ilc G, Giachin G, et al. Toward the molecular basis of inherited prion diseases: NMR structure of the human prion protein with V210I mutation. *J Mol Biol.* 2011;412:660–673.
22. Ilc G, Giachin G, Jaremko M, et al. NMR structure of the human prion protein with the pathological Q212P mutation reveals unique structural features. *PLoS One.* 2010;5(7): e11715.
23. Aguzzi A, Sigurdson C, Heikenwaelder M. Molecular mechanisms of prion pathogenesis. *Annu Rev Pathol.* 2008;3:11–40.
24. Calzolai L, Lysek DA, Guntert P, et al. NMR structures of three single-residue variants of the human prion protein. *Proc Natl Acad Sci USA.* 2000;97(15):8340–8345.
25. Christen B, Perez DR, Hornemann S, Wuthrich K. NMR structure of the bank vole prion protein at 20 degrees C contains a structured loop of residues 165-171. *J Mol Biol.* 2008;383(2):306–312.
26. Donne DG, Viles JH, Groth D, et al. Structure of the recombinant full-length hamster prion protein PrP(29-231): the N terminus is highly flexible. *Proc Natl Acad Sci USA.* 1997;94(25):13452–13457.
27. Gossert AD, Bonjour S, Lysek DA, Fiorito F, Wuthrich K. Prion protein NMR structures of elk and of mouse/elk hybrids. *Proc Natl Acad Sci USA.* 2005;102(3): 646–650.
28. James TL, Liu H, Ulyanov NB, et al. Solution structure of a 142-residue recombinant prion protein corresponding to the infectious fragment of the scrapie isoform. *Proc Natl Acad Sci USA.* 1997;94(19):10086–10091.
29. Lopez Garcia F, Zahn R, Riek R, Wuthrich K. NMR structure of the bovine prion protein. *Proc Natl Acad Sci USA.* 2000;97(15):8334–8339.

30. Perez DR, Damberger FF, Wuthrich K. Horse prion protein NMR structure and comparisons with related variants of the mouse prion protein. *J Mol Biol*. 2010;400(2): 121–128.
31. Wen Y, Li J, Yao W, et al. Unique structural characteristics of the rabbit prion protein. *J Biol Chem*. 2010;285(41):31682–31693.
32. Zahn R, Liu A, Luhrs T, et al. NMR solution structure of the human prion protein. *Proc Natl Acad Sci USA*. 2000;97(1):145–150.
33. Antonyuk SV, Trevitt CR, Strange RW, et al. Crystal structure of human prion protein bound to a therapeutic antibody. *Proc Natl Acad Sci USA*. 2009;106(8): 2554–2558.
34. Eghiaian F, Grosclaude J, Lesceu S, et al. Insight into the PrPC->PrPSc conversion from the structures of antibody-bound ovine prion scrapie-susceptibility variants. *Proc Natl Acad Sci USA*. 2004;101(28):10254–10259.
35. Haire LF, Whyte SM, Vasisht N, et al. The crystal structure of the globular domain of sheep prion protein. *J Mol Biol*. 2004;336(5):1175–1183.
36. Knaus KJ, Morillas M, Swietnicki W, Malone M, Surewicz WK, Yee VC. Crystal structure of the human prion protein reveals a mechanism for oligomerization. *Nat Struct Biol*. 2001;8(9):770–774.
37. Lee S, Antony L, Hartmann R, et al. Conformational diversity in prion protein variants influences intermolecular beta-sheet formation. *EMBO J*. 2010;29(1):251–262.
38. Bennett MJ, Sawaya MR, Eisenberg D. Deposition diseases and 3D domain swapping. *Structure*. 2006;14:811–824.
39. Surewicz WK, Apostol MI. Prion protein and its conformational conversion: a structural perspective. *Top Curr Chem*. 2011;305:135–167.
40. Caughey BW, Dong A, Bhat KS, Ernst D, Hayes SF, Caughey WS. Secondary structure analysis of the scrapie-associated protein PrP 27-30 in water by infrared spectroscopy. *Biochemistry*. 1991;30(31):7672–7680.
41. Pan KM, Baldwin M, Nguyen J, et al. Conversion of alpha-helices into beta-sheets features in the formation of the scrapie prion proteins. *Proc Natl Acad Sci USA*. 1993;90(23):10962–10966.
42. Diaz-Espinoza R, Soto C. High-resolution structure of infectious prion protein: the final frontier. *Nat Struct Mol Biol*. 2012;19(4):370–377.
43. Govaerts C, Wille H, Prusiner SB, Cohen FE. Evidence for assembly of prions with left-handed beta-helices into trimers. *Proc Natl Acad Sci USA*. 2004;101(22): 8342–8347.
44. DeMarco ML, Daggett V. From conversion to aggregation: protofibril formation of the prion protein. *Proc Natl Acad Sci USA*. 2004;101(8):2293–2298.
45. Cobb NJ, Sonnichsen FD, McHaourab H, Surewicz WK. Molecular architecture of human prion protein amyloid: a parallel, in-register β-structure. *Proc Natl Acad Sci USA*. 2007;104:18946–18951.
46. Smirnovas V, Baron GS, Offerdahl DK, Raymond GJ, Caughey B, Surewicz WK. Structural organization of brain-derived mammalian prions examined by hydrogen-deuterium exchange. *Nat Struct Mol Biol*. 2011;18(4):504–506.
47. Cavanagh J, Fairbrother WF, Palmer AG, Rance M, Skelton NJ. *Protein NMR Spectroscopy: Principles and Practice*. 2nd ed. San Diego: Elsevier Academic Press; 2007.
48. Sharma D, Rajarathnam K. 13C NMR chemical shifts can predict disulfide bond formation. *J Biomol NMR*. 2000;18(2):165–171.
49. Shen Y, Delaglio F, Cornilescu G, Bax A. TALOS+: a hybrid method for predicting protein backbone torsion angles from NMR chemical shifts. *J Biomol NMR*. 2009;44:213–223.
50. Guntert P, Mumenthaler C, Wuthrich K. Torsion angle dynamics for NMR structure calculation with the new program DYANA. *J Mol Biol*. 1997;273(1):283–298.

51. Herrmann T, Guntert P, Wuthrich K. Protein NMR structure determination with automated NOE assignment using the new software CANDID and the torsion angle dynamics algorithm DYANA. *J Mol Biol*. 2002;319(1):209–227.
52. Krieger E, Koraimann G, Vriend G. Increasing the precision of comparative models with YASARA NOVA—a self-parameterizing force field. *Proteins*. 2002;47:477–486.
53. Laskowski RA, Rullmann JAC, MacArthur MW, Kaptein R, Thornton JM. AQUA and PROCHECK-NMR: programs for checking the quality of protein structures solved by NMR. *J Biomol NMR*. 1996;8(4):477–486.
54. Vriend G. A molecular modelling and drug design program. *J Mol Graph*. 1990;9:52–56.
55. Zhang Y, Swietnicki W, Zagorski MG, Surewicz WK, Sonnichsen FD. Solution structure of the E200K variant of human prion protein. Implications for the mechanism of pathogenesis in familial prion diseases. *J Biol Chem*. 2000;275(43):33650–33654.
56. Pocchiari M, Salvatore M, Cutruzzola F, et al. A new point mutation of the prion protein gene in Creutzfeldt-Jakob disease. *Ann Neurol*. 1993;34(6):802–807.
57. Kovacs GG, Puopolo M, Ladogana A, et al. Genetic prion disease: the EUROCJD experience. *Hum Genet*. 2005;118(2):166–174.
58. Adrover M, Pauwels K, Prigent S, et al. Prion fibrillization is mediated by a native structural element that comprises helices H2 and H3. *J Biol Chem*. 2010;285(27): 21004–21012.
59. Riek R, Wider G, Billeter M, Hornemann S, Glockshuber R, Wuthrich K. Prion protein NMR structure and familial human spongiform encephalopathies. *Proc Natl Acad Sci USA*. 1998;95(20):11667–11672.
60. Van der Kamp MW, Daggett V. Pathogenic mutations in the hydrophobic core of the human prion protein can promote structural instability and misfolding. *J Mol Biol*. 2010;404(4):732–748.
61. Cohen FE, Pan KM, Huang Z, Baldwin M, Fletterick RJ, Prusiner SB. Structural clues to prion replication. *Science*. 1994;264(5158):530–531.
62. Kuwata K, Li H, Yamada H, et al. Locally disordered conformer of the hamster prion protein: a crucial intermediate to PrPSc? *Biochemistry*. 2002;41(41):12277–12283.
63. Nicholson EM, Mo H, Prusiner SB, Cohen FE, Marqusee S. Differences between the prion protein and its homolog Doppel: a partially structured state with implications for scrapie formation. *J Mol Biol*. 2002;316(3):807–815.
64. Apetri AC, Surewicz WK. Kinetic intermediate in the folding of human prion protein. *J Biol Chem*. 2002;277:44589–44592.
65. Aguzzi A, Sigurdson CJ. Antiprion immunotherapy: to suppress or to stimulate? *Nat Rev Immunol*. 2004;4(9):725–736.
66. Apetri AC, Maki K, Roder H, Surewicz WK. Early intermediate in human prion protein folding as evidenced by ultrarapid mixing experiments. *J Am Chem Soc*. 2006;128(35):11673–11678.
67. Booth DR, Sunde M, Bellotti V, et al. Instability, unfolding and aggregation of human lysozyme variants underlying amyloid fibrillogenesis. *Nature*. 1997;385:787–793.
68. Canet D, Last AM, Tito P, et al. Local cooperativity in the unfolding of an amyloidogenic variant of human lysozyme. *Nat Struct Biol*. 2002;9:308–315.
69. Kelly JW. The alternative conformations of amyloidogenic proteins and their multi-step assembly pathways. *Curr Opin Struct Biol*. 1998;8:101–106.
70. Gerum C, Schlepckow K, Schwalbe H. The unfolded state of the murine prion protein and properties of single-point mutants related to human prion diseases. *J Mol Biol*. 2010;401(1):7–12.
71. Gerum C, Silvers R, Wirmer-Bartoschek J, Schwalbe H. Unfolded-state structure and dynamics influence the fibril formation of human prion protein. *Angew Chem Int Ed Engl*. 2009;48(50):9452–9456.

72. Kumar J, Sreeramulu S, Schmidt TL, et al. Prion protein amyloid formation involves structural rearrangements in the C-terminal domain. *Chembiochem*. 2010;11(9):1208–1213.
73. Chiovitti K, Corsaro A, Thellung S, et al. Intracellular accumulation of a mild-denatured monomer of the human PrP fragment 90-231, as possible mechanism of its neurotoxic effects. *J Neurochem*. 2007;103(6):2597–2609.
74. Corsaro A, Thellung S, Bucciarelli T, et al. High hydrophobic amino acid exposure is responsible of the neurotoxic effects induced by E200K or D202N disease-related mutations of the human prion protein. *Int J Biochem Cell Biol*. 2011;43(3):372–382.
75. Thellung S, Corsaro A, Villa V, et al. Human PrP90-231-induced cell death is associated with intracellular accumulation of insoluble and protease-resistant macroaggregates and lysosomal dysfunction. *Cell Death Dis*. 2011;2: e138.
76. Meli M, Gasset M, Colombo G. Dynamic diagnosis of familial prion diseases supports the beta2-alpha2 loop as a universal interference target. *PLoS One*. 2011;6(4): e19093.
77. Rossetti G, Giachin G, Legname G, Carloni P. Structural facets of disease-linked human prion protein mutants: a molecular dynamic study. *Proteins*. 2010;78(16): 3270–3280.
78. Van der Kamp MW, Daggett V. Molecular dynamics as an approach to study prion protein misfolding and the effect of pathogenic mutations. *Top Curr Chem*. 2011;305:169–197.
79. Christen B, Hornemann S, Damberger FF, Wuthrich K. Prion protein NMR structure from tammar wallaby (*Macropus eugenii*) shows that the beta2-alpha2 loop is modulated by long-range sequence effects. *J Mol Biol*. 2009;389(5):833–845.
80. Campana V, Sarnataro D, Zurzolo C. The highways and byways of prion protein trafficking. *Trends Cell Biol*. 2005;15(2):102–111.
81. Casey JR, Grinstein S, Orlowski J. Sensors and regulators of intracellular pH. *Nat Rev Mol Cell Biol*. 2010;11(1):50–61.
82. Godsave SF, Wille H, Kujala P, et al. Cryo-immunogold electron microscopy for prions: toward identification of a conversion site. *J Neurosci*. 2008;28(47):12489–12499.
83. Pimpinelli F, Lehmann S, Maridonneau-Parini I. The scrapie prion protein is present in flotillin-1-positive vesicles in central- but not peripheral-derived neuronal cell lines. *Eur J Neurosci*. 2005;21(8):2063–2072.
84. Marijanovic Z, Caputo A, Campana V, Zurzolo C. Identification of an intracellular site of prion conversion. *PLoS Pathog*. 2009;5(5): e1000426.
85. Alonso DOV, de Armond SJ, Cohen FE, Daggett V. Mapping the early steps in the pH-induced conformational conversion of the prion protein. *Proc Natl Acad Sci USA*. 2001;98:2985–2989.
86. DeMarco ML, Daggett V. Molecular mechanism for low pH triggered misfolding of the human prion protein. *Biochemistry*. 2007;46:3045–3054.
87. Hornemann S, Glockshuber R. A scrapie-like unfolding intermediate of the prion protein domain PrP(121–231) induced by acidic pH. *Proc Natl Acad Sci USA*. 1998;95:6010–6014.
88. Langella E, Improta R, Barone V. Checking the pH-induced conformational transition of prion protein by molecular dynamics simulations: effect of protonation of histidine residues. *Biophys J*. 2004;87:3623–3632.
89. Van der Kamp MW, Daggett V. Influence of pH on the human prion protein: insights into the early steps of misfolding. *Biophys J*. 2010;99:2289–2298.
90. Swietnicki W, Petersen RB, Gambetti P, Surewicz K. pH-dependent stability and conformation of the recombinant human prion protein PrP(90-231). *J Biol Chem*. 1997;272:27517–27520.
91. Calzolai L, Zahn R. Influence of pH on NMR structure and stability of the human prion protein globular domain. *J Biol Chem*. 2003;278:35592–35596.

92. Eghiaian F, Daubenfeld T, Quenet Y, et al. Diversity in prion protein oligomerization pathways results from domain expansion as revealed by hydrogen/deuterium exchange and disulfide linkage. *Proc Natl Acad Sci USA*. 2007;104:7414–7419.
93. Singh J, Udgaonkar JB. Structural effects of multiple pathogenic mutations suggest a model for the initiation of misfolding of the prion protein. *Angew Chem Int Ed*. 2015;54:1–6.
94. Watzlawik J, Skora L, Frense D, et al. Prion protein helix 1 promotes aggregation but is not converted into beta-sheet. *J Biol Chem*. 2006;281:30242–30250.
95. Singh J, Udgaonkar JB. Unraveling the molecular mechanism of pH-induced misfolding and oligomerization of the prion protein. *J Mol Biol*. 2016;428(6):1345–1355.
96. Langella E, Improta R, Crescenzi O, Barone V. Assessing the acid-base and conformational properties of histidine residues in human prion protein (125-228) by means of pK(a) calculations and molecular dynamics simulations. *Proteins*. 2006;64:167–177.
97. Alpérovitch A, Zerr I, Pocchiari M, et al. Codon 129 prion protein genotype and sporadic Creutzfeldt–Jakob disease. *Lancet*. 1999;353:1673–1674.
98. Collinge J, Palmer MS, Dryden AJ. Genetic predisposition to iatrogenic Creutzfeldt-Jakob disease. *Lancet*. 1991;337:1441–1442.
99. Nozaki I, Hamaguchi T, Sanjo N, et al. Prospective 10-year surveillance of human prion diseases in Japan. *Brain*. 2010;133:3043–3057.
100. Palmer MS, Dryden AJ, Hughes JT, Collinge J. Homozygous prion protein genotype predisposes to sporadic Creutzfeldt–Jakob disease. *Nature*. 1991;352:340–342.
101. Cervenáková L, Goldfarb LG, Garruto R, Lee HS, Gajdusek DC, Brown P. Phenotype-genotype studies in kuru: implications for new variant Creutzfeldt–Jakob disease. *Proc Natl Acad Sci USA*. 1998;95:13239–13241.
102. Kobayashi A, Hizume M, Teruya K, Mohri S, Kitamoto T. Heterozygous inhibition in prion infection: the stone fence model. *Prion*. 2009;3:27–30.
103. Goldfarb LG, Petersen RB, Tabaton M, et al. Fatal familial insomnia and familial Creutzfeldt–Jakob disease: disease phenotype determined by a DNA polymorphism. *Science*. 1992;258:806–808.
104. Hosszu LLP, Jackson GS, Trevitt CR, et al. The residue 129 polymorphism in human prion protein does not confer susceptibility to Creutzfeldt-Jakob disease by altering the structure or global stability of PrPC. *J Biol Chem*. 2004;279:28515–28521.
105. Shibuya S, Higuchi J, Shin RW, Tateishi J, Kitamoto T. Codon 219 Lys allele of PRNP is not found in sporadic Creutzfeldt-Jakob disease. *Ann Neurol*. 1998;43: 826–828.
106. Soldevila M, Calafell F, Andres AM, et al. Prion susceptibility and protective alleles exhibit marked geographic differences. *Hum Mutat*. 2003;22:104–105.
107. Perrier V, Kaneko K, Safar J, et al. Dominant-negative inhibition of prion replication in transgenic mice. *Proc Natl Acad Sci USA*. 2002;99:13079–13084.
108. Crozet C, Lin YL, Mettling C, et al. Inhibition of PrPSc formation by lentiviral gene transfer of PrP containing dominant negative mutations. *J Cell Sci*. 2004;117: 5591–5597.
109. Hizume M, Kobayashi A, Teruya K, et al. Human prion protein (PrP) 219K is converted to PrPSc but shows heterozygous inhibition in variant Creutzfeldt–Jakob disease infection. *J Biol Chem*. 2009;284:3603–3609.

CHAPTER FIVE

Structural Modeling of Human Prion Protein's Point Mutations

Giulia Rossetti[*,†,‡], **Paolo Carloni**[*,§,1]
[*]Institute of Neuroscience and Medicine (INM-9) and Institute for Advanced Simulation (IAS-5), Jülich, Germany
[†]Jülich Supercomputing Centre, Jülich, Germany
[‡]University Hospital Aachen, RWTH-Aachen, Aachen, Germany
[§]RTWH-Aachen, Aachen, Germany
[1]Corresponding author: e-mail address: p.carloni@fz-juelich.de

Contents

Abstract

Prion diseases, or transmissible spongiform encephalopathies, are a group of rare fatal neurodegenerative maladies that affect humans and animals. The main event in disease progression is the posttranslational conversion of the ubiquitously expressed cellular form of the prion protein (PrP^{C}) into its misfolded and pathogenic isoform, known as prion or PrP^{Sc}. In the presence of specific disease-linked mutations, the conversion may occur spontaneously. Molecular simulation studies on human PrP^{C} wild-type and variants from several research groups, including ours, have provided a consistent picture of the effect of such mutations. In particular, the calculations have pinpointed "hot spots" for the conversion across several disease-linked variants. They have also identified a region of the protein containing two helices (Helix 2 and Helix 3) as a key structural element most prone to the conversion, consistently with a wealth of experimental data. Some of these findings are summarized here.

Progress in Molecular Biology and Translational Science, Volume 150
ISSN 1877-1173
http://dx.doi.org/10.1016/bs.pmbts.2017.07.001

1. INTRODUCTION

Prion diseases or transmissible spongiform encephalopathies (TSEs) are rare neurodegenerative diseases, affecting humans and animals.[1,2] They exhibit symptoms of both cognitive and motor dysfunction, vacuolation of the gray matter in the human central nervous system, neuronal loss, astrogliosis.[3] The hallmark of TSEs is the misfolding of the extracellular, membrane-anchored prion protein (PrP^{C}) into "prion" (proteinaceous infectious particles, called also PrP^{Sc} or scrapie).[a]

According to the *"Protein only hypothesis"* by Prusiner,[11] prions self-propagate by converting PrP^{C} into the scrapie form, the PrP^{Sc}. This may lead to neurodegeneration without using any nucleic acids. TSEs may share similar pathogenesis with common neurodegenerative syndromes such as Alzheimer's disease and Parkinson's disease[12,13],[b] for which there are currently no cure. Therefore, today, PrP^{C} is one of the most studied models for protein misfolding mechanism, and TSE serves as an excellent model for studying many other neurodegenerative diseases.[15]

One of the key arguments in support of the protein-only hypothesis is the evidence linking familial prion diseases with mutations in the gene coding for human prion protein[1,16,17]: a significant (~15%) fraction of human (Hu) TSE are genetic and related to several disease-linked mutations (DLMs) of $HuPrP^{C}$.[18] Genetic forms of HuTSEs are transmitted as autosomal dominant traits and cosegregate with missense or insertional/deletional mutations in *PRNP*. DLMs may cause spontaneous TSE, while artificial mutations in transgenic mice can determine the susceptibility to the infection of different prion strains.[19–21] Nonpathogenic, naturally occurring mutations (polymorphisms) also exist in the PrP gene.[18] They influence the etiology and neuropathology of the disease in both humans[22] and sheep.[23]

Deciphering mutations' effect on the structural stability of the protein is both pharmaceutically relevant and it provides hints on the molecular basis

[a] The tendency to go to induced or spontaneous misfolding and eventually neurodegeneration depends highly on PrP^{C} intrinsic features including the amino acid sequence[4,5] as well as secondary structure elements,[6–8] the highly flexible amino terminal region of the protein[9] and posttranslational modification elements.[10]

[b] The WHO estimated ca. 7.7 million new cases of Alzheimer's disease in the year 2010 and increasing numbers in the next decades due to global aging.[14] The European Parkinson's Disease Association (EPDA) estimated ca. 6.3 million people living with Parkinson's disease in the year 2011 (www.epda.eu.com).

of the $PrP^C \rightarrow PrP^{Sc}$ conversion. Computations—especially those which make contact with experiments[24]—are advancing our understanding of the effect of the above-mentioned mutations. Here, after reporting on computational studies on $HuPrP^C$ structure and stability, we review recent investigations, at the computational level, on most of the DLMs in humans and artificial mutations in mice.

The chapter is organized as follows. Section 2 describes the wild-type (WT) protein, which serves here as a reference. Section 3 reports structural modeling on protein variants, including point mutations (PMs) associated with diseases and polymorphisms in the human protein.[c] Section 4 reports on artificial mutation in mice. Some concluding remarks are drawn at the end of the paper.

2. WILD-TYPE HUMAN PRION PROTEIN

Human PrP^C is a mid-size protein, highly conserved across mammals both in terms of sequence[26,27] and structure.[28–32] It consists of a glycosylphosphatidylinositol (GPI) anchoring domain, that keeps the rest of the protein close to the outer surface of the membrane, a largely unordered N-terminal domain (residues 23–124), which might acquire a defined structure when PrP^C is present within membrane rafts,[33] and a globular domain (GD, residues 125–231),[31] whose structural determinants are conserved even in the absence of the other domain[26,27,31] (Fig. 1).

2.1 Globular Domain

A variety of X-ray and NMR studies[26,27,31] show that GD consists of three α-helices (residues 144–154, 173–194, and 200–228, respectively, H1–H3 hereafter) and a short antiparallel β-sheets comprising residues 128–131 and 161–164. A single disulfide bond, Cys^{179}–Cys^{214}, connects H2 and H3 (Fig. 1). Experimental studies show that the H2–H3 region of PrP^C is an independent α-helical unit able alone to form fibrils[8,31,34–36] and to reproduce the oligomerization pathway observed in the full-length protein.[8,37] Accordingly, H2 and H3 are part of the prions' amyloid fibril core,[8,36] while H1 seems not essential for prion infectivity,[1] and it retains its α-helical conformation in $HuPrP^C$ under a wide range of denaturing experimental conditions.[38]

[c] DLMs whose clinicopathological classification is missing[25] are not discussed here.

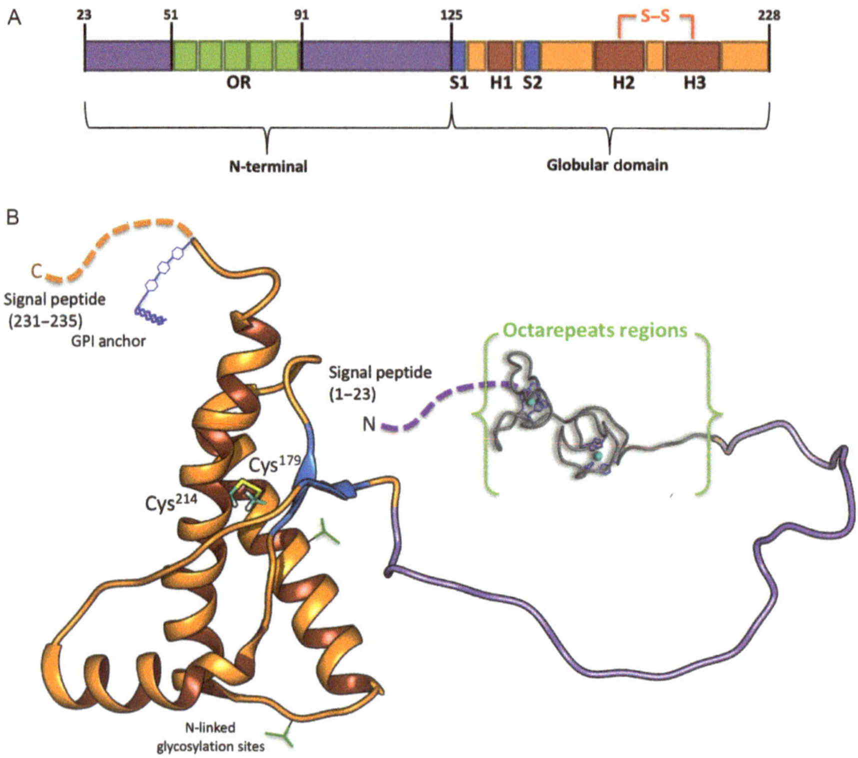

Fig. 1 Scheme (A) and tridimensional view (B) of HuPrP^C. It features two signal peptides (1–23 and 231–235), an octapeptide repeat region (OR, involved in binding Cu^{2+} ions), two β-sheets (S1 and S2), three α-helixes (H1, H2, and H3), one disulfide bond (SS) between cysteine residues 179 and 214, and two potential sites for N-linked glycosylation (*green* forks at residues 181 and 197). H2 and H3 helices linked by the SS bond constitute the H2–H3 domain. A glycosylphosphatidylinositol anchor (GPI, in *blue*) is attached to the C-terminus, which anchors the protein on the outside cellular membrane.

Consistently with experiment, early bioinformatics studies, structural analysis, and MD simulations suggested that (i) GD is a "frustrated" domain[6,7]: some of the residues in H2 and H3 are incompatible with their natural propensities.[6,7] (ii) H2 and H3 helices have both a high intrinsic β propensity and the ability to adopt multiple secondary structure conformations[4,5]. (iii) Residues at the H2 C-terminal may promote a partial α-helix-to-β-sheet conversion.[39] Molecular simulation investigations confirm these suggestions. MD simulations, using experimental hydrogen exchange factors as restraints, showed that the unfolding of the C-terminus of H2 may lead to the formation of β-strands and that H2 and H3 are crucial to the conversion of $HuPrP^C$ to $HuPrP^{Sc}$.[40] In addition, advanced enhanced sampling simulations, i.e., replica exchange molecular dynamics[41,42] and well-tempered metadynamics simulations,[43] showed that the origin of metastability of the $HuPrP^C$ form is due mainly to H2 and H3, which have the tendency to form either a random coil or a β-structure.

Further interesting findings emerged by molecular simulation studies concerning $HuPrP^C$ stability. MD simulations suggested that GPI anchors facilitate continuous interactions between $HuPrP^C$ and the membrane, decreasing the solvent accessibility of H2–H3 domain[44,45] and therefore stabilizing this domain. In addition, other MD-based studies explored the role of the Cys^{179}–Cys^{214} disulfide bridge, connecting H2 and H3, for the stabilization of the protein folding.[46,47] They showed that the breakage of the S—S bond is associated with misfolding propensity of the protein,[47,48] consistently with what found in denaturation experiments.[49] Moreover, the mechanical response of the protein is different in the absence of the disulfide bond.[46] However, the relevance of the stability of this disulfide bond for the actual conversion is under debate, because it has not been established whether the bond remains intact[50–52] or is reduced and then reformed[53–56] during the $HuPrP^C \rightarrow HuPrP^{Sc}$ conversion.

The intrinsic conformational instability of $HuPrP^C$ is evident at higher temperature[57–59] and particularly significant at lower pH.[60–69] Consistently with experimental evidence, MD of $HuPrP^C$ simulations performed at acidic pH revealed a reduced structural stability particularly prominent the C-terminus region of H2[60–65] and H1.[60–62,65–69] Specifically, the GB turns out to lose α-content, and it is further destabilized via disruption of critical intramolecular interactions between subdomains S1–H1–S2 and H2–H3. Those include salt bridges (SBs),[60,63] and hydrogen bonds.[64] Simulations at high temperature combined with acidic pH pointed out a H2 misfolding and the consequent destabilizations of H2–H3 domain were observed,[64,70] consistently with what already observed in previous computational studies.

2.2 N-Terminal Domain

The N-terminal region carries out important physiological functions.[71] It contains characteristic glycine-rich octapeptide repeats (PHGGGWGQ human sequence) which enables the protein to coordinate Cu(II) ions[72,73] (Figure 1); moreover, it also contains a major part of the so-called transmembrane domain (termed TM1, comprising roughly residues 112–135) and the preceding "stop transfer effector" (STE, a hydrophilic region containing roughly residues 104–111).[74,75] STE and TM1 act in concert to control the cotranslational translocation at the endoplasmic reticulum (ER) during the biosynthesis of HuPrP^{C}. Modified structural features of TM1 and STE can increase the proportion of the atypical forms of PrP, namely: PrP that integrate into the lipid bilayer with either the N-terminus or the C-terminus in the ER lumen (^{Ntm}PrP and ^{Ctm}PrP, respectively), the ones retained in the cytoplasm (^{cy}PrP), the one secreted (^{sec}PrP) (Figure 2). Alteration of these proportions can be detrimental in vivo.[74,76,77]

Moreover, the domain is a broad-spectrum molecular sensor.[78] Indeed, it interacts with chemicals including: (i) sulfated glycosaminoglycans,[79] which bind to the four histidines in the OR, as well as several lysines at the N-terminus[80]; (ii) Cu^{2+} ions,[81] which may bind at six histidine residues,[82] including those located in the OR. In addition, the Mo protein has been shown to interact with vitronectin,[83] the stress-inducible protein 1 (STI1)[84] and amyloid-β (Aβ) multimers.[85–87]

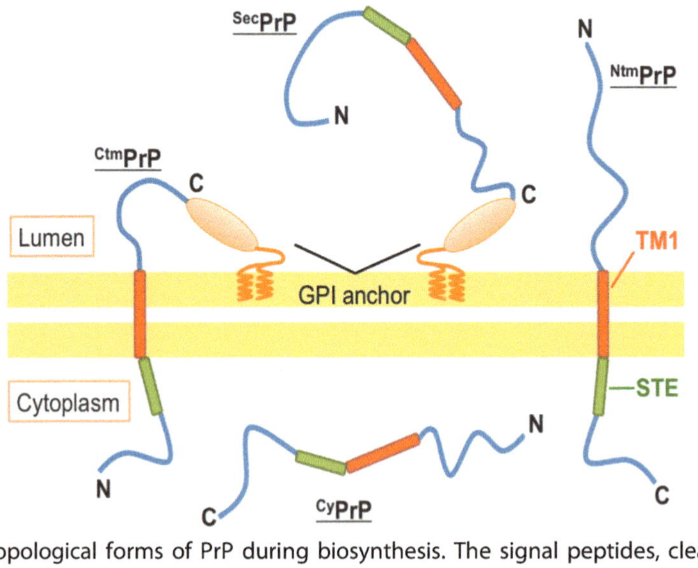

Fig. 2 Topological forms of PrP during biosynthesis. The signal peptides, cleaved off during the biosynthesis of the proteins, are not shown in this figure *Readapted from Cong X, Casiraghi N, Rossetti G, et al. Role of prion disease-linked mutations in the intrinsically disordered N-terminal domain of the prion protein. J Chem Theory Comput. 2013;9 (11):5158–5167.*

At present, the structural ensemble of the full-length N-terminal HuPrP^C domain has not been determined experimentally, and almost all computational studies have focuses on short fragments of it (see reviews[24,25]). These are not discussed here. The only atomistic study on the entire domain is based on a Monte Carlo simulation approach, PROFASI.[88] The latter suggest that the domain's structural ensemble can be roughly divided into six subregions with distinct structural behaviors. Among those, we mention here: (i) The so-called the amyloidogenic region (residues 105–125) that features transient helical structure[89]; (ii) one of the octave-tide repeats (residues 61–68) that features a stable loop/β-turn conformation.[89] The results of the simulation are consistent with published CD, NMR, and FTIR studies.[90–93]

3. PRION PROTEIN VARIANTS

58 Missense or insertional/deletional mutations have been identified so far in the gene coding for HuPrP^C (*PRNP*).[25]

Fourteen insertional or deletional mutations are localized in the OR[94]: from one to nine OR insertions have been reported. Most of the OR insertions lead to familial Creutzfeldt–Jakob disease (fCJD), except for an 8-OR insertion whose phenotype is unclassified. A 2-OR deletion in this region might cause fCJD,[95] albeit OR deletions are not yet firmly associated to Hu TSE. A methionine/valine polymorphism at codon 129 (129M/V) influences Hu TSE phenotype when coupled with certain PMs, although the polymorphism itself does not promote the disease.[96] Another polymorphism, E219K, presents in ~6% of the Japanese population, appears to be protective against sCJD.[97] A polymorphism at codon 127 (G127V), reported only in the natives of Papua New Guinea, is protective against Kuru.[98]

Missense mutations include 44 nonsynonymous codon substitutions, or PMs and five nonsense (or "stop") mutations (Y145stop, Q160stop, Y163stop, Y225stop, and Y226stop),[25,99,100] which code to prematurely terminate the protein sequence at the mutation position.[101,102] All these mutations may be associated with neurodegeneration and give rise to abnormal forms of HuPrP^C in the brain.[99] PMs are located all over the protein, from the disordered N-terminal domain (N-term_HuPrP^C hereinafter) to the folded C-terminal GD. They are mostly found in the GD,[25] and most of them have been studied by molecular simulation.

3.1 Globular Domain

In vitro PMs in the GB may increase the likelihood of misfolding by thermodynamic destabilization of $HuPrP^C$.[103–105] This is true for some studied mutants.[104–106] But PMs may also alter the prion protein surface, triggering in turn abnormal interactions with other not yet identified cofactors[107,108] or causing aberrant cellular trafficking and accumulation inside the cells.[109]

Several molecular simulation studies have been conducted on $HuPrP^C$ variants containing PMs[24,110–118] associated with diseases and, to a lesser extent, with polymorphisms. They have used both the AMBER[119] and CHARMM[120] force fields and run at different temperatures, from ~298 to ~500 K.

Overall, these studies suggest that PMs share common structural features which are different from those of the WT protein and from the protective dominant negative polymorphism $HuPrP^C$(E219K).[110,111] Indeed, almost all the mutations, independently of their location, trigger common structural features: these include loss of aromatic, hydrophobic, and salt bridges[121] which destabilize the proper GD fold (Figure 3).

The loss of aromatic interactions occurs mainly in the S2–H2 loop[110,111,122] causing an increase of loop flexibility with respect the WT. This is consistent with solution NMR studies showing that this loop could be highly flexible in some mammalian species (sharing 90% sequence identity or higher[123]) developing prion diseases (including Hu[31,124] and Mo[32,125]; and sheep[126]) or well structured in some other species such as (Syrian hamster,[127] elk,[32] bank vole,[29] wallaby,[28] rabbit,[128] and horse[129]). Importantly, no occurrence of TSE has been reported in rabbit, horse, or any marsupial species,[130] suggesting that the flexibility of the S2–H2 loop impact of prion resistance. The increased flexibility of the loop causes a larger solvent exposure of Tyr169 in the PMs than in WT $HuPrP^C$ and $HuPrP^C$(E219K-129M).[110,111] This could affect the interaction of the protein with cellular partners[107,131] and, as recently suggested, favor aggregation prone configurations of the protein.[132]

The weakened/loss of hydrophobic interactions are the ones mostly linking the S2–H2 loop with C-terminal part of H3.[110,111] The C-terminal part of H3 might modulate the S2–H2 loop conformation by long-range interactions,[28] based on the observation by NMR[28,133] that the high structural definition of the S2–H2 loop in tammar wallaby and horse PrP^C is entirely due to long-range interactions involving these regions.[28,29]

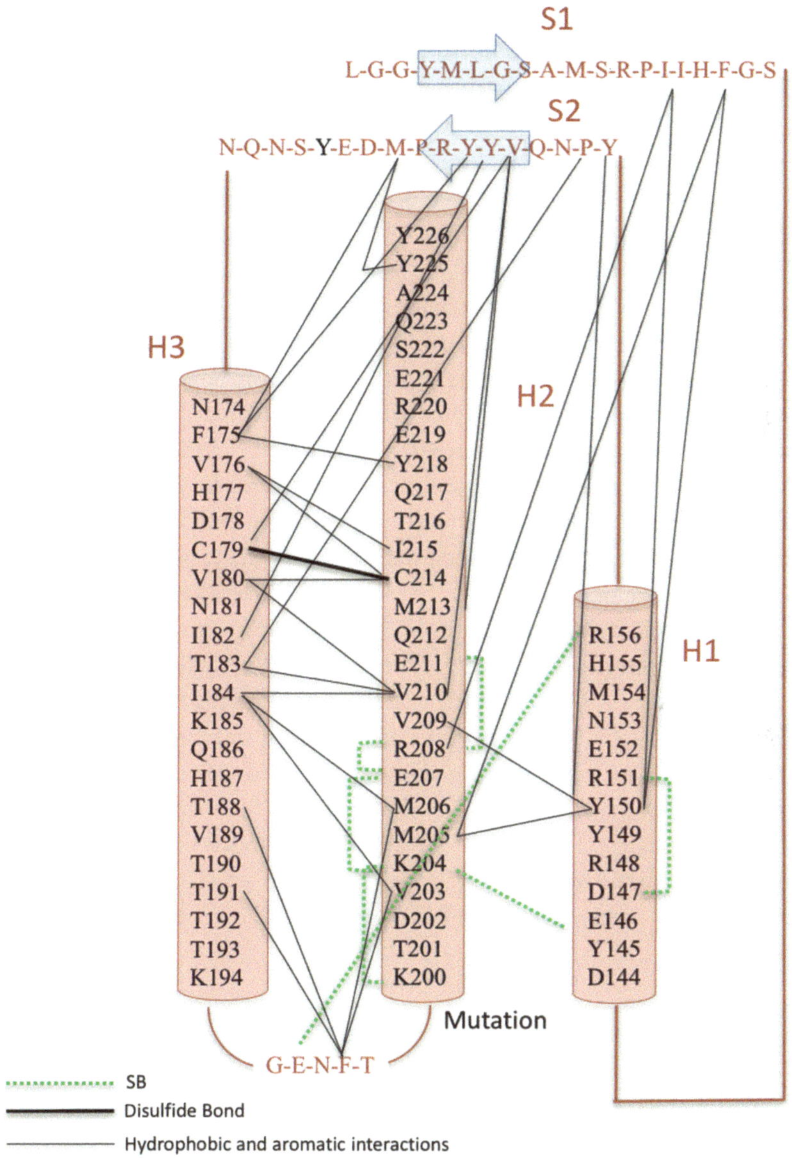

Fig. 3 Schematic representation of salt bridge (SB), aromatic, and hydrophobic interactions stabilizing $HuPrP^{C}$ fold. *The figure is readapted from Rossetti G, Cong XJ, Caliandro R, Legname G, Carloni P. Common structural traits across pathogenic mutants of the human prion protein and their implications for familial prion diseases. J Mol Biol. 2011;411 (3):700–712 and it is based on the structural information contained in Giachin G, Biljan I, Ilc G, Plavec J, Legname G. Probing early misfolding events in prion protein mutants by NMR spectroscopy. Molecules. 2013;18(8):9451–9476.*

The loss of SBs mainly occurs in the H2–H3 regions.[110–112,116,134,135] Notably, the residues involved in the SB network are highly conserved,[8] and their absence is linked with pathogenic mutants.[8]

3.2 N-Terminal Domain

Few PMs are located in N-terminal domain of $HuPrP^C$ (N-term_$HuPrP^C$). Q52P is likely associated with fCJD[100]; G114V has been reported with both fCJD[136] and Gerstmann–Sträussler–Scheinker syndrome (GSS).[99] P102L, P105L/S/T, and A117V are associated with GSS.[99] Our group simulated the structural ensembles on N-term from *Homo sapiens* and mouse,[89] using a Monte Carlo-based simulation approach. Our simulations show that the PMs do not change dramatically the conformational ensemble of the domain. This is consistent with in vitro experiments, which have shown that PMs do not affect the thermostability or misfolding kinetics of the protein molecule.[99,104–106] Rather, they mildly affect the intramolecular contacts in regions responsible for interactions with cellular partners, as well as the local structural features of STE/TM1. These in turn may alter the protein function[78] and biosynthesis.[76] Changes in the STE/TM1 may also affect the relative proportions of the ^{Ntm}PrP, ^{Ctm}PrP, and ^{Cy}PrP topological variants in human and mice.[76] This, in turn, could have repercussions on the disease, as elevated ^{Ctm}PrP level has been related to both familial (in GSS patients) and infectious (in Tg mice) prion diseases.[74,77] This result is consistent with the in vitro data that PMs P101L, P104L, and A116V increase the interactions between $MoPrP^C$ STE/TM1 and a membrane mimetic at pH 7.[137] The altered local features in STE/TM1 might also impact on the interactions of the protein with transacting factors in the cytosol and in the ER membrane.[138]

4. CHIMERIC PRION PROTEIN

Transmission of TSE across species is far less efficient than within species.[139] A striking example of the prion transmission across species came from several studies using transgenic (Tg) mice expressing chimeric Mo/$HuPrP^C$[19–21,140] (i.e., containing mutations from other species). Chimeric Mo/$HuPrP^C$ has been reported to exhibit transmission barriers against certain prion strains,[19] spontaneous conversion into prions,[20,21] abbreviated incubation times,[140] and diverse susceptibility or resistance to scrapie infection.[107] A recent MD study[141] from our group compared 11 different Chimeric Mo/$HuPrP^C$ structures with those of the WT protein. We found that the chimeras resistant to PrP^{Sc} infection show stronger specific interactions between the H1 and N-terminal of H3, and in turn, shorter intramolecular distances, than $HuPrP^C$, $MoPrP^C$, and the nonresistant chimeras. Therefore, the interactions between H1 helix and N-terminal of H3 helix are critical in prion propagation. On the other hand, we found that the S2–H2 loop exchanges between different conformations: 3_{10}-helix/turn pattern like in $MoPrP^C$ and a bend/turn pattern like in $HuPrP^C$. The dominant-negative effect of Mo PrP chimeras over WT $MoPrP^C$ occurs if the chimera not only resists PrP^{Sc} infection but also adopts the Mo-like pattern of exchanges between conformations in the S2–H2 loop. This suggests that the compatible loop conformation allows these dominant-negative chimeras to interfere with the conversion of MoPrP to PrP^{Sc}. Therefore, the S2–H2 loop conformation may help prevent the protein to be converted to the scrapie form.[141]

5. CONCLUSIONS

Molecular simulation studies with different force fields provide a rather unifying picture of the effect of PMs, consistent with NMR data. $HuPrP^C$ is characterized by an intrinsic instability of the structure, mainly in the H2–H3 region, and DLMs do affect this region. The S2–H2 loop is more flexible, and several of the hydrophobic and salt-bridges interactions in the H2–H3 domain are loss. Experiments and computations therefore provide a clear picture of H2–H3 region as the seeding element for the native-like misfolding mechanism. The resulting instability[5,142] may trigger and $\alpha \rightarrow \beta$ transition. Righteously, H2 and H3 were recently defined as the "Achilles heels" of prion protein stability.[41]

REFERENCES

1. Prusiner SB. Prions. *Proc Natl Acad Sci USA*. 1998;95(23):13363–13383.
2. Caughey B. Prion protein conversions: insight into mechanisms, TSE transmission barriers and strains. *Br Med Bull*. 2003;66:109–120.
3. Prusiner SB. Molecular biology and pathogenesis of prion diseases. *Trends Biochem Sci*. 1996;21(12):482–487.
4. Gendoo DMA, Harrison PM. Discordant and chameleon sequences: their distribution and implications for amyloidogenicity. *Protein Sci*. 2011;20(3):567–579.
5. Kuznetsov IB, Rackovsky S. Comparative computational analysis of prion proteins reveals two fragments with unusual structural properties and a pattern of increase in hydrophobicity associated with disease-promoting mutations. *Protein Sci*. 2004;13(12):3230–3244.
6. Dima R, Thirumalai D. Exploring the propensities of helices in PrPc to form beta sheet using NMR structures and sequence alignments. *Biophys J*. 2002;83(3):1268–1280.

7. Dima RI, Thirumalai D. Probing the instabilities in the dynamics of helical fragments from mouse PrPC. *Proc Natl Acad Sci USA*. 2004;101(43):15335–15340.
8. Adrover M, Pauwels K, Prigent S, et al. Prion fibrillization is mediated by a native structural element that comprises helices H2 and H3. *J Biol Chem*. 2010;285(27):21004–21012.
9. Priola SA, Meade-White K, Lawson VA, Chesebro B. Flexible N-terminal region of prion protein influences conformation of protease-resistant prion protein isoforms associated with cross-species scrapie infection in vivo and in vitro. *J Biol Chem*. 2004;279(14):13689–13695.
10. Rudd PM, Merry AH, Wormald MR, Dwek RA. Glycosylation and prion protein. *Curr Opin Struct Biol*. 2002;12(5):578–586.
11. Prusiner SB. Novel proteinaceous infectious particles cause scrapie. *Science*. 1982;216(4542):136–144.
12. Ashe KH, Aguzzi A. Prions, prionoids and pathogenic proteins in Alzheimer disease. *Prion*. 2013;7(1):55–59.
13. Mougenot AL, Nicot S, Bencsik A, et al. Prion-like acceleration of a synucleinopathy in a transgenic mouse model. *Neurobiol Aging*. 2012;33(9):2225–2228.
14. World Health Organization. Alzheimer's Disease International. In: *Dementia: A Public Health Priority*; 2012. Geneva.
15. Diack AB, Alibhai JD, Barron R, Bradford B, Piccardo P, Manson JC. Insights into mechanisms of chronic neurodegeneration. *Int J Mol Sci*. 2016;17(1):82, 1–18.
16. Collinge J. Prion diseases of humans and animals: their causes and molecular basis. *Annu Rev Neurosci*. 2001;24:519–550.
17. Weissmann C. Molecular biology of transmissible spongiform encephalopathies. *FEBS Lett*. 1996;389(1):3–11.
18. Prusiner SB. A unifying role for prions in neurodegenerative diseases. *Science*. 2012;336(6088):1511–1513.
19. Sigurdson CJ, Nilsson KPR, Hornemann S, et al. A molecular switch controls interspecies prion disease transmission in mice. *J Clin Invest*. 2010;120(7):2590–2599.
20. Sigurdson CJ, Nilsson KPR, Hornemann S, et al. De novo generation of a transmissible spongiform encephalopathy by mouse transgenesis. *Proc Natl Acad Sci USA*. 2009;106(1):304–309.
21. Sigurdson CJ, Joshi-Barr S, Bett C, et al. Spongiform encephalopathy in transgenic mice expressing a point mutation in the beta 2-alpha 2 loop of the prion protein. *J Neurosci*. 2011;31(39):13840–13847.
22. Bishop MT, Pennington C, Heath CA, Will RG, Knight RS. PRNP variation in UK sporadic and variant Creutzfeldt Jakob disease highlights genetic risk factors and a novel non-synonymous polymorphism. *BMC Med Genet*. 2009;10:146.
23. Bossers A, Schreuder BE, Muileman IH, Belt PB, Smits MA. PrP genotype contributes to determining survival times of sheep with natural scrapie. *J Gen Virol*. 1996;77(Pt 10):2669–2673.
24. Rossetti G, Bongarzone S, Carloni P. Computational studies on the prion protein. *Curr Top Med Chem*. 2013;13(19):2419–2431.
25. Giachin G, Biljan I, Ilc G, Plavec J, Legname G. Probing early misfolding events in prion protein mutants by NMR spectroscopy. *Molecules*. 2013;18(8):9451–9476.
26. Pan KM, Baldwin M, Nguyen J, et al. Conversion of alpha-helices into beta-sheets features in the formation of the scrapie prion proteins. *Proc Natl Acad Sci USA*. 1993;90(23):10962–10966.
27. Riek R, Hornemann S, Wider G, Billeter M, Glockshuber R, Wuthrich K. NMR structure of the mouse prion protein domain PrP(121 − 321). *Nature*. 1996;382(6587):180–182.
28. Christen B, Hornemann S, Damberger FF, Wuthrich K. Prion protein NMR structure from tammar wallaby (Macropus eugenii) shows that the beta2-alpha2 loop is modulated by long-range sequence effects. *J Mol Biol*. 2009;389(5):833–845.

29. Christen B, Perez DR, Hornemann S, Wuthrich K. NMR structure of the bank vole prion protein at 20 degrees C contains a structured loop of residues 165-171. *J Mol Biol.* 2008;383(2):306–312.
30. Lopez Garcia F, Zahn R, Riek R, Wuthrich K. NMR structure of the bovine prion protein. *Proc Natl Acad Sci USA.* 2000;97(15):8334–8339.
31. Zahn R, Liu A, Lührs T, et al. NMR solution structure of the human prion protein. *Proc Natl Acad Sci USA.* 2000;97(1):145–150.
32. Gossert AD, Bonjour S, Lysek DA, Fiorito F, Wüthrich K. Prion protein NMR structures of elk and of mouse/elk hybrids. *Proc Natl Acad Sci USA.* 2005;102(3): 646–650.
33. Naslavsky N, Stein R, Yanai A, Friedlander G, Taraboulos A. Characterization of detergent-insoluble complexes containing the cellular prion protein and its scrapie isoform. *J Biol Chem.* 1997;272(10):6324–6331.
34. Kuwata K, Kamatari YO, Akasaka K, James TL. Slow conformational dynamics in the hamster prion protein. *Biochemistry.* 2004;43(15):4439–4446.
35. Kuwata K, Li H, Yamada H, et al. Locally disordered conformer of the hamster prion protein: a crucial intermediate to PrPSc? *Biochemistry.* 2002;41(41):12277–12283.
36. Lu X, Wintrode PL, Surewicz WK. Beta-sheet core of human prion protein amyloid fibrils as determined by hydrogen/deuterium exchange. *Proc Natl Acad Sci USA.* 2007;104:1510–1515.
37. Chakroun N, Prigent S, Dreiss CA, et al. The oligomerization properties of prion protein are restricted to the H2H3 domain. *FASEB J.* 2010;24(9):3222–3231.
38. Ziegler J, Sticht H, Marx UC, Müller W, Rösch P, Schwarzinger S. CD and NMR studies of prion protein (PrP) helix 1. Novel implications for its role in the PrPC-> PrPSc conversion process. *J Biol Chem.* 2003;278(50):50175–50181.
39. Yamamoto N. Hot spot of structural ambivalence in prion protein revealed by secondary structure principal component analysis. *J Phys Chem B.* 2014;118(33): 9826–9833.
40. Khorvash M, Lamour G, Gsponer J. Long-time scale fluctuations of human prion protein determined by restrained MD simulations. *Biochemistry.* 2011;50(47): 10192–10194.
41. Chamachi NG, Chakrabarty S. Temperature-induced misfolding in prion protein: evidence of multiple partially disordered states stabilized by non-native hydrogen bonds. *Biochemistry.* 2017;56(6):833–844.
42. Baillod P, Garrec J, Tavernelli I, Rothlisberger U. Prion versus doppel protein misfolding: new insights from replica-exchange molecular dynamics simulations. *Biochemistry.* 2013;52(47):8518–8526.
43. Singh RK, Chamachi NG, Chakrabarty S, Mukherjee A. Mechanism of unfolding of human prion protein. *J Phys Chem B.* 2017;121(3):550–564.
44. Cheng CJ, Koldsø H, van der Kamp MW, Schiøtt B, Daggett V. Simulations of membrane-bound diglycosylated human prion protein reveal potential protective mechanisms against misfolding. *J Neurochem.* 2017.
45. Wu EL, Qi Y, Park S, et al. Insight into early-stage unfolding of gpi-anchored human prion protein. *Biophys J.* 2015;109(10):2090–2100.
46. Chen J, Thirumalai D. Helices 2 and 3 are the initiation sites in the PrP(C) → PrP(SC) transition. *Biochemistry.* 2013;52(2):310–319.
47. Tao W, Yoon G, Cao P, Eom K, Park HS. β-Sheet-like formation during the mechanical unfolding of prion protein. *J Chem Phys.* 2015;143(12):125101.
48. Ning L, Guo J, Jin N, Liu H, Yao X. The role of Cys179-Cys214 disulfide bond in the stability and folding of prion protein: insights from molecular dynamics simulations. *J Mol Model.* 2014;20(2):2106.
49. Benetti F, Biarnés X, Attanasio F, Giachin G, Rizzarelli E, Legname G. Structural determinants in prion protein folding and stability. *J Mol Biol.* 2014;426(22):3796–3810.

50. Pan KM, Stahl N, Prusiner SB. Purification and properties of the cellular prion protein from Syrian hamster brain. *Protein Sci.* 1992;1(10):1343–1352.
51. Herrmann LM, Caughey B. The importance of the disulfide bond in prion protein conversion. *Neuroreport.* 1998;9(11):2457–2461.
52. Muramoto T, Scott M, Cohen FE, Prusiner SB. Recombinant scrapie-like prion protein of 106 amino acids is soluble. *Proc Natl Acad Sci USA.* 1996;93(26):15457–15462.
53. Welker E, Raymond LD, Scheraga HA, Caughey B. Intramolecular versus intermolecular disulfide bonds in prion proteins. *J Biol Chem.* 2002;277(36):33477–33481.
54. Jackson GS, Hosszu LL, Power A, et al. Reversible conversion of monomeric human prion protein between native and fibrilogenic conformations. *Science.* 1999;283(5409): 1935–1937.
55. Feughelman M, Willis BK. Thiol-disulfide interchange a potential key to conformational change associated with amyloid fibril formation. *J Theor Biol.* 2000;206(2): 313–315.
56. Tompa P, Tusnády GE, Friedrich P, Simon I. The role of dimerization in prion replication. *Biophys J.* 2002;82(4):1711–1718.
57. el-Bastawissy E, Knaggs MH, Gilbert IH. Molecular dynamics simulations of wild-type and point mutation human prion protein at normal and elevated temperature. *J Mol Graph Model.* 2001;20(2):145–154.
58. Shamsir MS, Dalby AR. One gene, two diseases and three conformations: molecular dynamics simulations of mutants of human prion protein at room temperature and elevated temperatures. *Proteins.* 2005;59(2):275–290.
59. Blinov N, Berjanskii M, Wishart DS, Stepanova M. Structural domains and main-chain flexibility in prion proteins. *Biochemistry.* 2009;48(7):1488–1497.
60. Chen W, van der Kamp MW, Daggett V. Structural and dynamic properties of the human prion protein. *Biophys J.* 2014;106(5):1152–1163.
61. Sekijima M, Motono C, Yamasaki S, Kaneko K, Akiyama Y. Molecular dynamics simulation of dimeric and monomeric forms of human prion protein: insight into dynamics and properties. *Biophys J.* 2003;85(2):1176–1185.
62. Campos SRR, Machuqueiro M, Baptista AM. Constant-pH molecular dynamics simulations reveal a β-rich form of the human prion protein. *J Phys Chem B.* 2010;114(39):12692–12700.
63. Langella E, Improta R, Barone V. Checking the pH-induced conformational transition of prion protein by molecular dynamics simulations: effect of protonation of histidine residues. *Biophys J.* 2004;87(6):3623–3632.
64. Gu W, Wang T, Zhu J, Shi Y, Liu H. Molecular dynamics simulation of the unfolding of the human prion protein domain under low pH and high temperature conditions. *Biophys Chem.* 2003;104(1):79–94.
65. Langella E, Improta R, Crescenzi O, Barone V. Assessing the acid-base and conformational properties of histidine residues in human prion protein (125–228) by means of pKa calculations and molecular dynamics simulations. *Proteins.* 2006;64(1):167–177.
66. Singh J, Udgaonkar JB. Unraveling the molecular mechanism of pH-induced misfolding and oligomerization of the prion protein. *J Mol Biol.* 2016;428(6):1345–1355.
67. Alonso DO, DeArmond SJ, Cohen FE, Daggett V. Mapping the early steps in the pH-induced conformational conversion of the prion protein. *Proc Natl Acad Sci USA.* 2001;98(6):2985–2989.
68. Alonso DO, An C, Daggett V. Simulations of biomolecules: characterization of the early steps in the pH-induced conformational conversion of the hamster, bovine and human forms of the prion protein. *Philos Trans A Math Phys Eng Sci.* 2002;360(1795):1165–1178.
69. DeMarco ML, Daggett V. Molecular mechanism for low pH triggered misfolding of the human prion protein. *Biochemistry.* 2007;46(11):3045–3054.

70. Colacino S, Tiana G, Broglia RA, Colombo G. The determinants of stability in the human prion protein: insights into folding and misfolding from the analysis of the change in the stabilization energy distribution in different conditions. *Proteins*. 2005;62(3):698–707.
71. Shmerling D, Hegyi I, Fischer M, et al. Expression of amino-terminally truncated PrP in the mouse leading to ataxia and specific cerebellar lesions. *Cell*. 1998;93(2):203–214.
72. Viles JH, Cohen FE, Prusiner SB, Goodin DB, Wright PE, Dyson HJ. Copper binding to the prion protein: structural implications of four identical cooperative binding sites. *Proc Natl Acad Sci USA*. 1999;96(5):2042–2047.
73. Millhauser GL. Copper and the prion protein: methods, structures, function, and disease. *Annu Rev Phys Chem*. 2007;58:299–320.
74. Hegde RS, Mastrianni JA, Scott MR, et al. A transmembrane form of the prion protein in neurodegenerative disease. *Science*. 1998;279(5352):827–834.
75. Li AM, Christensen HM, Stewart LR, Roth KA, Chiesa R, Harris DA. Neonatal lethality in transgenic mice expressing prion protein with a deletion of residues 105-125. *EMBO J*. 2007;26(2):548–558.
76. Chakrabarti O, Ashok A, Hegde RS. Prion protein biosynthesis and its emerging role in neurodegeneration. *Trends Biochem Sci*. 2009;34(6):287–295.
77. Hegde RS, Tremblay P, Groth D, DeArmond SJ, Prusiner SB, Lingappa VR. Transmissible and genetic prion diseases share a common pathway of neurodegeneration. *Nature*. 1999;402(6763):822–826.
78. Beland M, Roucou X. The prion protein unstructured N-terminal region is a broad-spectrum molecular sensor with diverse and contrasting potential functions. *J Neurochem*. 2012;120(6):853–868.
79. Silva JL, Vieira TCRG, Gomes MPB, Rangel LP, Scapin SMN, Cordeiro Y. Experimental approaches to the interaction of the prion protein with nucleic acids and glycosaminoglycans: modulators of the pathogenic conversion. *Methods*. 2011;53(3): 306–317.
80. Taubner LM, Bienkiewicz EA, Copie V, Caughey B. Structure of the flexible amino-terminal domain of prion protein bound to a sulfated glycan. *J Mol Biol*. 2010;395(3):475–490.
81. Leal SS, Botelho HM, Gomes CM. Metal ions as modulators of protein conformation and misfolding in neurodegeneration. *Coord Chem Rev*. 2012;256(19–20):2253–2270.
82. Migliorini C, Porciatti E, Luczkowski M, Valensin D. Structural characterization of Cu2+, Ni2+ and Zn2+ binding sites of model peptides associated with neurodegenerative diseases. *Coord Chem Rev*. 2012;256(1–2):352–368.
83. Hajj GNM, Lopes MH, Mercadante AF, et al. Cellular prion protein interaction with vitronectin supports axonal growth and is compensated by integrins. *J Cell Sci*. 2007;120(11):1915–1926.
84. Zanata SM, Lopes MH, Mercadante AF, et al. Stress-inducible protein 1 is a cell surface ligand for cellular prion that triggers neuroprotection. *EMBO J*. 2002;21(13): 3307–3316.
85. Lauren J, Gimbel DA, Nygaard HB, Gilbert JW, Strittmatter SM. Cellular prion protein mediates impairment of synaptic plasticity by amyloid-beta oligomers. *Nature*. 2009;457(7233):1128–U1184.
86. Nicoll AJ, Panico S, Freir DB, et al. Amyloid-beta nanotubes are associated with prion protein-dependent synaptotoxicity. *Nat Commun*. 2013;4:2416.
87. Chen SG, Yadav SP, Surewicz WK. Interaction between human prion protein and amyloid-beta (a beta) oligomers role of n-terminal residues. *J Biol Chem*. 2010;285(34):26377–26383.
88. Irback A, Mohanty S. PROFASI: a Monte Carlo simulation package for protein folding and aggregation. *J Comput Chem*. 2006;27(13):1548–1555.

89. Cong X, Casiraghi N, Rossetti G, et al. Role of prion disease-linked mutations in the intrinsically disordered N-terminal domain of the prion protein. *J Chem Theory Comput.* 2013;9(11):5158–5167.
90. Degioia L, Selvaggini C, Ghibaudi E, et al. Conformational polymorphism of the amyloidogenic and neurotoxic peptide homologous to residues 106-126 of the prion protein. *J Biol Chem.* 1994;269(11):7859–7862.
91. Miura T, Yoda M, Takaku N, Hirose T, Takeuchi H. Clustered negative charges on the lipid membrane surface induce beta-sheet formation of prion protein fragment 106-126. *Biochemistry.* 2007;46(41):11589–11597.
92. Satheeshkumar KS, Jayakumar R. Conformational polymorphism of the amyloidogenic peptide homologous to residues 113-127 of the prion protein. *Biophys J.* 2003;85(1):473–483.
93. Zahn R. The octapeptide repeats in mammalian prion protein constitute a pH-dependent folding and aggregation site. *J Mol Biol.* 2003;334(3):477–488.
94. Croes EA, Theuns J, Houwing-Duistermaat JJ, et al. Octapeptide repeat insertions in the prion protein gene and early onset dementia. *J Neurol Neurosurg Psychiatry.* 2004;75(8):1166–1170.
95. Beck JA, Mead S, Campbell TA, et al. Two-octapeptide repeat deletion of prion protein associated with rapidly progressive dementia. *Neurology.* 2001;57(2):354–356.
96. Kovacs GG, Trabattoni G, Hainfellner JA, Ironside JW, Knight RSG, Budka H. Mutations of the prion protein gene—phenotypic spectrum. *J Neurol.* 2002;249(11): 1567–1582.
97. Soldevila M, Calafell F, Andres AM, et al. Prion susceptibility and protective alleles exhibit marked geographic differences. *Hum Mutat.* 2003;22(1):104–105.
98. Mead S, Whitfield J, Poulter M, et al. A novel protective prion protein variant that colocalizes with kuru exposure. *New Engl J Med.* 2009;361(21):2056–2065.
99. van der Kamp MW, Daggett V. The consequences of pathogenic mutations to the human prion protein. *Protein Eng Des Sel.* 2009;22(8):461–468.
100. Schelzke G, Stoeck K, Eigenbrod S, et al. Report about four novel mutations in the prion protein gene. *Dement Geriatr Cogn Disord.* 2013;35(3–4):229–237.
101. Jansen C, Parchi P, Capellari S, et al. Prion protein amyloidosis with divergent phenotype associated with two novel nonsense mutations in PRNP. *Acta Neuropathol.* 2010;119(2):189–197.
102. Revesz T, Holton JL, Lashley T, et al. Genetics and molecular pathogenesis of sporadic and hereditary cerebral amyloid angiopathies (vol 118, pg 115, 2009). *Acta Neuropathol.* 2009;118:115. 2009;118(2): 321–321.
103. Apetri AC, Vanik DL, Surewicz WK. Polymorphism at residue 129 modulates the conformational conversion of the D178N variant of human prion protein 90-231. *Biochemistry.* 2005;44(48):15880–15888.
104. Apetri AC, Surewicz K, Surewicz WK. The effect of disease-associated mutations on the folding pathway of human prion protein. *J Biol Chem.* 2004;279(17): 18008–18014.
105. Swietnicki W, Petersen RB, Gambetti P, Surewicz WK. Familial mutations and the thermodynamic stability of the recombinant human prion protein. *J Biol Chem.* 1998;273(47):31048–31052.
106. Liemann S, Glockshuber R. Influence of amino acid substitutions related to inherited human prion diseases on the thermodynamic stability of the cellular prion protein. *Biochemistry.* 1999;38(11):3258–3267.
107. Kaneko K, Zulianello L, Scott M, et al. Evidence for protein X binding to a discontinuous epitope on the cellular prion protein during scrapie prion propagation. *Proc Natl Acad Sci USA.* 1997;94(19):10069–10074.

108. Telling GC, Scott M, Mastrianni J, et al. Prion propagation in mice expressing human and chimeric PrP transgenes implicates the interaction of cellular PrP with another protein. *Cell*. 1995;83(1):79–90.
109. Ashok A, Hegde RS. Selective processing and metabolism of disease-causing mutant prion proteins. *PLoS Pathog*. 2009;5(6), e1000479.
110. Rossetti G, Cong XJ, Caliandro R, Legname G, Carloni P. Common structural traits across pathogenic mutants of the human prion protein and their implications for familial prion diseases. *J Mol Biol*. 2011;411(3):700–712.
111. Rossetti G, Giachin G, Legname G, Carloni P. Structural facets of disease-linked human prion protein mutants: a molecular dynamic study. *Proteins*. 2010;78(16): 3270–3280.
112. Bamdad K, Naderi-Manesh H. Contribution of a putative salt bridge and backbone dynamics in the structural instability of human prion protein upon R208H mutation. *Biochem Biophys Res Commun*. 2007;364(4):719–724.
113. van der Kamp MW, Daggett V. Pathogenic mutations in the hydrophobic core of the human prion protein can promote structural instability and misfolding. *J Mol Biol*. 2010;404(4):732–748.
114. Meli M, Gasset M, Colombo G. Dynamic diagnosis of familial prion diseases supports the beta 2-alpha 2 loop as a universal interference target. *PLoS One*. 2011;6(4), e19093.
115. Chebaro Y, Derreumaux P. The conversion of helix H2 to beta-sheet is accelerated in the monomer and dimer of the prion protein upon T183A mutation. *J Phys Chem B*. 2009;113(19):6942–6948.
116. De Simone A, Dodson GG, Verma CS, Zagari A, Fraternali F. Prion and water: tight and dynamical hydration sites have a key role in structural stability. *Proc Natl Acad Sci USA*. 2005;102(21):7535–7540.
117. Chen W, van der Kamp MW, Daggett V. Diverse effects on the native beta-sheet of the human prion protein due to disease-associated mutations. *Biochemistry*. 2010;49(45): 9874–9881.
118. Behmard E, Abdolmaleki P, Asadabadi EB, Jahandideh S. Prevalent mutations of human prion protein: a molecular modeling and molecular dynamics study. *J Biomol Struct Dyn*. 2011;29(2):379–389.
119. Cornell WD, Cieplak P, Bayly CI, et al. A second generation force field for the simulation of proteins, nucleic acids, and organic molecules. *J Am Chem Soc*. 1995;117(19):5179–5197.
120. Brooks BR, Bruccoleri RE, Olafson BD, States DJ, Swaminathan S, Karplus M. CHARMM: a program for macromolecular energy, minimization, and dynamics calculations. *J Comput Chem*. 1983;4(2):187–217.
121. Zuegg J, Gready JE. Molecular dynamics simulations of human prion protein: importance of correct treatment of electrostatic interactions. *Biochemistry*. 1999;38(42): 13862–13876.
122. Guo J, Ren H, Ning L, Liu H, Yao X. Exploring structural and thermodynamic stabilities of human prion protein pathogenic mutants D202N, E211Q and Q217R. *J Struct Biol*. 2012;178(3):225–232.
123. Wopfner F, Weidenhöfer G, Schneider R, et al. Analysis of 27 mammalian and 9 avian PrPs reveals high conservation of flexible regions of the prion protein. *J Mol Biol*. 1999;289(5):1163–1178.
124. Calzolai L, Zahn R. Influence of pH on NMR structure and stability of the human prion protein globular domain. *J Biol Chem*. 2003;278(37):35592–35596.
125. Riek R, Wider G, Billeter M, Hornemann S, Glockshuber R, Wuthrich K. Prion protein NMR structure and familial human spongiform encephalopathies. *Proc Natl Acad Sci USA*. 1998;95(20):11667–11672.

126. Lysek DA, Schorn C, Nivon LG, et al. Prion protein NMR structures of cats, dogs, pigs, and sheep. *Proc Natl Acad Sci USA*. 2005;102(3):640–645.
127. Donne DG, Viles JH, Groth D, et al. Structure of the recombinant full-length hamster prion protein PrP(29-231): the N terminus is highly flexible. *Proc Natl Acad Sci USA*. 1997;94(25):13452–13457.
128. Wen Y, Li J, Yao WM, et al. Unique structural characteristics of the rabbit prion protein. *J Biol Chem*. 2010;285(41):31682–31693.
129. Perez DR, Damberger FF, Wuthrich K. Horse prion protein NMR structure and comparisons with related variants of the mouse prion protein. *J Mol Biol*. 2010;400(2):121–128.
130. Agrimi U, Nonno R, Dell'Omo G, et al. Prion protein amino acid determinants of differential susceptibility and molecular feature of prion strains in mice and voles. *PLoS Pathog*. 2008;4(7): e1000113.
131. Korth C, Stierli B, Streit P, et al. Prion (PrPSc)-specific epitope defined by a monoclonal antibody. *Nature*. 1997;390(6655):74–77.
132. Huang D, Caflisch A. Evolutionary conserved Tyr169 stabilizes the β2-α2 loop of the prion protein. *J Am Chem Soc*. 2015;137(8):2948–2957.
133. Damberger FF, Christen B, Perez DR, Hornemann S, Wuthrich K. Cellular prion protein conformation and function. *Proc Natl Acad Sci USA*. 2011;108(42):17308–17313.
134. Zhang Y, Dai L, Iwamoto M, Ouyang Z. Molecular dynamics study on the conformational transition of prion induced by the point mutation: F198S. *Thin Solid Films*. 2006;499(1–2):224–228.
135. Guo J, Ning L, Ren H, Liu H, Yao X. Influence of the pathogenic mutations T188K/R/A on the structural stability and misfolding of human prion protein: insight from molecular dynamics simulations. *Biochim Biophys Acta Gen Subj*. 2012;1820(2): 116–123.
136. Liu Z, Jia L, Piao Y, et al. Creutzfeldt-Jakob disease with PRNP G114V mutation in a Chinese family. *Acta Neurol Scand*. 2010;121(6):377–383.
137. Hornemann S, von Schroetter C, Damberger FF, Wuthrich K. Prion protein-detergent micelle interactions studied by NMR in solution. *J Biol Chem*. 2009;284(34): 22713–22721.
138. Hegde RS, Kang SW. The concept of translocational regulation. *J Cell Biol*. 2008;182(2):225–232.
139. Gajdusek DC, Gibbs Jr CJ. Slow, latent and temperate virus infections of the central nervous system. *Res Publ Assoc Res Nerv Ment Dis*. 1968;44:254–280.
140. Telling GC, Scott M, Hsiao KK, et al. Transmission of Creutzfeldt-jakob-disease from humans to transgenic mice expressing chimeric human-mouse prion protein. *Proc Natl Acad Sci USA*. 1994;91(21):9936–9940.
141. Cong X, Bongarzone S, Giachin G, Rossetti G, Carloni P, Legname G. Dominant-negative effects in prion diseases: insights from molecular dynamics simulations on mouse prion protein chimeras. *J Biomol Struct Dyn*. 2012;1–12.
142. Kallberg Y, Gustafsson M, Persson B, Thyberg J, Johansson J. Prediction of amyloid fibril-forming proteins. *J Biol Chem*. 2001;276(16):12945–12950.

CHAPTER SIX

Prion Protein and Genetic Susceptibility to Diseases Caused by Its Misfolding

George A. Carlson[1,2]
McLaughlin Research Institute, Great Falls, MT, United States
[2]Corresponding author: e-mail address: george.carlson@ucsf.edu

Contents

Abstract

Early genetic studies on scrapie, an infectious neurodegenerative disease of sheep that was adapted to mice, provided evidence in support of the hypothesis that the agent was a slow virus with a nucleic acid genome independent of the host. Particularly compelling support for an independent genome came from the existence of strains of

[1] Present address: Institute for Neurodegenerative Diseases, University of California San Francisco, 675 Nelson Rising Lane, San Francisco, CA 94143, United States.

Progress in Molecular Biology and Translational Science, Volume 150
ISSN 1877-1173
http://dx.doi.org/10.1016/bs.pmbts.2017.06.007

scrapie agent, some of which were true breeding, while others appeared to mutate under selective pressure. Kuru, a neurodegenerative disease in the remote highlands of Papua New Guinea, had pathological changes similar to those in scrapie and also proved to be transmissible. Genetic studies with the tools of molecular biology and transgenic mice forced a reevaluation of earlier work and supported the prion hypothesis of a novel pathogen devoid of nucleic acid. In this chapter, I discuss the contributions of classical and molecular genetics to understanding PrP prion diseases and to determining that heritable information is enciphered in protein conformation.

1. INTRODUCTION

While classifying Alzheimer's (AD) and Parkinson's (PD) diseases, along with other neurodegenerative brain diseases, as prion diseases is not universally accepted, there is little disagreement that these and other neurodegenerative diseases involve misfolding, aggregation, and spread within the brain of pathological misfolded assemblies of the particular protein or proteins associated with each disorder. This perspective on neurodegenerative diseases derived from work on exceedingly rare transmissible human disorders including Kuru, Creutzfeldt–Jakob disease (CJD) and Gerstmann–Sträussler–Scheinker syndrome (GSS) and on scrapie, an agriculturally significant, transmissible neurodegenerative disease of sheep and goats. The prion hypothesis put forth by Stanley Prusiner argued that the causative infectious agent for scrapie was a novel proteinaceous pathogen distinct from viruses, viroids, and bacteria.[1] Ultimately, misfolded aggregates of a host-encoded protein, designated prion protein (PrP) were demonstrated to be the cause of scrapie, CJD, and GSS.[2] Genetic analyses of prion diseases caused by PrP misfolding and aggregation played an important role in understanding prion replication and disease susceptibility, not only for the rare diseases caused by PrP misfolding, but also for common neurodegenerative diseases involving the amyloid precursor protein (AD), tau (AD, frontotemporal dementia, chronic traumatic encephalopathy (CTE), Pick's disease, and others), α-synuclein (PD, multiple system atrophy, dementia with Lewy bodies), TDP-43 (amyotrophic lateral sclerosis), Huntingtin (Huntington's disease), among others.[3] With the exception of Huntington's disease, which is only familial, neurodegenerative prion diseases have sporadic, genetic, and acquired or infectious forms. PrP prions were once thought to be the only neurodegenerative disease with an infectious etiology, but while Alzheimer's and other dementias are not contagious, transmission of self-replicating misfolded protein aggregates from both mice and humans has been achieved in mouse models and cell cultures.[4–6] Propagation of prions from patients with CTE argues that the trauma can induce conformations that start tau prion propagation.[7] Recent evidence also suggests that human-to-human transmission of Alzheimer's Aß amyloid pathology may have resulted from treatment with contaminated growth hormone.[8] The shared features of protein misfolding diseases argue that genetic approaches used to study prototypical PrP prion diseases, including scrapie, may have broader applicability to other neurodegenerative diseases.

2. SCRAPIE, KURU, AND SLOW VIRUSES

The name scrapie derives from a common clinical sign of the disease—compulsive scraping against fences and other objects leading to loss of areas of fleece. Scrapie was known as a contagious disease by the agricultural community in Europe by the 18th century and likely earlier[9] with transmission by inoculation clearly established by the first half of the 20th century.[10,11] Therefore, studies on scrapie and human diseases took a microbiological perspective and focused on isolation and characterization of the peculiar infectious agents that caused spongiform degeneration of the brain. A long incubation period, the interval between infection and clinical signs, is a feature of the disease, leading to use of the term slow virus to describe the infectious agent.

The similarities in the pathology of scrapie and that of individuals dying of Kuru, an incurable neurodegenerative disorder endemic among the Fore people in the remote highlands of Papua New Guinea, was noted by Hadlow who suggested inoculation of nonhuman primates with brain homogenates from patients dying of Kuru.[12] Successful transmission of neurodegenerative disease from Kuru patients firmly established an infectious etiology for this disease.[13] The epidemiology and natural history of Kuru suggested cannibalism associated with funeral rituals as the mode of spread within the population with incubation periods from time of ingestion of infected tissue to disease onset that can be decades long.[14] The similar spongiform encephalopathies (TSEs) CJD and GSS also proved transmissible, establishing wide acceptance of "slow viruses" as the etiological agents for TSEs.[15,16]

3. PRIONS AND PrP

Several features, including resistance to ionizing radiation, autoclaving, or nuclease treatment, and lack of an adaptive immune response to the infectious agent distinguished the causative agent for scrapie from viruses, leading to a variety of proposals for the nature of this unconventional agent ranging from a naked RNA to a particle consisting only of protein.[2] Spurred by the unusual nature of the infectious agent, Prusiner and his colleagues began biochemical fractionation to identify molecules or particles that purified with infectivity. This approach was facilitated by development of an incubation time assay to assess the titer of scrapie agents in Syrian hamsters, which have much shorter incubation times, as well as larger brains, than mice (~60 days for hamsters compared to more than 100 days for mice).[17] Each fraction of the purification process, which included a limited proteinase K digestion step, was inoculated intracerebrally into groups of hamsters to determine the incubation time. Use of incubation time, rather than end point titration to estimate the titer of infectivity was a major advance that made this approach feasible. A single protein of 27–30 kDa purified with hamster scrapie infectivity and was designated PrP 27–30.[18] The prion hypothesis was published and stated that a proteinaceous infectious particle, called a prion, was novel and distinct from all other known infectious agents.[1] This proposal met with skepticism and intense resistance from the scrapie research community.

A partial amino acid sequence for PrP 27–30[19] enabled cloning of hamster[20] and mouse[21] PrP cDNAs. PrP was encoded in the host genome and present in both infected and uninfected animals. While to some, this indicated that PrP was not a functional component of the infectious agent,[21] use of rabbit antisera specific for PrP revealed that both infected and uninfected animals had a 32–33 kDa isoform of PrP that was completely digested in uninfected animals by proteinase K (PK) concentrations that produced PrP 27–30 in scrapie-infected animals.[20] The normal cellular isoform was designated PrP^{C} while the PK-resistant isoform, found only in scrapie prion-infected animals, was called PrP^{Sc}.

4. GENETIC LINKAGE OF SCRAPIE INCUBATION TIME AND THE PrP GENE

DNA probes for the PrP gene provided a new tool to reassess earlier genetic studies on mouse-adapted scrapie agent. Two inbred strains of mice that differ widely in incubation time for the Rocky Mountain Laboratory (RML) strain of prions were used to demonstrate genetic linkage of this key biological phenotype of scrapie to the PrP gene (*Prnp*). I/LnJ mice have very long incubation periods after inoculation with scrapie prions

(200–385 days) and NZW/LacJ mice have shorter ones (113 days ± 2.8 SEM). (NZW X I/Ln)F1 hybrid mice had incubation times of 223 ± 2.8 days indicating that longer incubation times were dominant. Incubation periods in offspring of a backcross of long incubation time (NZW X I/Ln)F1 mice to NZW/LacJ mice segregated into two groups (one averaging 130 days and the other 195 days) demonstrating that a single Mendelian gene controlled incubation time.[22] NZW and most other inbred strains carry the *Prnp*^{a} allele, which was originally defined by restriction fragment length polymorphisms that distinguished it from the less common *Prnp*^{b} allele.[22,23] Genotyping the backcross mice showed that the incubation time gene (designated *Prni*) was tightly linked to Prnp, the structural gene for PrP, whose entire open reading frame is encoded by a single exon. DNA sequencing found that the *a* and *b* alleles of *Prnp* differ at codons 108 (Leu/Phe) and 189 (Thr/Val).[24] This demonstration of linkage, together with the possibility that *Prnp* itself, rather than a linked gene, controlled prion incubation time provided unbiased biological evidence supporting the relevance of the biochemical association of PrP^{Sc} with the infectious prion particle. The concordance between incubation time and *Prnp* genotype was not absolute, however, raising the possibility that *Prni* and *Prnp* were distinct genes. That the two alleles were found to differ at codons 108 and 189[24] and the failure to identify genetic recombination events that separated the incubation time from *Prnp* genotype argued that incubation time was controlled by PrP itself.[23,25] This was not formally proven, however, until 1998 when one Prnp allele was converted to the other using knock-in technology.[26] Earlier, pre-PrP analyses had already demonstrated the power of genetics to help reveal fundamental features of the scrapie agent and its transmission.

5. SCRAPIE STRAINS

In 1964, Dickson and Mackay reported incubation periods ranging from approximately 120 to nearly 300 days in different inbred mouse strains that were inoculated with the same ME 7 strain of scrapie brain homogenate.[27] Backcrosses and intercrosses among short incubation time and long incubation time mouse strains demonstrated that a single autosomal gene, designated *Sinc* for Scrapie incubation, was responsible. *Sinc* accounted for most of the variance and had two alleles (*p*7 for prolonged ME7 and *s*7 for short ME7 strain incubation times). Only two closely related mouse strains in Edinburgh, VM/Dk and IM/Dk, carried the *p*7 allele. Importantly, long and short incubation times were not an absolute property of each of the two *Sinc* alleles, but were determined by interaction between genotype of the host and the particular strain of scrapie agent. For example, VM/Dk and IM/Dk mice homozygous for the *Sinc*^{p7} allele had much shorter incubation times for the 87 V scrapie strain than mice with the *Sinc*^{s7} allele—the opposite of the results obtained with the ME7 strain.[28] Multiple strains of the presumed scrapie virus were defined according to their incubation times and distribution of pathological lesions in *Sinc*^{s7} and *Sinc*^{p7} mice.[29,30] Unfortunately, the scrapie prion strains and the *Sinc*^{p7} mice that defined them, were not widely available slowing progress toward determining whether *Sinc* and *Prni* were the same gene or even whether *Sinc* was linked to *Prnp*, which had been mapped to mouse Chr 2.[31] For decades, the very existence of scrapie strains provided the strongest and most compelling argument against an infectious agent devoid of functional nucleic acid, since what else other than nucleic acid could encode heritable information and be subject to mutation and selection.[28–30]

Mouse scrapie strains could be classified into at least three general categories: true-breeding prion strains whose properties remained constant regardless of passage history through *Sinc*^{s7} or *Sinc*^{p7} mice; those whose incubation times shortened dramatically in a second passage (for example, *s*7 to *p*7 to *p*7) reminiscent of the species barrier to scrapie transmission, but maintained their patterns of relative incubation times in *s*7, *p*7, and *s*7/*p*7 mice; and unstable strains whose passage through *Sinc* mismatched mice effected a new pattern of interaction among the scrapie agent strain and host genotype. Given the assumption that the scrapie agent was a slow virus, it was entirely reasonable to conclude that agent strain properties reflected a nucleic acid genome in the scrapie agent and to use the very existence of strains as evidence against the prion hypothesis.

5.1 PrP, Passage History, and Prion Strains

However, if PrP^{Sc} were indeed a component of the scrapie agent, prions produced in *Prnp*^{a} mice would contain PrP-A molecules that differ in amino acid sequence from PrP-B made in *Prnp*^{b} mice. For example, incubation times in mice expressing PrP-B had longer incubation times when inoculated with RML prions maintained in mice expressing PrP-A than did RML prions passaged through PrP-B mice.[32] Although passage of prions through PrP-B mice shortened incubation time in PrP-B mice, passage history did not change the property of the RML prion strain that incubation times in PrP-A mice were always shorter than in PrP-B mice regardless of passage history. *Prnp^b^* or *Sinc^p7^* behave as dominant alleles in prolonging incubation time of RML or ME7 prion isolates. However, expression *Prnp^b^* transgenes in *Prnp^a^* mice shortened, rather than prolonging RML prion incubation time as might be expected for a dominant allele.[33] This result was due to the dramatically increased amount of PrP-B substrate, which, though less efficiently converted to PrP^{Sc} than PrP-A, supports prion replication and causes disease. Thus, the dominance of *Prnp^b^* in *Prnp^a^/Prnp^b^* heterozygotes reflects the reduced supply of PrP^{C}-A produced from the single copy of *Prnp^a^* rather than an inhibitory effect of PrP-B on prion replication.[34]

5.2 Species Barrier to Prion Transmission

In general, prions that share PrP primary structure with the host have shorter incubation times than mismatched prions. The species barrier to prion transmission provides a particularly compelling illustration.[35–38] Syrian hamster-adapted scrapie prions have incubation times that range from approximately 60–150 days depending on the particular strain of agent. When inoculated into mice the incubation times exceeded 300 days and in some cases the mice did not show clinical signs at all. However, inoculation of brain homogenates from the mice that eventually became ill following infection with hamster prions caused disease in mice with incubation times approximating those of scrapie prions maintained in mice. With the advent of transgenesis came the demonstration that the species barrier between hamsters and mice was due to PrP. Mice transgenic for Syrian hamster PrP (SHaPrP) were susceptible to SHa prions with incubation times as short as 50 days in high copy number transgenics, with incubation time inversely proportional to transgene copy number and the level of SHaPrP.[37,38] As expected, PrP^{Sc} produced by the Tg(SHaPrP) mice was hamster, not mouse; the distribution of pathology and presence of PrP amyloid plaques also were more similar to hamster than mouse scrapie. The prions produced in Tg(SHaPrP) mice readily transmitted to hamsters but not to mice. These Tg(SHaPrP) mice also expressed endogenous mouse PrP (MoPrP) and were susceptible to infection with mouse scrapie prions. The expression of SHaPrP transgenes slightly prolonged the incubation time for mouse prions. The converse also was shown to be true after knockouts of *Prnp* were produced.

6. *Prnp* KNOCKOUTS

The first targeted deletion of the mouse *Prnp* gene was produced in Zurich (this null allele is designated $Prnp^{Zrch1}$) and mice homozygous for this null allele were viable, did not develop disease after inoculation with prions, and failed to replicate infectivity or produce PrP^{Sc}.[39,40] In addition to providing strong evidence that PrP^{C} is essential for prion disease and replication of the infectious agent, the viability of $Prnp^{Zrch1}/Prnp^{Zrch1}$ mice indicated that the pathology in prion disease is not due to loss of PrP function. PrP null mice are often abbreviated $Prnp^{0/0}$ or $Prnp^{-/-}$ but it can be important to note the specific null allele. A PrP null allele later produced in Edinburgh ($Prnp^{Edbg}$) gave very similar results to the $Prnp^{Zrch1}$ allele.[41] However, three lines of mice homozygous for other *Prnp* null alleles, *Ngsk*, *Rcm0*, and *Zrch2*, developed late onset ataxia and Purkinje cell neurodegeneration.[42–44] These mice were produced with targeting constructs that abolished the splice acceptor for *Prnp*'s third exon that contains the entire PrP open reading frame; in contrast, the lines without ataxia had the splice acceptor intact while replacing downstream PrP coding sequences. Sequencing genomic clones of *Prnp* had revealed a PrP-related gene designated *Prnd*.[45] The Doppel protein encoded by this gene is highly expressed in the testis, but only minimally in the brain and central nervous system. The PrP knockout lines that developed ataxia used *Prnd*'s splice acceptor to produce a *Prnp–Prnd* chimeric transcript resulting in inappropriate expression of Doppel in the central nervous system.[44] The neurodegenerative effects of *Prnd* misexpression or overexpression in the CNS could be prevented by expression of PrP.[46]

A small number of humans heterozygous for loss-of-function *PRNP* alleles have been identified in a large population and individuals with only a single copy of *PRNP* are apparently healthy.[47] Whether homozygosity for null alleles causes disease or detrimental phenotypes in humans is unknown. However, if the reduced levels of PrP that are expected in null allele heterozygotes truly have no ill effects, therapies to reduce PrP levels could prove effective in slowing disease progression. Mice heterozygous for $Prnp^{0}$ express roughly half the levels of PrP as wild-type homozygotes have greatly prolonged incubation times in spite of production of levels of PrP^{Sc} and infectious prions comparable to those in infected wild-type mice.[48] The presence of high levels of PrP^{Sc} during a long asymptomatic period points out a central role of PrP^{C} in neurodegeneration and disease. This was further illustrated by experiments transplanting brain tissue from PrP overexpressing mice into PrP-deficient mice and then infecting with scrapie prions. Only the graft showed any evidence of neurodegeneration and spongiform change even though large amounts of PrP^{Sc} migrated into the parenchyma of the host brain, accumulated, and formed amyloid deposits.[49] This is strong evidence that expression of PrP^{C} is required not only for prion replication, but also for exogenous prions to exert their neurotoxic effects.

6.1 PrP Function

The function of PrP has proved elusive. A variety of phenotypes, ranging from perturbations in circadian rhythms to gait abnormalities, have been attributed to lack of PrP expression in mice homozygous for *Prnp* null alleles.[50] The embryonic stem cells used for *Prnp* targeting were derived from sublines of 129 strain inbred mice and in most cases have been backcrossed to other strains to make congenic lines or simply outcrossed and maintained on a mixed genetic background. Differences in genetic backgrounds easily could account for differences in phenotypes, as could segregating alleles in outbred lines.

At least some of the diversity in phenotypes In $Prnp^{0/0}$ mice can be attributed to differences in congenicity among the *Prnp* null lines, often due to genes within the differential segment. B6 mice congenic for the $Prnp^{Zrch1}$ knockout allele have a deficiency in phagocytosis of effete or apoptotic cells by macrophages that was attributed to PrP.[51] Elegant genetic analyses demonstrated that the protein encoded by the 129 strain *Sirpa* allele was included within the *Prnp* congenic segment, indicating that PrP itself was not responsible for this phenotype.[52] The recent targeted deletion of *Prnp* using TALEN-induced disruption of the protein coding sequence, created a truly coisogenic PrP knockout allele ($Prnp^{Zrch3}$) on the C57BL/6J background.[53] This new allele, together with the $Prnp^{Edbg}$ allele that is maintained on its original 129 background, provide important tools for assessing the function of PrP. The *Zrch3* knockout mouse has been used to confirm earlier results suggesting that PrP ablation causes chronic demyelinating polyneuropathy.[54] This demyelinating disease in aged PrP null mice results from PrP^{C}'s promotion of peripheral myelin homeostasis through binding to the G protein-coupled receptor Adgrg6 and increasing cAMP levels in Schwann cells.[55]

While some phenotypes in $Prnp^{0/0}$ congenic mice can be accounted for by genes within the differential chromosomal segment on mouse

Chromosome 2, other putative PrP functions may be due to background-dependent modifier genes or even to environmental differences, such as the microbiological status of different colonies. Attribution of a phenotype to the selected gene in a congenic strain always should be tempered with caution, as clearly illustrated by the example of PrP. Not only do congenic strains carry donor-derived alleles within the differential segment, but also passenger donor alleles on different chromosomes also can be coselected. For example, despite 12 generations of backcrossing with intentional selection only for *H2* on Chromosome 17 to produce B10.A(18R) mice, the T-cell receptor alpha chain complex on Chr 14 was also found to be donor-strain derived, probably due to coselection for resistance to viral pathogens present in the colony where the mice were made.[56]

7. PRNP AND FAMILIAL PRION DISEASES

The discovery that Kuru was an infectious disease caused by ceremonial ingestion of contaminated tissue, favored the view that the disease had a viral etiology. The discovery that alleles of *Prnp* influenced susceptibility to mouse-adapted scrapie as reflected by incubation time led to studies to determine whether familial forms of GSS disease could be influenced by the nucleic acid sequence of the human PrP gene, *PRNP*. A missense mutation at codon 102 was discovered and linked to disease.[57] Other familial PrP prion diseases were genetically linked to mutations in *PRNP* and could be transmitted by inoculation into experimental animals. The explanation that the linkage of *PRNP* to familial prion disease reflected susceptibility to a ubiquitous virus was unsatisfying. Transgenic mice expressing high levels of *PRNP* with the mouse equivalent of the GSS mutation developed spontaneous neurodegenerative disease that could be transmitted with short incubation times to transgenic mice that expressed low levels of mutant PrP and developed disease only late in life.[58,59]

Prion disease is extremely rare in humans with only one or two cases per million people each year. An intriguing feature of PrP prion diseases is that they can have infectious, sporadic, or familial origin.[2] Infectious or iatrogenic prion disease, exemplified by Kuru and by CJD caused by prion contaminated cadaver-derived growth hormone inoculation,[60] is exceptionally rare. Sporadic prion disease is the most common, with less than 20% of prion disease cases associated with a family history. A variety of *PRNP* mutations have been reported in patients with prion disease. Assuming multiple multigenerational families are available, genetic linkage provides strong evidence that some *PRNP* mutations, exemplified by P102L, are causative and fully penetrant.[57,61,62] However, caution is necessary when attributing causation to mutations found in only one or a few patients afflicted with prion disease. More than 60 variants in the *PRNP* gene had been reported as pathogenic. Minikel and colleagues[47] used large-scale population data to assess the likelihood that *PRNP* variants are causative, based on the simple premise that a fully penetrant mutation should be no more frequent in the general population than the disease that it purportedly causes. Mutations with strong genetic evidence for Mendelian inheritance with complete penetrance were not found in a control population of 60,706 sequenced exomes; these mutations (A117V, D178N, P102L, and E200K) also produced spontaneous disease in mouse models. Together these four mutations account for more than 50% of familial prion disease. Variants that were frequent in the control population, but rarely found among patients with confirmed PrP prion disease are very unlikely to be causative. Some variant PrPs are not clearly causative, with a frequency in controls too high to be fully penetrant, but lower than the frequency in the patient population. These mutations (M232R, V180I, V210I, and others) are likely to be incompletely penetrant but may increase the risk of disease over the lifetime of the carriers.

7.1 Primary Structure Does Not Predict Infectivity

The propensity of variant PrPs for misfolding and aggregation is experimentally tractable both in culture and in mouse models, but how a particular primary structure of PrP affects conformation, let alone whether it can cause prion disease, cannot be predicted at present. In general, mismatching of amino acid sequences between the host PrP^{C} and PrP^{Sc} in the inoculum results in inefficient disease transmission or long incubation times, but prediction of whether a particular prion will cross a species barrier is not possible. Witness the ability of bovine spongiform encephalopathy (BSE) prions to infect humans, albeit at low efficiency, and cause a variant form of CJD (vCJD). In spite of primary structure differences, both vCJD and BSE transmit to mice and evidence indicates that they are the same strain of prions. Both isolates produce near identical profiles of pathology and patterns of PrP^{Sc} deposition within the brain. BSE and vCJD efficiently infect mice with the mouse strain distribution pattern of incubation times having RIIS shorter than B6, which is shorter than in VM/Dk.[63,64] The incubation time of RIIIS mice (~130 days) suggests the absence of a species barrier. RIIIS mice produce PrP-A, but the Prnp open reading frame is within a haplotype designated *Prnp^c* based on flanking restriction site polymorphisms.[23] It is not known whether the remarkably short incubation time of RIII mice after challenge with BSE and vCJD prions compared to other mice expressing PrP-A is due to genes other than *Prnp* or to levels of *Prnp* expression.

Bank vole PrP (BVPrP) differs from mouse PrP at eight positions, and bank voles are susceptible to prions from a variety species.[65–68] The promiscuous susceptibility of bank voles to prions is due to the structure of PrP rather than to other factors in the species since transgenic mice expressing BVPrP transgenes are similarly susceptible to a variety of prion strains from different species, including humans.[69,70] BVPrP also can serve as an exceptionally efficient substrate for PrP conversion assays in vitro.[71,72]

8. PrP CONFORMATION, PRION STRAINS, AND SUSCEPTIBILITY TO HUMAN PRION DISEASES

Effects of PRNP polymorphisms on susceptibility to human prion diseases. The distinct effects of *a* and *b* alleles of *Prnp* on experimental prion incubation time in the mouse is similar to the effects of polymorphisms in the human PRNP gene on susceptibility to sporadic, transmitted, and familial prion diseases. The variant in PrP most commonly associated with prion disease susceptibility is at position 129, which can be either a methionine or a valine.[73] In general, heterozygosity for codon 129 protects against prion diseases, possibly due to reduction in the concentration of the most efficiently converted PrP allotype that matches the codon 129 of the disease-inducing PrP^{Sc} and to inefficient interactions between the two allotypes. For example, heterozygosity in patients with sporadic CJD (sCJD) is much lower than expected for its frequency in the general population[74]; though frequencies differ among populations, heterozygosity and 129M homozygosity are more common than 129V homozygosity. In the United States and the United Kingdom populations, ~50% are M/V heterozygous with ~40% M/M homozygous.[74] Of the more than 200 individuals afflicted with vCJD all have been M/M homozygotes with the possible exception of a single M/V heterozygous case.[75] Evidence from genetically modified mice and from humans suggests, however, that all three codon 129 genotypes of PrP can replicate BSE or vCJD infectivity.[76]

8.1 Protein Conformation Enciphers Heritable Information

Although the species barrier ultimately results from mismatching of PrP amino acid sequence, the primary structure of PrP cannot explain all the differences in the efficiency of prion transmission from one host to another. For example, experimental transmission of transmissible mink encephalopathy to Syrian hamsters resulted in two distinct stable strains of prions designated hyper (HY) and drowsy (DY).[77] The strains differed in incubation time, distribution of pathological lesions within the brain, and clinical signs. Hamsters with the DY strain were lethargic and had longer incubation times (168 ± 2 days) compared to HY strain-infected hamsters (incubation times, 65 ± 1 days) that exhibited hyperesthesia and ataxia. The amino acid sequences of PrP in the HY and DY prions are identical, but the PrP^{Sc} in the strains differed in susceptibility to proteinase-K digestion and are cleaved at different amino-terminal sites, indicating conformational differences between the two prion strains.[78,79]

Similar evidence that the heritable information that distinguishes prions strains is encoded by protein conformation comes from transmission of two distinct human prion diseases into mice.[80] A point mutation at codon 178 (D178N) causes familial prion disease, but whether the mutation causes familial CJD (fCJD) or fatal familial insomnia (FFI) depends on the codon 129 genotype. FFI is produced when D178N is on PrP with methionine at position 129, while the same mutation on 129V PrP produces a subtype of fCJD.[81] As the name implies, the most striking features that distinguish FFI from CJD are insomnia, presumably due to severe atrophy of thalamic nuclei, and dysautonomia. Spongiform degeneration is less widespread in FFI than in CJD. The unglycosylated fragment sizes of PrP^{Sc} following limited PK digestion also differ according to whether the D178N mutation is on 129M PrP (FFI, 19 kDa) or 129V PrP (fCJD, 21 kDa).[82] This conformational difference influencing the PK cleavage site could be interpreted as being due to single amino acid difference distinguishing PrP_{D178N} in FFI patients from that in fCJD patients. However, strong experimental evidence indicates that conformation alone encrypts consistent heritable properties of different strains passaged in the same strain of mice that previously had been attributed to nucleic acid genomes.[83]

Transmission of human prions that produced distinct diseases and that were caused by different mutations in *PRNP* showed that the disease-specific inocula produced distinct prion strains within the same mouse strain and that their strain properties were maintained on repeated passage.[80,84] Brain homogenates from a patient who had died of FFI (D178N, M129) or of fCJD (E200K, M129) were inoculated into transgenic mice expressing a chimeric MoHu2Mo PrP transgene on $Prnp^{0/0}$ background. Unlike the case where the species barrier of mice to hamster prions is overcome by expressing hamster PrP transgenes,[38] simply expressing a human PrP transgene is not sufficient to overcome the species barrier to human prions. Eliminating expression of endogenous mouse *Prnp* or expression of mouse–human chimeric transgenes (or preferably both) is necessary for efficient infection of mice with human prions.[85,86] FFI transmitted to Tg(MHu2Mo) *Prnp*$^{0/0}$ mice with incubation times of 206 ± 7 (SEM) days and produced a PrP^{Sc} PK-resistant fragment of 19 kDa, while fCJD inoculation produced incubation times of 170 ± 2 days and a 21 kDa PrP^{Sc} fragment. The pathology of the human diseases was reflected in mice with FFI inoculated mice having very prominent disease in the thalamus with relative sparing of the hypothalamus in comparison to fCJD inoculated mice. While the primary structures of PrP^{Sc} that initiated the two diseases differed, the PrP amino acid sequence of the hosts was identical.[80] Even more striking is the fact that the strains' properties, including PK-resistant PrP fragment sizes, are maintained on subsequent passage in Tg(MHu2Mo) *Prnp*$^{0/0}$ mice.[84] The FFI incubation time on second passage fell to 136 ± 1 days, while the fCJD incubation time was stable at 167 ± 3 days. Thus, conformation alone is sufficient to encrypt strain properties, and allelic forms of *Prnp* and *PRNP* that determine incubation time or susceptibility to disease reflect the effect of primary structure on protein conformation.

Epidemiological evidence indicates that *PRNP* polymorphisms influence susceptibility to Kuru, the now extinct neurodegenerative disease that was endemic and restricted to a remote region of the Papua New Guinea highlands. Again, heterozygosity at codon 129 was associated with longer incubation times and survival following ingestion of tissue infected with Kuru prions as part of funerary rites. Ritual cannibalism ceased in the 1950s, and incubation times exceeding 50 years have been documented.[87] Virtually all elderly Kuru patients were heterozygous at codon 129, while those afflicted at a young age (putative short incubation times) were homozygous for the M or V allele, resulting in a dearth of heterozygotes in the population.[88]

Genotyping Papua New Guineans also revealed a novel allele of *PRNP* (G127V) that was common within the Kuru region but not found in any individuals with Kuru.[88a] Thus, the *PRNP* V127, which always was inherited with M129, appears to be a resistance allele recently selected during the Kuru epidemic; its presence was postulated to confer complete resistance to Kuru. Mice expressing both V127M129 and G127M129 human *PRNP* transgenes confirmed this.[89] The mice expressing both transgenes were completely resistant to Kuru and to sCJD, but were susceptible to vCJD. Mice expressing only the V127 variant *PRNP* transgene, however, were resistant to all human prion strains that were tested. Results from transgenic lines expressing various levels of PrP led to the conclusions that PrP-V127 is completely resistant to conversion by the prion strains that have been tested and can act as a concentration-dependent inhibitor of wild-type PrP replication. The effects of V127 and all other variants in PrP that affect disease susceptibility, incubation time, clinical signs, or distribution of pathological lesions most likely are mediated by their effects on PrP conformation.

8.2 Selection and Evolution of Prion Strains

Amino acid sequence clearly does not dictate a single conformation as illustrated, for example, by the formation of two distinct conformers by FFI and fCJD prions in mice expressing the same PrP transgene.[80] It is likely that multiple conformations of PrP^{Sc} exist in infected animals. Properties of prion strains can change when passaged in different cell lines, when going from mice to cell culture (and vice versa), or in the presence of drugs that prolong scrapie incubation time in mice.[90–92] Evolution of prion strains and acquisition of drug resistance either by induced conformational change or selection of preexisting conformers has serious implications for development of prion therapeutics. For example, although treatment of mice with the 2-amino thiazole IND24 doubles the incubation time of the RML strain of prions, the treated mice eventually develop disease and retain resistance to IND24 in subsequent passage.[92] Differences in the ability to infect distinct cell lines also were induced by the drug treatment. Since multiple conformers of PrP^{Sc} are responsible for selection of distinct "mutant" strains of prions by drug selection or passage in resistant hosts, therapeutics that target a single prion strain or particular site of PrP^{C}–PrP^{Sc} interaction may ultimately prove ineffective. Particularly relevant is the failure of IND24 and other drugs that are effective against mouse prions to have any effect on replication of CJD prions in humanized mice.

Spread of misfolded protein aggregates along neural pathways is a feature of animal models for neurodegenerative diseases such as PD and AD that are much more common than those caused by PrP misfolding.[3,93,94] Pathology also can be transmitted by inoculation of mice or cell cultures with aggregates of tau, Aß-, or α-synuclein.[95–99] It is clear that PrP is not the only protein driving neurodegenerative disease that can produce multiple conformers from a single primary structure.[100,101] As only one example,

Aß transmissions can also lead to distinct strains of Aß prions.[102] Multiple conformations of misfolded proteins subject to selective pressures leading to emergence of "mutant" forms with different properties presents a challenge comparable to that in cancer therapy with evolution of malignant cells through mutations in the cellular DNA. In cancer, identification of the genes involved in resistance to therapy offers new potential targets and pathways, rather than a fluid set of conformations that, at present, cannot be predicted from the amino acid sequence adding difficulties to the design of combination therapies.

9. GENES IN ADDITION TO THE PrP GENE THAT INFLUENCE SUSCEPTIBILITY TO PrP PRION DISEASES

Strains of mice that express identical PrP sequences can vary widely in prion incubation time, with the intervals from inoculation of RML prions to onset of disease ranging from 100 days in SJL/J to over 200 days.[23,24,103] Quantitative trait analysis using microsatellite markers scattered across the genome of offspring of intercrosses and backcrosses between long and short incubation time *Prnp*^{a} mouse strains identified several highly significant, but broad, chromosomal regions that harbored quantitative trait loci (QTL) affecting prion incubation time.[104–107] The same chromosomal regions on Chr11 and Chr 2 were identified in independent studies, two of which used crosses involving the wild-derived CAST/Ei strain. Significant QTL unique to the individual studies also were detected, highlighting the fact that different QTL may be detected in crosses of different progenitors, with only two mouse strains sampled in each conventional cross. In all cases, regions of chromosome that could be identified contained hundreds of genes, making identification of the genes underlying QTL problematic. The difficulty in going from a QTL region to the underlying modifier gene(s) is compounded by the relatively small contribution of a QTL to the total variance. Transferring a QTL to a different genetic background to produce congenic strains often results in loss of the modifier's effect.

Newer approaches were facilitated by the availability of genome sequence and high-density single-nucleotide polymorphisms to produce stocks of mice with a random mix of alleles that approximate a human population. Heterogeneous stock (HS) mice have been generated from multiple inbred and wild-derived mouse strains and bred over generations to produce a high-density mix of alleles.[108] Similarly, diversity outbred (DO) mice were a by-product of the collaborative cross that has produce hundreds of recombinant inbred lines from HS mice carrying alleles from eight strains of mice.[109] These stocks provide increased diversity and much higher mapping resolution than conventional crosses between two inbred strains. For example, use of HS mice mapped a locus affecting RML prion incubation time to a small 2.9 Mb region on Chromosome 19. Sequencing revealed a strong candidate, the E3 ubiquitin ligase *Hectd2*, whose human counterpart also proved to be involved in susceptibility to human prion disease.[110] Additional candidates for genes underlying QTL have been identified, but as yet, with the possible exception of HECTD2, none show great promise as targets for therapeutic intervention.

A limitation of all approaches to identifying QTL may be inherent to the biology of prion disease and its genetic modifiers; to date, each QTL region accounts for only a relatively small portion of the variance in incubation times, generally less than 20%. The small effect size of individual genes also was observed in a compilation of prion incubation times in 20 gene knockout mice and their controls.[111] Of the 20 mice, the three knockouts producing a significant effect (*App*, *Il1r1*, and *Sod1*) only slightly prolonged incubation times (13%, 16%, and 19%).

10. CONCLUDING REMARKS: GENETICS AND THE PROSPECTS FOR PrP PRION DISEASE THERAPEUTICS

The ultimate goal of analysis of genetic susceptibility to PrP prion diseases and to neurodegenerative diseases involving misfolding of other proteins is to identify genes and pathways that can serve as therapeutic targets. PRNP is clearly a target and ablation of this gene renders animals absolutely resistant to disease and to prion replication. While the consequences of reducing PrP concentrations in humans is unknown, hemizygous individuals have been identified and have no obvious defects.[47] Developing methods for specifically reducing expression of a single protein over a long period of time is not a trivial undertaking, but, at least theoretically, is feasible.

Both genetic and biochemical analyses point to the interaction between PrP^{C} and PrP^{Sc} as being central to prion replication and disease progression, regardless of whether the origin of misfolded PrP^{Sc} is exogenous, due to PRNP mutations that favor misfolding, or to sporadic events. The existence of prion strains with different (and unpredictable from the primary structure) conformations of PrP^{Sc}, together with the lack of a cell culture model for replication of human prions, have thus far proven to be insurmountable barriers to development of therapeutics for PrP prion diseases. In vitro assays for human tau and α-synuclein prions hold promise for screens for drugs against tauopathies, PD, and multiple system atrophy though the problem of distinct strains of misfolded proteins extends to these diseases.[4,7,96,97,112,113]

Modifier genes identified through a variety of approaches also may offer targets for therapeutic intervention even though such modifiers of PrP prion diseases individually account for only a small portion of the variance. On the positive side, HECTD2 has been shown to influence susceptibility to human PrP prion disease. It is likely that drugs targeting only single target will not suffice for effective prevention or treatment of prion diseases. Further investigation of disease susceptibility and genetic analysis of effects on endophenotypes of the disease will be required for success.

REFERENCES

1. Prusiner SB. Novel proteinaceous infectious particles cause scrapie. *Science*. 1982;216:136–144.
2. Prusiner SB. Prions. *Proc Natl Acad Sci USA*. 1998;95:13363–13383.
3. Jucker M, Walker LC. Self-propagation of pathogenic protein aggregates in neurodegenerative diseases. *Nature*. 2013;501:45–51.
4. Holmes BB, Diamond MI. Cellular models for the study of prions. In: Prusiner SB, ed. *Prion Diseases*. Long Island, NY: Cold Spring Harbor Laboratory Press; 2017.
5. Dugger BN, Perl DP, Carlson GA. Studies of neurodegenerative disease transmission and transgenesis in mice. In: Prusiner SB, ed. *Prion Biology*. Long Island, NY: Cold Spring Harbor Laboratory Press; 2017.
6. Clavaguera F, Tolnay M, Goedert M. The prion-like behavior of assembled tau in transgenic mice. In: Prusiner SB, ed. *Prion Diseases*. Long Island, NY: Cold Spring Harbor Laboratory Press; 2016.
7. Woerman AL, Aoyagi A, Patel S, et al. Tau prions from Alzheimer's disease and chronic traumatic encephalopathy patients propagate in cultured cells. *Proc Natl Acad Sci USA*. 2016;113:E8187–E8196.
8. Jaunmuktane Z, Mead S, Ellis M, et al. Evidence for human transmission of amyloid-β pathology and cerebral amyloid angiopathy. *Nature*. 2015;525:247–250.
9. Schneider K, Fangerau H, Michaelsen B, Raab WH-M. The early history of the transmissible spongiform encephalopathies exemplified by scrapie. *Brain Res Bull*. 2008;77:343–355.
10. Cuillé J, Chelle P-L. La maladie dite tremblante du mouton est-elle inoculable? *C R Acad Sci*. 1936;203:1552–1554.
11. Cuillé J, Chelle P-L. Transmission experimentale de la tremblante a la chevre. *C R Acad Sci*. 1939;208:1058–1060.
12. Hadlow WJ. Scrapie and Kuru. *Lancet*. 1959;274:289–290.
13. Gajdusek DC, Gibbs Jr CJ, Alpers M. Experimental transmission of a Kuru-like syndrome to chimpanzees. *Nature*. 1966;209:794–796.
14. Gajdusek DC. Unconventional viruses and the origin and disappearance of Kuru. *Science*. 1977;197:943–960.

15. Brownell B, Campbell MJ, Greenham LW, Peacock DB. Experimental transmission of Creutzfeldt–Jakob disease. *Lancet*. 1975;306:186–187.
16. Tateishi J, Ohta M, Koga M, Sato Y, Kuroiwa Y. Transmission of chronic spongiform encephalopathy with Kuru plaques from humans to small rodents. *Ann Neurol*. 1979;5:581–584.
17. Prusiner SB, Groth DF, Cochran SP, Masiarz FR, McKinley MP, Martinez HM. Molecular properties, partial purification, and assay by incubation period measurements of the hamster scrapie agent. *Biochemistry*. 1980;21:4883–4891.
18. Bolton DC, McKinley MP, Prusiner SB. Identification of a protein that purifies with the scrapie prion. *Science*. 1982;218:1309–1311.
19. Prusiner SB, Groth DF, Bolton DC, Kent SB, Hood LE. Purification and structural studies of a major scrapie prion protein. *Cell*. 1984;38:127–134.
20. Oesch B, Westaway D, Wälchli M, et al. A cellular gene encodes scrapie PrP 27–30 protein. *Cell*. 1985;40:735–746.
21. Chesebro B, Race R, Wehrly K, et al. Identification of scrapie prion protein-specific mRNA in scrapie-infected and uninfected brain. *Nature*. 1985;315:331–333.
22. Carlson GA, Kingsbury DT, Goodman PA, et al. Linkage of prion protein and scrapie incubation time genes. *Cell*. 1986;46:503–511.
23. Carlson GA, Goodman PA, Lovett M, et al. Genetics and polymorphism of the mouse prion gene complex: control of scrapie incubation time. *Mol Cell Biol*. 1988;8:5528–5540.
24. Westaway D, Goodman PA, Mirenda CA, McKinley MP, Carlson GA, Prusiner SB. Distinct prion proteins in short and long scrapie incubation period mice. *Cell*. 1987;51:651–662.
25. Carlson GA, Ebeling C, Torchia M, Westaway D, Prusiner SB. Delimiting the location of the scrapie prion incubation time gene on chromosome 2 of the mouse. *Genetics*. 1993;133:979–988.
26. Moore RC, Hope J, McBride PA, et al. Mice with gene targeted prion protein alterations show that *Prnp*, *Sinc* and *Prni* are congruent. *Nat Genet*. 1998;18:118–125.
27. Dickinson AG, Mackay JM. Genetical control of the incubation period in mice of the neurological disease, scrapie. *Heredity*. 1964;19:279–288.
28. Dickinson AG, Meikle VMH. Host-genotype and agent effects in scrapie incubation: change in allelic interaction with different strains of agent. *Mol Gen Genet*. 1971;112:73–79.
29. Dickinson AG, Meikle VMH. A comparison of some biological characteristics of the mouse-passaged scrapie agents, 22A and ME7. *Genet Res*. 1969;13:213–225.
30. Bruce ME, Fraser H. Scrapie strain variation and its implications. *Curr Top Microbiol Immunol*. 1991;172:125–138.
31. Sparkes RS, Simon M, Cohn VH, et al. Assignment of the human and mouse prion protein genes to homologous chromosomes. *Proc Natl Acad Sci USA*. 1986;83:7358–7362.
32. Carlson GA, Westaway D, DeArmond SJ, Peterson-Torchia M, Prusiner SB. Primary structure of prion protein may modify scrapie isolate properties. *Proc Natl Acad Sci USA*. 1989;86:7475–7479.
33. Westaway D, Mirenda CA, Foster D, et al. Paradoxical shortening of scrapie incubation times by expression of prion protein transgenes derived from long incubation period mice. *Neuron*. 1991;7:59–68.
34. Carlson GA, Ebeling C, Yang S-L, et al. Prion isolate specified allotypic interactions between the cellular and scrapie prion proteins in congenic and transgenic mice. *Proc Natl Acad Sci USA*. 1994;91:5690–5694.
35. Zlotnik I, Rennie JC. Experimental transmission of mouse passaged scrapie to goats, sheep, rats and hamsters. *J Comp Pathol*. 1965;75:147–157.

36. Kimberlin RH, Walker CA. Pathogenesis of scrapie: agent multiplication in brain at the first and second passage of hamster scrapie in mice. *J Gen Virol*. 1979;42:107–117.
37. Prusiner SB, Scott M, Foster D, et al. Transgenetic studies implicate interactions between homologous PrP isoforms in scrapie prion replication. *Cell*. 1990;63:673–686.
38. Scott M, Foster D, Mirenda C, et al. Transgenic mice expressing hamster prion protein produce species-specific scrapie infectivity and amyloid plaques. *Cell*. 1989;59:847–857.
39. Büeler H, Fisher M, Lang Y, et al. Normal development and behaviour of mice lacking the neuronal cell-surface PrP protein. *Nature*. 1992;356:577–582.
40. Büeler H, Aguzzi A, Sailer A, et al. Mice devoid of PrP are resistant to scrapie. *Cell*. 1993;73:1339–1347.
41. Manson JC, Clarke AR, Hooper ML, Aitchison L, McConnell I, Hope J. 129/Ola mice carrying a null mutation in PrP that abolishes mRNA production are developmentally normal. *Mol Neurobiol*. 1994;8:121–127.
42. Sakaguchi S, Katamine S, Nishida N, et al. Loss of cerebellar Purkinje cells in aged mice homozygous for a disrupted PrP gene. *Nature*. 1996;380:528–531.
43. Moore R. *Gene Targeting Studies at the Mouse Prion Protein Locus* [PhD thesis] Edinburgh, Scotland: University of Edinburgh; 1997.
44. Moore RC, Lee IY, Silverman GL, et al. Ataxia in prion protein (PrP)-deficient mice is associated with upregulation of the novel PrP-like protein Doppel. *J Mol Biol*. 1999;292:797–817.
45. Lee IY, Westaway D, Smit AFA, et al. Complete genomic sequence and analysis of the prion protein gene region from three mammalian species. *Genome Res*. 1998;8:1022–1037.
46. Moore RC, Mastrangelo P, Bouzamondo E, et al. Doppel-induced cerebellar degeneration in transgenic mice. *Proc Natl Acad Sci USA*. 2001;98:15288–15293.
47. Minikel EV, Vallabh SM, Lek M, et al. Quantifying prion disease penetrance using large population control cohorts. *Sci Transl Med*. 2016;8: 322ra329.
48. Büeler H, Raeber A, Sailer A, Fischer M, Aguzzi A, Weissmann C. High prion and PrP^{Sc} levels but delayed onset of disease in scrapie-inoculated mice heterozygous for a disrupted PrP gene. *Mol Med*. 1994;1:19–30.
49. Brandner S, Isenmann S, Raeber A, et al. Normal host prion protein necessary for scrapie-induced neurotoxicity. *Nature*. 1996;379:339–343.
50. Steele AD, Lindquist S, Aguzzi A. The prion protein knockout mouse: a phenotype under challenge. *Prion*. 2007;1:83–93.
51. de Almeida CJG, Chiarini LB, da Silva JP, PMR ES, Martins MA, Linden R. The cellular prion protein modulates phagocytosis and inflammatory response. *J Leukoc Biol*. 2005;77:238–246.
52. Nuvolone M, Kana V, Hutter G, et al. SIRPa polymorphisms, but not the prion protein, control phagocytosis of apoptotic cells. *J Exp Med*. 2013;210:2539–2552.
53. Nuvolone M, Hermann M, Sorce S, et al. Strictly co-isogenic C57BL/6J-*Prnp*−/− mice: a rigorous resource for prion science. *J Exp Med*. 2016;213:313–327.
54. Bremer J, Baumann F, Tiberi C, et al. Axonal prion protein is required for peripheral myelin maintenance. *Nat Neurosci*. 2010;13:310–318.
55. Küffer A, Lakkaraju AKK, Mogha A, et al. The prion protein is an agonistic ligand of the G protein–coupled receptor Adgrg6. *Nature*. 2016;536:464–468.
56. Nakajima PB, Di Vincenzo JP, Jameson SC, Gascoigne NRJ. Chromosome 14 in B10. A (18R) mice is recombinant and includes Tcr V-alpha[a] alleles. *Immunogenetics*. 1992;35:190–198.
57. Hsiao K, Baker HF, Crow TJ, et al. Linkage of a prion protein missense variant to Gerstmann–Sträussler syndrome. *Nature*. 1989;338:342–345.

58. Hsiao KK, Scott M, Foster D, Groth DF, DeArmond SJ, Prusiner SB. Spontaneous neurodegeneration in transgenic mice with mutant prion protein. *Science.* 1990;250:1587–1590.
59. Hsiao KK, Groth D, Scott M, et al. Genetic and transgenic studies of prion proteins in Gerstmann-Straussler-Scheinker disease. In: Prusiner SB, Collinge J, Powell J, Anderton B, eds. *Prion Diseases of Humans and Animals.* London: Ellis Horwood Ltd; 1992.
60. Brown P, Gajdusek DC, Gibbs Jr CJ, Asher DM. Potential epidemic of Creutzfeldt–Jakob disease from human growth hormone therapy. *N Engl J Med.* 1985;313:728–731.
61. Hsiao K, Meiner Z, Kahana E, et al. Mutation of the prion protein in Libyan Jews with Creutzfeldt–Jakob disease. *N Engl J Med.* 1991;324:1091–1097.
62. Medori R, Tritschler H-J, LeBlanc A, et al. Fatal familial insomnia, a prion disease with a mutation at codon 178 of the prion protein gene. *N Engl J Med.* 1992;326: 444–449.
63. Bruce M, Chree A, McConnell I, Foster J, Pearson G, Fraser H. Transmission of bovine spongiform encephalopathy and scrapie to mice: strain variation and the species barrier. *Philos Trans R Soc Lond B Biol Sci.* 1994;343:405–411.
64. Bruce ME, Will RG, Ironside JW, et al. Transmissions to mice indicate that 'new variant' CJD is caused by the BSE agent. *Nature.* 1997;389:498–501.
65. Chandler RL, Turfrey BA. Inoculation of voles, Chinese hamsters, gerbils and guinea-pigs with scrapie brain material. *Res Vet Sci.* 1972;13:219–224.
66. Di Bari MA, Chianini F, Vaccari G, et al. The bank vole (*Myodes glareolus*) as a sensitive bioassay for sheep scrapie. *J Gen Virol.* 2008;89:2975–2985.
67. Nonno R, Di Bari MA, Cardone F, et al. Efficient transmission and characterization of Creutzfeldt–Jakob disease strains in bank voles. *PLoS Pathog.* 2006;2: e12.
68. Pirisinu L, Di Bari MA, D'Agostino C, et al. Gerstmann–Sträussler–Scheinker disease subtypes efficiently transmit in bank voles as genuine prion diseases. *Sci Rep.* 2016;6: 20443.
69. Watts JC, Giles K, Stöhr J, et al. Spontaneous generation of rapidly transmissible prions in transgenic mice expressing wild-type bank vole prion protein. *Proc Natl Acad Sci USA.* 2012;109:3498–3503.
70. Watts JC, Giles K, Patel S, Oehler A, DeArmond SJ, Prusiner SB. Evidence that bank vole PrP is a universal acceptor for prions. *PLoS Pathog.* 2014;10: e1003990.
71. Cosseddu GM, Nonno R, Vaccari G, et al. Ultra-efficient PrP^{Sc} amplification highlights potentialities and pitfalls of PMCA technology. *PLoS Pathog.* 2011;7: e1002370.
72. Orrú CD, Groveman BR, Raymond LD, et al. Bank vole prion protein as an apparently universal substrate for RT-QuIC-based detection and discrimination of prion strains. *PLoS Pathog.* 2015;11: e1004983.
73. Collinge J, Palmer MS, Dryden AJ. Genetic predisposition to iatrogenic Creutzfeldt–Jakob disease. *Lancet.* 1991;337:1441–1442.
74. Palmer MS, Dryden AJ, Hughes JT, Collinge J. Homozygous prion protein genotype predisposes to sporadic Creutzfeldt–Jakob disease. *Nature.* 1991;352:340–342.
75. Kaski D, Mead S, Hyare H, et al. Variant CJD in an individual heterozygous for *PRNP* codon 129. *Lancet.* 2009;374:2128.
76. Diack AB, Head MW, McCutcheon S, et al. Variant CJD: 18 years of research and surveillance. *Prion.* 2014;8(3):1–10.
77. Bessen RA, Marsh RF. Identification of two biologically distinct strains of transmissible mink encephalopathy in hamsters. *J Gen Virol.* 1992;73:329–334.
78. Bessen RA, Marsh RF. Distinct PrP properties suggest the molecular basis of strain variation in transmissible mink encephalopathy. *J Virol.* 1994;68:7859–7868.

79. Bessen RA, Kocisko DA, Raymond GJ, Nandan S, Lansbury PT, Caughey B. Non-genetic propagation of strain-specific properties of scrapie prion protein. *Nature*. 1995;375:698–700.
80. Telling GC, Parchi P, DeArmond SJ, et al. Evidence for the conformation of the pathologic isoform of the prion protein enciphering and propagating prion diversity. *Science*. 1996;274:2079–2082.
81. Goldfarb LG, Petersen RB, Tabaton M, et al. Fatal familial insomnia and familial Creutzfeldt–Jakob disease: disease phenotype determined by a DNA polymorphism. *Science*. 1992;258:806–808.
82. Monari L, Chen SG, Brown P, et al. Fatal familial insomnia and familial Creutzfeldt–Jakob disease: different prion proteins determined by a DNA polymorphism. *Proc Natl Acad Sci USA*. 1994;91:2839–2842.
83. Bruce ME, McConnell I, Fraser H, Dickinson AG. The disease characteristics of different strains of scrapie in *Sinc* congenic mouse lines: implications for the nature of the agent and host control of pathogenesis. *J Gen Virol*. 1991;72:595–603.
84. Prusiner SB. Prion diseases and the BSE crisis. *Science*. 1997;278:245–251.
85. Telling GC, Scott M, Hsiao KK, et al. Transmission of Creutzfeldt–Jakob disease from humans to transgenic mice expressing chimeric human-mouse prion protein. *Proc Natl Acad Sci USA*. 1994;91:9936–9940.
86. Collinge J, Palmer MS, Sidle KCL, et al. Transmission of fatal familial insomnia to laboratory animals (Lett.). *Lancet*. 1995;346:569–570.
87. Collinge J, Whitfield J, McKintosh E, et al. Kuru in the 21st century—an acquired human prion disease with very long incubation periods. *Lancet*. 2006;367:2068–2074.
88. Mead S, Stumpf MP, Whitfield J, et al. Balancing selection at the prion protein gene consistent with prehistoric Kurulike epidemics. *Science*. 2003;300:640–643.
88a. Mead S, Whitfield J, Poulter M, et al. A novel protective prion protein variant that co-localizes with kuru exposure. *New Engl J Med*. 2009;361:2056–2061.
89. Asante EA, Smidak M, Grimshaw A, et al. A naturally occurring variant of the human prion protein completely prevents prion disease. *Nature*. 2015;522:478–481.
90. Li J, Browning S, Mahal SP, Oelschlegel AM, Weissmann C. Darwinian evolution of prions in cell culture. *Science*. 2010;327:869–872.
91. Oelschlegel AM, Weissmann C. Acquisition of drug resistance and dependence by prions. *PLoS Pathog*. 2013;9: e1003158.
92. Berry DB, Lu D, Geva M, et al. Drug resistance confounding prion therapeutics. *Proc Natl Acad Sci USA*. 2013;110:E4160–E4169.
93. de Calignon A, Polydoro M, Suarez-Calvet M, et al. Propagation of tau pathology in a model of early Alzheimer's disease. *Neuron*. 2012;73:685–697.
94. Olanow CW, Prusiner SB. Is Parkinson's disease a prion disorder? *Proc Natl Acad Sci USA*. 2009;106:12571–12572.
95. Watts JC, Giles K, Oehler A, et al. Transmission of multiple system atrophy prions to transgenic mice. *Proc Natl Acad Sci USA*. 2013;110:19555–19560.
96. Woerman AL, Stöhr J, Aoyagi A, et al. Propagation of prions causing synucleinopathies in cultured cells. *Proc Natl Acad Sci USA*. 2015;112:E4949–E4958.
97. Woerman AL, Watts JC, Aoyagi A, Giles K, Middleton LT, Prusiner SB. α-Synuclein: multiple system atrophy prions. In: Prusiner SB, ed. *Prion Diseases*. Long Island, NY: Cold Spring Harbor Laboratory Press; 2016.
98. Luk KC, Kehm VM, Zhang B, O'Brien P, Trojanowski JQ, Lee VMY. Intracerebral inoculation of pathological α-synuclein initiates a rapidly progressive neurodegenerative α-synucleinopathy in mice. *J Exp Med*. 2012;209:975–986.
99. Clavaguera F, Bolmont T, Crowther RA, et al. Transmission and spreading of tauopathy in transgenic mouse brain. *Nat Cell Biol*. 2009;11:909–913.

100. Prusiner SB. Biology and genetics of prions causing neurodegeneration. *Annu Rev Genet*. 2013;47:601–623.
101. Collinge J. Mammalian prions and their wider relevance in neurodegenerative diseases. *Nature*. 2016;539:217–226.
102. Watts JC, Condello C, Stöhr J, et al. Serial propagation of distinct strains of Aβ prions from Alzheimer's disease patients. *Proc Natl Acad Sci USA*. 2014;111:10323–10328.
103. Dickinson AG. Host–pathogen interactions in scrapie. *Genetics*. 1975;79(suppl):387–395.
104. Stephenson DA, Chiotti K, Ebeling C, et al. Quantitative trait loci affecting prion incubation time in mice. *Genomics*. 2000;69:47–53.
105. Lloyd SE, Onwuazor ON, Beck JA, et al. Identification of multiple quantitative trait loci linked to prion disease incubation period in mice. *Proc Natl Acad Sci USA*. 2001;98:6279–6283.
106. Manolakou K, Beaton J, McConnell I, et al. Genetic and environmental factors modify bovine spongiform encephalopathy incubation period in mice. *Proc Natl Acad Sci USA*. 2001;98:7402–7407.
107. Moreno CR, Lantier F, Lantier I, Sarradin P, Elsen J-M. Detection of new quantitative trait loci for susceptibility to transmissible spongiform encephalopathies in mice. *Genetics*. 2003;165:2085–2091.
108. Mott R, Flint J. Simultaneous detection and fine mapping of quantitative trait loci in mice using heterogeneous stocks. *Genetics*. 2002;160:1609–1618.
109. Svenson KL, Gatti DM, Valdar W, et al. High-resolution genetic mapping using the mouse diversity outbred population. *Genetics*. 2012;190:437–447.
110. Lloyd SE, Maytham EG, Pota H, et al. *HECTD2* is associated with susceptibility to mouse and human prion disease. *PLoS Genet*. 2009;5: e1000383.
111. Tamgüney G, Giles K, Glidden DV, et al. Genes contributing to prion pathogenesis. *J Gen Virol*. 2008;89:1777–1788.
112. Holmes BB, Diamond MI. Prion-like properties of tau protein: the importance of extracellular tau as a therapeutic target. *J Biol Chem*. 2014;289:19855–19861.
113. Sanders DW, Kaufman SK, DeVos SL, et al. Distinct tau prion strains propagate in cells and mice and define different tauopathies. *Neuron*. 2014;82:1271–1288.

CHAPTER SEVEN

The Prion Concept and Synthetic Prions

Giuseppe Legname[*,1], **Fabio Moda**[†,1]
[*]Scuola Internazionale Superiore di Studi Avanzati (SISSA), Trieste, Italy
[†]IRCCS Foundation Carlo Besta Neurological Institute, Milan, Italy
[1]Corresponding authors: e-mail address: legname@sissa.it; fabio.moda@istituto-besta.it

Contents

Abstract

Transmissible spongiform encephalopathies or prion diseases are a group of fatal neurodegenerative diseases caused by unconventional infectious agents, known as prions (PrP^{Sc}). Prions derive from a conformational conversion of the normally folded prion protein (PrP^{C}), which acquires pathological and infectious features. Moreover, PrP^{Sc} is able to transmit the pathological conformation to PrP^{C} through a mechanism that is still not well understood. The generation of synthetic prions, which behave like natural prions, is of fundamental importance to study the process of PrP^{C} conversion and to assess the efficacy of therapeutic strategies to interfere with this process. Moreover, the ability of synthetic prions to induce pathology in animals confirms that the pathological properties of the prion strains are all enciphered in abnormal conformations, characterizing these infectious agents.

1. THE PRION CONCEPT

Prion diseases or transmissible spongiform encephalopathies (TSEs) are devastating disorders which include scrapie in sheep,[1] bovine spongiform encephalopathy (BSE) in cattle,[2] chronic wasting disease in cervids,[3] transmissible mink encephalopathy in mink,[4] and Creutzfeldt–Jakob disease

Progress in Molecular Biology and Translational Science, Volume 150
ISSN 1877-1173
http://dx.doi.org/10.1016/bs.pmbts.2017.06.002

(CJD)[5] in humans. These diseases have sporadic or genetic origins and can be acquired by infectious route. One of the most important features of prions is their ability to be transmissible among the same species or between different species. Zoonotic transmission of BSE to humans was responsible for the emergence of a variant form of Creutzfeldt–Jakob disease (vCJD),[6] while cases of prion transmission between humans were also reported. In particular, the use of contaminated surgical instruments, or administration/transplant of infectious samples (dura mater grafts, growth hormone, etc.) to healthy subjects transmitted iatrogenic CJD,[7–9] while blood transfusions were found to efficiently transmit vCJD among humans.[10–15]

The first case of prion disease in animals was reported in 1732 in Lincolnshire in sheep affected by scrapie,[16] but the origin of this disease might date back even earlier. Initially, these types of diseases were thought to have a viral nature, but an accidental event led to the discovery that the infectious agent could not be a virus. Indeed, sheep treated with a vaccine against a common sheep virus, which was inadvertently prepared from formalin-fixed brain tissue collected from a scrapie-infected animal, start to develop typical scrapie pathology.

Subsequently, to confirm this observation, experiments of scrapie transmission were performed[17] either within the same species[18] or between different species, including common laboratory animals (mice).[19] In both cases, the infectious agent was efficiently transmitted, thus generating many hypothesis on its unique nature: (i) a subvirus,[20] (ii) a self-replicating membrane,[20,21] and (iii) a viroid or a retrovirus-like element,[22] but none of them were completely proved.

While studying different cases of atypical encephalitis occurring in the endogenous Fore population in Papua New Guinea, Carleton Gajdusek discovered that these diseases could be efficiently transmitted among the population through cannibalistic rituals.[23] These diseases were later found to be a prion disorder named kuru.

While comparing kuru and scrapie pathology, William Hadlow came to the conclusion that both diseases could have been caused by the same infectious agent. To better understanding this hypothesis, kuru was inoculated in chimpanzee and surprisingly induced pathology which was characterized by a long incubation time.[24] For this reason, the infectious agent was initially thought to be a slow virus,[25] but its resistance to UV treatment and ionizing radiations[26] ruled out this possibility. Subsequent studies have tried to support the hypothesis that the nucleic acid could be enveloped by a protein coat (virino) able to protect the genetic materials from denaturing agents.[27]

However, also in this case, the association of TSE infectivity with the presence of nucleic acid has never been demonstrated. Finally, the Nobel prize laureate Stanley B. Prusiner showed that the infectious agent was an abnormally folded protein that he named *prion* for *proteinaceous infectious particle*, which is able to self-replicate in the body[28] and formulated the so-called protein-only hypothesis.[29] The latter introduces prions as uncommon infectious agent able to transmit the disease. Further studies revealed that prions were abnormally folded and protease-resistant forms of the physiological cellular form of the prion protein (PrP^C), also known as PrP^{Sc}.[30] The discovery that the level of infectivity was reduced using either agents that destroy all the protein structures or anti-PrP antibodies[31] further supported the protein-only hypothesis, which is now well accepted in the prion field. In 1985, the gene that encodes for prion protein was identified (*PRNP*),[32] and genetic studies revealed that mutations in this gene were able to induce prion pathology. For instance, the Gerstmann–Sträussler–Scheinker syndrome was the first type of genetic prion disease caused by a point mutation (P102L) in the *PRNP* gene.[33] Additional studies revealed the presence of different mutations in *PRNP* which were associated with other genetic forms of prion disorders.[34] In vivo experiments revealed that transgenic mouse models devoid of PrP were not susceptible to prion disease, thus showing that the presence of PrP^C is sufficient to sustain prion pathology.[35]

2. NATURAL AND RECOMBINANT PrP CAN MISFOLD IN VITRO

Conversely to PrP^C, PrP^{Sc} is rich in β-sheet structures, and its partial proteolytic digestion leads to the formation of a PK-resistant core, PrP27–30, that can form amyloid fibrils. Notably, the incubation of PrP^C with two different prion strains generated two distinct types PrP^{res}.[36] In most cases, in vitro generated PrP^{res} failed to induce any prion pathology after inoculation in susceptible animals and confirmed that the acquisition of proteinase resistance was not enough for the propagation of infectivity.[37–40] Subsequent studies revealed that prions can be synthetically produced and can acquire infectious properties. For instance, intracerebral inoculation of a synthetic PrP peptide, carrying the P101L mutation and refolded into a β-sheet structure, into transgenic mice expressing the mutant PrP(P101L) induced severe prion pathology after about 200 days after injection.[41,42] Surprisingly, these animals developed a unique type of prion pathology characterized by the presence of PK-sensitive PrP^{Sc}.[43] This disease was not transmissible to wild-type animals.[44] The injection of peptides with the same sequence but lacking the β-sheet structures did not induce any pathological alteration. Therefore, these studies revealed that PrP^C can be induced to misfold in vitro under specific experimental conditions, but such abnormal conformations are not always associated to infectivity.[45]

3. SYNTHETIC PRIONS ARE INFECTIOUS

Experiments performed with the recombinant prion protein (recPrP) revealed that, under specific experimental conditions, the protein misfolds and aggregates to form amyloid structures. Using denaturing conditions, recPrP corresponding to the PrP27–30 (residues 89–230 of mouse sequence) was induced to misfold and acquired two different structures: β-oligomers (in the presence of acidic pH and urea) or fibrillar structures (neutral pH and low concentration of urea), the formation of which was measured by thioflavin-T.[37] The addition of preformed amyloid assemblies (referred to as seeds) to the fibrillization reaction considerably reduced the time of the aggregation, thus showing that the formation of fibrils can be promoted by seeding. To assess the infectivity of these structures, transgenic mice (Tg9949) overexpressing the N-terminally truncated mouse PrP (residues 89–230) were intracerebrally inoculated with (i) prefolded amyloid fibrils (unseeded) and (ii) seeded amyloid fibrils. In both cases, animals succumbed to prion disease but with different neuropathological alterations.[46] Particularly, the second group of mice exhibited shorter incubation time than the first one and contained more PrP^{res}. This isolate was designated as MoSP1 and, conversely to the peptide inoculated into P101L mice, the new isolate could be passaged to wild-type mice and acquired protease resistance. One of the most important observations, which was gathered from these experiments, was the fact that the stability of a specific prion isolate and the length of incubation time after inoculation in susceptible animals were directly correlated.[47,48] In other words, the most stable isolate induces the longest incubation times in animals, while unstable isolates accumulate more rapidly in the brain and kill the host faster. In order to better characterize the relationship between conformation, stability, and infectivity, several laboratories started to produce synthetic prions characterized by different conformations (by varying the biochemical conditions used for their formation, including pH, urea concentration, and temperature).[48] These amyloids were then challenged in transgenic mice and revealed that not all the amyloids were infectious and those which were infectious induced neuropathological changes characterized by unique pattern of PrP deposition, revealing that each inoculum was a unique isolate. The amyloids that were not able to produce any prion pathology were characterized by the lowest conformational stability and, likely, they might have been destroyed readily by proteases. Furthermore, these data support that the infectious properties are dictated by the conformation of the prion protein itself and the N-linked glycosylation seems to have a marginal role.[49–52] This concept is of fundamental importance for developing efficient therapeutic strategies to prevent intra- and interspecies prion transmission. Unfortunately, the effectiveness of different antiprion compounds seems to be closely related to particular strains of PrP^{Sc}, as no universal inhibitor for all strains exists.[53,54]

4. SYNTHETIC PRIONS UNDERGO A PROCESS OF ADAPTATION

One phenomenon recently described is the ability of prions to adapt to specific environments through a Darwinian process: the most aggressive prion strain generally dominates.[55–58] This is often observed when prions are transmitted to different host species, although their molecular mechanism still needs to be determined. Two hypothesis were made so far: (i) prions are a mixture of strains and adaptation appears when the more aggressive amplifies to sufficient titers to appear within the host or (ii) the adaptation may result from conformational changes during prion replication. An example of this phenomenon was observed when MoSP1 strain was serially transmitted to wild-type mice and revealed the presence of at least of two different prion isolates named: MoSP1(1) and MoSP1(2).[59,60] Both strains were characterized by two different molecular masses of the protease-resistant core, 21 and 19 kDa, respectively.[61] When inoculated in cells, MoSP1(1) and MoSP1(2) were able to replicate with high fidelity. However, when cells were infected with a mixture of both strains, MoSP1(1) preferentially replicates over MoSP1(2) which slowly disappeared. Similar results were obtained after inoculation of MoSP1(1) and MoSP1(2) in animals. Surprisingly, mice challenged with MoSP1(2) still contained small amount of MoSP1(1). Thus, MoSP1(1) replicates faster than MoSP1(2), and this isolate might be more effective at interacting with PrP^{C} (or other factors) competing with MoSP1(2) for the substrates required for its propagation. More recent evidences show that mammalian[56,62] and yeast prions[63] can adapt in response to environmental pressure.

5. ALTERNATIVE APPROACHES TO GENERATE SYNTHETIC PRIONS

An alternative approach to produce synthetic prions was based on a process named *annealing*. In particular, full-length hamster PrP was converted into amyloid fibrils which were incubated with normal hamster brain homogenate for 5 cycles alternating 1 min of incubation at 80°C and 1 min of incubation at 37°C (annealing).[64] Western blot analysis of the final products revealed the presence of a PK-resistant band migrating at around 16 kDa. Inoculation of this material into the brain of hamsters was able to produce a new type of prion disease characterized by a very long incubation time. The newly generated synthetic prion was named Synthetic Strain Leading to OverWeight (SSLOW). Another strategy to generate synthetic prion was based on a technique developed by Claudio Soto's group named Protein Misfolding Cyclic Amplification (PMCA).[65] In this case infectious PrP^{Sc} is mixed with an excess of PrP^{C}. Cycles of incubation and sonication are repeated in a cyclic manner. The sonication disaggregates PrP^{Sc} fibrils, thus generating different seeds that can recruit and convert normal PrP^{C}. Using this technique[66] it was shown that although not absolutely required, the replication of infectious prions was significantly increased by the presence of polyanions, such as nucleic acids, glycosaminoglycans, and chaperone proteins.[67–69] Moreover, after several rounds of PMCA using recPrP that has never been exposed to any prion, a de novo generation of PrP^{res} was observed. In particular, two types of self-perpetuating PrP^{res} characterized by two different PK-resistant cores of 17 and 14 kDa, respectively, were generated.[70] However, only PrP^{res} of 17 kDa was highly infectious to mice. Another approach to produce infectious prion using PMCA was shown in 2009 by Barria and collaborators.[71] In particular, through serial rounds of PMCA, PrP^{Sc} was generated in the absence of preexisting PrP^{Sc}. The de novo generated PrP^{Sc} was infectious when inoculated into wild-type hamsters, producing a new disease phenotype with unique neuropathological features. Through PMCA, full-length mouse recombinant PrP was able to produce a new prion strain using different combination of synthetic anionic phospholipid and RNA. The generated PrP^{Sc} was infectious, and the infectivity was maintained after serial passages in wild-type mice.[72] Surprisingly, even after replacing RNA with a synthetic polyriboadenylic acid, infectious prions could be generated, thus indicating that RNA does not contain specific information that might facilitate the conversion of PrP^{C} in its infectious isoform.[73] All these data confirm that there are several approaches able to efficiently generate infectious prion in vitro and provide important evidence in support of the protein-only hypothesis.

6. CONCLUDING REMARKS

We have recently performed a study aimed at verifying the presence of any correlation between the conformational diversity of misfolded recPrP and their pathological activities. In particular, using a precise mixture of common chemicals, we have generated a large amount of misfolded recPrP characterized by different conformational structures. Neither infected brain extracts nor amplified PrP^{Sc} was used for this analysis. Atomic force microscopy assessment indicated that many preparations had the typical amyloid shape, while other preparations showed a more spherical conformation, similar to β-oligomers. Many of these amyloids were infectious when challenged in mouse hypothalamic GT1 and neuroblastoma N2a cell lines[74] and produced a PK-resistant PrP. The most infectious preparations were inoculated in wild-type mice but failed to induce any prion pathology. The same preparations were subjected to several rounds of PMCA, and one of them was able to amplify and produce a PK-resistant PrP. The latter was then inoculated in wild-type mice and induced a severe prion pathology with a 100% attack rate. Through PMCA we were able to detect the prion isolate in the blood (plasma fraction) of symptomatic animals as earlier as 140 days postinoculation. Therefore, we have defined a strategy to generate putative infectious material in vitro using solely recPrP, common chemicals, and controlled and well-defined biophysical and biochemical conditions, without employing any external source of PrP^{Sc}. Working with natural prions is not always simple since they are complex and heterogeneous. Synthetic prions seem easier to control than natural strains, being more homogeneous and structurally defined. These features can be exploited to (i) study different aspects of prion biology, (ii) analyze the pathological mechanisms which drive the process of misfolding, (iii) study the efficacy of specific therapeutic compounds to interfere with this pathological process, and finally (iv) develop personalized treatments for each patient.

REFERENCES

1. Wood JL, Lund LJ, Done SH. The natural occurrence of scrapie in moufflon. *Vet Rec.* 1992;130(2):25–27.
2. Wells GA, Scott AC, Johnson CT, et al. A novel progressive spongiform encephalopathy in cattle. *Vet Rec.* 1987;121(18):419–420.

3. Williams ES, Young S. Spongiform encephalopathy of Rocky Mountain elk. *J Wildl Dis.* 1982;18(4):465–471.
4. Barlow RM. Transmissible mink encephalopathy: pathogenesis and nature of the aetiological agent. *J Clin Pathol Suppl.* 1972;6:102–109.
5. Gibbs Jr CJ, Gajdusek DC, Asher DM, et al. Creutzfeldt-Jakob disease (spongiform encephalopathy): transmission to the chimpanzee. *Science.* 1968;161(3839):388–389.
6. Bruce ME, Will RG, Ironside JW, et al. Transmissions to mice indicate that 'new variant' CJD is caused by the BSE agent. *Nature.* 1997;389(6650):498–501.
7. Weissmann C, Enari M, Klohn PC, Rossi D, Flechsig E. Transmission of prions. *Proc Natl Acad Sci USA.* 2002;99(suppl 4):16378–16383.
8. Andreoletti O, Litaise C, Simmons H, et al. Highly efficient prion transmission by blood transfusion. *PLoS Pathog.* 2012;8(6): e1002782.
9. Denkers ND, Seelig DM, Telling GC, Hoover EA. Aerosol and nasal transmission of chronic wasting disease in cervidized mice. *J Gen Virol.* 2010;91(pt 6):1651–1658.
10. Davidson LR, Llewelyn CA, Mackenzie JM, Hewitt PE, Will RG. Variant CJD and blood transfusion: are there additional cases? *Vox Sang.* 2014;107(3):220–225.
11. Hewitt PE, Llewelyn CA, Mackenzie J, Will RG. Creutzfeldt-Jakob disease and blood transfusion: results of the UK Transfusion Medicine Epidemiological Review study. *Vox Sang.* 2006;91(3):221–230.
12. Turner ML, Ironside JW. New-variant Creutzfeldt-Jakob disease: the risk of transmission by blood transfusion. *Blood Rev.* 1998;12(4):255–268.
13. Diack AB, Head MW, McCutcheon S, et al. Variant CJD: 18 years of research and surveillance. *Prion.* 2014;8(4):286–295.
14. Concha-Marambio L, Pritzkow S, Moda F, et al. Detection of prions in blood from patients with variant Creutzfeldt-Jakob disease. *Sci Transl Med.* 2016;8(370): 370ra183.
15. Bougard D, Brandel JP, Belondrade M, et al. Detection of prions in the plasma of presymptomatic and symptomatic patients with variant Creutzfeldt-Jakob disease. *Sci Transl Med.* 2016;8(370):370ra182.
16. Liberski PP. Historical overview of prion diseases: a view from afar. *Folia Neuropathol.* 2012;50(1):1–12.
17. Aad G, Abajyan T, Abbott B, et al. Measurement of top quark polarization in top-antitop events from proton-proton collisions at radicals = 7 TeV using the ATLAS detector. *Phys Rev Lett.* 2013;111(23):232002.
18. Aaij R, Adeva B, Adinolfi M, et al. Precision measurement of the Lambdab(0) baryon lifetime. *Phys Rev Lett.* 2013;111(10):102003.
19. Chandler RL. Encephalopathy in mice produced by inoculation with scrapie brain material. *Lancet.* 1961;1(7191):1378–1379.
20. Adams DH. The nature of the scrapie agent. A review of recent progress. *Pathol Biol.* 1970;18(9):559–577.
21. Gibbons RA, Hunter GD. Nature of the scrapie agent. *Nature.* 1967;215(5105): 1041–1043.
22. Bastian FO. Spiroplasma as a candidate agent for the transmissible spongiform encephalopathies. *J Neuropathol Exp Neurol.* 2005;64(10):833–838.
23. Liberski PP. Kuru and D. Carleton Gajdusek: a close encounter. *Folia Neuropathol.* 2009;47(2):114–137.
24. Gajdusek DC, Gibbs CJ, Alpers M. Experimental transmission of a Kuru-like syndrome to chimpanzees. *Nature.* 1966;209(5025):794–796.
25. Cho HJ. Is the scrapie agent a virus? *Nature.* 1976;262(5567):411–412.
26. Alper T, Cramp WA, Haig DA, Clarke MC. Does the agent of scrapie replicate without nucleic acid? *Nature.* 1967;214(5090):764–766.
27. Kimberlin RH. Scrapie agent: prions or virinos? *Nature.* 1982;297(5862):107–108.
28. Griffith JS. Self-replication and scrapie. *Nature.* 1967;215(5105):1043–1044.

29. Prusiner SB. Novel proteinaceous infectious particles cause scrapie. *Science*. 1982;216(4542):136–144.
30. Bolton DC, McKinley MP, Prusiner SB. Identification of a protein that purifies with the scrapie prion. *Science*. 1982;218(4579):1309–1311.
31. Gabizon R, McKinley MP, Groth D, Prusiner SB. Immunoaffinity purification and neutralization of scrapie prion infectivity. *Proc Natl Acad Sci USA*. 1988;85(18):6617–6621.
32. Chesebro B, Race R, Wehrly K, et al. Identification of scrapie prion protein-specific mRNA in scrapie-infected and uninfected brain. *Nature*. 1985;315(6017):331–333.
33. Hsiao K, Baker HF, Crow TJ, et al. Linkage of a prion protein missense variant to Gerstmann-Straussler syndrome. *Nature*. 1989;338(6213):342–345.
34. Collinge J. Prion diseases of humans and animals: their causes and molecular basis. *Annu Rev Neurosci*. 2001;24:519–550.
35. Bueler H, Aguzzi A, Sailer A, et al. Mice devoid of PrP are resistant to scrapie. *Cell*. 1993;73(7):1339–1347.
36. Bessen RA, Kocisko DA, Raymond GJ, Nandan S, Lansbury PT, Caughey B. Non-genetic propagation of strain-specific properties of scrapie prion protein. *Nature*. 1995;375(6533):698–700.
37. Baskakov IV, Legname G, Baldwin MA, Prusiner SB, Cohen FE. Pathway complexity of prion protein assembly into amyloid. *J Biol Chem*. 2002;277(24):21140–21148.
38. Baskakov IV, Legname G, Prusiner SB, Cohen FE. Folding of prion protein to its native alpha-helical conformation is under kinetic control. *J Biol Chem*. 2001;276 (23):19687–19690.
39. Baskakov IV, Aagaard C, Mehlhorn I, et al. Self-assembly of recombinant prion protein of 106 residues. *Biochemistry*. 2000;39(10):2792–2804.
40. Hill AF, Antoniou M, Collinge J. Protease-resistant prion protein produced in vitro lacks detectable infectivity. *J Gen Virol*. 1999;80(pt 1):11–14.
41. Kaneko K, Ball HL, Wille H, et al. A synthetic peptide initiates Gerstmann-Straussler-Scheinker (GSS) disease in transgenic mice. *J Mol Biol*. 2000;295(4):997–1007.
42. Hsiao KK, Groth D, Scott M, et al. Serial transmission in rodents of neurodegeneration from transgenic mice expressing mutant prion protein. *Proc Natl Acad Sci USA*. 1994;91(19):9126–9130.
43. Nazor KE, Kuhn F, Seward T, et al. Immunodetection of disease-associated mutant PrP, which accelerates disease in GSS transgenic mice. *EMBO J*. 2005;24(13):2472–2480.
44. Tremblay P, Ball HL, Kaneko K, et al. Mutant PrPSc conformers induced by a synthetic peptide and several prion strains. *J Virol*. 2004;78(4):2088–2099.
45. Benetti F, Legname G. De novo mammalian prion synthesis. *Prion*. 2009;3(4):213–219.
46. Legname G, Baskakov IV, Nguyen HO, et al. Synthetic mammalian prions. *Science*. 2004;305(5684):673–676.
47. Peretz D, Scott MR, Groth D, et al. Strain-specified relative conformational stability of the scrapie prion protein. *Protein Sci*. 2001;10(4):854–863.
48. Colby DW, Giles K, Legname G, et al. Design and construction of diverse mammalian prion strains. *Proc Natl Acad Sci USA*. 2009;106(48):20417–20422.
49. Collinge J, Clarke AR. A general model of prion strains and their pathogenicity. *Science*. 2007;318(5852):930–936.
50. Collinge J, Sidle KC, Meads J, Ironside J, Hill AF. Molecular analysis of prion strain variation and the aetiology of 'new variant' CJD. *Nature*. 1996;383(6602):685–690.
51. DeArmond SJ, Sanchez H, Yehiely F, et al. Selective neuronal targeting in prion disease. *Neuron*. 1997;19(6):1337–1348.
52. Tuzi NL, Cancellotti E, Baybutt H, et al. Host PrP glycosylation: a major factor determining the outcome of prion infection. *PLoS Biol*. 2008;6(4): e100.
53. McKenzie D, Kaczkowski J, Marsh R, Aiken J. Amphotericin B delays both scrapie agent replication and PrP-res accumulation early in infection. *J Virol*. 1994;68(11): 7534–7536.

54. Supattapone S, Wille H, Uyechi L, et al. Branched polyamines cure prion-infected neuroblastoma cells. *J Virol.* 2001;75(7):3453–3461.
55. Bartz JC, Bessen RA, McKenzie D, Marsh RF, Aiken JM. Adaptation and selection of prion protein strain conformations following interspecies transmission of transmissible mink encephalopathy. *J Virol.* 2000;74(12):5542–5547.
56. Li J, Browning S, Mahal SP, Oelschlegel AM, Weissmann C. Darwinian evolution of prions in cell culture. *Science.* 2010;327(5967):869–872.
57. Makarava N, Baskakov IV. The evolution of transmissible prions: the role of deformed templating. *PLoS Pathog.* 2013;9(12): e1003759.
58. Baskakov IV. The many shades of prion strain adaptation. *Prion.* 2014;8(2):169–172.
59. Legname G, Nguyen HO, Baskakov IV, Cohen FE, Dearmond SJ, Prusiner SB. Strain-specified characteristics of mouse synthetic prions. *Proc Natl Acad Sci USA.* 2005;102(6):2168–2173.
60. Legname G, Nguyen HO, Peretz D, Cohen FE, DeArmond SJ, Prusiner SB. Continuum of prion protein structures enciphers a multitude of prion isolate-specified phenotypes. *Proc Natl Acad Sci USA.* 2006;103(50):19105–19110.
61. Ghaemmaghami S, Watts JC, Nguyen HO, Hayashi S, DeArmond SJ, Prusiner SB. Conformational transformation and selection of synthetic prion strains. *J Mol Biol.* 2011;413(3):527–542.
62. Ghaemmaghami S, Ahn M, Lessard P, et al. Continuous quinacrine treatment results in the formation of drug-resistant prions. *PLoS Pathog.* 2009;5(11): e1000673.
63. Chen B, Bruce KL, Newnam GP, Gyoneva S, Romanyuk AV, Chernoff YO. Genetic and epigenetic control of the efficiency and fidelity of cross-species prion transmission. *Mol Microbiol.* 2010;76(6):1483–1499.
64. Makarava N, Kovacs GG, Bocharova O, et al. Recombinant prion protein induces a new transmissible prion disease in wild-type animals. *Acta Neuropathol.* 2010;119(2):177–187.
65. Saborio GP, Permanne B, Soto C. Sensitive detection of pathological prion protein by cyclic amplification of protein misfolding. *Nature.* 2001;411(6839):810–813.
66. Deleault NR, Harris BT, Rees JR, Supattapone S. Formation of native prions from minimal components in vitro. *Proc Natl Acad Sci USA.* 2007;104(23):9741–9746.
67. Gonzalez-Montalban N, Lee YJ, Makarava N, Savtchenko R, Baskakov IV. Changes in prion replication environment cause prion strain mutation. *FASEB J.* 2013;27(9): 3702–3710.
68. Geoghegan JC, Valdes PA, Orem NR, et al. Selective incorporation of polyanionic molecules into hamster prions. *J Biol Chem.* 2007;282(50):36341–36353.
69. Wang F, Yang F, Hu Y, et al. Lipid interaction converts prion protein to a PrPSc-like proteinase K-resistant conformation under physiological conditions. *Biochemistry.* 2007;46(23):7045–7053.
70. Zhang Z, Zhang Y, Wang F, et al. De novo generation of infectious prions with bacterially expressed recombinant prion protein. *FASEB J.* 2013;27(12):4768–4775.
71. Barria MA, Mukherjee A, Gonzalez-Romero D, Morales R, Soto C. De novo generation of infectious prions in vitro produces a new disease phenotype. *PLoS Pathog.* 2009;5(5): e1000421.
72. Wang F, Wang X, Yuan CG, Ma J. Generating a prion with bacterially expressed recombinant prion protein. *Science.* 2010;327(5969):1132–1135.
73. Wang F, Zhang Z, Wang X, et al. Genetic informational RNA is not required for recombinant prion infectivity. *J Virol.* 2012;86(3):1874–1876.
74. Bosque PJ, Prusiner SB. Cultured cell sublines highly susceptible to prion infection. *J Virol.* 2000;74(9):4377–4386.

CHAPTER EIGHT

Gene Targeted Transgenic Mouse Models in Prion Research

Abigail B. Diack*[,1], James D. Alibhai[†], Jean C. Manson*

*The Roslin Institute & R(D)SVS, University of Edinburgh, Easter Bush, United Kingdom
[†]The National CJD Research and Surveillance Unit, Centre for Clinical Brain Sciences, University of Edinburgh, Edinburgh, United Kingdom
[1]Corresponding author: e-mail address: abigail.diack@roslin.ed.ac.uk

Contents

Abstract

The production of transgenic mice expressing different forms of the prion protein (PrP) or devoid of PrP has enabled researchers to study the role of PrP in the infectious process of a prion disease and its normal function in the healthy individual. A wide range of transgenic models have been produced ranging from PrP null mice, normal expression levels to overexpression models, models expressing different species of the *Prnp* gene and different mutations and polymorphisms within the gene. Using this range of transgenic models has allowed us to define the influence of PrP expression on disease susceptibility and transmission, assess zoonotic potential, define strains of human prion diseases, elucidate the function of PrP, and start to unravel the mechanisms involved in chronic neurodegeneration. This chapter focuses mainly on the use of the gene targeted transgenic models and summarizes the ways in which they have allowed us to study the role of PrP in prion disease and the insights they have provided into the mechanisms of neurodegenerative diseases.

Progress in Molecular Biology and Translational Science, Volume 150
ISSN 1877-1173
http://dx.doi.org/10.1016/bs.pmbts.2017.06.008

1. INTRODUCTION

Transgenic mice have been important tools in research of prion disease or transmissible spongiform encephalopathy (TSE) research since 1989. They have been used for a wide range of studies including identifying infectivity, defining the critical role of the *Prnp* gene and its mutations and polymorphisms in host susceptibility, and studying the normal function of PrP^C. The models produced have ranged from normal expression levels to high levels of expression of the *Prnp* gene, using different species of *Prnp* and different mutations and polymorphisms within the gene.

The first models produced by Scott *et al.* were transgenic mice which overexpressed hamster prion protein (PrP) via random insertion of the hamster *Prnp* gene into the murine genome.[1] This approach to transgenic production has since been used to produce a great number of models in which PrP from a different species is ectopically expressed and at varying levels of expression. Overexpression models have been particularly useful in identifying infectivity and producing rapid disease models. However the expression patterns and cellular localization in the central nervous system (CNS) and periphery differ between models and are dependent on the promoter driving expression thus direct comparison between models can prove difficult.

An alternative approach has been to produce gene targeted mice in which homologous recombination is used to replace the endogenous PrP gene with the gene of interest.[2] The mice are genetically identical to each other apart from the mutant PrP gene. Expression of the gene is at physiological levels and has the normal expression pattern both in the CNS and peripheral tissues thus allowing direct comparison between the transgenic models and their wild-type controls. Gene targeted models expressing human, bovine, and ovine PrP[3,4] and mice with specific mutations and polymorphisms of the *Prnp* gene have all been produced. In addition, a number of different gene targeted transgenic mice in which PrP has been abolished have been created to study the function of the PrP in health and disease. It is outwith the scope of this chapter to review all the transgenic studies in prion research, our focus will therefore be the use of the gene targeted transgenic models and the ways in which they have furthered our understanding of these diseases.

2. THE MOUSE AS A MODEL OF PRION DISEASE

The mouse is an important in vivo model of prion diseases showing the full spectrum of the natural disease in terms of the clinical, biochemical (proteinase K (PK) resistance and PK-sensitive PrP), and neuropathological (including TSE vacuolation, abnormal PrP deposition, and gliosis) phenotype. The prion strain of interest is inoculated by the intracerebral (i.c.) and/or peripheral route into the chosen mouse line. The i.c. route is the most efficient route for transmitting disease but the peripheral route with extended incubation times is important for some particular studies. Following an incubation period, clinical signs such as ataxia, kyphosis, gait, and behavioral abnormalities can be observed, and the animal is culled at a set clinical endpoint. The incubation times are dependent on the host/agent combination and are highly reproducible. A TSE vacuolation profile, which plots the distribution and severity of vacuolation in defined regions of the brain, can then be plotted producing a characteristic "lesion profile" for each host/agent combination. PrP deposition patterns can be assessed in a similar manner. The consistency of these models allows for rigorous time course studies to also be conducted examining the early events during the neurodegenerative process. These studies include assessing neuronal and synaptic loss, behavioral abnormalities, gliosis, and transcriptomic studies and have led to well-established time lines for the early and late pathological events, providing a basis for investigating disease mechanism and intervention strategies (Fig. 1).

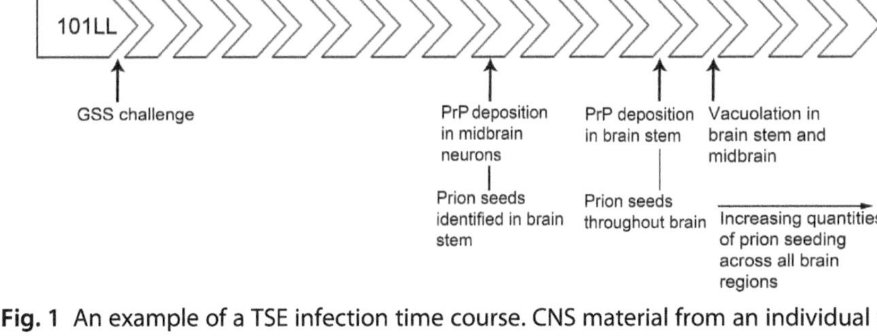

Fig. 1 An example of a TSE infection time course. CNS material from an individual with GSS is intracerebrally inoculated into gene targeted 101LL mice. Throughout the approximately 290-day time course of the study, specific events such as abnormal PrP deposition, astrogliosis, vacuolation, and presence of prion seeds can be detected.[5,6]

3. THE INFLUENCE OF HOST PrP EXPRESSION ON DISEASE SUSCEPTIBILITY

Scott *et al.* inserted a hamster *Prnp* gene into the mouse genome by random integration. These mice were susceptible to hamster scrapie exhibiting clinical signs of disease and with a significantly shorter incubation period than wild-type mice.[1] PrP expression levels were shown to influence incubation periods with mice expressing the highest levels of hamster *Prnp* demonstrating showing the shortest incubation periods.[1,7] This relationship between expression levels and incubation time has also been demonstrated for mice overexpressing other *Prnp* genes.[8]

The relationship between PrP expression levels and incubation periods of disease was further explored when the first PrP null mice were produced. A number of mouse lines in which PrP has been abolished have been produced including *Prnp*$^{-/-}$ (Edinburgh),[9] *Prnp*$^{0/0}$ (Zurich I),[10] and *Prnp*$^{-/-}$ (Nagasaki).[11] Each line used different targeting vectors and consequently different regions of the *Prnp* gene were altered. Each of these lines has proved to be resistant to prion infection when inoculated with a number of different prion agents.[12,13] Moreover, heterozygous null mice with only one copy of *Prnp*[9,12,14] which expressed lower levels of PrP (50%–70% of wild-type levels) had consistently longer incubation periods when inoculated with prion agents than wild-type controls and showed a longer disease progression.[12,13] These studies demonstrate that host PrP is essential for the development of disease and that incubation periods and disease progression are inversely correlated with the amount of PrP expressed in the mice. This knowledge has led to the production of a vast array of mouse lines overexpressing different *Prnp* genes to produce rapid disease models.[15–17] While these lines are important for identifying infection some caution should be taken in their use for an understanding transmission potential and defining a prion strain between since overexpression appears to influence both these criteria.[18]

4. PrP MUTATIONS AND DISEASE TRANSMISSION

Host PrP is a fundamental requirement for the transmission of prion diseases. Alterations in the *Prnp* gene such as mutations leading to amino acid changes which may change the stability of the PrP protein structure, or post-translational modifications such as glycosylation can have a significant influence on transmission characteristics such as disease susceptibility, pathology and incubation periods. The consequences of such alterations can be unpredictable and the mechanisms behind this are not yet fully understood.

Early transmission studies using scrapie prions inoculated into wild-type mice suggested that the PrP gene was linked to incubation periods. Classical genetic analysis identified *Sinc/Prni* as the major gene controlling mouse scrapie incubation time and *Sinc/Prni* were shown to be linked to *Prnp*.[19–21] *Prnp* alleles express distinct PrP protein variants, PrP-A and PrP-B, which arise from codon 108 leucine/phenylalanine and 189 threonine/valine dimorphisms.[22] *Prnp* genotype segregates with incubation time suggesting, but not proving, that incubation time is controlled by PrP dimorphisms, and that the *Sinc/Prni* and *Prnp* loci are congruent. Moore *et al.* used gene targeting to construct mice in which the endogenous *Prnp* allele had been modified to express PrP-B instead of PrP-A. When challenged with the mouse-adapted bovine spongiform encephalopathy (BSE) strain 301V, these mice exhibited dramatically shortened incubation times (133 ± 1 vs 244 ± 2 days in PrP-A mice) and demonstrated that PrP dimorphisms at codon 108 and/or 189 control incubation time, and that *Sinc/Prni* and *Prnp* are congruent.[23]

The inherited human prion disease, Gerstmann–Sträussler–Scheinker (GSS), is most commonly associated with a proline (P) to leucine (L) substitution at codon 102 (P102L) in *PRNP*.[24] In order to investigate this mutation, gene targeted mice with a P to L mutation at codon 101 (101LL) of the murine PrP sequence (equivalent to the 102LL mutation in *PRNP*) were produced.[25] No spontaneous disease was observed in these mice but when inoculated with brain homogenate from a human case of P102L GSS clinical disease with 100% susceptibility was observed in 288 ± 4 days compared with 456 days in wild-type mice; this suggested that sequence homology between the host and donor is important in determining transmissibility.[25] However when these mice were inoculated with strains of prion disease from different species and carrying a proline at the equivalent codon 101 position, the incubation periods were also dramatically reduced when compared to wild-type mice. Both natural sheep scrapie (SSBP/1) (346 ± 5 vs over 400 days) and hamster-passaged scrapie (263K) (464 ± 18 vs over 600 days) exhibited this phenomenon.[26] In contrast, murine scrapie strains, ME7, 139A, and 79A which also carry PrP with proline at codon 101 exhibit longer incubation periods in 101LL mice compared to wild-type mice.[27,28] These studies demonstrate that host/donor sequence homology is not the only criterion that determines disease transmission and other factors must play a role.

Unlike the gene targeted 101LL mice, transgenic mice which over express P101L at a high level develop a spontaneous neurodegenerative disease in which abnormal PrP including PrP amyloid plaques can be observed in the brain but with an absence of detectable PK-resistant PrP.[29,30] Despite the lack of PK-resistant PrP, brain material from these mice can transmit disease to mice which over express P101L and develop spontaneous disease but not to wild-type control mice or mice overexpressing wild-type murine PrP.[30,31] Later studies suggested that transmission to the overexpressing P101L mice was better described as an acceleration of an existing disease rather than true infectious transmission.[32]

A central aspect of the prion hypothesis that misfolding of PrP is sufficient to cause infectivity and therefore models of familial prion disease which cause spontaneous disease should also generate infectivity.[33] A gene targeted model of fatal familial insomnia (FFI) carrying the mouse equivalent of the aspartate (D) to asparagine (N) mutation at codon 178 (D178N in human, D177N in mice) was produced in which additional amino acid substitutes provided a transmission barrier against mouse scrapie.[34] The mice, designated ki-3F4-FFI, spontaneously developed a disease which included behavioral abnormalities, gliosis, neuronal loss, and low amounts of PK-resistant PrP (PrP^{res}) which is similar to the disease phenotype in humans. This disease was able to be transmitted and serially passaged in mice expressing physiological levels of wild-type PrP with the same human epitope (ki-3F4-WT) and in mice overexpressing mouse PrP (Tga20). The mice were highly resistant to mouse scrapie confirming that the infectivity was a consequence of the mutation and not contamination from other prion agents.[34] A similar approach was used to produce a model of familial CJD caused by a lysine (K) to glutamate (E) mutation at codon 199.[35] Again, these mice produced a spontaneous disease containing the hallmark features of the human disease which was able to be transmitted to both transgenic and wild-type mice, confirming the generation of infectivity in this model.

5. POSTTRANSLATIONAL MODIFICATIONS CAN INFLUENCE DISEASE TRANSMISSION

Posttranslational modifications such as glycosylation have been shown both in vitro and in vivo to influence disease transmission and susceptibility to prion disease. PrP^C is variably glycosylated at two highly conserved sites (amino acids 180 and 196 in mice). N-glycan attachments at these sites give rise to four glycosylated forms of PrP; unglycosylated, monoglycosylated at position 180, monoglycosylated at position 196, and diglycosylated. In vitro studies have demonstrated that glycosylation can influence cross-species transmission.[36] Consequently, a number of glycosylation-deficient mouse lines have been developed including conventional transgenics,[37] 3F4 tagged murine transgenics,[38] and gene targeted transgenics.[39] Cancellotti *et al.* produced three lines of gene targeted mice in which the first, second, or both glycosylation sites were abolished: G1, N180T; G2, N196T; and G3, N180T and N196T, respectively.[39] Glycosylation of host PrP was shown to have a major influence on the host's susceptibility to disease. G2 mice were susceptible to 79A, ME7, and 301C with incubation periods similar to that of wild-type mice inoculated with 79A and ME7 but longer than wild-type mice inoculated with 301C.[40] In contrast, G1 and G3 mice appeared resistant to ME7 and 301C prion strains and exhibited very long incubation times with 79A. More recent studies investigated the role of glycosylation in cross-species transmission and demonstrated that G2 mice were susceptible to 263K, variant Creutzfeldt–Jakob disease (vCJD) and sCJDMM2. In particular, G2 mice showed a shorter incubation time and a higher susceptibility rate than wild-type mice when inoculated with sCJDMM2. In contrast, the absence of glycosylation at the first (G1) or both (G3) N-glycan attachment sites resulted in almost complete resistance to the three agents.[41] These studies demonstrate that glycosylation is a key factor in determining disease transmission between species and suggest that glycosylation at the second N-glycan site may protect against disease transmission.

6. CROSSING THE SPECIES BARRIER AND ASSESSING ZOONOTIC POTENTIAL

As discussed earlier, the inability of a prion agent from one species to infect a host of another species has been attributed to a lack of sequence homology between the PrP of the host and agent and can be influenced by different factors including glycosylation and specific mutations. This inefficient transmission of a prion agent to a new species is often described as the species barrier and is associated with low attack rates and having a long incubation period upon primary passage which may decrease upon subsequent passage in the new host.[42,43] This phenomenon is frequently observed in transmissions of prion agents to wild-type mouse lines. In order to overcome the species barrier and understand more about the transmission characteristics of specific diseases, a number of conventional and gene targeted mice expressing PrP of different species have been produced. Current models range from mice expressing different polymorphisms of human,[4,44] deer,[45,46] and ovine PrP genes.[16,47]

Assessing the zoonotic potential of prion diseases, particularly of emerging prion diseases, is important and challenging. To achieve this, both conventional transgenic and gene targeted transgenic mice expressing human PrP have been used in conjunction with in vitro assay systems to assess zoonotic potential. Two polymorphisms in the *PRNP* gene; methionine (M) or valine (V) at codon 129 and glutamic acid (E) or lysine (K) at codon 219 have been associated with susceptibility to prion diseases.[48] The *PRNP* codon 129 genotype has been shown to affect the phenotype of disease in CJD and FFI and is associated with susceptibility to vCJD (reviewed in Ref.[48]).

The need to test zoonotic potential of prion diseases has arisen due to the ability of BSE to transmit between species with the BSE strain identified in humans, ungulates, and felids.[49–51] Experimental challenges utilizing gene targeted mice expressing human PrP did not show any evidence of BSE transmission (HuM, HuV, and HuMV)[4] or very limited transmission (Ki-Hu129M/M).[44] Conventional transgenic mice overexpressing human PrP 129MM (Tg35) have an overall transmission rate of 29% when challenged with BSE isolates and incubation periods of between 338 and 492 days.[52] Similar results have been observed using different transgenic mouse lines.[53] Transgenic mice overexpressing human PrP 129VV (Tg-152) also show evidence of transmission at a low transmission rate (38% vs 88% in wild-type FVB mice) and with longer incubation periods (602 ± 50 days vs 466 ± 26 days) than wild-type mice.[54] Thus the overexpression clearly lowers the barrier to transmission but may over exaggerate the transmission potential between species. The gene targeted mice demonstrate that there is a significant species barrier present between BSE and human PrP and may represent a truer reflection of the situation between cattle and humans and could explain why despite the large number of BSE cases reported and high exposure of the UK population (www.oie.int), there has been a limited number of cases of vCJD; 178 as of February 2017 (www.cjd.ed.ac.uk).

Although there is a large species barrier between BSE and human PrP, transmission of BSE through an intermediate species such as sheep may alter transmission efficiency. Both gene targeted and mice overexpressing human PrP demonstrate an increased susceptibility to sheep BSE compared to classical BSE. In the overexpression models, reduced incubation periods and increased susceptibility of up to 100% are observed[53] whereas in gene targeted mice nearly 70% of HuM mice show evidence of disease compared to 0% challenged with cattle BSE.[55] Goat BSE has also been shown to transmit more efficiently to overexpressing mice than cattle BSE.[53] These results indicate that transmission through an intermediate species could increase the zoonotic risk of BSE compared to cattle BSE.

Although scrapie has been endemic in the sheep population for many years, there appears to be no evidence of transmission to humans. However studies involving the challenge of mice overexpressing human PrP (fourfold more compared with that found in normal human brain tissue) with natural scrapie have recently demonstrated disease transmission. Upon primary passage mice expressing human 129MV showed disease-associated PrP^{res} accumulation in their brains, however no clinical signs were observed. Following second passage of sheep scrapie in the same transgenic lines, clinical signs, and PrP^{res} accumulation in the brains of a proportion of mice expressing either human PrP129MM or 129MV were observed.[56] Studies in gene targeted mice have proved negative to date[55] and in vitro analysis using human brain homogenate as a substrate in the protein misfolding cyclic amplification assay showed no conversion of human PrP[57] thus providing conflicting evidence. In a similar manner, transmission studies of chronic wasting disease (CWD) have shown no evidence of positive transmissions in gene targeted mice[58] and a number of overexpressing mouse lines.[59,60] However, a recent reports suggest evidence of transmission of CWD in overexpressing mice at subclinical levels[61] and in vitro analysis provides evidence that CWD can convert human PrP.[57] Thus there is some evidence both in vivo and in vitro that CWD has the potential for cross-species transmission and the risks of zoonotic transmission cannot therefore be dismissed.

Assessing zoonotic potential of a prion agent is important for public health and there is a variety of both in vivo and in vitro methods. The transgenic mouse is the most common tool but as has been discussed above, there are a great many different lines of transgenic mice, each with different copy numbers and different sites of integration into the mouse genome. This can make it difficult to compare between mouse lines and overexpression of PrP can make interpretation of transmission potential more challenging. Combined with the number of different strains of disease such as scrapie[62,63] and CWD[64,65] which may hold different zoonotic properties, interpretation of the resulting data requires careful thought. The use of a combination of different in vivo models and in vitro assays provides a larger evidence base to make assessments of zoonotic potential and comparisons between different agents.

7. MODELING HUMAN PRION DISEASE TRANSMISSION

The production of transgenic mice which express PrP of a different species allows the study of intraspecies transmission and can be used in modeling the effects of host genetics such as *PRNP* codon 129 in humans. This continues to be of importance when assessing the risks of human-to-human transmission of disease and understanding disease pathogenesis. In particular, vCJD has been studied extensively in mouse models expressing the different codon 129 genotypes.

Variant CJD has been shown to transmit to both mice overexpressing human PrP and gene targeted mice with varying degrees of susceptibility dependent upon codon 129 genotype. Utilizing gene targeted mice which allowed direct comparison between the mouse lines, Bishop *et al.* demonstrated that although all three genotypes were susceptible, susceptibility to vCJD decreased in the order MM > MV > VV with a decrease in attack rates and longer incubation periods in the HuMV and HuV mice; thus predicting a lengthy preclinical phase in the 129MV and 129VV genotypes.[4] Indeed to date this prediction has borne out in the human population as all definite/probable cases of clinical vCJD which have been genotyped have been of the 129MM genotype with the exception of one clinical 129MV case which was reported in 2016 (www.cjd.ed.ac.uk).[66] Furthermore, these results suggest that there may be a significant proportion of individuals carrying *PRNP* codon 129VV and codon 129MV who may be susceptible to vCJD infection albeit with prolonged incubation periods or in the absence of clinical symptoms. Retrospective analyses of anonymized appendix samples have suggested that up to 1 in 2000 individuals in the UK may carry abnormal PrP deposition in their appendices.[67,68] The positive results were observed in all codon 129 genotypes and in a range of birth cohorts which could suggest a large asymptomatic proportion of the United Kingdom population. To date, four probable cases of vCJD infection through transmission of infected red blood cells[69–71] and one likely case through infected UK factor VIII concentrates[72] have been identified indicating that individuals are infectious during the preclinical or asymptomatic stage and that blood is an efficient means of transmitting infectivity. Studies using both wild-type and gene targeted mice demonstrate that material from a blood transfusion-associated case of vCJD showed similar transmission characteristics to that of a primary case of vCJD[73] and that human-to-human transmission had not caused significant alterations to the transmission efficiency of the agent.

Moreover, spleen material from an asymptomatic heterozygous individual in which abnormal PrP had only been identified in the spleen and a lymph node was also able to transmit infectivity to HuM mice.[74]

Understanding the relationship between presence of abnormal deposition and infectivity, the role of the periphery in transmission of infection, and the implications of transmission between individuals of different genotypes is essential for public health both at an individual and population level. The recent identification of a 36-year-old man who was heterozygote at codon 129 with pathologically confirmed clinical vCJD has now led to fresh concerns over a second wave of vCJD cases in individuals of the MV genotype.[66] Thus the use of transgenic mice in understanding the impact of the codon 129 genotype on transmission potential and strain characteristics of disease will aid in assessing the risks of further cases of disease.

8. DEFINING HUMAN PRION DISEASE STRAINS AND TRANSMISSIBILITY POTENTIAL

Prion disease strains can be defined using a fixed set of agent and host parameters.[75] Commonly applied to the experimental transmission of an isolate into panels of wild-type mice of different *Prnp* genotypes, the phenotypic properties include a combination of incubation period of disease and incubation period rankings of different mouse lines, the distribution and intensity of TSE-associated vacuoles in the brain, distribution of abnormal PrP deposition in the brain and the biochemical profile of PrP^{res} as assessed by PK digestion and Western blot analysis[19,76,77] (Fig. 2). To allow a full characterization of the strain of interest, the characteristics of the isolate must stabilize which can require serial passage, however some strains such as BSE/vCJD appear uniquely stable upon primary passage.[51,77] In some instances, the species barrier is too great or there is an agent/host incompatibility and the strain of agent does not transmit well to wild-type mice, i.e., sCJD.[51] The use of transgenic mice expressing human PrP can be useful in this respect, in particular gene targeted mice in which only the polymorphism of interest differs between mouse lines.

Sporadic CJD is a rare human prion disease and is thought to occur as a consequence of the random conversion of the normal PrP to the abnormal form. A range of subtypes of sCJD have been classified using a combination of their *PRNP* codon 129 genotype and their protease resistant (PrP^{res}) type on Western blot.[78] sCJDMM1 appears to be the most common form (57%) whereas thalamic sCJDMM2 appears to be less common (<1%).[79] Utilizing

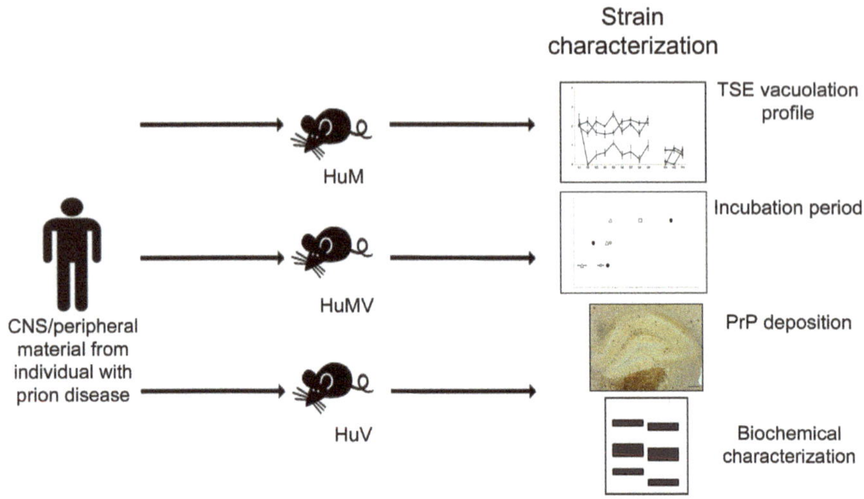

Fig. 2 Strain characterization of human prion agents in vivo. Gene targeted mice are challenged with human prion disease and strain characteristics including TSE vacuolation profiles, PrP deposition patterns, and incubation period rankings are determined and compared against standard datasets.

a traditional strain typing approach in gene targeted mice expressing human PrP, the range of phenotypes in the subtypes can now be attributed to different prion strains. Six typical sCJD cases, each representing a subgroup, i.e., MM1, cortical MM2, MV1, MV2, VV1, and VV2, were inoculated into HuM, HuMV, and HuV mice. The resulting data indicate that four strains of sCJD exist: $M1^{CJD}$ consisting of sCJDMM1 and sCJDMV1 which show similar incubation periods, vacuolation profiles, and PrP deposition staining in each mouse lines. The $V2^{CJD}$ strain comprising MV2 and VV2 which again share similar characteristics including a relatively short incubation period of ~280 days in HuV mice, the shortest of any incubation period observed in this transmission set. The remaining two strains are $M2^{CJD}$, the $sCJD^{MM2}$ cortical subtype which gave no evidence of clinical disease or vacuolation in the challenged mice and PrP deposition was only observed in HuMV and HuM mice; and $V1^{CJD}$, sCJDVV1 with clinical signs and vacuolation only observed in HuMV and HuV mice and limited numbers of mice with PrP deposition in each mouse line.[80] The same four strains of agent were also identified using in vitro techniques.[81] A fifth potential strain was described by Moda *et al.* originating from MM2 thalamic CJD (sometimes referred to as sporadic familial insomnia). Unlike MM2-cortical, MM2-thalamic transmitted disease to the HuM mice with vacuolation and PrP deposition observed.[82] These studies have provided a standard dataset against which other atypical or sporadic human prion diseases can be compared against and also provide a means of assessing transmission potential.

As novel forms of human prion diseases are identified it is important that we not only identify the strain of agent but also assess its transmissibility potential. Identification of the strain can indicate if diagnostic criteria need to be adapted or updated whereas as has been demonstrated with the transmission of vCJD through blood and CJD through human growth hormone, it is essential that we assess the risk of ongoing transmission. In 2008, a novel prion disease, variably protease-sensitive prionopathy (VPSPr) was identified.[83] This is thought to be a sporadic disease with clinical features similar to atypical dementias such as frontotemporal dementia. However the defining feature of this disease are the biochemical properties of PrP^{res}. Compared to sCJD PrP^{res}, $VPSPr^{res}$ is variably resistant to PK digestion and has a ladder-like electrophoretic profile on Western blot with a prominent low molecular weight fragment of approximately 8 kDa. Like sCJD it can occur in all codon 129 genotypes but there appears to be an overrepresentation of the 129VV genotype.[83–85] Two complementary transmission studies have now been carried out to assess the risks of transmission and to define the strain of this novel disease. Both transmission studies used transgenic mice expressing human PrP; one used overexpression models[86] and the other, gene targeted mouse lines.[87] In both studies, no evidence of clinical disease with pathological confirmation of prion disease was observed. Although the conventional transgenic mice (expressing human PrP at 2-, 3-, and 8-fold normal levels) showed evidence of vacuolation and or presence of abnormal PrP,[86] the only evidence of disease in the gene targeted mice was abnormal PrP deposition in mice challenged with VPSPr brain isolates from two 129VV individuals; a third 129MV isolate did not transmit disease.[87] Susceptibility to VPSPr from the 129VV individuals appeared to be in the order HuV > HuMV > HuM with mice showing evidence of abnormal PrP deposition predominantly in the form of plaque-like deposits (Fig. 3). Interestingly, one HuV mouse showed similar PrP deposits to those found in the human cases of VPSPr with reactive astrocytosis within the vicinity of the deposits. This same mouse also exhibited the same pattern of differential immunoreactivity with four PrP antibodies as human cases of VPSPr indicating that this represents a truly infectious disease albeit at extremely low levels. Comparison of the strain characteristics against sCJD transmissions in the same mice revealed some similarities with sCJDMM2, however

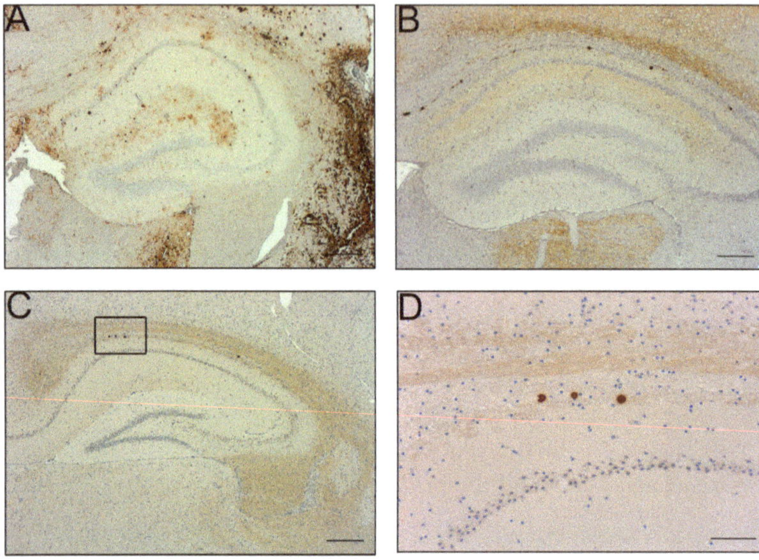

Fig. 3 Examples of abnormal PrP deposition in transgenic mice expressing human *PRNP* codon 129 VV genotype challenged with human prion diseases. (A) HuV mouse challenged with vCJD, (B) HuV mouse challenged with sCJD VV2, (C and D) HuV mouse challenged with UK case of VPSPr, F is magnification of area indicated in *box*. Immunohistochemistry was performed with (A–B) 6H4 anti-PrP antibody or (C–D) 3F4 anti-PrP antibody. Representative sections are shown through the hippocampus/thalamus. Scale bars: (A–C) 200 μm and (D) 50 μm.

PrP deposition differed in brain region and form between the two studies.[87] Taken in combination with the data from the Notari *et al.* study, the transmission studies indicate that VPSPr is a unique prion strain with limited transmission potential in humans. The two studies also highlight the need for surveillance for novel or atypical forms of human prion disease.

9. UNDERSTANDING THE FUNCTION OF PrP

PrP^C is an essential requirement for prion disease and has been implicated in the development of other neurodegenerative diseases such as Alzheimer's disease,[88,89] yet the physiological role of PrP^C has not been fully established. Using mice in which the gene of interest is knocked out and determining their phenotype under both normal and physiologically stressed conditions is a means of determining a proteins function. A number of PrP knockout mouse lines have been produced in order to ascertain the function of PrP^C, however four have been produced in such a way that ectopic expression of the prion-related protein Doppel (Dpl) occurred becoming highly expressed in the brain.[11,90–92] Ectopic expression of Dpl leads to the development of late-onset ataxia thus confounding the effects of the PrP^C knockout phenotype. Only those lines without ectopic Dpl expression will be discussed here.

PrP is highly expressed within the nervous system although there are regional, cellular, and neuronal population differences. PrP is also expressed at varying levels within the immune system, i.e., blood, bone marrow, spleen, and in other tissues such muscles and reproductive tracts.[93] Despite this ubiquitous expression, PrP knockout mice appear to have no overt phenotype and develop and reproduce normally.[9,10] Similarly, goats which are naturally devoid of PrP[94] and transgenic cattle lacking PrP[95] appear to have no overt phenotype. However more detailed studies of these animals, in particular, the mouse models have revealed a number of deficits in neuronal function. Of particular interest was that $Prnp^{-/-}$ (Edinburgh) mice demonstrated differences in their circadian system with significant increases in period length of activity between the $Prnp^{-/-}$ compared with wild-type mice and differences in sleep fragmentation and responses to sleep deprivation were observed. This was one of the first times a null mutation had been shown to affect sleep regulation while the phenotype was similar to the sleep alterations in FFI.[96]

PrP^C is also thought to play a role in peripheral neuropathies. Ablation of PrP^C in four different mouse strains led to chronic demyelinating polyneuropathy (CDP).[97] Furthermore, Bremer *et al.* demonstrated that it was the depletion of PrP^C in neurons rather than Schwann cells that triggered CDP thus showing that neuronal PrP^C is essential for peripheral myelin maintenance. These models may aid in identifying therapeutic targets for peripheral neuropathies. The neuronal loss of PrP^C is also associated with deficits in hippocampal spatial learning and hippocampal synaptic plasticity; symptoms of human prion diseases.[98] The further study of the PrP knockout models has led to the identification of PrP^C role in olfactory system,[99] oxidative homeostasis,[100] mitochondria,[101] and iron homeostasis[102] among others.

10. MECHANISMS OF NEURODEGENERATION

Pathological features common to protein misfolding neurodegeneration in the CNS include, disease-specific protein deposition, gliosis, and neuronal loss. Recent studies have suggested that the spread of misfolded protein species between distinct brain regions in a "prion-like" mechanism occurs not only in the prion diseases but also in other diseases such as Alzheimer's and Parkinson's disease.[103,104] These studies have led to the conclusion that accumulation, spread, and distribution of misfolded proteins predicts regions that ultimately undergo neurodegeneration, and thus defines disease outcome. The appearance of detectable misfolded proteins and their relationship to neurodegeneration has therefore been the major focus for defining disease mechanisms. However, a number of findings question the direct relationship between protein misfolding and neurodegeneration. For example, misfolded PrP accumulates in the brain in some situations unaccompanied by other typical neuropathological changes or any clinical signs of disease.[105–107] Thus it remains unclear what precise role misfolded proteins have in disease pathogenesis.

Over several decades, the neuron has been subject to the majority of research into protein misfolding diseases, but it is now apparent that glial cells are important players in the neurodegenerative process. Many protein misfolding diseases, including Alzheimer's disease, Parkinson's disease, and prion diseases, demonstrate activation of glial cells in the brain during the course of disease alongside accumulation of misfolded protein, but the precise role of the glial cells in the disease process is not known.[108–111]

Vincenti *et al.* examined the role of microglia, the major resident immune cell population of the brain, in a murine prion disease model of chronic neurodegeneration.[112] This study demonstrates that genes upregulated throughout the disease process are expressed predominantly by microglia. A disease-specific, innate immune response was observed in addition to an upregulation of genes associated with metabolism and respiratory stress.[112] This study thus strongly supports the growing recognition of the important contribution of microglia to a chronic neurodegenerative disease process.

A transgenic model of prion disease was used to further investigate the relationship between misfolded protein, gliosis, and neurodegeneration.[5] The GSS/101LL model shows restricted and specific regions of neurodegeneration namely the thalamus, midbrain, and brainstem whereas other regions remain apparently healthy.[106] This model therefore allowed examination of both resilient and susceptible regions of the brain during a neurodegenerative process (Fig. 1). Using a sensitive assay for detection (RT-QuIC) misfolded PrP was observed widespread throughout the brain, accumulating in all brain regions examined irrespective of neurodegeneration demonstrating active mechanisms of protein misfolding in all brain regions. Neither time of exposure nor amount of misfolded protein determined regions of neurodegeneration. The global gene expression differences between brain regions which were susceptible or resilient to neurodegeneration demonstrated two distinct microglia responses in prion-infected brains: a novel homeostatic response in all regions and an innate immune response restricted to sites of neurodegeneration.[5] Alibhai *et al.* concluded that protein misfolding events alone do not define targeting of neurodegeneration, which instead manifests only when misfolded PrP accompanies a specific innate immune response. Understanding the mechanisms underlying susceptibility and resilience might provide new opportunities to develop therapeutic interventions which, rather than aiming to prevent mechanisms of neurodegeneration, could be targeted to upregulate pathways to enhance neuroprotection.

11. CONCLUSION

The use of transgenic models in prion disease research has allowed researchers to define the role of PrP in normal homeostasis and in disease. The identifications of associations of PrP genotypes and expression levels with disease susceptibility and the ability to predict transmission of disease between species, in particular to and between humans can have a direct impact on public health and risk management. Moreover, transgenic mouse models have aided in recognizing the role that PrP^C can play in other neurodegenerative disease. As new transgenic technologies become available and new mouse models are produced, our understanding of the prion diseases will increase and aid in identifying therapeutic targets and strategies for treating and/or controlling the spread of these fatal diseases. The transgenic mouse remains a valuable tool in reaching these targets.

REFERENCES

1. Scott M, Foster D, Mirenda C, et al. Transgenic mice expressing hamster prion protein produce species-specific scrapie infectivity and amyloid plaques. *Cell.* 1989;59(5):847–857.
2. Capecchi MR. Gene targeting in mice: functional analysis of the mammalian genome for the twenty-first century. *Nat Rev Genet.* 2005;6(6):507–512.
3. Kitamoto T, Mohri S, Ironside JW, et al. Follicular dendritic cell of the knock-in mouse provides a new bioassay for human prions. *Biochem Biophys Res Commun.* 2002;294(2):280–286.
4. Bishop MT, Hart P, Aitchison L, et al. Predicting susceptibility and incubation time of human-to-human transmission of vCJD. *Lancet Neurol.* 2006;5(5):393–398.
5. Alibhai J, Blanco RA, Barria MA, et al. Distribution of misfolded prion protein seeding activity alone does not predict regions of neurodegeneration. *PLoS Biol.* 2016;14(11). e1002579.

6. Alibhai JD. *Role of Misfolded Prion Protein in Neurodegeneration*. PhD thesis, Edinburgh: The University of Edinburgh; 2015.
7. Prusiner SB, Scott M, Foster D, et al. Transgenetic studies implicate interactions between homologous PrP isoforms in scrapie prion replication. *Cell*. 1990;63(4):673–686.
8. Westaway D, Mirenda CA, Foster D, et al. Paradoxical shortening of scrapie incubation times by expression of prion protein transgenes derived from long incubation period mice. *Neuron*. 1991;7(1):59–68.
9. Manson JC, Clarke AR, Hooper ML, Aitchison L, McConnell I, Hope J. 129/Ola mice carrying a null mutation in PrP that abolishes mRNA production are developmentally normal. *Mol Neurobiol*. 1994;8(2–3):121–127.
10. Bueler H, Fischer M, Lang Y, et al. Normal development and behaviour of mice lacking the neuronal cell-surface PrP protein. *Nature*. 1992;356(6370):577–582.
11. Sakaguchi S, Katamine S, Nishida N, et al. Loss of cerebellar Purkinje cells in aged mice homozygous for a disrupted PrP gene. *Nature*. 1996;380(6574):528–531.
12. Bueler H, Aguzzi A, Sailer A, et al. Mice devoid of PrP are resistant to scrapie. *Cell*. 1993;73(7):1339–1347.
13. Manson J. Prnp gene dosage, allelic specificity and gene regulation in the transmissible spongiform encephalopathies. In: Court L, Dodet B, eds. *Transmissible Subacute Spongiform Encephalopathies: Prion Diseases*. Paris, France: Elsevier; 1996:239–245.
14. Bueler H, Raeber A, Sailer A, Fischer M, Aguzzi A, Weissmann C. High prion and PrPSc levels but delayed onset of disease in scrapie-inoculated mice heterozygous for a disrupted PrP gene. *Mol Med*. 1994;1(1):19–30.
15. Fischer M, Rulicke T, Raeber A, et al. Prion protein (PrP) with amino-proximal deletions restoring susceptibility of PrP knockout mice to scrapie. *EMBO J*. 1996;15(6):1255–1264.
16. Vilotte JL, Soulier S, Essalmani R, et al. Markedly increased susceptibility to natural sheep scrapie of transgenic mice expressing ovine prp. *J Virol*. 2001;75(13):5977–5984.
17. Beringue V, Le Dur A, Tixador P, et al. Prominent and persistent extraneural infection in human PrP transgenic mice infected with variant CJD. *PLoS One*. 2008;3(1): e1419.
18. Le Dur A, Lai TL, Stinnakre MG, et al. Divergent prion strain evolution driven by PrPC expression level in transgenic mice. *Nat Commun*. 2017;8: 14170.
19. Dickinson AG, Meikle VMH, Fraser H. Identification of a gene which controls the incubation period of some strains of scrapie agent in mice. *J Comp Pathol*. 1968;78(3):293–299.
20. Carlson GA, Kingsbury DT, Goodman PA, et al. Linkage of prion protein and scrapie incubation time genes. *Cell*. 1986;46(4):503–511.
21. Carlson GA, Goodman PA, Lovett M, et al. Genetics and polymorphism of the mouse prion gene complex: control of scrapie incubation time. *Mol Cell Biol*. 1988;8(12):5528–5540.
22. Westaway D, Goodman PA, Mirenda CA, McKinley MP, Carlson GA, Prusiner SB. Distinct prion proteins in short and long scrapie incubation period mice. *Cell*. 1987;51(4):651–662.
23. Moore RC, Hope J, McBride PA, et al. Mice with gene targeted prion protein alterations show that Prnp, Sinc and Prni are congruent. *Nat Genet*. 1998;18(2):118–125.
24. Hsiao K, Baker HF, Crow TJ, et al. Linkage of a prion protein missense variant to Gerstmann–Straussler syndrome. *Nature*. 1989;338(6213):342–345.
25. Manson JC, Jamieson E, Baybutt H, et al. A single amino acid alteration (101L) introduced into murine PrP dramatically alters incubation time of transmissible spongiform encephalopathy. *EMBO J*. 1999;18(23):6855–6864.
26. Barron RM, Thomson V, Jamieson E, et al. Changing a single amino acid in the N-terminus of murine PrP alters TSE incubation time across three species barriers. *EMBO J*. 2001;20(18):5070–5078.

27. Manson JC, Barron R, Jamieson E, et al. A single amino acid alteration in murine PrP dramatically alters TSE incubation time. *Arch Virol Suppl.* 2000;16:95–102.
28. Barron RM, Thomson V, King D, Shaw J, Melton DW, Manson JC. Transmission of murine scrapie to P101L transgenic mice. *J Gen Virol.* 2003;84(pt 11):3165–3172.
29. Hsiao K, Scott M, Foster D, Groth D, DeArmond S, Prusiner S. Spontaneous neurodegeneration in transgenic mice with mutant prion protein. *Science.* 1990;250(4987):1587–1590.
30. Telling GC, Haga T, Torchia M, Tremblay P, DeArmond SJ, Prusiner SB. Interactions between wild-type and mutant prion proteins modulate neurodegeneration in transgenic mice. *Genes Dev.* 1996;10(14):1736–1750.
31. Hsiao KK, Groth D, Scott M, et al. Serial transmission in rodents of neurodegeneration from transgenic mice expressing mutant prion protein. *Proc Natl Acad Sci USA.* 1994;91(19):9126–9130.
32. Nazor KE, Kuhn F, Seward T, et al. Immunodetection of disease-associated mutant PrP, which accelerates disease in GSS transgenic mice. *EMBO J.* 2005;24(13):2472–2480.
33. Prusiner SB. Novel proteinaceous infectious particles cause scrapie. *Science.* 1982;216(4542):136–144.
34. Jackson WS, Borkowski AW, Faas H, et al. Spontaneous generation of prion infectivity in fatal familial insomnia knockin mice. *Neuron.* 2009;63(4):438–450.
35. Jackson WS, Borkowski AW, Watson NE, et al. Profoundly different prion diseases in knock-in mice carrying single PrP codon substitutions associated with human diseases. *Proc Natl Acad Sci USA.* 2013;110(36):14759–14764.
36. Priola SA, Lawson VA. Glycosylation influences cross-species formation of protease-resistant prion protein. *EMBO J.* 2001;20(23):6692–6699.
37. DeArmond SJ, Sanchez H, Yehiely F, et al. Selective neuronal targeting in prion disease. *Neuron.* 1997;19(6):1337–1348.
38. Neuendorf E, Weber A, Saalmueller A, et al. Glycosylation deficiency at either one of the two glycan attachment sites of cellular prion protein preserves susceptibility to bovine spongiform encephalopathy and scrapie infections. *J Biol Chem.* 2004;279(51):53306–53316.
39. Cancellotti E, Wiseman F, Tuzi NL, et al. Altered glycosylated PrP proteins can have different neuronal trafficking in brain but do not acquire scrapie-like properties. *J Biol Chem.* 2005;280(52):42909–42918.
40. Tuzi NL, Cancellotti E, Baybutt H, et al. Host PrP glycosylation: a major factor determining the outcome of prion infection. *PLoS Biol.* 2008;6(4): e100.
41. Wiseman FK, Cancellotti E, Piccardo P, et al. The glycosylation status of PrPC is a key factor in determining transmissible spongiform encephalopathy transmission between species. *J Virol.* 2015;89(9):4738–4747.
42. Kimberlin RH, Walker CA. Pathogenesis of scrapie: agent multiplication in brain at the first and second passage of hamster scrapie in mice. *J Gen Virol.* 1979;42(1):107–117.
43. Kimberlin RH, Cole S, Walker CA. Temporary and permanent modifications to a single strain of mouse scrapie on transmission to rats and hamsters. *J Gen Virol.* 1987;68(7):1875–1881.
44. Asano M, Mohri S, Ironside JW, Ito M, Tamaoki N, Kitamoto T. vCJD prion acquires altered virulence through trans'species infection. *Biochem Biophys Res Commun.* 2006;342(1):293–299.
45. Browning SR, Mason GL, Seward T, et al. Transmission of prions from mule deer and elk with chronic wasting disease to transgenic mice expressing Cervid PrP. *J Virol.* 2004;78(23):13345–13350.
46. Tamguney G, Giles K, Bouzamondo-Bernstein E, et al. Transmission of elk and deer prions to transgenic mice. *J Virol.* 2006;80(18):9104–9114.

47. Crozet C, Flamant F, Bencsik A, Aubert D, Samarut J, Baron T. Efficient transmission of two different sheep scrapie isolates in transgenic mice expressing the ovine PrP gene. *J Virol.* 2001;75(11):5328–5334.
48. Kobayashi A, Teruya K, Matsuura Y, et al. The influence of PRNP polymorphisms on human prion disease susceptibility: an update. *Acta Neuropathol.* 2015;130(2):159–170.
49. Bruce M, Chree A, McConnell I, Foster J, Pearson G, Fraser H. Transmission of bovine spongiform encephalopathy and scrapie to mice: strain variation and the species barrier. *Philos Trans R Soc Lond B Biol Sci.* 1994;343(1306):405–411.
50. Fraser H, Pearson GR, McConnell I, Bruce ME, Wyatt JM, Gruffydd-Jones TJ. Transmission of feline spongiform encephalopathy to mice. *Vet Rec.* 1994;134(17):449.
51. Bruce ME, Will RG, Ironside JW, et al. Transmissions to mice indicate that 'new variant' CJD is caused by the BSE agent. *Nature.* 1997;389(6650):498–501.
52. Asante EA, Linehan JM, Desbruslais M, et al. BSE prions propagate as either variant CJD-like or sporadic CJD-like prion strains in transgenic mice expressing human prion protein. *EMBO J.* 2002;21(23):6358–6366.
53. Padilla D, Beringue V, Espinosa JC, et al. Sheep and goat BSE propagate more efficiently than cattle BSE in human PrP transgenic mice. *PLoS Pathog.* 2011;7(3): e1001319.
54. Hill AF, Desbruslais M, Joiner S, et al. The same prion strain causes vCJD and BSE. *Nature.* 1997;389(6650):448–450526.
55. Plinston C, Hart P, Chong A, et al. Increased susceptibility of human–PrP transgenic mice to bovine spongiform encephalopathy infection following passage in sheep. *J Virol.* 2011;85(3):1174–1181.
56. Cassard H, Torres J-M, Lacroux C, et al. Evidence for zoonotic potential of ovine scrapie prions. *Nat Commun.* 2014;5:5821.
57. Barria MA, Balachandran A, Morita M, et al. Molecular barriers to zoonotic transmission of prions. *Emerg Infect Dis.* 2014;20(1):88–97.
58. Wilson R, Plinston C, Hunter N, et al. Chronic wasting disease and atypical forms of bovine spongiform encephalopathy and scrapie are not transmissible to mice expressing wild-type levels of human prion protein. *J Gen Virol.* 2012;93(pt 7):1624–1629.
59. Kong Q, Huang S, Zou W, et al. Chronic wasting disease of elk: transmissibility to humans examined by transgenic mouse models. *J Neurosci.* 2005;25(35):7944–7949.
60. Sandberg MK, Al-Doujaily H, Sigurdson CJ, et al. Chronic wasting disease prions are not transmissible to transgenic mice overexpressing human prion protein. *J Gen Virol.* 2010;91(pt 10):2651–2657.
61. Kong Q, Cali I, Qing L, et al. *Zoonotic Potential of CWD Prions: An Update Prion 2016 Tokyo;* Tokyo: Prion (conference abstract); 2016.
62. Bruce ME, Boyle A, Cousens S, et al. Strain characterization of natural sheep scrapie and comparison with BSE. *J Gen Virol.* 2002;83(3):695–704.
63. Moore SJ, Smith JD, Greenlee MH, Nicholson EM, Richt JA, Greenlee JJ. Comparison of two US sheep scrapie isolates supports identification as separate strains. *Vet Pathol.* 2016;53(6):1187–1196.
64. Raymond GJ, Raymond LD, Meade-White KD, et al. Transmission and adaptation of chronic wasting disease to hamsters and transgenic mice: evidence for strains. *J Virol.* 2007;81(8):4305–4314.
65. Perrott MR, Sigurdson CJ, Mason GL, Hoover EA. Evidence for distinct chronic wasting disease (CWD) strains in experimental CWD in ferrets. *J Gen Virol.* 2012;93(pt 1):212–221.
66. Mok T, Jaunmuktane Z, Joiner S, et al. Variant Creutzfeldt–Jakob disease in a patient with heterozygosity at PRNP codon 129. *N Engl J Med.* 2017;376(3):292–294.
67. Hilton DA, Ghani AC, Conyers L, et al. Prevalence of lymphoreticular prion protein accumulation in UK tissue samples. *J Pathol.* 2004;203(3):733–739.

68. Gill ON, Spencer Y, Richard-Loendt A, et al. Prevalent abnormal prion protein in human appendixes after bovine spongiform encephalopathy epizootic: large scale survey. *BMJ*. 2013;347:f5675.
69. Peden AH, Head MW, Ritchie DL, Bell JE, Ironside JW. Preclinical vCJD after blood transfusion in a PRNP codon 129 heterozygous patient. *Lancet*. 2004;364(9433):527–529.
70. Hewitt PE, Llewelyn CA, Mackenzie J, Will RG. Three reported cases of variant Creutzfeldt–Jakob disease transmission following transfusion of labile blood components. *Vox Sang*. 2006;91(4):348.
71. Wroe SJ, Pal S, Siddique D, et al. Clinical presentation and pre-mortem diagnosis of variant Creutzfeldt–Jakob disease associated with blood transfusion: a case report. *Lancet*. 2006;368(9552):2061–2067.
72. Peden A, McCardle L, Head MW, et al. Variant CJD infection in the spleen of a neurologically asymptomatic UK adult patient with haemophilia. *Haemophilia*. 2010;16(2):296–304.
73. Bishop MT, Ritchie DL, Will RG, et al. No major change in vCJD agent strain after secondary transmission via blood transfusion. *PLoS One*. 2008;3(8): e2878.
74. Bishop MT, Diack AB, Ritchie DL, Ironside JW, Will RG, Manson JC. Prion infectivity in the spleen of a PRNP heterozygous individual with subclinical variant Creutzfeldt–Jakob disease. *Brain*. 2013;136(pt 4):1139–1145.
75. Bartz JC. Prion strain diversity. *Cold Spring Harb Perspect Med*. 2016;6(12). pii: a024349.
76. Bruce MB, Will RG, Fraser H. Comparison of the biological characteristics of BSE and CJD in mice. In: Iqbal K, Swaab DF, Winblad B, Wisniewski HM, eds. *Alzheimer's Disease and Related Disorders*. Chichester: John Wiley; 1999:553–560.
77. Ritchie DL, Boyle A, McConnell I, Head MW, Ironside JW, Bruce ME. Transmissions of variant Creutzfeldt–Jakob disease from brain and lymphoreticular tissue show uniform and conserved bovine spongiform encephalopathy-related phenotypic properties on primary and secondary passage in wild-type mice. *J Gen Virol*. 2009;90(pt 12):3075–3082.
78. Parchi P, Giese A, Capellari S, et al. Classification of sporadic Creutzfeldt–Jakob disease based on molecular and phenotypic analysis of 300 subjects. *Ann Neurol*. 1999;46(2):224–233.
79. Will RG, Ironside JW. Sporadic and Infectious Human Prion Diseases. *Cold Spring Harb Perspect Med*. 2017;7(1). pii: a024364.
80. Bishop MT, Will RG, Manson JC. Defining sporadic Creutzfeldt–Jakob disease strains and their transmission properties. *Proc Natl Acad Sci USA*. 2010;107(26):12005–12010.
81. Uro-Coste E, Cassard H, Simon S, et al. Beyond PrP res type 1/type 2 dichotomy in Creutzfeldt–Jakob disease. *PLoS Pathog*. 2008;4(3): e1000029.
82. Moda F, Suardi S, Di Fede G, et al. MM2-thalamic Creutzfeldt–Jakob disease: neuropathological, biochemical and transmission studies identify a distinctive prion strain. *Brain Pathol*. 2012;22(5):662–669.
83. Gambetti P, Dong Z, Yuan J, et al. A novel human disease with abnormal prion protein sensitive to protease. *Ann Neurol*. 2008;63(6):697–708.
84. Zou WQ, Puoti G, Xiao X, et al. Variably protease-sensitive prionopathy: a new sporadic disease of the prion protein. *Ann Neurol*. 2010;68(2):162–172.
85. Gambetti P, Puoti G, Zou WQ. Variably protease-sensitive prionopathy: a novel disease of the prion protein. *J Mol Neurosci*. 2011;45(3):422–424.
86. Notari S, Xiao X, Espinosa JC, et al. Transmission characteristics of variably protease-sensitive prionopathy. *Emerg Infect Dis*. 2014;20(12):2006–2014.
87. Diack AB, Ritchie DL, Peden AH, et al. Variably protease-sensitive prionopathy, a unique prion variant with inefficient transmission properties. *Emerg Infect Dis*. 2014;20(12):1969–1979.

88. Parkin ET, Watt NT, Hussain I, et al. Cellular prion protein regulates beta-secretase cleavage of the Alzheimer's amyloid precursor protein. *Proc Natl Acad Sci USA*. 2007;104(26):11062–11067.
89. Salazar SV, Strittmatter SM. Cellular prion protein as a receptor for amyloid-β oligomers in Alzheimer's disease. *Biochem Biophys Res Commun*. 2017;483(4):1143–1147.
90. Moore RC, Lee IY, Silverman GL, et al. Ataxia in prion protein (PrP)-deficient mice is associated with upregulation of the novel PrP-like protein doppel. *J Mol Biol*. 1999;292(4):797–817.
91. Rossi D, Cozzio A, Flechsig E, et al. Onset of ataxia and Purkinje cell loss in PrP null mice inversely correlated with Dpl level in brain. *EMBO J*. 2001;20(4):694–702.
92. Yokoyama T, Kimura KM, Ushiki Y, et al. Vivo conversion of cellular prion protein to pathogenic isoforms, as monitored by conformation-specific antibodies. *J Biol Chem*. 2001;276(14):11265–11271.
93. Linden R, Martins VR, Prado MAM, Cammarota M, Izquierdo I, Brentani RR. Physiology of the prion protein. *Physiol Rev*. 2008;88(2):673–728.
94. Benestad SL, Austbø L, Tranulis MA, Espenes A, Olsaker I. Healthy goats naturally devoid of prion protein. *Vet Res*. 2012;43(1):87.
95. Richt JA, Kasinathan P, Hamir AN, et al. Production of cattle lacking prion protein. *Nat Biotechnol*. 2007;25(1):132–138.
96. Tobler I, Gaus SE, Deboer T, et al. Altered circadian activity rhythms and sleep in mice devoid of prion protein. *Nature*. 1996;380(6575):639–642.
97. Bremer J, Baumann F, Tiberi C, et al. Axonal prion protein is required for peripheral myelin maintenance. *Nat Neurosci*. 2010;13(3):310–318.
98. Criado JR, Sanchez-Alavez M, Conti B, et al. Mice devoid of prion protein have cognitive deficits that are rescued by reconstitution of PrP in neurons. *Neurobiol Dis*. 2005;19(1–2):255–265.
99. Le Pichon CE, Valley MT, Polymenidou M, et al. Olfactory behavior and physiology are disrupted in prion protein knockout mice. *Nat Neurosci*. 2009;12(1):60–69.
100. Brown DR, Nicholas RSJ, Canevari L. Lack of prion protein expression results in a neuronal phenotype sensitive to stress. *J Neurosci Res*. 2002;67(2):211–224.
101. Miele G, Jeffrey M, Turnbull D, Manson J, Clinton M. Ablation of cellular prion protein expression affects mitochondrial numbers and morphology. *Biochem Biophys Res Commun*. 2002;291(2):372–377.
102. Singh A, Kong Q, Luo X, Petersen RB, Meyerson H, Singh N. Prion protein (PrP) knock-out mice show altered iron metabolism: a functional role for PrP in iron uptake and transport. *PLoS One*. 2009;4(7): e6115.
103. Jucker M, Walker LC. Pathogenic protein seeding in Alzheimer disease and other neurodegenerative disorders. *Ann Neurol*. 2011;70(4):532–540.
104. Walker LC, Jucker M. Neurodegenerative diseases: expanding the prion concept. *Annu Rev Neurosci*. 2015;38:87–103.
105. Lasmezas CI, Deslys JP, Robain O, et al. Transmission of the BSE agent to mice in the absence of detectable abnormal prion protein. *Science*. 1997;275(5298):402–405.
106. Barron RM, Campbell SL, King D, et al. High titres of TSE infectivity associated with extremely low levels of PrPSc in vivo. *J Biol Chem*. 2007.
107. Jeffrey M, McGovern G, Chambers EV, et al. Mechanism of PrP-amyloid formation in mice without transmissible spongiform encephalopathy. *Brain Pathol*. 2012;22(1):58–66.
108. Williams AE, van Dam AM, Man-A-Hing WKH, Berkenbosch F, Eikelenboom P, Fraser H. Cytokines, prostaglandins and lipocortin-1 are present in the brains of scrapie-infected mice. *Brain Res*. 1994;654(2):200–206.
109. Betmouni S, Perry VH, Gordon JL. Evidence for an early inflammatory response in the central nervous system of mice with scrapie. *Neuroscience*. 1996;74(1):1–5.

110. Ye X, Scallet AC, Kascsak RJ, Carp RI. Astrocytosis and amyloid deposition in scrapie-infected hamsters. *Brain Res.* 1998;809(2):277–287.
111. Jebelli J, Su W, Hopkins S, Pocock J, Garden GA. Glia: guardians, gluttons, or guides for the maintenance of neuronal connectivity? *Ann N Y Acad Sci.* 2015;1351(1):1–10.
112. Vincenti JE, Murphy L, Grabert K, et al. Defining the microglia response during the time course of chronic neurodegeneration. *J Virol.* 2015;90(6):3003–3017.

CHAPTER NINE

Transmission and Replication of Prions

Alba Marín-Moreno*, Natalia Fernández-Borges*, Juan C. Espinosa*, Olivier Andréoletti†, Juan M. Torres*,[1]

*Centro de Investigación en Sanidad Animal, CISA-INIA, Madrid, Spain
†UMR INRA-ENVT 1225, Interactions Hôte Agent Pathogène, Ecole Nationale Vétérinaire de Toulouse, Toulouse, France
[1]Corresponding author: e-mail address: jmtorres@inia.es

Contents

Abstract

Transmissible spongiform encephalopathies (TSEs) are a group of progressive, invariably fatal diseases that affect the nervous system of many mammals including humans.

The key molecular event in the pathogenesis of TSEs is the conversion of the cellular prion protein PrP^{C} into a disease-associated isoform PrP^{Sc}. The "protein-only hypothesis" argues that PrP^{Sc} itself is the infectious agent. In effect, PrP^{Sc} can adopt several structures that represent different prion strains.

The interspecies transmission of TSEs is difficult because of differences between the host and donor primary PrP sequence. However, transmission is not impossible as this occurred when bovine spongiform encephalopathy spread to humans causing variant Creutzfeldt–Jakob disease (vCJD).

This event determined a need for a thorough understanding of prion replication and transmission so that we could be one step ahead of further threats for human health. This chapter focuses on these concepts and on new insights gained into prion propagation mechanisms.

Progress in Molecular Biology and Translational Science, Volume 150
ISSN 1877-1173
http://dx.doi.org/10.1016/bs.pmbts.2017.06.014

1. PRION REPLICATION

1.1 The "Protein-Only" Hypothesis

The longest known transmissible spongiform encephalopathy (TSE), scrapie, was first thought to be a muscle disease caused by parasites[1] or by "filterable" virus of slow progression.[2] Soon, however, there was more and more evidence to suggest that the scrapie agent was something different from a virus. It was noted to be resistant to treatments that destroy nucleic acids[2,3] but susceptible to treatments that disrupt protein structure.[4–8] In parallel, attempts to identify a scrapie-specific nucleic acid were unsuccessful.[2,3,9] Thus, it was proposed the revolutionary suggestion by Griffith[10] that it could be a self-replicating protein. Later, Stanley Prusnier proposed the term "prion" to describe the proteinaceous infectious particles responsible for TSEs, and this led to the so-called "protein-only" hypothesis to explain prion diseases.[9] This hypothesis states that the main, if not the only, constituent of the infectious agent is an alternative isoform of the normal cellular prion protein PrP^{C}, called PrP^{Sc}. PrP^{Sc} self-catalyzes its formation by recruiting and transforming PrP^{C}.[11]

PrP^{Sc} was soon noticed to show different properties to PrP^{C} (Table 1). One such difference is that it is partially resistant to protease digestion, and treatment with proteinase K (PK) releases a C-terminally resistant core called PrP^{res}. Effectively, PrP^{res} is extensively used as a biochemical marker for prion detection.[12] In addition, PrP^{Sc} has a high tendency to aggregate to form amyloids, is insoluble in nonionic detergents and is strikingly resistant to common sterilization procedures. In contrast, PrP^{C} is monomeric, protease-sensitive, and soluble in nonionic detergents.[13] In support of the "protein-only" hypothesis, both PrP isoforms were found to be encoded

Table 1 Differential Properties of PrP^{C} and Prp^{Sc} Arising From Their Different Conformation

	PrP^{C}	PrP^{Sc}
Resistance to protease digestion	×	✓
Tendency to aggregate	×	✓
Solubility in nonionic detergents	✓	×
Resistance to common sterilization protocols	×	✓
Predominant secondary structure	α-Helices and turns	β-Sheet strands

by the same host gene[14] and no apparent differences were detected in their mRNA or primary protein sequence.[15] All this evidence supported that the difference between PrP^{Sc} and PrP^{C} could be purely conformational.[15,16] The structure of recombinant PrP^{C} consists of an unfolded N-terminal domain and a mostly α-helical C-terminal domain which contains three α-helices and two short β-sheet strands.[17,18] The insoluble nature and aggregation properties of PrP^{Sc} make it unsuitable for most structural analysis techniques, though it is known that PrP^{Sc} is mostly composed of β-sheets, showing a reduced content of helices and turns when compared to PrP^{C}.[19] These data are consistent with several prediction models.[20]

Further evidence supporting the "protein-only" hypothesis is provided by the first PrP knockout transgenic mouse models. These mice develop normally in the absence of PrP[21] but are resistant to prion infection.[22] Accordingly, prion replication seems to be an exponential process with the prerequisite that PrP^{C} has to be present to be converted into PrP^{Sc}.[23,24]

The final proof for this hypothesis has been provided the protein misfolding cyclic amplification technique (PMCA). PMCA generates infectious PrP^{Sc} amplified through the incubation with healthy brain homogenates and sonication cycles.[25] Instead of healthy brain homogenate, brain-derived purified PrP^{C} or even recombinant PrP^{C} can be used as substrate for PMCA, but this reduces amplification efficiency.[26,27] Finally, in 2010 de novo infectious prions were generated by PMCA using recombinant PrP^{C}.[28]

1.2 General Aspects of Prion Replication

Within the framework of the "prion-only" hypothesis, early transmission studies in transgenic mice led to several general conclusions about prion replication: the replication process is specific, the produced PrP^{Sc} is homologous to that in the inoculum and determines the neurological disease observed in the transgenic mice.[24] For conversion, it is thought that PrP^{C} binds PrP^{Sc} generating an intermediate complex before forming nascent PrP^{Sc}.[24] Through in vitro assays, binding and conversion were found to be kinetically separate steps.[29–31]

Once prions enter an organism, an exponential replication process begins generating more and more PrP^{Sc} particles. However, prion persistence without detectable amplification has been reported.[32] When a critical threshold level of particles is reached, animals show clinical signs of neurologic dysfunction and eventually die. The time between inoculation and the appearance of clinical signs is denominated incubation time. This incubation time is inversely proportional to the levels of PrP expressed in the brains of transgenic mice (Fig. 1).[24,33] However, just a small proportion of total PrP becomes protease resistant.[24,34] Furthermore, much of the protease-resistant PrP isolated from infected brains is not N-terminally truncated.[35] So PrP^{res}, the most used surrogate marker for prion presence, is not as robust as was believed. Indeed, there have been several reports of prion replication and infectivity in the absence of PrP^{res}.[36–38] Moreover, the first evidence for PK-sensitive PrP^{Sc} ($sPrP^{Sc}$) was obtained in infected hamsters[39] and could account for 90% of total PrP^{Sc} in some prion diseases like CJD.[40] PK-sensitive PrP^{Sc} can be amplified by PMCA[41] and is as infectious as its resistant counterpart.[42] Its basic structure seems to be the same as that of

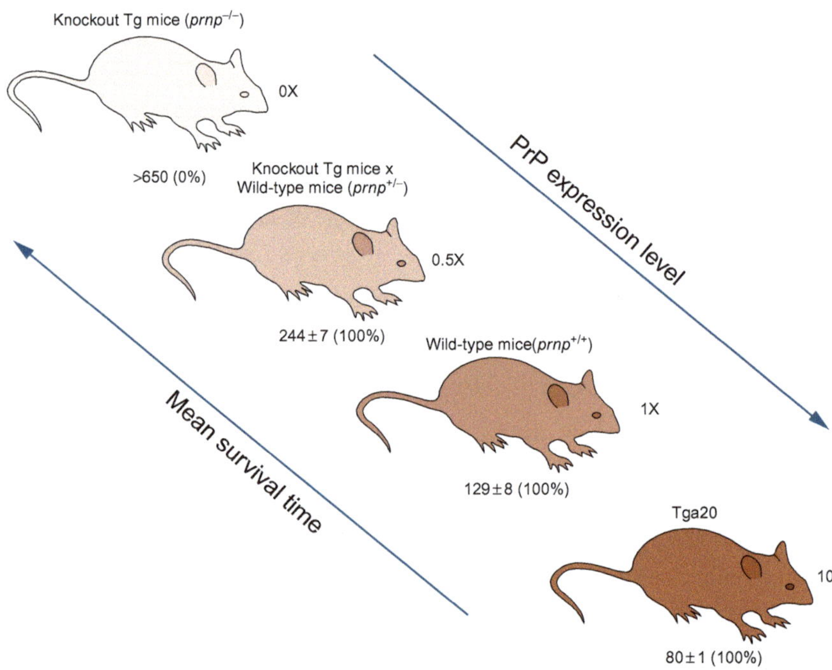

Fig. 1 Incubation time is inversely proportional to the levels of PrP expressed in the brains of transgenic mice. Experimental data obtained through 22 L inoculation in several rodent models expressing different levels of the PrP protein. The rodent models used were *prnp* knockout mice, wild-type mice, hemizygous wild-type mice, or homozygous murine PrP transgenic mice (Tga20). Mice were challenged intracerebrally with 20 μL of 10% brain homogenate from cows naturally infected with BSE. Mean survival times ± standard deviation recorded in each group are provided along with attack rates in *brackets*.

classic PrP^{Sc},[43] so one extended theory argues it might be forming less compact oligomers more exposed to the actions of PK.

The complicated nature of prion replication is also reflected by dissociation between prion propagation and toxicity. Studies measuring prion titers during the incubation period have reached variable conclusions about the kinetics of prion propagation. Thus, some authors have reported low or undetectable levels during the incubation period followed by an exponential rise in the terminal stage of disease.[43,44] However, a prion titer plateau has been described for the late incubation period by others.[45] More recently, infectivity and toxicity have been proposed to be separated in two different processes: a clinically silent exponential phase not rate limited by PrP^{C} level which rapidly reaches a maximal prion titer, followed by a distinct switch to a plateau phase.[46] The latter determines time to clinical onset in a manner inversely proportional to PrP^{C} level. The existence of two different particles for each stage, infectious and toxic, has also been suggested.[46] With regards to the nature of the infectious particle, it has been reported that the most infectious prion particles are nonfibrillar aggregates equivalent to 14–28 PrP molecules.[47] The toxic particle, also referred to as lethal (PrP^{L}), is thought to comprise PrP^{Sc} oligomers generated as an intermediate or side-product during prion propagation.[48]

1.3 Prion Replication Models

A complete understanding of the conversion process is still pending. Traditionally, two theoretical models were used to explain the conversion of PrP^{C} into PrP^{Sc}.

The first model to explain prion replication is known as the "template-assisted conversion model" and was proposed by Prusiner.[11] Assuming that spontaneous PrP^{C} conversion into PrP^{Sc} is thermodynamically unfavorable, this model proposes that a PrP^{Sc} molecule joins with a PrP^{C} molecule to form a heterodimer, assisting its refolding. The strong energetic barrier that separates both states are abrogated and PrP^{C} is transformed into a PrP^{Sc} molecule. Both molecules can now act as templates for the generation of more PrP^{C} molecules in an exponential process (Fig. 2A). Instead of PrP^{C}, a metastable or partially unfolded intermediate isoform called PrP* could be the particle that is transformed by PrP^{Sc}. PrP* is thought to be formed by stochastic fluctuations in the structure of PrP^{C}. However, no physical evidence for the existence of PrP* has been obtained.[49] How prion replication is triggered by the first PrP^{Sc} molecule is also explained by this model according to

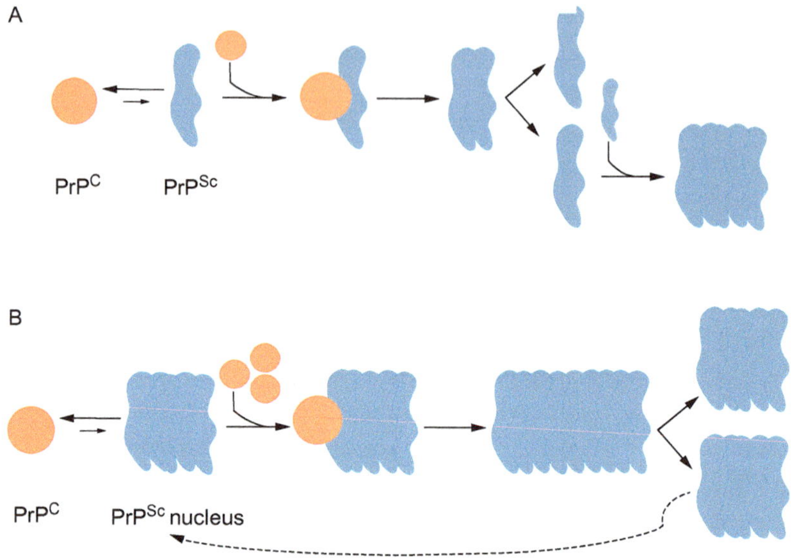

Fig. 2 Diagram of the two models proposed for the conversion of PrP^C into PrP^{Sc}. The "template-assisted conversion model" proposes that a PrP^{Sc} molecule forms a heterodimer with a PrP^C molecule and helps its refolding to produce two PrP^{Sc} molecules (A). The "nucleated-polymerization model" assumes that PrP^{Sc} and PrP^C are in equilibrium and that PrP^C is the predominant isoform. Eventually, the PrP^{Sc} molecules produced form a nucleus that later acts as a seed (B).

the different TSE etiologies. In the case of infectious TSEs like scrapie and Creutzfeldt–Jakob disease (vCJD), infection with exogenous PrP^{Sc} would act as a template to promote the conversion of PrP^C or PrP* into PrP^{Sc}.[24] In contrast, sporadic prion diseases like sCJD in humans would be the outcome of the sufficient build-up of PrP* to produce PrP^{Sc}. Finally, genetic mutations that destabilize PrP^C would promote its conversion into PrP* and later in PrP^{Sc} causing genetic TSEs like fatal familial insomnia in humans.[50,51]

Based on this model, there is some evidence pointing to a cofactor involved in PrP^{Sc} generation.[49,52] This presumed molecule is a cellular protein, commonly known as protein X, which could act as chaperone for PrP^C facilitating its conversion into PrP^{Sc}.[53] This species specific protein would be responsible for the absence of prion transmission to new hosts. Currently, the identity of protein X is not known and the generation of de novo infectious prions through the PMCA of recombinant PrP^C in the absence of other proteins does not support the existence of protein X.[28]

The second model is called the "nucleated-polymerization model" and was proposed by Jarret and Lansbury.[53] This model assumes that PrP^{Sc} and PrP^{C} are in a simple equilibrium whereby PrP^{C} is the predominant isoform and eventually formed PrP^{Sc} molecules would form a nucleus. The formation of this structure is not thermodynamically favorable; it is slow and is the rate determining phase of prion propagation. Once the nucleus has formed it acts as a seed, recruiting and transforming more and more PrP^{C} molecules in a more thermodynamically optimal manner, finally creating amyloid aggregates (Fig. 2B). The sporadic etiology of TSE is explained by the slow nucleus formation process. An exogenous prion infection would accelerate this process.[53] Similarly, PrP mutations that alter the PrP^{C}/PrP^{Sc} equilibrium could reduce the nucleation time, giving rise to genetic TSEs.[53]

In addition, the PMCA technique has provided interesting information about prion replication by revealing the minimal components necessary to ensure the generation of bona fide prions in vitro. Several studies using the PMCA technique showed that lipids and polyanions like RNA were the most successful PMCA cofactors.[28,54] The role of cofactors in prion infectivity remains to be clarified, but they seem to play an important role. The absence of lipids and polyanions during PMCA of recombinant or purified PrP^{C} generated prions with lower infectivity compared to supplemented PMCA-generated prions. The precise mechanism by which cofactors improve the PMCA remains unclear. They may act binding PrP^{C} altering the normal conformation to increase β-sheet content and PrP aggregation, may facilitate PrP conversion by concentrating PrP^{Sc} and PrP^{C} on the surface of a single lipid vesicle or a single polyanion molecule or may stabilize the infectious PrP^{Sc} conformation by forming a complex with PrP^{Sc} being a part of the infectious particle.[55]

1.4 Prion Strains

One of the main problems with the "protein-only" hypothesis is explaining the existence of multiple prion strains. Prion strains are defined as infectious isolates which, when transmitted to identical hosts, exhibit distinct prion disease phenotypes. Such phenotypes include differences in incubation time, distribution of prion deposits in the brain, clinical symptoms in vivo, and biochemical features like glycosylation patterns, denaturation profiles, and PrP^{res} electrophoretic mobility.[56] Most of these traits have proved stable across serial passages in animal models. Sometimes new strains can be observed after prion transmission in a different host or in the same host expressing polymorphisms of the prion gene.[57]

The strain phenomenon was first used to argue against the "protein-only hypothesis." As the differential phenotypic traits that characterize each strain could be serially propagated in inbred mice with the same PrP genotype, the different information cannot be encoded by differences in the PrP amino acid sequence or by differences in the PrP nucleotide sequence.[58] Hence, the most accepted theory to explain prion strains today is that PrP^{Sc} itself is able to transmit different information by adopting different tertiary and quaternary structures. An alternative hypothesis states that PrP^{Sc} is capable of transmitting the disease but its characteristics are modified by a cellular nucleic acid.[59] Unfortunately, there is currently no experimental evidence to support this theory.

The first hints that prion strains existed were obtained in goats inoculated with "hyper" or "drowsy" isolates.[60] In these experiments, isolates from goats with a "drowsy" syndrome transmitted a similar disease to inoculated recipients while isolates from goats with a "hyper" or ataxic syndrome transmitted the same phenotypic characteristics to recipient goats. Soon, numerous scrapie strains were identified on the basis of the extent of spongiform degeneration and distribution of brain lesions produced in the host.[60,61]

This evidence was followed by data suggesting that different PrP^{Sc} strains also showed variable stability when placed in contact with chaotropic salts.[62] Such evidence along with their different glycosylation state clearly revealed that different strains exposed different epitopes as a result of their distinct tridimensional structure. In this same line of investigation, it has been observed through infrared spectroscopy that several PrP^{Sc} molecules associated with hamster strains could have different conformations[63] reinforcing the idea of different prion strains having different structures although with the same primary sequence.

In subsequent work, PrP glycosylation state and relative proportions of unglycosylated, monoglycosylated, and diglycosylated isoforms were shown to contribute to prion strain diversity.[64] In some cases, robust differences in glycoform ratio exist and these differences can therefore be used for prion strain typing. According to these criteria, three distinct patterns of $PrP^{Sc/Res}$ have been identified in cases of sporadic (sCJD) and iatrogenic (iCJD) CJD (Fig. 3).[65,66] In contrast, all vCJD cases show a novel pattern, widely known as type 4 (Fig. 3).[57] How differences in PrP^{Sc} glycosylation state affect the transmitted disease remains unclear. It has been suggested that the prevalence of distinct glycoforms determines the structure of infectious

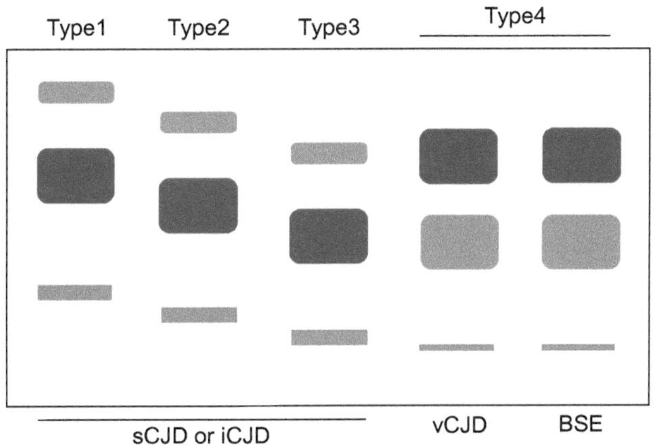

Fig. 3 Diagram of the Western blotting signatures of existing human prion strains. Human prion strains differ in their electrophoretic mobility due to the different sizes of the PK-resistant cores released after PK treatment. Strains also display differences in PrP glycoform proportions. Note that vCJD and BSE show the same signature as they come from the same prion agent.

seeds and strain properties,[67] although weak evidence for this theory has only recently emerged.

In the past few years, new prion strains have been identified in non-human diseases. In chronic wasting disease (CWD), a TSE affecting several cervid species widely extended across North America[68] and recently detected in Europe,[69] two strains CWD1 and CWD2 have been identified in transgenic mice expressing cervid PrP^C.[70] Besides classic bovine spongiform encephalopathy (C-type BSE), which caused a major food safety crisis in the late 1990s,[71] two new BSE strains have been identified. These strains, L-type[72] and H-type[73] (low or high electrophoretic mobility compared to that of the C-type) are also known as atypical BSE agents. Classic scrapie shows a vast diversity of strains[71] and atypical forms.[74]

2. PRION TRANSMISSION

2.1 Prion Transmissibility and the Species Barrier

One of the main characteristics that differentiates TSEs from other neurodegenerative diseases caused by protein misfolding and aggregation is their transmissibility among individuals of the same species and other species. Transmission between different species is an especially worrying characteristic since a given prion disease could "jump" from its natural host to other species including humans.

Fortunately, because of the so-called species barrier, interspecies transmission is not so straightforward.[75] Prions synthesized de novo always reflect the sequence of the host PrP gene and not that of the PrP^{Sc} molecules in the inoculum.[76] Typically, on first passage of prions from one species to another, not all inoculated animals succumb to disease and incubation times are usually longer and more variable than those reported for prion transmission within the same species (Fig. 4A). However, if subsequent serial passages are performed in the new host, incubation times become shorter and more uniform, and finally stable passage features will be achieved. This process is called adaptation (Fig. 4A). In other cases, animals of the new species will not be affected at all on first passage but if the brains of these animals are inoculated in susceptible hosts, these animals will succumb to disease, indicative of a subclinical infection (Fig. 4A). The final features of the prion may resemble the ones of the initial inoculum, maintaining the original prion strain as occurred with BSE and vCJD.[77] However, after interspecies transmission, a new strain with new characteristics may arise[62] (Fig. 4A), and this is commonly seen in in vitro models of prion replication like PMCA.[78] Furthermore, full transmission or complete resistance to infection can also happen when dealing with species barrier (Fig. 4A).

In effect, although the species barrier is mostly influenced by donor–host differences in the PrP gene sequence,[24,33] prion strains also play an important role (Fig. 4B). Hence, the efficiency of interspecies transmission and incubation period length are determined by the prion strain. For instance, two different strains of transmissible mink encephalopathy (TME) were found to give rise to different incubation periods and brain deposition patterns when inoculated into Syrian hamsters although the strains came from the same prion agent.[79] Hence, to unify the species barrier and strain concepts, the conformational selection model proposes that only some of the many theoretically possible prion strains exist in thermodynamic terms and that only a subset of these would be compatible with a given PrP sequence (Fig. 5). Only overlap between "permissible" conformations for PrP^{Sc} derived from two different species would ensure easy transmission between species. Conversely, two species with no "permissible" PrP^{Sc} conformations in common, would have a large transmission barrier and any transmission of infectivity between such species would require a strain change.[48]

Other factors beside PrP sequence and prion strain may affect interspecies transmission. A recent study has shown that deglycosylation of PrP^{C}

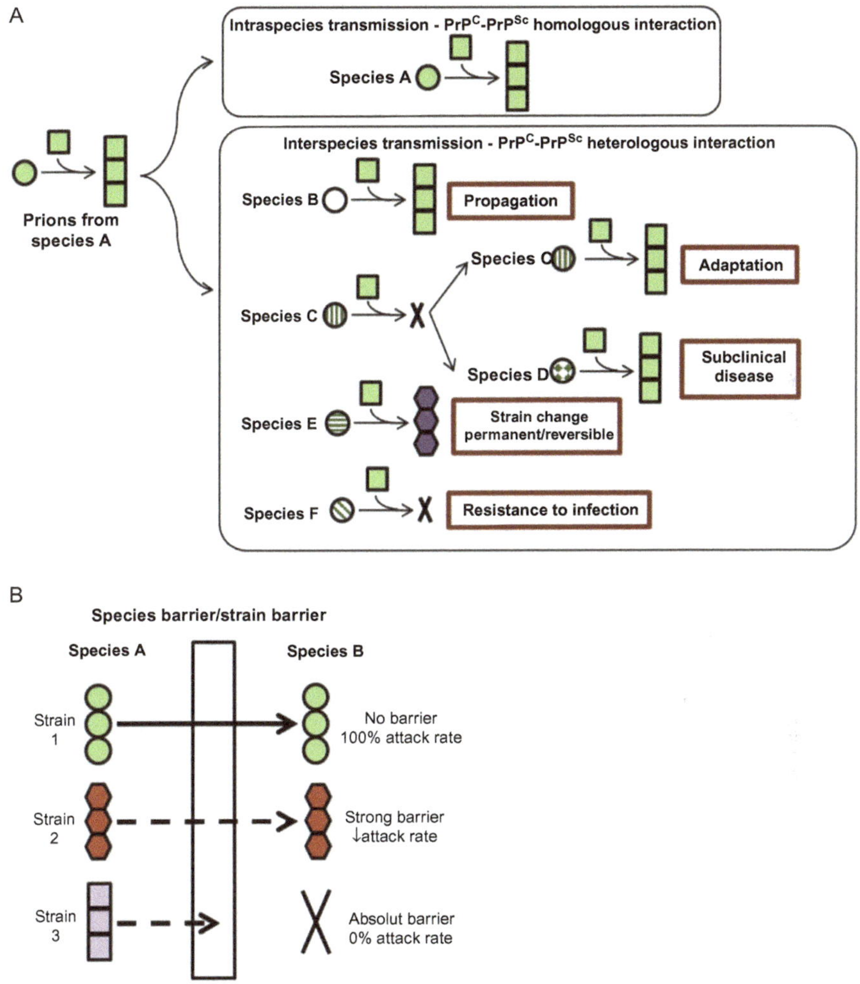

Fig. 4 Species barrier and strain barrier for interspecies prion transmission. The species barrier prevents prion transmission from one species to another. This phenomenon is mainly due to different PrP sequences between host and donor. The main outcomes observed following interspecies transmission are summarized in the figure (A). Nevertheless, prion strains also modulate the species barrier. Different strains of the same prion agent could display different abilities to infect other species, which is known as strain barrier (B).

reduces or even eliminates the species barrier for prion amplification in PMCA, suggesting that the negative charge of sialic acid controls the energy barrier of homologous and heterologous prion replication.[80] Moreover, in mice challenged with human and hamster prion strains, the absence of PrP^{C}

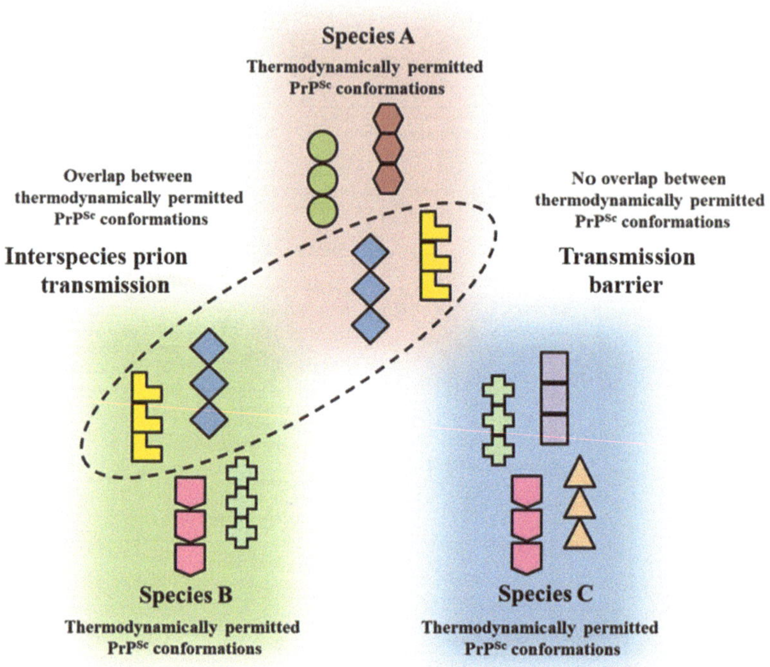

Fig. 5 Conformational selection model. Not all theoretical PrP^{Sc} conformations are thermodynamically possible for a given PrP^{C} protein. Only overlap of the allowed conformations for PrP^{Sc} between two species results in prion transmission. If the two species do not share "permissible" PrP^{Sc} conformations, there will be a high transmission barrier.

glycosylation results in almost complete resistance to disease.[81] Thus, when the glycosylation state of a PrP^{Sc} inoculum seed was eliminated none of the inoculated animals became infected, but when sialylation was restored all animals developed the disease.[82]

In addition to glycosylation state, host-specific cell factors seem to regulate the interspecies transmission of prions and the prion strain diversity obtained from a single prion isolate. In an isolate of CWD transmitted from Syrian golden hamsters to transgenic mice expressing PrP, two CWD strains were identified. However, interspecies transmission to Syrian golden hamster caused drastic changes in strain phenotypes.[83] Similarly, when a panel of diverse prions of different origins was propagated in bank voles and transgenic mice expressing physiological levels of bank vole

PrP^C, similar transmission properties were recorded in both animals but remarkable differences were detected in PrP^{Sc} propagated by the voles and the corresponding transgenic mice.[84] The findings of both these studies suggest that host cell factors other than PrP^C can modulate prion strain features.

2.2 Polymorphic Variants and the Species Barrier

Sheep and goats share the same PrP^C amino acid sequence but with a variety of different polymorphism that modify the susceptibility or resistance to scrapie. Sheep genotypes $V_{136}R_{154}Q_{171}$ and $A_{136}R_{154}Q_{171}$ are associated with a high susceptibility to classical scrapie, while the $A_{136}R_{154}R_{171}$ genotype is linked to resistance.[71] Interestingly, atypical scrapie more frequently affects sheep carrying classical scrapie-resistant genotypes than susceptible animals. In goats, around 50 prion gene polymorphisms have been described.[85] Although several polymorphisms are shared by sheep and goats, little information is available about the goat PrP^C and susceptibility to TSEs. A low susceptibility to classical scrapie has been linked to the R/Q_{211} and Q/K_{222} variants of the goat PrP^C.[71,86] Indeed, K_{222} variant is strongly protective against classical scrapie and goat BSE[86,87] but not against atypical scrapie.[88]

In the human PrP^C amino acid sequence, the major determinant of susceptibility/resistance to prion agents is the M/V_{129} dimorphism. In the general population, M/V_{129} heterozygosity is the most common genotype.[89] Although this dimorphism also plays a role in the development of sporadic prion diseases in combination with other mutations,[90] it is the principal factor responsible for the human species barrier against BSE and BSE-related prions. BSE was first diagnosed in the United Kingdom in 1986[91] and soon spread world-wide reaching epidemic proportions in the late 1990s.[85] The first cases of vCJD were later reported in 1996 in young individuals.[92] The etiology of vCJD was linked to exposure to BSE-infected meat products through epidemiological studies and experimental transmission experiments to macaques and mice and biochemical strain typing.[77,92] Despite intense exposure of United Kingdom and other European residents to BSE, only 228 cases of vCJD have been reported to date.[93] All of these cases are M_{129} homozygotes except for a single case of an M/V_{129} heterozygote.[93] These data are supported by experimental transmission in several human PrP transgenic mouse lines.[57,94,95]

2.3 Routes of Infection

In transmission experiments, a commonly used prion infection route is the intracerebral inoculation. Prions replicate directly in their target organ, the brain, where PrP^C expression level is higher.[96,97] However, this is not what happens in nature. Exceptions are iCJD cases due to prion-contaminated neurosurgery material, electrode probes, or to cadaveric dura mater transplantation.[98] Typically, prions access their hosts via peripheral routes where they overcome numerous hurdles to reach the brain. A summary of the different routes of prion infection can be found in Table 2.

In acquired prion diseases, infection is often induced by oral exposure, i.e., the consumption of prion-contaminated food products. This is true for TSEs like Kuru,[99] BSE,[100] vCJD,[101] or TME.[102] After prion entry via oral infection, there is an increase in prion infectivity in what it is known as gut-associated lymphatic tissue (GALT). Next, prions reach the nervous system via the peripheral nerves initiating their path to the brain.[103] The route from the gastrointestinal tract to peripheral nerves has been studied in detail in some prion strains and is known to involve three main actors

Table 2 Summary of Prion Infection Routes and Corresponding Prion Diseases

Infection Route	Source of Prions	Prion Disease	Affected Hosts
Intracerebral	Experimental	Several TSEs	Several animal models
	Prion-contaminated neurosurgery material, electrode probes, and dura mater grafts	iCJD	Human
Oral	Prion-contaminated food products	Kuru, vCJD, BSE, TME	Human, cattle, mink
Ocular	Experimental	Several TSEs	Several animal models
	Prion-contaminated corneal grafts	iCJD	Human
Intravenous	Experimental	Several TSEs	Several animal models
	Prion-contaminated blood products	vCJD	Human
Horizontal transmission	Environmental contamination	Scrapie, CWD	Sheep, goat, cervid species

as part of GALT. These actors are Peyer's patches (PPs), membranous epithelial cells (M cells), and mesenteric lymph nodes. Prions replicate in these areas long before they can be detected in the brain.[104,105] Although the possibility of prion uptake by M cells has been established,[106–108] another two options are dendritic cells (DC) uptake, and endocytosis in vesicular structures via a ferritin-dependent mechanism[109] or via laminin receptor binding.[110] Gastrointestinal tract alterations like those caused by microbial infections or idiopathic inflammatory diseases could affect prion entry and spread.[111] DCs, localized within the PPs acquire antigens that have crossed from the gut and deliver them to peripheral nerves in lymphoid organs.[112] Using two scrapie strains, it has been shown that DC depletion in GALT and spleen before oral exposure blocks early prion accumulation in these tissues.[113]

The eye can also be an efficient inoculation site to trigger prion diseases as reflected by reports of iCJD cases following the transplant of prion-contaminated corneal grafts.[98] Thus, it has been possible to inoculate prions in animal models through intraocular injection[114] or conjunctival instillation.[115] Intravenous and intraperitoneal injection[116] can also be experimentally used for prion inoculation, although with less efficiency than intracerebral injection but this always depends on the prion strain.[44] Accidental prion transmission via intravenous routes was long underestimated due to the low presence of prions in human blood although in other animals like sheep the efficiency is comparable to that obtained with the intracerebral route.[117] However, reports exist of vCJD cases due to blood transfusion.[118] This has determined a need for strict surveillance of blood products.

Some TSEs like scrapie or CWDs are maintained within their animal populations through natural horizontal transmission because of their persistence in the environment.[119] This transmission route has long been attributed to prion excretion because of the important role played by lymphoid tissue in prion dissemination and pathogenesis.[119] From lymphoid tissue, the spread to other tissues and secretion of prions in placenta, saliva, urine, and feces to the environment is more feasible.[112,120] However, lymphoid tissue only plays a minor role in the pathogenesis of BSE in cattle and BSE prions are mostly confined to the central nervous system.[121,122] By contrast, BSE prions in sheep accumulate in the lymphoid organs.[123] It is not known why some prion strains are capable of such dissemination outside the central nervous system while others remain in the brain even if elicited by oral challenge but it seems to be a consequence of complex prion strain–host interactions.

3. CONCLUDING REMARKS

Advances in the study of prion transmission and replication have paralleled the development of appropriate tools and technologies shedding some light on the still obscure pieces of the complicated puzzle of TSEs.

The molecular mechanisms of prion replication have not been clearly defined. Neither is it known how strain information is maintained and transmitted or which mechanisms define the tropisms of prion strains. Resolving these issues may also significantly impact other protein misfolding disorders that share some features of TSEs such as Alzheimer's disease or α-synucleinopathies.

Finally, the zoonotic potential of newly emerged prions like atypical BSE strains also remains unclear stressing the importance of an in-depth understanding of prion replication and transmission in preparation for possible future threats to human health.

REFERENCES

1. M'Fadyean JM. Scrapie. *J Comp Pathol.* 1918;31:102–131.
2. Kimberlin RH. Scrapie agent: prions or virinos? *Nature.* 1982;297:107–108.
3. Alper T, Haig DA, Clarke MC. The exceptionally small size of the scrapie agent. *Biochem Biophys Res Commun.* 1966;22:278–284.
4. Prusiner SB, McKinley MP, Groth DF, et al. Scrapie agent contains a hydrophobic protein. *Proc Natl Acad Sci USA.* 1981;78:6675–6679.
5. Hunter GD, Gibbons RA, Kimberlin RH, Millson GC. Further studies of the infectivity and stability of extracts and homogenates derived from scrapie affected mouse brains. *J Comp Pathol.* 1967;77:301–307.
6. Prusiner SB, Groth DF, Cochran SP, Masiarz FR, McKinley MP, Martinez HM. Molecular properties, partial purification, and assay by incubation period measurements of the hamster scrapie agent. *Biochemistry.* 1980;19:4883–4891.
7. Hunter GD, Millson GC. Attempts to release the scrapie agent from tissue debris. *J Comp Pathol.* 1967;88:301–307.
8. Prusiner SB, Groth DF, McKinley MP, Cochran SP, Bowman KA, Kasper KC. Thiocyanate and hydroxyl ions inactivate the scrapie agent. *Proc Natl Acad Sci USA.* 1981;78:4606–4610.
9. Prusiner SB. Novel proteinaceous infectious particles cause scrapie. *Science.* 1982;216:136–144.
10. Griffith JS. Self-replication and scrapie. *Nature.* 1967;215:1043–1044.
11. Prusiner SB. Molecular biology of prion diseases. *Science.* 1991;252:1515–1522.
12. Gielbert A, Davis LA, Sayers AR, Hope J, Gill AC, Sauer MJ. High-resolution differentiation of transmissible spongiform encephalopathy strains by quantitative n-terminal amino acid profiling (n-taap) of pk-digested abnormal prion protein. *J Mass Spectrom.* 2009;44:384–396.
13. Prusiner SB. Prions. *Proc Natl Acad Sci USA.* 1998;95:13363–13383.

14. Basler K, Oesch B, Scott M, et al. Scrapie and cellular PrP isoforms are encoded by the same chromosomal gene. *Cell.* 1986;46:417–428.
15. Hope J, Morton LJD, Farquhar CF, Multhaup G, Beyreuther K, Kimberlin RH. The major polypeptide of scrapie-associated fibrils (SAF) has the same size, charge distribution and N-terminal protein sequence as predicted for the normal brain protein (PrP). *EMBO J.* 1986;5:2591–2597.
16. Stahl N, Baldwin MA, Teplow DB, et al. Structural studies of the scrapie prion protein using mass spectrometry and amino acid sequencing. *Biochemistry.* 1993;32:1991–2002.
17. Riek R, Hornemann S, Wider G, Glockshuber R, Wüthrich K. NMR characterization of the full-length recombinant murine prion protein, mPrP (23 – 231). *FEBS Lett.* 1997;413:282.
18. Knaus KJ, Morillas M, Swietnicki W, Malone M, Surewicz WK, Yee VC. Crystal structure of the human prion protein reveals a mechanism for oligomerization. *Nat Struct Biol.* 2001;8:770–774.
19. Vázquez-Fernández E, Vos MR, Afanasyev P, et al. The structural architecture of an infectious mammalian prion using electron cryomicroscopy. *PLoS Pathog.* 2016;8:12.
20. Requena JR, Wille H. The structure of the infectious prion protein: experimental data and molecular models. *Prion.* 2014;8:60–66.
21. Bueler H, Fischer M, Lang Y, et al. Normal development and behaviour of mice lacking the neuronal cell-surface PrP protein. *Nature.* 1992;356:577–582.
22. Bueler H, Aguzzi A, Sailer A, et al. Mice devoid of PrP are resistant to scrapie. *Cell.* 1993;13:1339–1347.
23. Borchelt DR, Scott M, Taraboulos A, Stahl N, Prusiner SB. Scrapie and cellular prion proteins differ in their kinetics of synthesis and topology in cultured cells. *J Cell Biol.* 1990;110:743–752.
24. Prusiner SB, Scott M, Foster D, et al. Transgenetic studies implicate interactions between homologous PrP isoforms in scrapie prion replication. *Cell.* 1990;63:673–686.
25. Saa P, Castilla J, Soto C. Ultra-efficient replication of infectious prions by automated protein misfolding cyclic amplification. *J Biol Chem.* 2006;281:35245–35252.
26. Deleault NR, Geoghegan JC, Nishina K, Kascsak R, Williamson RA, Supattapone S. Protease-resistant prion protein amplification reconstituted with partially purified substrates and synthetic polyanions. *J Biol Chem.* 2005;280:26873–26879.
27. Atarashi R, Moore RA, Sim VL, et al. Ultrasensitive detection of scrapie prion protein using seeded conversion of recombinant prion protein. *Nat Methods.* 2007;4:645–650.
28. Wang F, Wang X, Yuang CG, Ma J. Generating a prion with bacterially expressed recombinant prion protein. *Science.* 2010;327:1132–1135.
29. Horiuchi M, Chabry J, Caughey B. Specific binding of normal prion protein to the scrapie form via a localized domain initiates its conversion to the protease-resistant state. *EMBO J.* 1999;18:3193–3203.
30. Chabry J, Priola SA, Wehrly K, Nishio J, Hope J, Chesebro B. Species-independent inhibition of abnormal prion protein (PrP) formation by a peptide containing a conserved PrP sequence. *J Virol.* 1999;73:6245–6250.
31. Horiuchi M, Priola SA, Chabry J, Caughey B. Interactions between heterologous forms of prion protein: binding, inhibition of conversion, and species barriers. *Proc Natl Acad Sci USA.* 2000;97:5836–6841.
32. Race R, Raines A, Raymond GJ, Caughey B, Chesebro B. Long-term subclinical carrier state precedes scrapie replication and adaptation in a resistant species: analogies to bovine spongiform encephalopathy and variant Creutzfeldt–Jakob disease in humans. *J Virol.* 2001;75:10106–10112.
33. Scott M, Foster D, Mirenda C, et al. Transgenic mice expressing hamster prion protein produce species-specific scrapie infectivity and amyloid plaques. *Cell.* 1989;59:847–857.

34. Caughey B, Raymond GJ. The scrapie-associated form of PrP is made from a cell surface precursor that is both protease- and phospholipase-sensitive. *J Biol Chem*. 1991;266:18217–18223.
35. Bolton DC, Bendheim PE, Marmostein AD, Potempska A. Isolation and structural studies of the intact scrapie agent protein. *Arch Biochem Biophys*. 1987;258:579–590.
36. Lasmezas CI, Deslys JP, Robain O, et al. Transmission of the BSE agent to mice in the absence of detectable abnormal prion protein. *Science*. 1997;275:402–405.
37. Barron RM, Campbell SL, King D, et al. High titers of transmissible spongiform encephalopathy infectivity associated with extremely low levels of PrPSc in vivo. *J Biol Chem*. 2007;282:35878–35886.
38. Dobie K, Barron R. Dissociation between transmissible spongiform encephalopathy (TSE) infectivity and proteinase K-resistant PrP(Sc) levels in peripheral tissue from a murine transgenic model of TSE disease. *J Virol*. 2013;87:5895–5903.
39. Safar JG, Wille H, Itri V, et al. Eight prion strains have PrP(Sc) molecules with different conformations. *Nat Med*. 1998;4:1157–1165.
40. Safar JG, Geschwind MD, Deering C, et al. Diagnosis of human prion disease. *Proc Natl Acad Sci USA*. 2005;102:3501–3506.
41. Pastrana MA, Sajnani G, Onisko B, et al. Isolation and characterization of a proteinase K-sensitive PrPSc fraction. *Biochemistry*. 2006;45:15710–15717.
42. Sajnani G, Silva CJ, Ramos A, et al. PK-sensitive PrPSc is infectious and shares basic structural features with PK-resistant PrPSc. *PLoS Pathog*. 2012;8: e1002547.
43. Kimberlin RH, Walker CA. Pathogenesis of mouse scrapie: dynamics of agent replication in spleen, spinal cord and brain after infection by different routes. *J Comp Pathol*. 1979;89:551–562.
44. Kimberlin RH, Walker CA. Pathogenesis of scrapie (strain 263k) in hamsters infected intracerebrally, intraperitoneally or intraocularly. *J Gen Virol*. 1986;67:255–263.
45. Czub M, Braig HR, Diringer H. Pathogenesis of scrapie: study of the temporal development of clinical symptoms, of infectivity titers and scrapie-associated fibrils in brains of hamsters infected intraperitoneally. *J Gen Virol*. 1986;67:2005–2009.
46. Sandberg MK, Al-Doujaily H, Sharps B, Clarke AR, Collinge J. Prion propagation and toxicity in vivo occur in two distinct mechanistic phases. *Nature*. 2011;470:540–542.
47. Silveira JR, Raymond GJ, Hughson AG, et al. The most infectious prion protein particles. *Nature*. 2005;437:257–261.
48. Collinge J, Clarke A. A general model of prion strains and their pathogenicity. *Science*. 2007;318:930–936.
49. Cohen FE, Pan KM, Huang Z, Baldwin M, Fletterick RJ, Prusiner SB. Structural clues to prion replication. *Science*. 1994;264:530–531.
50. Westaway D, DeArmond SJ, Cayetano-Canlas J, et al. Degeneration of skeletal muscle, peripheral nerves, and the central nervous system in transgenic mice overexpressing wild-type prion proteins. *Cell*. 1994;16:117–129.
51. Westaway D, Cooper C, Turner S, Da Costa M, Carlson GA, Prusiner SB. Structure and polymorphism of the mouse prion protein gene. *Proc Natl Acad Sci USA*. 1994;91:6418–6422.
52. Telling GC, Scott M, Mastrianni J, et al. Prion propagation in mice expressing human and chimeric PrP transgenes implicates the interaction of cellular PrP with another protein. *Cell*. 1995;83:79–90.
53. Jarrett JT, Lansbury PT. Seeding "one-dimensional crystallization" of amyloid: a pathogenic mechanism in Alzheimer's disease and scrapie? *Cell*. 1993;13:1055–1058.
54. Deleault NR, Harris BT, Rees JR, Supattapone S. Formation of native prions from minimal components in vitro. *Proc Natl Acad Sci USA*. 2007;104:9741–9746.
55. Ma J. The role of cofactors in prion propagation and infectivity. *PLoS Pathog*. 2012;8(4): e1002589.

56. Scott M, Peretz D, Ridley RM, Baker HF, DeArmond SJ, Prusiner SB. *Transgenetic investigations of the species barrier and prion strains. Prion Biology and Diseases.* 2nd ed. Cold Spring Harbor, NY: Cold Spring Harbor Laboratory Press; 2004.
57. Wadsworth JD, Asante EA, Desbruslais M, et al. Human prion protein with valine 129 prevents expression of variant CJD phenotype. *Science.* 2004;306:1793–1796.
58. Bruce M, Chree A, McConnell I, Foster J, Pearson G, Fraser H. Transmission of bovine spongiform encephalopathy and scrapie to mice: strain variation and the species barrier. *Phil Trans R Soc Lond.* 1994;343:405–411.
59. Weissmann C. A "unified theory" of prion propagation. *Nature.* 1991;352:679–683.
60. Pattison IH, Millson GC. Scrapie produced experimentally in goats with special reference to the clinical syndrome. *J Comp Pathol.* 1961;71:101–108.
61. Fraser H, Dickinson AG. Scrapie in mice. Agent-strain differences in the distribution and intensity of grey matter vacuolation. *J Comp Pathol.* 1973;83:29–40.
62. Peretz D, Williamson RA, Legname G, et al. A change in the conformation of prions accompanies the emergence of a new prion strain. *Neuron.* 2002;34:921–932.
63. Caughey B, Raymond GJ, Bessen RA. Strain-dependent differences in beta-sheet conformations of abnormal prion protein. *J Biol Chem.* 1998;273:32230–32235.
64. Parchi P, Capellari S, Chen SG, et al. Typing prion isoforms. *Nature.* 1997;386: 232–234.
65. Collinge J, Sidle KC, Meads J, Ironside J, Hill AF. Molecular analysis of prion strain variation and the aetiology of "new variant" CJD. *Nature.* 1996;383:685–690.
66. Parchi P, Giese A, Capellari S, et al. Classification of sporadic Creutzfeldt–Jakob disease based on molecular and phenotypic analysis of 300 subjects. *Ann Neurol.* 1999;46: 224–233.
67. Collinge J. Molecular neurology of prion disease. *J Neurol Neurosurg Psychiatry.* 2005;76:906–919.
68. Haley NJ, Hoover EA. Chronic wasting disease of cervids: current knowledge and future perspectives. *Annu Rev Anim Biosci.* 2015;3:305–325.
69. Benestad SL, Mitchell G, Simmons M, Ytrehus B, Vikøren T. First case of chronic wasting disease in Europe in a Norwegian free-ranging reindeer. *Vet Res.* 2016;47(1):88.
70. Angers RC, Kang HE, Napier D, et al. Prion strain mutation determined by prion protein conformational compatibility and primary structure. *Science.* 2010;328:1154–1158.
71. Aguilar-Calvo P, García C, Espinosa JC, Andreoletti O, Torres JM. Prion and prion-like diseases in animals. *Virus Res.* 2015;207:82–93.
72. Casalone C, Zanusso G, Acutis P, et al. Identification of a second bovine amyloidotic spongiform encephalopathy: molecular similarities with sporadic Creutzfeldt–Jakob disease. *Proc Natl Acad Sci USA.* 2004;101:3065–3070.
73. Biacabe AG, Laplanche JL, Ryder S, Baron T. Distinct molecular phenotypes in bovine prion diseases. *EMBO Rep.* 2004;5:110–115.
74. Benestad SL, Sarradin P, Thu B, Schönheit J, Tranulis MA, Bratberg B. Cases of scrapie with unusual features in Norway and designation of a new type, Nor98. *Vet Rec.* 2003;153:202–208.
75. Pattison IH. Experiments with scrapie with special reference to the nature of the agent and the pathology of the disease. *Res Publ Assoc Res Nerv Ment Dis.* 1968;44:254–280.
76. Bockman JM, Prusiner SB, Tateishi J, Kingsbury DT. Immunoblotting of Creutzfeldt–Jakob disease prion proteins: host species-specific epitopes. *Ann Neurol.* 1987;21:589–595.
77. Bruce ME, Will RG, Ironside JW, et al. Transmissions to mice indicate that 'new variant' CJD is caused by the BSE agent. *Nature.* 1997;389:498–501.
78. Barria MA, Telling GC, Gambetti P, Mastrianni JA, Soto C. Generation of a new form of human PrP(Sc) in vitro by interspecies transmission from cervid prions. *J Biol Chem.* 2011;286:7490–7495.

79. Bessen RA, Marsh RF. Biochemical and physical properties of the prion protein from two strains of the transmissible mink encephalopathy agent. *J Virol.* 1992;66: 2096–2101.
80. Katorcha E, Makarava N, Savtchenko R, D'Azzo A, Baskakov IV. Sialylation of prion protein controls the rate of prion amplification, the cross-species barrier, the ratio of PrPSc glycoform and prion infectivity. *PLoS Pathog.* 2014;10: e1004366.
81. Wiseman FK, Cancellotti E, Piccardo P, et al. The glycosylation status of PrPC is a key factor in determining transmissible spongiform encephalopathy transmission between species. *J Virol.* 2015;89:4738–4747.
82. Katorcha E, Daus ML, Gonzalez-Montalban N, et al. Reversible off and on switching of prion infectivity via removing and reinstalling prion sialylation. *Sci Rep.* 2016;6: 33119.
83. Crowell J, Hughson A, Caughey B, Bessen RA. Host determinants of prion strain diversity independent of prion protein genotype. *J Virol.* 2015;89:10427–10441.
84. Espinosa JC, Nonno R, Di Bari M, et al. PrP^{C} governs susceptibility to prion strains in bank vole, while other host factors modulate strain features. *J Virol.* 2016;90:10660–10669.
85. Acín C, Martín-Burriel I, Monleón E, et al. Prion protein gene variability in spanish goats. Inference through susceptibility to classical scrapie strains and pathogenic distribution of peripheral prp(sc.). *PLoS One.* 2013;8: e61118.
86. Aguilar-Calvo P, Espinosa JC, Pintado B, et al. Role of the goat K222-PrP(C) polymorphic variant in prion infection resistance. *J Virol.* 2014;88:2670–2676.
87. Aguilar-Calvo P, Fast C, Tauscher K, et al. Effect of Q211 and K222 PRNP polymorphic variants in the susceptibility of goats to oral infection with goat bovine spongiform encephalopathy. *J Infect Dis.* 2015;212:664–672.
88. Aguilar-Calvo P, Espinosa JC, Andréoletti O, et al. Goat K222-PrPC polymorphic variant does not provide resistance to atypical scrapie in transgenic mice. *Vet Res.* 2016;47:96.
89. Ironside JW. Variant Creutzfeldt–Jakob disease: an update. *Folia Neuropathol.* 2012;50:50–56.
90. Lloyd S, Mead S, Collinge J. Genetics of prion disease. *Top Curr Chem.* 2011;305:1–22.
91. Wells GA, Scott AC, Johnson CT, et al. A novel progressive spongiform encephalopathy in cattle. *Vet Rec.* 1987;121:419–420.
92. Will RG, Ironside JW, Zeidler M, et al. A new variant of Creutzfeldt–Jakob disease in the UK. *Lancet.* 1996;347:921–925.
93. Mok T, Jaunmuktane Z, Joiner S, et al. Variant Creutzfeldt–Jakob disease in a patient with heterozygosity at PRNP codon 129. *N Engl J Med.* 2017;376:292–294.
94. Takeuchi A, Kobayashi A, Ironside JW, Mohri S, Kitamoto T. Characterization of variant Creutzfeldt–Jakob disease prions in prion protein-humanized mice carrying distinct codon 129 genotypes. *J Biol Chem.* 2013;288:21659–21666.
95. Padilla D, Béringue V, Espinosa JC, et al. Sheep and goat BSE propagate more efficiently than cattle BSE in human PrP transgenic mice. *PLoS Pathog.* 2011;7: e1001319.
96. Brown HR, Goller NL, Rudelli RD, et al. The mRNA encoding the scrapie agent protein is present in a variety of non-neuronal cells. *Acta Neuropathol.* 1990;80:1–6.
97. Kretzschmar HA, Prusiner SB, Stowring LE, DeArmond SJ. Scrapie prion proteins are synthesized in neurons. *Am J Pathol.* 1986;122:1–5.
98. Bonda DJ, Manjila S, Mehndiratta P, et al. Human prion diseases: surgical lessons learned from iatrogenic prion transmission. *Neurosurg Focus.* 2016;41:E10. http://dx.doi.org/10.3171/2016.5.FOCUS15126.
99. Collinge J, Whitfield J, McKintosh E, et al. Kuru in the 21st century—an acquired human prion disease with very long incubation periods. *Lancet.* 2006;367:2068–2074.
100. Anderson RM, Donnelly CA, Ferguson NM, et al. Transmission dynamics and epidemiology of BSE in British cattle. *Nature.* 1996;382:779–788.
101. Hilton DA. Pathogenesis and prevalence of variant Creutzfeldt–Jakob disease. *J Pathol.* 2006;208:134–141.

102. Marsh RF, Hadlow WJ. Transmissible mink encephalopathy. *Rev Sci Tech.* 1992;11:539–550.
103. McBride PA, Schulz-Schaeffer WJ, Donaldson M, et al. Early spread of scrapie from the gastrointestinal tract to the central nervous system involves autonomic fibers of the splanchnic and vagus nerves. *J Virol.* 2001;75:9320–9327.
104. Bons N, Mestre-Frances N, Belli P, Cathala F, Gajdusek DC, Brown P. Natural and experimental oral infection of nonhuman primates by bovine spongiform encephalopathy agents. *Proc Natl Acad Sci USA.* 1999;96:4046–4051.
105. Andréoletti O, Berthon P, Marc D, et al. Early accumulation of PrP(Sc) in gut-associated lymphoid and nervous tissues of susceptible sheep from a Romanov flock with natural scrapie. *J Gen Virol.* 2000;81:3115–3126.
106. Heppner FL, Christ AD, Klein MA, et al. Transepithelial prion transport by M cells. *Nat Med.* 2001;7:976–977.
107. Neutra MR, Pringault E, Kraehenbuhl JP. Antigen sampling across epithelial barriers and induction of mucosal immune responses. *Annu Rev Immunol.* 1996;14:275–300.
108. Donaldson DS, Kobayashi A, Ohno H, Yagita H, Williams IR, Mabbott NA. M cell-depletion blocks oral prion disease pathogenesis. *Mucosal Immunol.* 2012;5:216–225.
109. Mishra RS, Basu S, Gu Y, et al. Protease-resistant human prion protein and ferritin are cotransported across Caco-2 epithelial cells: implications for species barrier in prion uptake from the intestine. *J Neurosci.* 2004;24:11280–11290.
110. Morel E, Andrieu T, Casagrande F, et al. Bovine prion is endocytosed by human enterocytes via the 37 kDa/67 kDa laminin receptor. *Am J Pathol.* 2005;167:1033–1042.
111. Sigurdson CJ, Heikenwalder M, Manco G, et al. Bacterial colitis increases susceptibility to oral prion disease. *J Infect Dis.* 2008;199:243–252.
112. Mabbott NA, MacPherson GG. Prions and their lethal journey to the brain. *Nat Rev Microbiol.* 2006;4:201–211.
113. Raymond CR, Aucouturier P, Mabbott NA. In vivo depletion of CD11c+ cells impairs scrapie agent neuroinvasion from the intestine. *J Immunol.* 2007;179:7758–7766.
114. Fraser H. Neuronal spread of scrapie agent and targeting of lesions within the retino-tectal pathway. *Nature.* 1982;295:149–150.
115. Scott JR, Foster JD, Fraser H. Conjunctival instillation of scrapie in mice can produce disease. *Vet Microbiol.* 1993;34:305–309.
116. Kimberlin RH, Walker CA. Pathogenesis of mouse scrapie: effect of route of inoculation on infectivity titers and dose–response curves. *J Comp Pathol.* 1978;88:39–47.
117. Andréoletti O, Litaise C, Simmons H, et al. Highly efficient prion transmission by blood transfusion. *PLoS Pathog.* 2012;8: e1002782.
118. Llewelyn CA, Hewitt PE, Knight RS, et al. Possible transmission of variant Creutzfeldt–Jakob disease by blood transfusion. *Lancet.* 2004;363:417–421.
119. Gough KC, Maddison BC. Prion transmission: prion excretion and occurrence in the environment. *Prion.* 2010;4:275–282.
120. Press CM, Heggebø R, Espenes A. Involvement of gut-associated lymphoid tissue of ruminants in the spread of transmissible spongiform encephalopathies. *Adv Drug Deliv Rev.* 2004;56:885–899.
121. Buschmann A, Groschup MH. Highly bovine spongiform encephalopathy-sensitive transgenic mice confirm the essential restriction of infectivity to the nervous system in clinically diseased cattle. *J Infect Dis.* 2005;192:934–942.
122. Espinosa JC, Morales M, Castilla J, Rogers M, Torres JM. Progression of prion infectivity in asymptomatic cattle after oral bovine spongiform encephalopathy challenge. *J Gen Virol.* 2007;88:1379–1383.
123. Lezmi S, Ronzon F, Bencsik A, et al. PrP(d) accumulation in organs of ARQ/ARQ sheep experimentally infected with BSE by peripheral routes. *Acta Biochim Pol.* 2006;53:399–405.

CHAPTER TEN

Immunology of Prion Protein and Prions

Neil A. Mabbott[1]
The Roslin Institute & Royal (Dick) School of Veterinary Sciences, University of Edinburgh, Midlothian, United Kingdom
[1]Corresponding author: e-mail address: neil.mabbott@roslin.ed.ac.uk

Contents

Abstract

Many natural prion diseases are acquired peripherally, such as following the oral consumption of contaminated food or pasture. After peripheral exposure many prion isolates initially accumulate to high levels within the host's secondary lymphoid tissues. The replication of prions within these tissues is essential for their efficient spread to the brain where they ultimately cause neurodegeneration. This chapter describes our current understanding of the critical tissues, cells, and molecules which the prions exploit to mediate their efficient propagation from the site of exposure (such as the intestine) to the brain. Interactions between the immune system and prions are not only restricted to the secondary lymphoid tissues. Therefore, an account of how the activation status of the microglial in the brain can also influence progression of prion disease pathogenesis is provided. Prion disease susceptibility may also be influenced by additional factors such as chronic inflammation, coinfection with other pathogens, and

Progress in Molecular Biology and Translational Science, Volume 150
ISSN 1877-1173
http://dx.doi.org/10.1016/bs.pmbts.2017.06.004

aging. Finally, the potential for immunotherapy to provide a means of safe and effective prophylactic or therapeutic intervention in these currently untreatable diseases is considered.

1. INTRODUCTION

Prion diseases are infectious chronic neurodegenerative diseases that affect humans and several domestic and free-ranging animal species. During prion disease, aggregations of PrP^{Sc}, abnormally folded isoforms of the host cellular prion protein (PrP^{C}), accumulate in affected tissues. Prion infectivity copurifies with PrP^{Sc} and appears to constitute the major component of the infectious agent.[1,2] Host cells must express PrP^{C} in order to replicate prions.[3,4] According to the prion hypothesis proposed by Prof. Stanley Prusiner in 1980s,[5] PrP^{Sc} proteins can self-propagate by recruiting cellular PrP^{C} and, in doing so, confer the conformational changes into the disease-causing PrP^{Sc} isoform. In support of this hypothesis Prusiner and independent scientists elsewhere have shown that recombinant mouse PrP, when refolded into the disease-specific form in vitro, can transmit disease to recipient animals[2,6] (see chapter "The prion concept and synthetic prions" by Legname and Moda).

Cellular PrP^{C} is a 30–35-kDa glycoprotein and is encoded by the *PRNP* gene. This protein is inserted into the outer leaflet of the cell membrane via its glycosylphosphatidylinositol (GPI) anchor.[7] The secondary structure of PrP^{C} is rich in α-helical content with relatively little β-pleated sheet.[8,9] During prion disease changes occur in the secondary, tertiary, and quaternary structures of the PrP molecule, increasing the amount of β-pleated sheet.[8,10] These changes dramatically affect the physicochemical and biological characteristics of the PrP molecule. As a consequence, disease-specific PrP^{Sc} is relatively resistant to proteinase digestion, accumulates in affected tissues in insoluble aggregates, and is neurotoxic (see chapter "The structure of the infectious prion protein and its propagation" by Requena and Wille). How these conformational and biological changes occur is unknown, but may require additional chaperone factors such as RNA, proteoglycans, and lipids.[11,12] In this chapter, several terms are used to describe the presence of prions. PrP^{Sc} is used where evidence of proteinase resistance was demonstrated. Disease-specific PrP, or PrP^{d}, is used where it was not possible to confirm the proteinase resistance of the PrP accumulations. Infectious prions are only used when the disease transmission was demonstrated.

Within the central nervous system (CNS) PrP^{C} is highly expressed in nerves and plays an important role in promoting myelin homeostasis.[13]

Within the immune system PrP^C is expressed on lymphocytes, leukocytes, granulocytes, and stromal cells where it may play a role in cell activation,[14–16] T cell differentiation,[17] intercellular interactions,[18] and phagocytosis.[19,20] Despite these potential roles, PrP^C-deficient mice display no obvious immune deficit in vivo.[21] Data obtained from some of these studies have occasionally been contradictory, as careful attention must be made to ensure that any effects observed in certain PrP^C-deficient mouse lines are not due to effects on other linked gene loci.[22] See chapters "Functions of the prion protein" by Hirsch et al. and "Cell biology of prion protein" by Sarnataro for discussion on the cell biology of PrP.

2. PRIONS INFECT SECONDARY LYMPHOID ORGANS (SLO) BEFORE THE CNS

Table 1 concisely lists many of the different forms of spontaneous, genetic, or acquired prion diseases that have been described in humans and animals. Many of these diseases such as natural sheep scrapie and chronic wasting disease (CWD) in cervids are acquired, for example, after oral exposure, or lesions to the skin or mucous membranes. The consumption of feed contaminated with bovine spongiform encephalopathy (BSE) prions was most likely involved in the spread of this devastating disease among cattle.[23] This disease also has zoonotic potential demonstrated by the emergence of variant Creutzfeldt–Jakob disease (vCJD) in the UK human population.[24,25] Cases of iatrogenic CJD have also occurred by the accidental transmission of disease through the use of prion-contaminated surgical instruments, stereotactic EEG electrodes, transplantation of tissues (cornea, dura mater), injection of pituitary-derived growth hormones from sporadic CJD-affected cadavers, or transfusion of blood or blood products from vCJD-affected donors.

After peripheral infection many prion agent strains accumulate first within the secondary lymphoid organs (SLO) before they establish disease in the CNS. For example, after oral exposure of mice to scrapie prions,[26–28] mule deer (*Odocoileus hemionus*) to CWD,[29] and some strains of natural scrapie in sheep[30–32] the prions accumulate first in the gut-associated lymphoid tissues (GALT) such as the Peyer's patches in the small intestine. In humans, evidence of vCJD prion accumulation within the GALT before the onset of clinical signs has also been described.[33] Sporadic CJD, in contrast, has not been detected within the SLO during the preclinical phase of the disease, even after iatrogenic infection.[34,35] BSE prions also have little involvement in SLO during the preclinical phase after oral infection in cattle,[36–39] but

Table 1 Prion Diseases of Humans and Animals

Prion Disease	Affected Species	Route of Transmission
Iatrogenic Creutzfeldt–Jakob disease	Human	Accidental medical exposure to CJD-contaminated tissues or tissue products
Sporadic Creutzfeldt–Jakob disease	Human	Unknown. Theories include somatic mutation or spontaneous conversion of PrP^c to PrP^{Sc}
Variant Creutzfeldt–Jakob disease	Human	Ingestion of BSE-contaminated food or transfusion of blood or blood products from CJD-infected blood donor
Familial Creutzfeldt–Jakob disease	Human	Germ-line mutations of the *PRNP* gene
Gerstmann–Straussler–Scheinker syndrome	Human	Germ-line mutations of the *PRNP* gene
Kuru	Human	Ritualistic cannibalism
Fatal familial insomnia	Human	Germ-line mutations of the *PRNP* gene
Bovine spongiform encephalopathy	Cattle	Ingestion of contaminated food
Scrapie	Sheep, goats, mouflon	Acquired. Ingestion, horizontal transmission, vertical transmission unclear
Chronic wasting disease	Elk, deer, moose	Acquired. Ingestion, horizontal transmission, vertical transmission unclear
Transmissible mink encephalopathy	Mink	Acquired (ingestion) source unknown
Feline spongiform encephalopathy	Domestic and zoological cats	Ingestion of BSE-contaminated food
Exotic ungulate encephalopathy	Nyala, kudu	Ingestion of BSE-contaminated food

BSE, bovine spongiform encephalopathy; *CJD*, Creutzfeldt–Jakob disease; *PRNP*, the gene that encodes the cellular prion protein PrP^C.

accumulate to high levels within SLO after their transmission to other host species including humans (in the form of vCJD), sheep, and mice.[40–42]

The early accumulation of prions within the SLO is essential for the efficient transmission of the prions from the site of exposure to the brain (termed *neuroinvasion*). For example, survival times after intraperitoneal injection with prions are significantly extended in splenectomized mice.[43,44] Subsequent studies have since shown that prion neuroinvasion is similarly impeded in the absence of the SLO which drain the site of infection, such as the Peyer's patches in the small intestine after oral exposure,[27,28,45–47] or the local lymph node following infection via skin lesions[48] or the footpad.[49] This requirement for SLO in the efficient transmission of prions to the CNS may depend on the route of exposure. Studies in hamsters show that efficient prion neuroinvasion after exposure via the tongue or nasal cavity can occur independently of SLO, implying direct infection of the nervous system.[50,51]

Prions also accumulate with the GALT in the large intestine such as the cecal and colonic patches during the preclinical phase.[33,52–56] However, studies in mice show that the large intestinal GALT, in contrast to the Peyer's patches in the small intestine, are not early sites of prion accumulation or neuroinvasion from the intestine.[46] The large intestinal GALT are also not considered to be early sites of prion accumulation in sheep with natural scrapie.[57–60] Data from the analyses of large intestinal GALT biopsies (rectoanal mucosal-associated lymphoid tissue; RAMALT) from scrapie-infected goats[61] and CWD-infected cervids[54,62–65] are consistent with the conclusion that the large intestinal GALT become infected with prions much later in the disease process in other natural host species.

2.1 Prion Replication Upon Follicular Dendritic Cells Within SLO Is Essential to Achieve Neuroinvasion

After infection via a peripheral route, some prion strains accumulate first within the B cell follicles and germinal centers of SLO.[29,30,33,66,67] However, the absence of T cells[44,68–70] or PrP^C expression on B cells[70–72] has no effect on the accumulation of prions in SLO or disease susceptibility. This demonstrates that B cells and T cells themselves are not the essential sites of prion accumulation and replication in these tissues.

Within the SLO of mice exposed to prions, sheep with natural scrapie, cervids with CWD, and patients with vCJD,[29,30,33,66,67] abundant prion accumulation is evident upon the surfaces of follicular dendritic cells (FDC). These important stromal cells differentiate from ubiquitous perivascular precursor cells (pericytes)[73] and are essential for the maintenance of B cell follicles and germinal center responses. A large body of data shows that the early accumulation of prions upon FDC is essential for the efficient transmission of disease to the CNS. For example, B cells produce the cytokines TNFα or lymphotoxin (LT) $\alpha_1\beta_2$ which help maintain the FDC in their differentiated state.[74] Coincident with the absence of FDC, prion accumulation in SLO and neuroinvasion are each blocked in mice which lack B cells[66,69,72,75,76] or the expression of TNFα, $LT\alpha1\beta_2$, or their receptors.[28,77,78] Prion accumulation in SLO and neuroinvasion are also impeded when FDC are temporarily depleted following treatment with soluble receptors which block LT- or TNFα-mediated signaling between B cells and FDC.[26,79–83]

As you can see in Fig. 1, FDC express PrP^C highly, but the ectopic expression of high levels of PrP^C specifically in B cells[84] or T cells[85] (20 × *Prnp* copies) does enable prion replication within them. FDC characteristically trap and retain native antigens on their surfaces for long periods which they display to B cells within the germinal center. This long-term retention of antigens by FDC helps to promote immunoglobulin (antibody) isotype class switching, affinity maturation of naïve B cells, and the maintenance of immunological memory.[86] Whether FDC also simply trapped and accumulated prions after their replication by other infected cell populations, or were sites of prion replication themselves was uncertain. To address this issue, compound transgenic mice were created which enabled PrP^C expression to be "switched on" or "switched off" only in FDC.[87] Experiments using these mice showed that after intraperitoneal injection with mouse-passaged ME7 scrapie prions, the expression of PrP^C only on FDC was sufficient to sustain high levels of prion replication in the spleen. Conversely, prion replication in the spleen was blocked when PrP^C expression was specifically ablated only on the FDC. Thus, data from a wide range of independent studies show that FDC are the essential sites of prion replication in SLO.

FDC have many slender and long dendritic processes which they use to trap and retain native antigens in the form of immune complexes, comprising antigen–antibody and/or opsonizing complement components. Complement component C1q can bind to PrP^{Sc} (Refs. 88–90) and may mediate the uptake of prions by immune cells.[91] Furthermore, in the absence of systemic opsonizing complement components (C1q, C2, C3, and C4) or complement receptor (CR) expression on FDC, the accumulation of prions within SLO is impaired and neuroinvasion delayed.[92–96] This suggests that soon after the prions infect the host they are bound by soluble opsonizing complement components and subsequently acquired by FDC in a CR-dependent manner.[92–96]

After being acquired by FDC, immune complexes are internalized and then undergo cyclical rounds of display on the cell surface.[97] This cyclical mode of immune complex expression may help to protect the native

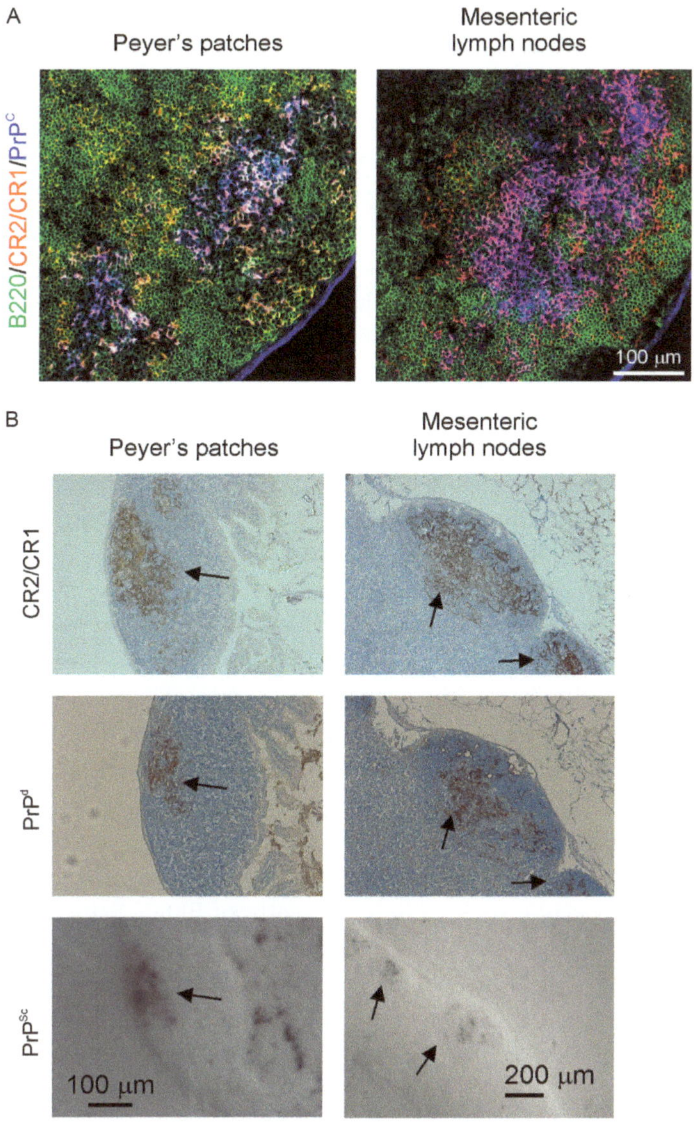

Fig. 1 Follicular dendritic cells (FDC) in the B cell follicles of secondary lymphoid organs (SLO) express high levels of PrP^{C}, and accumulate prions (PrP^{Sc}) on their surfaces during oral prion disease. (A) Peyer's patches and mesenteric lymph node sections were immunostained to detect B cells (B220, *green*), FDC (CR2/CR1, *red*), and PrP^{C} (*blue*). (B) After oral exposure of mice to scrapie prions high levels of disease-specific PrP (PrP^{d}, *brown, middle rows*) are detected in association with FDC (CR2/CR1, *brown, upper panels*) in the Peyer's patches and mesenteric lymph nodes. Analysis of adjacent sections by PET immunoblot analysis confirmed the presence of prion-specific, relatively proteinase K-resistant PrP^{Sc} (*blue/black*). *Elements of this figure are reproduced from Donaldson DS, Sehgal A, Rios D, Williams IR, Mabbott NA. Increased abundance of M cells in the gut epithelium dramatically enhances oral prion disease susceptibility.* PLoS Pathog. *2016;12:e1006075. http://doi.org/10.1371/journal.ppat.1006075 under the terms of the Creative Commons Attribution 4.0 International (CC BY 4.0) License.*

antigens from degradation, enabling the FDC to retain the antigens for long durations.[86] However, prion accumulation and replication seems to occur only on the FDC surface as intracellular PrP^{Sc} has not been detected by high-resolution electron microscopy.[98–101] This contrasts prion infection nerves, where intracellular PrP^{Sc} conversion can occur within their endosomal recycling compartments.[102]

Studies in experimental mice show that prions can establish infection within the enteric nervous system of the gut wall as soon as 21 days after oral exposure.[101] Consequently, the temporary depletion of FDC around this time after prion exposure has no effect on neuroinvasion as infection has already been established in the peripheral nervous system.[26,79,81] Thus although high levels of prions are retained upon FDC throughout the disease, their role in prion disease pathogenesis appears to be limited to the amplification of the prions above the threshold that is required for them to infect the nervous system and achieve neuroinvasion.

After their initial replication within the Peyer's patches, the prions subsequently establish infection upon FDC in other SLO such as the mesenteric lymph nodes and spleen, most likely due to their dissemination from the Peyer's patches (see Section 4). However, the absence of the mesenteric lymph nodes or spleen does not affect oral prion disease susceptibility.[28,46,47,103,104] This indicates that prion neuroinvasion initially occurs directly from the local lymphoid tissue draining the site of exposure.

Examples have however been described where some prion strains can establish infection within SLO independently of FDC. Whereas the replication of mouse-adapted ME7 scrapie prions is blocked in the absence of FDC in the SLO of TNFα-deficient mice,[77] the mouse-adapted RML prion isolate can establish infection in association with the high endothelial venules in the lymph nodes of these mice.[78,105] Under certain circumstances the RML prion isolate can also establish infection in the spleen in the absence of PrP^C expression in FDC.[106]

2.2 Prions Exploit M Cells to Enter Peyer's Patches and Establish Host Infection

In the gastrointestinal tract a single layer of tightly connected epithelial cells (enterocytes) provides a protective barrier against the microbial burden within the gut lumen. However, the epithelium which covers the lumenal surfaces of the GALT, including the Peyer's patches, contains a unique population of epithelial cells known as M cells. Unlike the surrounding enterocytes, these cells are highly phagocytic and specialized for the transcytosis of pathogens and particulate antigens across the gut epithelium into the GALT.[107] The sampling of particulate antigens by M cells is important for the initiation of efficient mucosal immune responses against pathogenic[108] and commensal bacteria.[109] Some pathogenic microorganisms, including the bacteria *Salmonella* Typhimurium,[110] *Yersinia enterocolitica*, *Listeria monocytogenes*,[111] and *Mycobacterium avium* subsp. *paratuberculosis*[112] as well as norovirus,[113] HIV-1,[114] influenza virus,[115] polio virus,[116] and reovirus,[117] are considered to exploit the transcytotic activities of M cells to establish host infection. Data from a range of independent in vivo and in vitro studies indicate that prions also utilize M cells to enter the GALT and establish infection.[47,118–122] Indeed, as Fig. 2 shows both the early replication of prions upon FDC in the GALT and disease susceptibility are blocked in mice that lack M cells in the gut epithelium.[47,122]

The density and distribution of M cells in the gut epithelium can be dramatically increased after exposure to certain proinflammatory stimuli.[123] Some gastrointestinal pathogens such as *S.* Typhimurium can also enhance the abundance of M cells to establish host infection.[110] The differentiation and maturation of M cells is regulated by the production of the cytokine RANKL by subepithelial stromal cells, and systemic treatment with RANKL increases M cell density throughout the gut epithelium.[109,124] As shown in Fig. 2, factors that increase M cell density can also enhance the uptake of prions from the gut lumen and, by doing so, increase disease susceptibility.[47] M cells express a diverse range of proteins on their apical surfaces which act as uptake receptors for certain pathogenic microorganisms.[107] Although M cells express high levels of PrP^C,[125] the initial uptake of prions from the intestine is not influenced by the absence of PrP expression in the gut epithelium.[101,121]

Data on the role of M cells in the uptake of prions from the gut lumen have occasionally been conflicting. High-resolution cryo-immunogold electron microscopy studies have shown that immediately after oral infection disease-specific PrP is predominantly detected within the large multivesicular endosomes of enterocytes overlying the Peyer's patches.[101] However, if prions predominantly exploit enterocytes, rather than M cells, to establish host infection,[101,126] one would not expect oral prion disease susceptibility to be blocked in the specific absence of M cells.[47,122] M cells are also present in the epithelia covering the nasal-associated lymphoid tissue,[127,128] but studies in hamsters indicate that after exposure via the nasal cavity the majority of the prions had crossed the nasal epithelium by paracellular transport.[129]

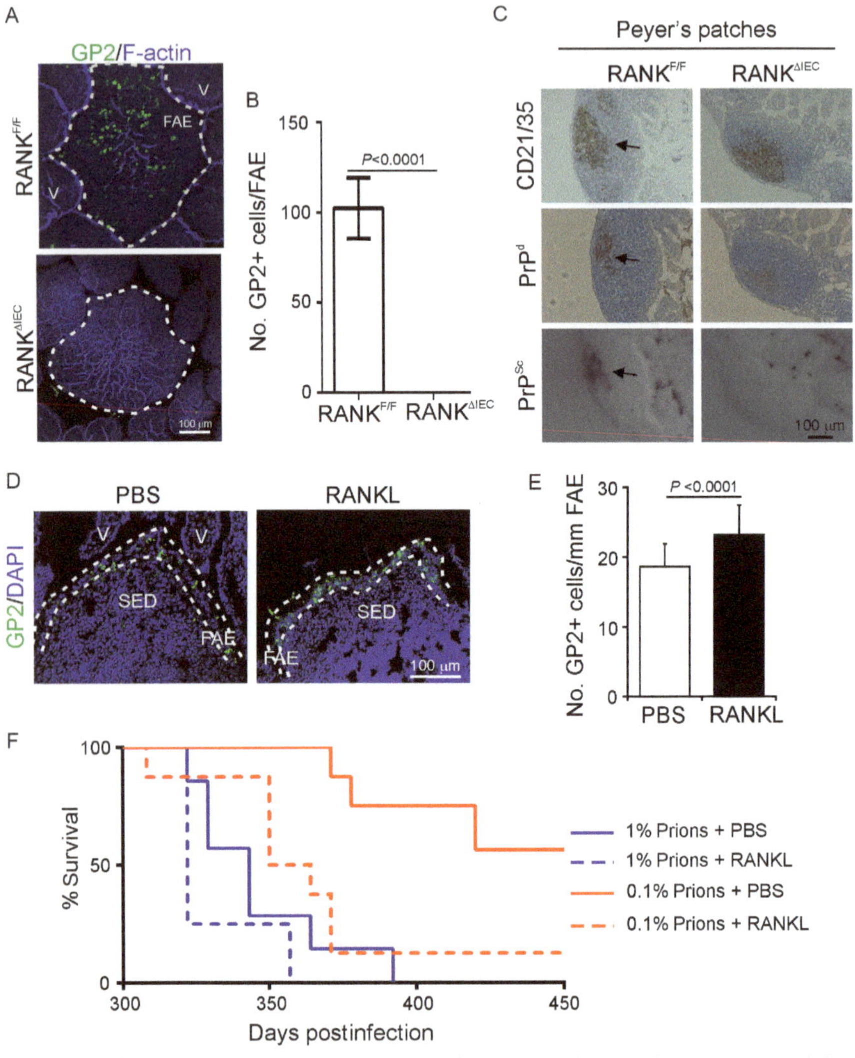

Fig. 2 M cell density in the gut epithelium influences oral prion disease susceptibility. (A) Peyer's patches from M cell-deficient RANK$^{\Delta IEC}$ mice and RANK$^{F/F}$ (control) mice were whole-mount immunostained to detect M cells (GP2^{+} cells, *green*) and F-actin (*blue*) as a counterstain. The *broken line* indicates the boundary of the follicle-associated epithelium (FAE) overlying the Peyer's patches. *V*, villi. This immunohistochemical (IHC) analysis indicated an absence of GP2^{+} M cells in the FAE of RANK$^{\Delta IEC}$ mice. (B) Morphometric analysis confirmed that the number of GP2^{+} cells/FAE was significantly reduced in RANK$^{\Delta IEC}$ mice. (C) PrP^{Sc} accumulation upon Peyer's patches FDC after oral prion exposure is blocked in M cell-deficient RANK$^{\Delta IEC}$ mice. Mice were orally exposed to ME7 scrapie prions and tissues collected at intervals afterward at 105 days postinfection. High levels of disease-specific PrP (PrP^{d}, *brown*, *middle rows*) were detected in association

3. THE CONTRASTING ROLES OF MONONUCLEAR PHAGOCYTES

The mononuclear phagocytes are a diverse group of hematopoietically derived phagocytic cells which includes conventional dendritic cells (DC), tissue macrophages, Langerhans cells in the epidermal layer of the skin, and the microglia in the CNS. As described below, data suggest that specific populations of these cells may play contrasting roles during prion disease, either mediating the propagation of prions to or within SLO, or by protecting the host by phagocytosing and destroying them.

3.1 Conventional Dendritic Cells Propagate Prions Toward and Within SLO

Classical DC sample their environments for pathogens and their antigens and are a distinct cell lineage from the stromal-derived FDC.[73,130,131] After antigen uptake, classical DC undergo maturation and migrate toward the local SLO to initiate immune responses. Although a major role of conventional

---

with FDC ($CD21/35^+$ cells, *brown, upper panels*) in the Peyer's patches of $RANK^{F/F}$ control mice. Analysis of adjacent sections by PET immunoblot analysis confirmed the presence of prion-specific PK-resistant PrP^{Sc} (*blue/black*). However, no PrP^{Sc} was detected in any of the Peyer's patches from the orally exposed M cell-deficient $RANK^{\Delta IEC}$ mice. (D and E) The density of functionally mature M cells in the intestine can be promoted in mice through exogenous administration of RANKL. C57BL/6 mice were treated daily with RANKL (or PBS as a control) to induce M cell differentiation. Peyer's patches and intestines were collected on day 3 of treatment. (D) Representative distribution of M cells ($GP2^+$ cells, *green*) in the Peyer's patches of mice from each group. *Broken lines* indicate the boundary of the FAE. *SED*, subepithelial dome; *V*, villi. (E) Morphometric analysis confirmed a significant increase in the number of $GP2^+$ M cells in the Peyer's patches of RANKL-treated mice. (F) C57BL/6 mice were treated daily for 4 days with RANKL (or PBS as a control) to induce M cell differentiation, and orally exposed to a moderate (1% scrapie brain homogenate) or limiting (0.1%) dose of ME7 scrapie prions between the 3rd and 4th treatments. RANKL treatment significantly increased disease susceptibility following oral exposure to a limiting dose of prions. Clinical prion disease survival curves for PBS- (*solid lines*) or RANKL-treated mice (*broken lines*) orally exposed to either a 1% (*blue lines*) or 0.1% (*red lines*) dose of prions (PBS/1% vs RANKL/1%, $P=0.120$; PBS/0.1% vs RANKL/0.1%, $P<0.0078$; PBS/1% vs RANKL/0.1%, $P=0.205$; log-rank [Mantel–Cox] test). *This figure is reproduced from Donaldson DS, Sehgal A, Rios D, Williams IR, Mabbott NA. Increased abundance of M cells in the gut epithelium dramatically enhances oral prion disease susceptibility.* PLoS Pathog. *2016;12:e1006075. http://doi.org/10.1371/journal.ppat.1006075 under the terms of the Creative Commons Attribution 4.0 International (CC BY 4.0) License.*

DC is to present processed (partially digested) antigens to T cells, some are capable of delivering unprocessed (native) antigens[132,133] to B cells.[134–136] The migratory characteristics of conventional DC are exploited by some pathogens to facilitate their delivery to the SLO.[137–140] Prion disease pathogenesis is delayed in the temporary absence of $CD11c^+$ or $CD8^+CDll1c^+$ conventional DC at the time of intraperitoneal prion exposure, indicating that prions may also exploit conventional DC to establish infection with SLO.[141,142]

In the GALT, particulate antigens that have been transported across the gut epithelium by M cells are subsequently released into the basolateral pocket beneath them and sampled by the mononuclear phagocytes within it.[143] Consistent with this, disease-specific PrP is similarly detectable within mononuclear phagocytes immediately beneath the epithelium overlying the Peyer's patches soon after oral exposure.[101] Furthermore, the initial accumulation of prions upon FDC in the Peyer's patches was impeded when conventional DC were transiently depleted at the time of oral exposure.[144] Chemokines help to attract lymphocytes and conventional DC to SLO and mediate their positioning within them. The chemokine CXCL13 is expressed by FDC and other stromal cells in the B cell follicles and specifically recruits CXCR5-expressing cells toward them.[145,146] In the absence of CXCR5-expressing conventional DC the early accumulation of prions upon FDC in Peyer's patches is impaired, and disease susceptibility reduced.[104] Deficiency in CCR7-CCL19/CCL21 signaling in contrast, which coordinates the localization of conventional DC to the interfollicular T cell regions of Peyer's patches,[147] does not influence oral prion disease pathogenesis.[148] These data suggest that once the prions have been transferred across the gut epithelium into the Peyer's patches by M cells they are subsequently acquired by conventional DC[101,144] and propagated by them in a CXCL13-CXCR5-dependent manner toward FDC.[104]

A diverse variety of DC subsets have been described, but their precise contributions to prion disease pathogenesis are uncertain and may be dependent on the route of exposure. For example, the depletion of $CD8^+CDllc^+$ conventional DC delayed disease pathogenesis when mice were infected with prions via the peritoneal cavity, but had no effect after oral exposure.[141] Similarly, prion disease pathogenesis after infection via epidermal skin lesions was impeded in the specific absence of $CD11c^+langerin^-$ dermal DC, but the depletion of epidermal Langerhan's cells or $langerin^+$ dermal DC had no effect.[149] Infectious prions have also been detected in association with splenic plasmacytoid DC,[150] but their role in disease pathogenesis is uncertain. Of course, a fraction of the prions may also initially reach the SLO in a cell-free manner.[129,149,151,152] For example, a highly detailed tracing study showed that some of the prions initially arrived at the draining SLO in a cell-autonomous manner, with cell-associated prions arriving in a subsequent wave a few hours later.[151]

The uptake of prions by mononuclear phagocytes is not PrP^{C} dependent as disease pathogenesis is unaffected by the absence of PrP^{C} expression in hematopoietic cells.[70–72,87,151,153] Two studies have shown that prion uptake by conventional DC is enhanced in the presence of complement components C1q and C3,[91,151] implying that this may be CR mediated. Conventional DC may also simply acquire prions nonspecifically as they constitutively sample their microenvironment by constitutive fluid-phase micropinocytosis.

3.2 Macrophages Can Engulf and Destroy Prions

Current theory suggests that the tissue macrophages provide a host-protective role by scavenging and degrading prions. This is supported by data which show that macrophages can acquire and degrade prions after extended in vitro exposure,[154,155] and that their depletion in vivo enhances the accumulation of PrP^{Sc} within SLO.[156,157]

3.3 Distinguishing Between Conventional DC and Macrophages

The majority of the studies which have attempted to determine the role of conventional DC in prion disease pathogenesis have defined them based on their expression of the integrin CD11c. However, CD11c expression is not exclusively restricted to these cells as it is also expressed by some macrophage and monocyte populations.[158–160] Many studies have described the interactions between prions and bone marrow-derived DC cultivated in the presence of GM-CSF, but transcriptional analyses show that these cells more closely resemble macrophages than conventional DC.[158,160] Treatments that ablate tissue macrophages[156,157] can also affect the abundance of other mononuclear phagocyte populations. Clearly more refined model systems are required to definitively determine the role of the different mononuclear phagocyte populations in prion disease pathogenesis.

4. B CELLS DISSEMINATE PRIONS BETWEEN SLO

The influence of B cells in prion disease pathogenesis can extend beyond the provision of homeostatic support for the maintenance of FDC. A common feature of many acquired prion diseases is that after the initial accumulation and replication of the prions in the draining SLO, they subsequently spread to most other SLO including the spleen.[26,28,59,81] B cells may play an important role in this dissemination.[161] B cells can recirculate via the blood stream and lymphatics for several weeks between SLO[162] and often strip antigens from the surfaces of FDC as they travel through the B cell follicles and germinal centers within them.[163] As Fig. 3 shows, when the recirculation of B cells between SLO was specifically blocked, the dissemination of prions between the SLO of infected mice was also blocked.[161] This indicates that recirculating B cells most likely acquire prions from infected FDC in the draining SLO and transfer them to FDC in other SLO as they recirculate around the body via the blood and lymph. Prions have also been detected in association with lymphocytes[164,165] and B cells in the peripheral blood of sheep with scrapie[166] and deer with CWD,[167] implying a similar role during natural prion diseases. The recognition that B cells and SLO played an important role in the pathogenesis of certain prion diseases[69] correctly raised the concern that vCJD (which also accumulates in the SLO[33,41]) might also contaminate the blood stream of infected individuals. In response to this concern, leukodepletion was introduced in the United Kingdom to attempt to reduce the potential burden of prions in donated blood or blood products. As a consequence of this and other preventative measures, the number of vCJD cases which have been linked to infection via transfusion of blood or blood products has fortunately been rare.[168–170]

B cells may also play an important role in the initial propagation of prions to FDC in the draining lymph nodes. Lymph fluid draining from local tissues enters into the subcapsular sinus region of the draining lymph node. The specialized macrophages which line this region acquire lymph-borne immune complexes and pass them through the floor of the subcapsular sinus to the underlying follicular B cells.[171] These B cells then acquire the immune complexes via their CR and transport them to the FDC.[171,172] Detailed in vivo PrP^{Sc} tracing studies indicate B cells may play a similar role in the initial uptake and delivery of complement-opsonized prions from the subcapsular sinus to FDC in the draining lymph node.[151]

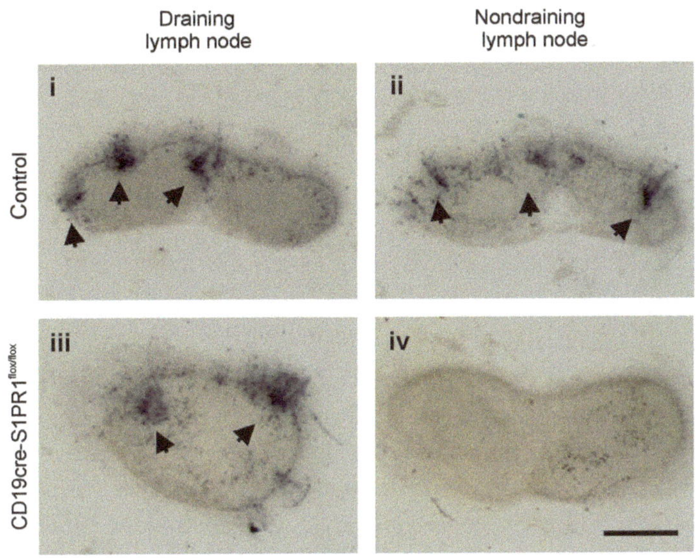

Fig. 3 B cells disseminate prions between secondary lymphoid organs (SLO). Prion dissemination between SLO is blocked in mice in which B cell recirculation via the blood stream and lymphatics is reduced. Panels (i) and (ii) show PrP^{Sc} accumulation (*blue/black*) upon follicular dendritic cells (*arrows*) within the draining and nondraining inguinal lymph nodes of control mice (Cre-deficient) at 105 days after infection by lesions in the skin epidermis. However, in mice in which B cell recirculation is specifically blocked (CD19cre $S1PR1^{flox/flox}$ mice; panels (iii, iv)), the dissemination of prions to the nondraining lymph nodes was blocked. Scale bar, 0.5 mm. *Originally published in Mok SW, Proia RL, Brinkmann V, Mabbott NA. B cell-specific S1PR1 deficiency blocks prion dissemination between secondary lymphoid organs.* J Immunol. *2012;188:5032–5040. Copyright 2012 The American Association of Immunologists, Inc.*

5. PRION NEUROINVASION FROM SLO OCCURS VIA THE PERIPHERAL NERVOUS SYSTEM

The SLO are highly innervated.[173] Following their replication upon FDC in the GALT, detailed immunohistochemical tracing studies show that orally acquired prions subsequently infect enteric nerves and spread along efferent fibers of both the sympathetic (e.g., splanchnic) and parasympathetic (e.g., vagus) nervous systems to establish CNS infection.[101,119,174–178] In support of these observations, the specific depletion of sympathetic nerves impairs prion neuroinvasion from SLO.[179] Conversely, prion disease pathogenesis is exacerbated in mice in which sympathetic nerve density in SLO is increased,[179] or in which the distance between FDC and sympathetic nerves is reduced.[82,180]

Some studies have proposed that prions may also enter the CNS by direct hematogenous transfer, as low/trace levels of PrP^{Sc} may be initially detected in the blood stream and brain soon after exposure using extremely sensitive assays.[181] In some scrapie-affected sheep disease-specific PrP has been detected in the circumventricular organs supporting the potential for hematogenous prion transfer into the CNS.[182] However, this may not be common to other host species as a similar distribution of disease-specific PrP is not evident in the brains of vCJD patients.[183] Studies in mice also show that CNS prion disease pathogenesis is not affected in the absence of recruitment of circulating monocytes.[184] Thus although the potential for some hematogenous transfer cannot be entirely excluded, strong evidence supports the conclusion that prions predominantly exploit synaptically connected autonomic ganglia and efferent fibers of the sympathetic and parasympathetic nervous systems to spread from SLO to the CNS.[178]

How the prions initially spread from the FDC to the peripheral nerves is uncertain. Mononuclear phagocytes are closely associated with enteric nerves in the intestine,[185] and a role for conventional DC in this transfer has been proposed.[186–189] Tunneling nanotubes (TNT) are slender membrane-bound cylinders of cytoplasm which can connect cells to enable cell-to-cell communication and the intercellular transfer of plasma membrane or cytoplasmic components. Infectious prions can be transferred between conventional DC and nerves via endolysosomal vesicles within TNT.[187–190] Whether such direct TNT-mediated DC:nerve interactions mediate the transfer of infectious prions in vivo remains to be determined. Infectious prions can also be released from infected cells such as conventional DC in association with small endosomal-derived vesicles termed exosomes which could transfer infection to nearby cells.[150,191] Whatever the mechanism of prion transfer, the presence of the GPI anchor moiety of the PrP molecule is essential as neuroinvasion is substantially impaired in mice that express PrP lacking the GPI anchor.[192]

6. MICROGLIA PROVIDE A HOST-PROTECTIVE ROLE IN THE CNS

Once the prions enter the CNS they amplify exponentially and cause extensive neuropathology which is typically characterized by the activation of microglia and astrocytes, accumulations of PrP^{Sc}, and neurodegeneration (see chapter "Neuropathology of human prion diseases" by Ritchie and Ironside for a detailed account of the neuropathology during human prion diseases). The resident mononuclear phagocytes of the CNS are the microglia which contribute to the maintenance of neuronal homeostasis, remodeling of synapses, and clearance of apoptotic cells, as well as providing a first line of defense against pathogens.[193–196] A change in microglial phenotype from resting to activated is an early neuropathological sign in the prion-affected CNS and precedes the onset of the neuropathology.[197] During prion disease a TGF-β- and PGE_2-dominated cytokine response is induced in the brain which stimulates the microglia to display a predominantly antiinflammatory profile.[198–203] This antiinflammatory profile is considered to be host protective as CNS prion disease pathogenesis is exacerbated in mice deficient in the antiinflammatory cytokines interleukin (IL)-4, IL-10, IL-13.[204,205] Conversely, the selective inhibition of CSF1R signaling during CNS prion disease enhances the expression of certain antiinflammatory microglial markers, slowing the development of the neuropathology.[206]

Proinflammatory mediators, in contrast, appear to play a little role in CNS prion disease during the steady state. Disease pathogenesis and neuropathology are unaltered in the absence of TNFα and TNFR1,[69,77] complement components,[92,93,95,96,207] the chemokine receptors CCR2, CCR5, and CX3CR1,[205,208] IL-6, IL-12p35, and IL-12p40,[77,209] IL1-Ra,[210] interferon neutralization,[211] the NLRP3 inflammasome,[212] and MyD88-mediated Toll-like receptor signaling.[213]

The depletion of microglia in in vitro cultivated cerebellar organotypic slices enhances prion accumulation,[214] and CNS prion disease pathogenesis is exacerbated in vivo in the absence of microglia.[215] The engulfment of apoptotic cells by mononuclear phagocytes is mediated by milk fat globule epidermal growth factor 8 (MFGE8). In the brains of mice deficient in astrocyte-derived MFGE8, CNS prion disease is accelerated due to the reduced ability of the microglia to clear apoptotic cell remnants.[193] These data clearly show that during CNS prion disease the microglia attempt to provide a general neuroprotective role by engulfing prions and prion-affected cells. Neurodegeneration appears to occur when PrP^{Sc} accumulation is associated by an inflammatory response in the microglia.[216] Interestingly, triggering receptor expressed on myeloid cells 2 (TREM2) can also mediate the engulfment of apoptotic nerves.[217] Although TREM2 expression is significantly elevated in microglia in the prion-affected brain, CNS prion disease pathogenesis is unaltered in the brains of TREM2-deficient mice.[218]

The mammalian intestine is host to large burden of commensal microorganisms, especially bacteria, referred to as the microbiota. These microorganisms provide many beneficial effects to host including the metabolism of essential nutrients, regulating tissue and cell development, and providing protection against pathogens by competing for habitat niches. Microglial function and maturation is impaired in germ-free mice, as microbiota-derived bacterial fermentation products (short-chain fatty acids) are important for the constitutive regulation of microglial homeostasis.[219] This suggests that dramatic modifications to commensal bacterial microbiota, such as following prolonged use of broad-spectrum antibiotics, might affect the progression of CNS prion disease.[220] Systemic concurrent infection with pathogenic microorganisms may also exacerbate CNS prion disease by influencing microglial phenotype. For example, systemic exposure of mice with CNS prion infection to the proinflammatory components of certain pathogenic microorganisms such as bacterial lipopolysaccharide can trigger a predominantly proinflammatory phenotype in the microglia, leading to exacerbated neuropathology.[221,222]

7. INFLUENCE OF AGING

The developmental status of the SLO can profoundly affect prion disease susceptibility. For example, the FDC in the spleens of neonatal mice are underdeveloped and have a reduced ability to replicate prions.[223] Although the majority of the clinical cases of sporadic CJD have occurred in elderly individuals (median age at onset of disease = 67 years), the majority of the clinical cases of vCJD have occurred in young adults (median age at onset of disease = 26 years). Aging is accompanied by a significant decline in immune function, termed immunosenescence. As a consequence of this immunosenescence, the elderly have decreased responses to vaccination and an increased susceptibility to cancer and infectious diseases. As Fig. 4 shows, immunosenescence also adversely affects the microarchitecture of the SLO.[224] As a consequence of these disturbances the ability of prions to establish infection upon the FDC within aged SLO is dramatically impaired. The functional maturation of M cells is adversely affected in the Peyer's patches of aged mice and reduces their ability to transfer particulate antigens across the gut epithelium.[225] Since the density of M cells in the gut epithelium can directly influence the efficiency of oral prion infection,[47] the effects of aging on their functional maturation are most

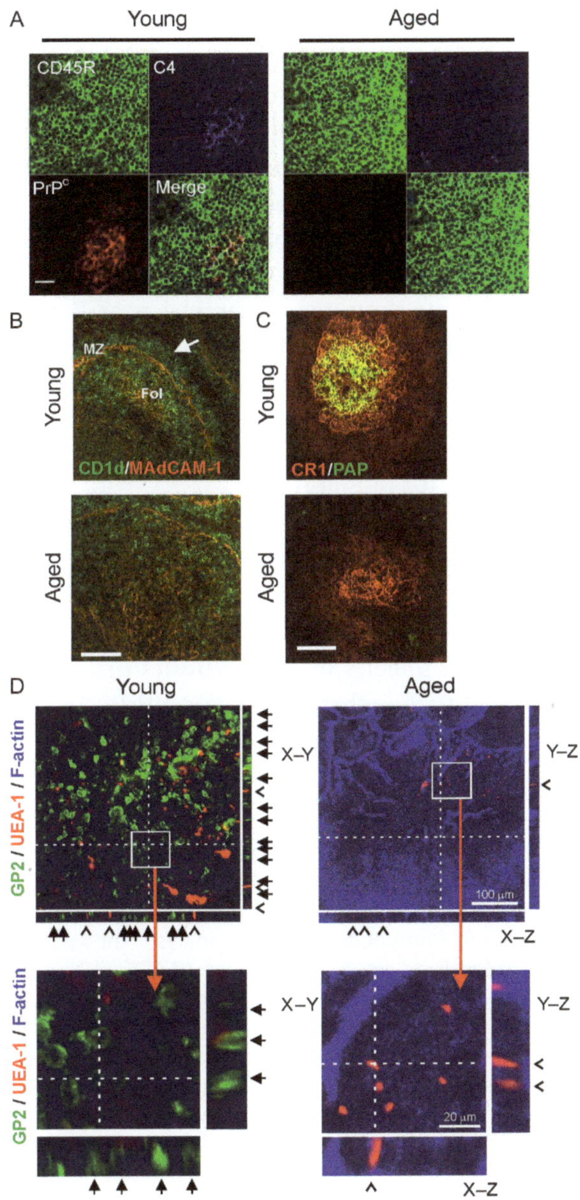

Fig. 4 Aging adversely affects the microarchitecture of secondary lymphoid organs. (A) Immunofluorescent analysis shows that follicular dendritic cells (FDC) in the spleens of aged mice (600 days old) have reduced retention of complement component C4 (*blue*) and expression of PrP^{C} (*red*) when compared to young mice (6–8 weeks old). Sections were counterstained with anti-CD45R monoclonal antibody to detect B cells (*green*). (B) In the spleens of young mice, abundant CD1d-expressing marginal zone

(Continued)

likely responsible for the dramatically reduced susceptibility of aged mice to oral prion disease.[226]

In the spleen, the marginal zone forms a barrier around the lymphocyte-containing white pulp regions and plays an important role in the removal of antigens and immune complexes from the blood stream. It is within this region that immune complexes are removed from the blood stream by marginal zone B cells and delivered by them to the FDC in the B cell follicles.[227,228] The splenic marginal zone is grossly disturbed in aged mice which adversely affects the shuttling of marginal zone B cells to the B cell follicles and the delivery of blood-borne immune complexes to FDC[229,230] (see also our animation video describing these effects: https://tinyurl.com/m2g2ej4). These aging-related impairments to the marginal zone also appear to affect the initial delivery of blood-borne prions to splenic FDC, reducing disease susceptibility.[42,229]

In natural host species age-related effects on prion disease pathogenesis and susceptibility have also been described. For example, the incidence of clinical prion disease is greater in lambs orally exposed to BSE prions during the preweaning stage than those exposed postweaning.[231] Aging effects on the abundance of B cell follicles in GALT tissues may also have practical implications, by reducing the reliability to detect PrP^{Sc} in RAMALT specimens from aged CWD-affected cervids.[232,233]

Fig. 4—Cont'd B cells (*green, arrow*) are present within the marginal zone (MZ) and B cell follicles (Fol). However, in aged mice, the distribution of marginal zone B cells is disturbed. Scale bar, 50 μm. (C) Young and aged mice were passively immunized with preformed immune complexes (PAP, *green*), and 24 h later, their presence upon FDC (CR1/CD35, *red*) assessed. Immunofluorescence analysis showed that the level of immune complexes colocalized upon the surfaces of FDC in spleens of aged mice was substantially reduced when compared to FDC in the spleens of young mice. Scale bar, 50 μm. (D) M cell density is dramatically reduced in the follicle-associated epithelium covering the Peyer's patches of aged mice. Peyer's patches were whole-mount immunostained to detect M cells (GP2; *green*), goblet cells (UEA-1; *red*), and F-actin (*blue*). The positions of the X–Z and Y–Z projections are indicated by the *broken line* in the main X–Y images. *Closed arrows* indicate $GP2^+$ M cells with characteristic basolateral pockets. *Open arrowheads* indicate $GP2^-UEA^-$ goblet cells. The *boxed areas* in each of the main X–Y images are shown below at higher magnification. *Data in (A–C) were reproduced with permission from Brown KL, Gossner A, Mok S, Mabbott NA. The effects of host age on the transport of complement-bound complexes to the spleen and the pathogenesis of intravenous scrapie infection.* J Virol. *2012;86:1228–1237. Data in (D) were reproduced with permission from Kobayashi A, Donaldson DS, Erridge C, et al. The functional maturation of M cells is dramatically reduced in the Peyer's patches of aged mice.* Mucosal Immunol. *2013;6:1027–1037.*

8. INFLUENCE OF INFLAMMATION AND PATHOGEN COINFECTION

Although a huge number of BSE-infected cattle are estimated to have entered the food chain during the BSE epidemic (~500,000; Refs. 234,235), the number of clinical cases of vCJD in humans has fortunately been disproportionally low (178 definite or probable cases, as of April 3, 2017; www.cjd.ed.ac.uk). This indicates that additional risk factors may have influenced disease susceptibility in the individuals that succumbed to clinical disease. Host susceptibility to prion infections is influenced by a range of factors including host species and *PRNP* genotype and prion agent strain (see chapter "Prion protein and genetic susceptibility to diseases caused by its misfolding" by Carlson). Various studies have also shown that the activation status of the host immune system can impact on disease pathogenesis and susceptibility. Treatment with mitogens around the time of peripheral prion exposure[236] or repetitive immunization (two injections with CpG, followed by repeated BSA/alum injections) can each enhance susceptibility to a low dose of prions.[237] Repetitive immunization coincided with increased FDC network size and an increased abundance of marginal zone B cells in the spleen, which may have enhanced the efficiency of prion replication and neuroinvasion from this tissue.[237]

Certain bacterial or viral infections or autoimmunity can trigger chronic inflammation in nonlymphoid tissues and the development of tertiary lymphoid tissues within them. Since these tertiary lymphoid tissues can contain FDC and B cell follicles, their induction in the kidneys, liver, and pancreas can facilitate the replication of prions within peripheral tissues which would otherwise be devoid of infectivity.[238,239] The presence of these tertiary lymphoid tissues in natural prion-affected hosts can have important implications. Their presence in the mammary glands of sheep with lentiviral mastitis may lead to enhanced transmission of prions to suckling lambs.[240] Nephritis in prion-affected hosts can enhance the excretion of prions in urine,[238] which could increase the contamination of the environment and the horizontal disease transmission. Chronic inflammation in the intestine, through the induction of FDC-containing isolated lymphoid follicles, can also enhance the distribution of prions throughout the gastrointestinal tract and provide additional sites of neuroinvasion after oral exposure.[28,46] Mouse studies show that within soft tissue granulomas an LTβR-dependent inflammatory stromal cell population distinct from FDC can also replicate prions.[241]

However, it is uncertain whether granulomatous lesions are also important additional sites of prion accumulation in natural host species such as sheep.[242]

Concurrent infection with the enteroinvasive bacterium *S.* Typhimurium around the time of oral prion exposure has also been shown to significantly increase prion disease susceptibility.[243] The original study proposed that this effect was primarily due to the pathology (colitis) caused within the large intestine by the *S.* Typhimurium infection. However, a subsequent study using the large intestine-restricted helminth pathogen *Trichuris muris* did not observe an effect on oral prion disease susceptibility despite also causing significant pathology within the cecum.[46] In addition to causing colitis in the large intestine, *S.* Typhimurium can also enhance the density of M cells in the GALT.[110] These effects on M cell density may have significantly increased the uptake of prions into Peyer's patches in the small intestine, reducing disease susceptibility.[47] Lesions to the mucosal surfaces in the oral[244] and nasal cavities[245] can also increase the initial uptake of prions at these sites and reduce disease susceptibility.

Transmissions of cattle (primary) BSE prions to experimental mice indicate that after interspecies prion exposure, the processing and amplification of prions upon FDC in SLO is important for their adaptation to the new host.[42,246] Inflammation at the site of exposure, or the pathology caused by a congruent pathogen infection, might enhance the initial uptake and delivery of prions to the draining SLO. This could reduce the transmission barrier by providing a greater opportunity for prion quasi-species with zoonotic potential to be selected, undergo host adaptation and amplification upon FDC,[95] instead of being phagocytosed and destroyed by macrophages and/or excreted before this could be achieved.[156,157]

Retrospective analyses of human tonsil and appendix specimens indicate a higher incidence of individuals with signs of vCJD accumulation in their SLO (~1/2000) than the clinical data suggest.[247] This implies that many individuals/animals may harbor detectable levels of prions in their tissues in the absence of clinical signs, and the potential existence of a subclinical carrier state. Using transgenic mice expressing the ovine or human form of PrP^C, one study has compared the ability of the CNS and SLO to replicate prions derived from other host species such as BSE and CWD. In these experiments the SLO were consistently more permissive than the brain to cross-species prion infection. These data illustrate how continued surveillance of SLO will be important to provide reliable estimate of the incidence of preclinical prion disease within populations. The factors which influence whether such individuals remain as subclinical "silent carriers" or progress to develop clinical disease remain to be determined.

9. PROSPECTS FOR IMMUNOTHERAPY AGAINST PRIONS

Many of the studies described above show how experimental treatments that block prion accumulation in SLO can also significantly delay the progression of clinical disease.[26,79–81,83,122,142,144,149] However, these treatments were only effective when applied immediately before or after infection (prion to neuroinvasion). Many of these approaches are also immunosuppressive and may enhance susceptibility to other pathogens. Due to the ubiquitous cellular expression of PrP^{C} this protein is tolerated by the host's immune system preventing the development of specific immune responses against prion-specific PrP^{Sc}, even though high levels of prions often accumulate and persist within the SLO throughout the infection.[248,249] However, some anti-PrP antibodies may be elicited during the clinical (terminal) phase.[250] Since the emergence of vCJD in humans intensive effort has been focused on the development of potential antiprion vaccines. Initial studies in experimental mice showed that this was plausible as both the passive administration of anti-PrP monoclonal antibodies[251] and vaccination using recombinant PrP in complete Freund's adjuvant[252] could each delay the development of prion disease after intraperitoneal exposure.

Many attempts have been made to overcome the host's tolerance to PrP in order to improve vaccine efficacy. Elegant experimental approaches have been described, including adoptive transfers of PrP peptide-loaded conventional DC,[253] PrP-sensitized $CD4^{+}$ T cells from PrP-deficient donor mice,[254] or transgenic T cells with PrP-specific T cell receptors.[255] However, these costly and technically challenging approaches are unlikely to be feasible for widespread application.

Although significant challenges remain to be addressed, mucosal vaccination (using an attenuated *Salmonella* vector expressing recombinant PrP) can provide some prophylactic protection against orally acquired prion infections in mice.[256] This vaccination method elicits a strong mucosal (IgA-dominated) anti-PrP antibody response. In humans the risk of accidental iatrogenic vCJD transmission via contaminated blood or blood products, tissues, or contaminated surgical instruments is a current concern. However, mucosal vaccination may confer limited protection, if any, against a systemic prion infection where induction of a strong anti-PrP IgG response is desirable. Intramuscular vaccination can induce a strong anti-PrP IgG response in mice, and impede prion disease pathogenesis after intraperitoneal exposure, but a similar approach was unable to provide protection against oral CWD infection in mule deer.[257] It is therefore desirable that an effective antiprion vaccine should be able to induce strong mucosal and systemic anti-PrP antibody responses.

Of course, prophylactic vaccines are only likely to be effective during the early stages of the infection, by preventing the prions from establishing infection in SLO and spreading to the CNS. Since large molecules such immunoglobulins are unable to cross the blood–brain barrier in significant amounts, mucosal or systemic vaccination is unlikely to be effective once disease is established in the CNS. To attempt to address this issue, one study has shown that camelid single-domain PrP-specific antibodies can cross the blood–brain barrier,[258] but their potential to attenuate CNS prion disease is not yet known. Studies have shown that the intracerebroventricular administration of anti-PrP antibodies directly into the brains of prion disease-affected mice can prolong survival times.[259] At the time of writing clinical trials were being planned to determine whether a similar administration of anti-PrP antibodies in the CNS might also offer therapeutic protection to patients with clinical signs of sporadic CJD.

However, irrespective of the approaches used it is essential that the anti-PrP antibodies do not recognize host PrP^C and, by doing so, cause autoimmunity or neurotoxicity. Data show that neurotoxicity occurred in the mouse brain after exposure to PrP-specific antibodies which bound to the $\alpha 1$ and $\alpha 3$ helices of the PrP^C globular domain.[260] This helps to provide important safety information to enable any novel antiprion vaccines or PrP-specific monoclonal antibodies to be screened for potential neurotoxic effects.[261]

10. CONCLUDING REMARKS

Fig. 5 summarizes the significant progress that has been made in recent years in identifying the key tissues, cells, and molecules that play mediate the propagation of prions from the periphery to the CNS. Promising progress has also been made in the development of potential immunotherapy-based prophylactic and therapeutic treatments. The identification of a neuroprotective role for the microglia during CNS prion disease reveals a novel opportunity for therapeutic intervention, by designing strategies to enhance this activity to delay or prevent the onset of further neuropathology. Despite the substantial advances that have been made, important knowledge gaps

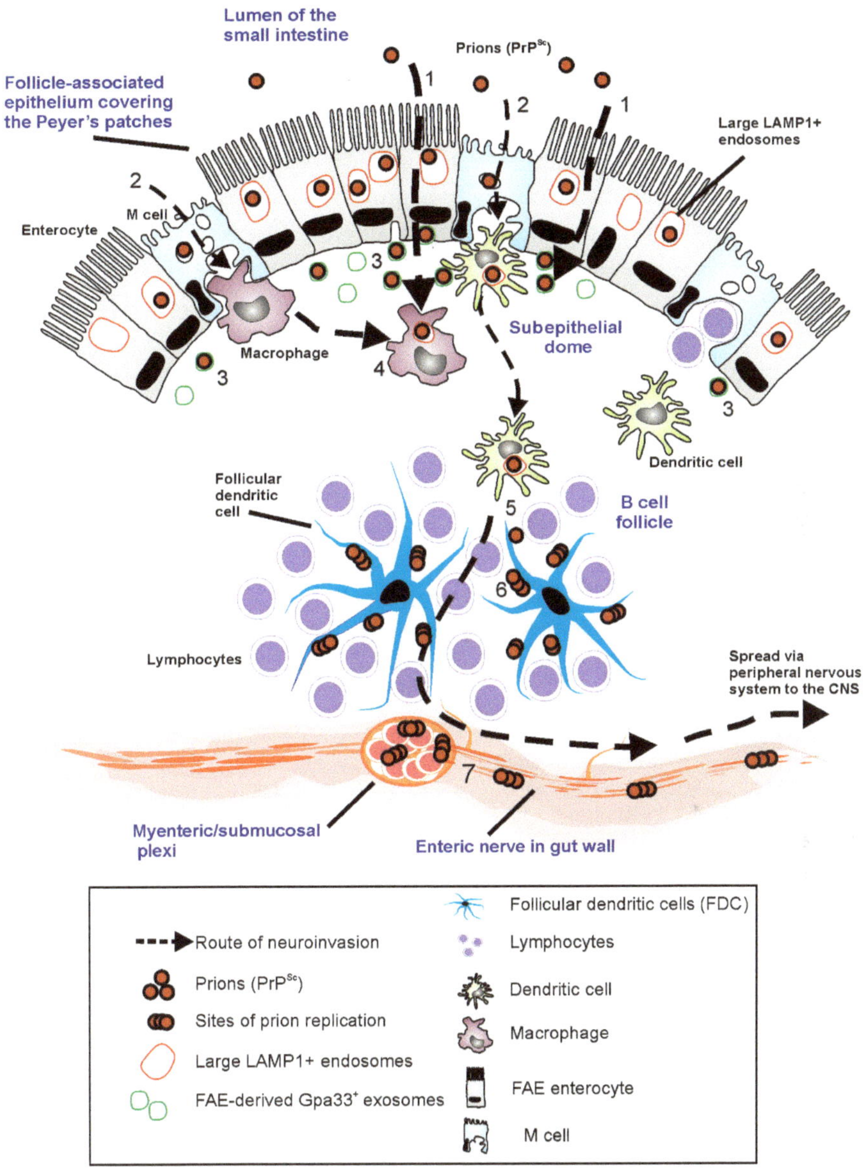

Fig. 5 Prions exploit an elegant cellular relay to establish host infection and spread from the gut lumen to the brain. (1) Although prion uptake into Peyer's patches from gut lumen can occur via large LAMP1-positive endosomes of enterocytes,[101] (2) oral prion disease susceptibility is blocked in the absence of M cells indicating that prions exploit M cells to infect Peyer's patches.[47,122] (3) The prions which are transported by the enterocytes are released into subepithelial dome region[101] and (4) most likely engulfed by macrophages.[156] (5) Conventional dendritic cells are considered to acquire prions after their transcytosis by M cells and deliver them to the follicular dendritic cells (FDC) in the B cell follicles.[104] (6) The prions then replicate on the surface of FDC.[26,87,101] (7) Once the prions have been amplified on the surfaces of FDC they then infected enteric nerves close to Peyer's patches[101] and spread through the peripheral nervous system to establish disease within the CNS.[178] *This figure is adapted from Kujala P, Raymond C, Romeijn M, et al. Prion uptake in the gut: identification of the first uptake and replication sites.* PLoS Pathog. *2011;7:e1002449. http://dx.doi.org/10.1371/journal.ppat.1002449 under the terms of the Creative Commons Attribution License.*

remain (for detailed discussion, see Ref. 262). Any future outbreaks of novel and emerging prion diseases with significant consequence for human and animal health are likely to be orally acquired. It is therefore essential that we continue to study the factors that influence disease susceptibility and the risk of cross-species transmission and zoonosis.

REFERENCES

1. Bolton DC, McKinley MP, Prusiner SB. Identification of a protein that purifies with the scrapie prion. *Science*. 1982;218:1309–1311.
2. Legname G, Baskakov IV, Nguyen H-OB, et al. Synthetic mammalian prions. *Science*. 2004;305:673–676.
3. Manson JC, Clarke AR, McBride PA, McConnell I, Hope J. PrP gene dosage determines the timing but not the final intensity or distribution of lesions in scrapie pathology. *Neurodegeneration*. 1994;3:331–340.
4. Prusiner SB, Groth D, Serban A, et al. Ablation of the prion protein (PrP) gene in mice prevents scrapie and facilitates production of anti-PrP antibodies. *Proc Natl Acad Sci USA*. 1993;90:10608–10612.
5. Prusiner SB. Novel proteinaceous infectious particles cause scrapie. *Science*. 1982;216:136–144.
6. Wang F, Wang X, Yuan CG, Ma J. Generating a prion with bacterially expressed recombinant prion protein. *Science*. 2010;327:1132–1135.
7. Stahl N, Borchelt DR, Hsiao K, Prusiner SB. Scrapie prion protein contains a phosphatidylinositol glycolipid. *Cell*. 1987;51:229–240.
8. Pan K-M, Baldwin M, Nguyen J, et al. Conversion of alpha-helices into beta-sheets features in the formation of the scrapie prion protein. *Proc Natl Acad Sci USA*. 1993;90:10962–10966.
9. Riek R, Hornemann S, Wider G, Glockshuber R, Wuthrich K. NMR characterization of the full-length recombinant murine prion protein, *m*PrP. *FEBS Lett*. 1997;413:282–288.
10. Caughey BW, Dong A, Bhat KS, Ernst D, Hayes SF, Caughey WS. Secondary structure analysis of the scrapie-associated protein PrP 27-30 in water by infrared spectroscopy. *Biochemistry*. 1991;30:7672–7680.
11. Deleault NR, Lucassen RW, Supattapone S. RNA molecules stimulate prion protein conversion. *Nature*. 2003;425:717–720.
12. Ma J. The role of cofactors in prion propagation and infectivity. *PLoS Pathog*. 2012;8: e1002589.
13. Küffer A, Lakkraju AK, Mogha A, et al. The prion protein is an agonistic ligand of the G protein-coupled receptor Adgrg6. *Nature*. 2016;536:464–468.
14. Cashman NR, Loertscher R, Nalbantoglu J, et al. Cellular isoform of the scrapie agent protein participates in lymphocyte activation. *Cell*. 1990;61:185–192.
15. Mabbott NA, Brown KL, Manson J, Bruce ME. T lymphocyte activation and the cellular form of the prion protein. *Immunology*. 1997;92:161–165.
16. Martínez del Hoyo G, López-Bravo M, Metharom P, Ardavín C, Aucouturier P. Prion protein expression by mouse dendritic cells is restricted to the nonplasmacytoid subsets and correlates with the maturation state. *J Immunol*. 2006;177:6137–6142.
17. Jouvin-Marche E, Attuli-Audenis V, Aude-Garcia C, et al. Overexpression of cellular prion protein induces and antioxidant environment altering T cell development in the thymus. *J Immunol*. 2006;176:3490–3497.

18. Ballerini C, Gourdain P, Bachy V, et al. Functional implication of cellular prion protein in antigen-driven interactions between T cells and dendritic cells. *J Immunol.* 2006;176:7254–7262.
19. Nakato G, Hase K, Suzuki M, et al. Cutting edge: *Brucella abortus* exploits a cellular prion protein on intestinal M cells as an invasive receptor. *J Immunol.* 2012;189: 1540–1544.
20. de Almeida CJG, Chiarini LB, da Silva JP, E Silva PMR, Martins MA, Linden R. The cellular prion protein modulates phagocytosis and inflammatory response. *J Leukoc Biol.* 2005;77:238–246.
21. McCulloch L, Brown KL, Mabbott NA. Ablation of the cellular prion protein, PrP^C, specifically on follicular dendritic cells has no effect on their maturation or function. *Immunology.* 2013;138:246–257.
22. Nuvolone M, Kana V, Hutter G, et al. SIRPα polymorphisms, but not the prion protein, control phagocytosis of apoptotic cells. *J Exp Med.* 2013;210:2539–2552.
23. Wilesmith JW, Ryan JB, Atkinson MJ. Bovine spongiform encephalopathy: epidemiological studies of the origin. *Vet Rec.* 1991;218:199–203.
24. Bruce ME, Will RG, Ironside JW, et al. Transmissions to mice indicate that 'new variant' CJD is caused by the BSE agent. *Nature.* 1997;389:498–501.
25. Hill AF, Desbruslais M, Joiner S, Sidle KCL, Gowland I, Collinge J. The same prion strain causes vCJD and BSE. *Nature.* 1997;389:448–450.
26. Mabbott NA, Young J, McConnell I, Bruce ME. Follicular dendritic cell dedifferentiation by treatment with an inhibitor of the lymphotoxin pathway dramatically reduces scrapie susceptibility. *J Virol.* 2003;77:6845–6854.
27. Prinz M, Huber G, Macpherson AJS, et al. Oral prion infection requires normal numbers of Peyer's patches but not of enteric lymphocytes. *Am J Pathol.* 2003;162: 1103–1111.
28. Glaysher BR, Mabbott NA. Role of the GALT in scrapie agent neuroinvasion from the intestine. *J Immunol.* 2007;178:3757–3766.
29. Sigurdson CJ, Williams ES, Miller MW, Spraker TR, O'Rourke KI, Hoover EA. Oral transmission and early lymphoid tropism of chronic wasting disease PrP^{res} in mule deer fawns (*Odocoileus hemionus*). *J Gen Virol.* 1999;80:2757–2764.
30. Andreoletti O, Berthon P, Marc D, et al. Early accumulation of PrP^{Sc} in gut-associated lymphoid and nervous tissues of susceptible sheep from a Romanov flock with natural scrapie. *J Gen Virol.* 2000;81:3115–3126.
31. Heggebø R, Press CM, Gunnes G, et al. Distribution of prion protein in the ileal Peyer's patch of scrapie-free lambs and lambs naturally and experimentally exposed to the scrapie agent. *J Gen Virol.* 2000;81:2327–2337.
32. van Keulen LJM, Schreuder BEG, Vromans MEW, Langeveld JPM, Smits MA. Scrapie-associated prion protein in the gastro-intestinal tract of sheep with scrapie. *J Comp Pathol.* 1999;121:55–63.
33. Hilton D, Fathers E, Edwards P, Ironside J, Zajicek J. Prion immunoreactivity in appendix before clinical onset of variant Creutzfeldt-Jakob disease. *Lancet.* 1998; 352(9129):703–704.
34. Glatzel M, Abela E, Maissen M, Aguzzi A. Extraneural pathological prion protein in sporadic Creutzfeldt-Jakob disease. *N Engl J Med.* 2003;349:1812–1820.
35. Hill AF, Butterworth RJ, Joiner S, et al. Investigation of variant Creutzfeldt-Jakob disease and other prion diseases with tonsil biopsy samples. *Lancet.* 1999;353:183–189.
36. Somerville RA, Birkett CR, Farquhar CF, et al. Immunodetection of PrP^{Sc} in spleens of some scrapie-infected sheep but not BSE-infected cows. *J Gen Virol.* 1997;78:2389–2396.
37. Terry LA, Marsh S, Ryder SJ, Hawkins SAC, Wells GAH, Spencer YI. Detection of disease-specific PrP in the distal ileum of cattle exposed orally to the agent of bovine spongiform encephalopathy. *Vet Rec.* 2003;152:387–392.

38. Wells GAH, Dawson M, Hawkins SAC, et al. Infectivity in the ileum of cattle challenged orally with bovine spongiform encephalopathy. *Vet Rec.* 1994;135:40–41.
39. Hoffmann C, Eiden M, Kaatz M, et al. BSE infectivity in jejunum, ileum and ileocaecal junction of incubating cattle. *Vet Res.* 2011;42:21.
40. Foster JD, Parnham DW, Hunter N, Bruce M. Distribution of the prion protein in sheep terminally affected with BSE following experimental oral transmission. *J Gen Virol.* 2001;82:2319–2326.
41. Hill AF, Zeidler M, Ironside J, Collinge J. Diagnosis of new variant Creutzfeldt-Jakob disease by tonsil biopsy. *Lancet.* 1997;349:99–100.
42. Brown KL, Mabbott NA. Evidence of subclinical prion disease in aged mice following exposure to bovine spongiform encephalopathy. *J Gen Virol.* 2014;95:231–243.
43. Fraser H, Dickinson AG. Pathogenesis of scrapie in the mouse: the role of the spleen. *Nature.* 1970;226:462–463.
44. Fraser H, Dickinson AG. Studies on the lymphoreticular system in the pathogenesis of scrapie: the role of spleen and thymus. *J Comp Pathol.* 1978;88:563–573.
45. Horiuchi M, Furuoka H, Kitamura N, Shinagawa M. Alymphoplasia mice are resistant to prion infection via oral route. *Jpn J Vet Res.* 2006;53:149–157.
46. Donaldson DS, Else KJ, Mabbott NA. The gut-associated lymphoid tissues in the small intestine, not the large intestine, play a major role in oral prion disease pathogenesis. *J Virol.* 2015;15:9532–9547.
47. Donaldson DS, Sehgal A, Rios D, Williams IR, Mabbott NA. Increased abundance of M cells in the gut epithelium dramatically enhances oral prion disease susceptibility. *PLoS Pathog.* 2016;12: e1006075.
48. Glaysher BR, Mabbott NA. Role of the draining lymph node in scrapie agent transmission from the skin. *Immunol Lett.* 2007;109:64–71.
49. Kratzel C, Kruger D, Beekes M. Relevance of the regional lymph node in scrapie pathogenesis after peripheral infection of hamsters. *BMC Vet Res.* 2007;3:22.
50. Bessen RA, Martinka S, Kelly J, Gonzales D. Role of the lymphoreticular system in prion neuroinvasion from the oral and nasal mucosa. *J Virol.* 2009;83:6435–6445.
51. Haybaeck J, Heikenwalder M, Klevenz B, et al. Aerosols transmit prions to immunocompetent and immunodeficient mice. *PLoS Pathog.* 2011;7: e1001257.
52. González L, Dagleish MP, Bellworthy SJ, et al. Postmortem diagnosis of preclinical and clinical scrapie in sheep by the detection of disease-associated PrP in their rectal mucosa. *Vet Rec.* 2006;158:325–331.
53. Espenes A, Press CM, Landsverk T, et al. Detection of PrP^{Sc} in rectal biopsy and necropsy samples from sheep with experimental scrapie. *J Comp Pathol.* 2006;134:115–125.
54. Spraker TR, Gidlewski TL, Balachandran A, VerCauteren KC, Creekmore L, Munger RD. Detection of PrP^{CWD} in postmortem rectal lymphoid tissues in Rocky Mountain elk (*Cervus elaphus nelsoni*) infected with chronic wasting disease. *J Vet Diagn Invest.* 2006;18:553–557.
55. Wolfe LL, Spraker TR, González L, et al. PrP^{CWD} in rectal lymphoid tissue of deer (*Odocoileus* spp.). *J Gen Virol.* 2007;88:2078–2082.
56. Dennis MM, Thomsen BV, Marshall KL, et al. Evaluation of immunohistochemical detection of prion protein in rectoanal mucosa-associated lymphoid tissue for diagnosis of scrapie in sheep. *Am J Vet Res.* 2009;70:63–72.
57. van Keulen LJ, Schreuder BE, Vromans ME, Langeveld JP, Smits MA. Pathogenesis of natural scrapie in sheep. *Arch Virol Suppl.* 2000;16:57–71.
58. van Keulen LJM, Vromans MEW, van Zijderveld FG. Early and late pathogenesis of natural scrapie infection in sheep. *APMIS.* 2002;110:23–32.
59. van Keulen LJM, Bossers A, Van Zijderveld FG. TSE pathogenesis in cattle and sheep. *Vet Res.* 2008;39:24.

60. Tabouret G, Lacroux C, Lugan S, et al. Relevance of oral experimental challenge with classical scrapie in sheep. *J Gen Virol.* 2010;91:2139–2144.
61. Gonzalez L, Martin S, Siso S, et al. High prevalence of scrapie in a dairy goat herd: tissue distribution of disease-associated PrP and effect of *PRNP* genotype and age. *Vet Res.* 2009;40:65.
62. Keane D, Barr D, Osborn R, et al. Validation of use of rectoanal mucosa-associated lymphoid tissue for immunohistochemical diagnosis of chronic wasting disease in white-tailed deer (*Odocoileus virginianus*). *J Clin Microbiol.* 2009;47:1412–1417.
63. Thomsen BV, Schneider DA, O'Rourke KI, et al. Diagnostic accuracy of rectal mucosa biopsy testing for chronic wasting disease within white-tailed deer (*Odocoileus virginianus*) herds in North America: effects of age, sex, polymorphism at *PRNP* codon 96, and disease progression. *J Vet Diagn Invest.* 2012;24:878–887.
64. Monello RJ, Powers JG, Hobbs NT, Spraker TR, O'Rourke KI, Wild MA. Efficacy of antemortem rectal biopsies to diagnose and estimate prevalence of chronic wasting disease in free-ranging cow elk (*Cervus elaphus nelsoni*). *J Wildl Dis.* 2013;49:270–278.
65. Spraker TR, VerCauteren KC, Gidlewski T, et al. Antemortem detection of PrPCWD in preclinical, ranch-raised Rocky Mountain elk (*Cervus elaphus nelsoni*) by biopsy of the rectal mucosa. *J Vet Diag Invest.* 2009;21:15–24.
66. Kitamoto T, Muramoto T, Mohri S, Doh-Ura K, Tateishi J. Abnormal isoform of prion protein accumulates in follicular dendritic cells in mice with Creutzfeldt-Jakob disease. *J Virol.* 1991;65:6292–6295.
67. McBride P, Eikelenboom P, Kraal G, Fraser H, Bruce ME. PrP protein is associated with follicular dendritic cells of spleens and lymph nodes in uninfected and scrapie-infected mice. *J Pathol.* 1992;168:413–418.
68. McFarlin DE, Raff MC, Simpson E, Nehlsen SH. Scrapie in immunologically deficient mice. *Nature.* 1971;233:336.
69. Klein MA, Frigg R, Flechsig E, et al. A crucial role for B cells in neuroinvasive scrapie. *Nature.* 1997;390:687–691.
70. Klein MA, Frigg R, Raeber AJ, et al. PrP expression in B lymphocytes is not required for prion neuroinvasion. *Nat Med.* 1998;4:1429–1433.
71. Brown KL, Stewart K, Ritchie D, et al. Scrapie replication in lymphoid tissues depends on PrP-expressing follicular dendritic cells. *Nat Med.* 1999;5:1308–1312.
72. Mohan J, Brown KL, Farquhar CF, Bruce ME, Mabbott NA. Scrapie transmission following exposure through the skin is dependent on follicular dendritic cells in lymphoid tissues. *J Dermatol Sci.* 2004;35:101–111.
73. Krautler NJ, Kana V, Kranich J, et al. Follicular dendritic cells emerge from ubiquitous perivascular precursors. *Cell.* 2012;150:194–206.
74. Mackay F, Browning JL. Turning off follicular dendritic cells. *Nature.* 1998;395:26–27.
75. Fraser H, Brown K. Peripheral pathogenesis of scrapie in normal and immunocompromised mice. *Anim Technol.* 1994;45:21–22.
76. Fraser H, Brown KL, Stewart K, McConnell I, McBride P, Williams A. Replication of scrapie in spleens of SCID mice follows reconstitution with wild-type mouse bone marrow. *J Gen Virol.* 1996;77:1935–1940.
77. Mabbott NA, Williams A, Farquhar CF, Pasparakis M, Kollias G, Bruce ME. Tumor necrosis factor-alpha-deficient, but not interleukin-6-deficient, mice resist peripheral infection with scrapie. *J Virol.* 2000;74:3338–3344.
78. Prinz M, Montrasio F, Klein MA, et al. Lymph nodal prion replication and neuroinvasion in mice devoid of follicular dendritic cells. *Proc Natl Acad Sci USA.* 2002;99:919–924.
79. Mabbott NA, Mackay F, Minns F, Bruce ME. Temporary inactivation of follicular dendritic cells delays neuroinvasion of scrapie. *Nat Med.* 2000;6:719–720.

80. Montrasio F, Frigg R, Glatzel M, et al. Impaired prion replication in spleens of mice lacking functional follicular dendritic cells. *Science*. 2000;288:1257–1259.
81. Mohan J, Bruce ME, Mabbott NA. Follicular dendritic cell dedifferentiation reduces scrapie susceptibility following inoculation via the skin. *Immunology*. 2005;114:225–234.
82. Prinz M, Heikenwalder M, Junt T, et al. Positioning of follicular dendritic cells within the spleen controls prion neuroinvasion. *Nature*. 2003;425:957–962.
83. Mabbott NA, McGovern G, Jeffrey M, Bruce ME. Temporary blockade of the tumour necrosis factor signaling pathway impedes the spread of scrapie to the brain. *J Virol*. 2002;76:5131–5139.
84. Montrasio F, Cozzio A, Flechsig E, et al. B-lymphocyte-restricted expression of the prion protein does not enable prion replication in PrP knockout mice. *Proc Natl Acad Sci USA*. 2001;98:4034–4037.
85. Raeber AJ, Sailer A, Hegyi I, et al. Ectopic expression of prion protein (PrP) in T lymphocytes or hepatocytes of PrP knockout mice is insufficient to sustain prion replication. *Proc Natl Acad Sci USA*. 1999;96:3987–3992.
86. Heesters BA, Myers RC, Carroll MC. Follicular dendritic cells: dynamic antigen libraries. *Nat Rev Immunol*. 2014;14:495–504.
87. McCulloch L, Brown KL, Bradford BM, et al. Follicular dendritic cell-specific prion protein (PrP^C) expression alone is sufficient to sustain prion infection in the spleen. *PLoS Pathog*. 2011;7: e1002402.
88. Blanquet-Grossard F, Thielens NM, Vendrely C, Jamin M, Arlaud GJ. Complement protein C1q recognizes a conformationally modified form of the prion protein. *Biochemistry*. 2005;44:4349–4356.
89. Sim RB, Kishore U, Villiers CL, Marche PN, Mitchell DA. C1q binding and complement activation by prions and amyloid. *Immunobiology*. 2007;212:355–362.
90. Mitchell DA, Kirby L, Paulin SM, Villiers CL, Sim RB. Prion protein activates and fixes complement directly via the classical pathway: implications for the mechanism of scrapie agent propagation in lymphoid tissue. *Mol Immunol*. 2007;44:2997–3004.
91. Flores-Lagnarica A, Sebti Y, Mitchell DA, Sim RB, MacPherson GG. Scrapie pathogenesis: the role of complement C1q in scrapie agent uptake by conventional dendritic cells. *J Immunol*. 2009;182:1305–1313.
92. Klein MA, Kaeser PS, Schwarz P, et al. Complement facilitates early prion pathogenesis. *Nat Med*. 2001;7:488–492.
93. Mabbott NA, Bruce ME, Botto M, Walport MJ, Pepys MB. Temporary depletion of complement component C3 or genetic deficiency of C1q significantly delays onset of scrapie. *Nat Med*. 2001;7:485–487.
94. Zabel MD, Heikenwalder M, Prinz M, et al. Stromal complement receptor CD21/35 facilitates lymphoid prion colonization and pathogenesis. *J Immunol*. 2007;179: 6144–6152.
95. Michel B, Ferguson A, Johnson T, et al. Genetic depletion of complement receptors CD21/35 prevents terminal prion disease in a mouse model of chronic wasting disease. *J Immunol*. 2012;189:4520–4527.
96. Michel B, Ferguson A, Johnson T, et al. Complement protein C3 exacerbates prion disease in a mouse model of chronic wasting disease. *Int Immunol*. 2013;25:697–702.
97. Heesters BA, Chatterjee P, Kim YA, Kuligowski MP, Kirchhausen T, Carroll MC. Endocytosis and recycling of immune complexes by follicular dendritic cells enhances B cell antigen binding and activation. *Immunity*. 2013;38:1164–1175.
98. Jeffrey M, McGovern G, Goodsir CM, Brown KL, Bruce ME. Sites of prion protein accumulation in scrapie-infected mouse spleen revealed by immuno-electron microscopy. *J Pathol*. 2000;191:323–332.

99. McGovern G, Brown KL, Bruce ME, Jeffrey M. Murine scrapie infection causes an abnormal germinal centre reaction in the spleen. *J Comp Pathol.* 2004;130:181–194.
100. McGovern G, Mabbott NA, Jeffrey M. Scrapie affects the maturation cycle and immune complex trapping by follicular dendritic cells in mice. *PLoS One.* 2009;4: e8186.
101. Kujala P, Raymond C, Romeijn M, et al. Prion uptake in the gut: identification of the first uptake and replication sites. *PLoS Pathog.* 2011;7: e1002449.
102. Marijanovic Z, Caputo A, Campana V, Zurzolo C. Identification of an intracellular site of prion conversion. *PLoS Pathog.* 2009;5: e1000426.
103. Kimberlin RH, Walker CA. Pathogenesis of scrapie in mice after intragastric infection. *Virus Res.* 1989;12:213–220.
104. Bradford BM, Reizis B, Mabbott NA. Oral prion disease pathogenesis is impeded in the specific absence of CXCR5-expressing dendritic cells. *J Virol.* 2017;91. http://dx.doi.org/10.1128/JVI.00124-17. pii: e00124-17.
105. O'Connor T, Frei N, Sponarova J, Schwarz P, Heikenwalder M, Agguzi A. Lymphotoxin, but not TNF, is required for prion invasion of lymph nodes. *PLoS Pathog.* 2012;8: e1002867.
106. Blättler T, Brandner S, Raeber AJ, et al. PrP-expressing tissue required for transfer of scrapie infectivity from spleen to brain. *Nature.* 1997;389:69–73.
107. Mabbott NA, Donaldson DS, Ohno H, Williams IR, Mahajan A. Microfold (M) cells: important immunosurveillance posts in the intestinal epithelium. *Mucosal Immunol.* 2013;6:666–677.
108. Hase K, Kawano K, Nochi T, et al. Uptake through glycoprotein 2 of FimH$^+$ bacteria by M cells initiates mucosal immune responses. *Nature.* 2009;462:226–231.
109. Rios D, Wood MB, Li J, Chassaing B, Gewirtz AT, Williams IR. Antigen sampling by intestinal M cells is the principal pathway initiating mucosal IgA production to commensal enteric bacteria. *Mucosal Immunol.* 2016;9:907–916.
110. Tahoun A, Mahajan S, Paxton E, et al. Salmonella transforms follicle-associated epithelial cells into M cells to promote intestinal invasion. *Cell Host Microbe.* 2012;12:645–666.
111. Chiba S, Nagai T, Hayashi T, Baba Y, Nagai S, Koyasu S. Listerial invasion protein internalin B promotes entry into ileal Peyer's patches in vivo. *Microbiol Immunol.* 2011;55:123–129.
112. Ponnusamy D, Periasamy S, Tripathi BN, Pal A. Mycobacterium Avium subsp. paratuberculosis invades through M cell and enterocytes across ileal and jejunal mucosa of lambs. *Res Vet Sci.* 2013;94:306–312.
113. Kolawole AO, Gonzalez-Hernandez MB, Turula H, Yu C, Elftman MD, Wobus CE. Oral norovirus infection is blocked in mice lacking Peyer's patches and mature M cells. *J Virol.* 2015;90:1499–1506.
114. Amerongen HM, Weltzin R, Farnet CM, Michetti P, Haseltine WA, Neutra MR. Transepithelial transport of HIV-1 by intestinal M cells: a mechanism for transmission of AIDS. *J Acquir Immune Defic Syndr.* 1991;4:760–765.
115. Fujimura Y, Takeda M, Ikai H, et al. The role of M cells of human nasopharyngeal lymphoid tissue in influenza virus sampling. *Virchows Arch.* 2004;444:36–42.
116. Takahashi Y, Misumi S, Muneoka A, et al. Nonhuman primate intestinal villous M-like cells: an effective poliovirus entry site. *Biochem Biophys Res Commun.* 2008;368:501–507.
117. Wolf JL, Rubin DH, Finberg R, et al. Intestinal M cells: a pathway for entry of reovirus into the host. *Science.* 1981;212:471–472.
118. Heppner FL, Christ AD, Klein MA, et al. Transepithelial prion transport by M cells. *Nat Med.* 2001;7:976–977.

119. Beekes M, McBride PA. Early accumulation of pathological PrP in the enteric nervous system and gut-associated lymphoid tissue of hamsters orally infected with scrapie. *Neurosci Lett.* 2000;278:181–184.
120. Miyazawa K, Kanaya T, Takakura I, et al. Transcytosis of murine-adapted bovine spongiform encephalopathy agents in an in vitro bovine M cell model. *J Virol.* 2010;84:12285–12291.
121. Takakura I, Miyazawa K, Kanaya T, et al. Orally administered prion protein is incorporated by M cells and spreads to lymphoid tissues with macrophages in prion protein knockout mice. *Am J Pathol.* 2011;179:1301–1309.
122. Donaldson DS, Kobayashi A, Ohno H, Yagita H, Williams IR, Mabbott NA. M cell depletion blocks oral prion disease pathogenesis. *Mucosal Immunol.* 2012;5:216–225.
123. Terahara K, Yoshida M, Igarashi O, et al. Comprehensive gene expression profiling of Peyer's patch M cells, villous M-like cells, and intestinal epithelial cells. *J Immunol.* 2008;180:7840–7846.
124. Knoop KA, Kumar N, Butler BR, et al. RANKL is necessary and sufficient to initiate development of antigen-sampling M cells in the intestinal epithelium. *J Immunol.* 2009;183:5738–5747.
125. Nakato G, Fukuda S, Hase K, Goitsuka R, Cooper MD, Ohno H. New approach for M-cell-specific molecules by screening comprehensive transcriptome analysis. *DNA Res.* 2009;16:227–235.
126. Jeffrey M, González L, Espenes A, et al. Transportation of prion protein across the intestinal mucosa of scrapie-susceptible and scrapie-resistant sheep. *J Pathol.* 2006;209:4–14.
127. Mutoh M, Kimura S, Takashi-Iwanaga H, Hisamoto M, Iwanaga T, Iida J. RANKL regulates differentiation of microfold cells in mouse nasopharynx-associated lymphoid tissue (NALT). *Cell Tissue Res.* 2016;364:175–184.
128. Nair VR, Franco LH, Zacharia VM, et al. Microfold cells actively translocate *Mycobacterium tuberculosis* to initiate infection. *Cell Rep.* 2016;16:1253–1258.
129. Kincaid AE, Hudson KF, Richey MW, Bartz JC. Rapid transepithelial transport of prions following inhalation. *J Virol.* 2012;86:12731–12740.
130. Shortman K, Liu Y-J. Mouse and human dendritic cell subtypes. *Nat Rev Immunol.* 2002;2:151–161.
131. Mabbott NA, Bailie JK, Kobayashi A, et al. Expression of mesenchyme-specific gene signatures by follicular dendritic cells: insights from the meta-analysis of microarray data from multiple mouse cell populations. *Immunology.* 2011;133:482–498.
132. Wykes M, Pombo A, Jenkins C, MacPherson GG. Dendritic cells interact directly with naive B lymphocytes to transfer antigen and initiate class switching in a primary T-dependent response. *J Immunol.* 1998;161:1313–1319.
133. Macpherson AJ, Uhr T. Induction of protective IgA by intestinal dendritic cells carrying commensal bacteria. *Science.* 2004;303:1662–1665.
134. Saeki H, Wu M, Olasz E, Hwang ST. A migratory population of skin-derived dendritic cells expresses CXCR5, responds to B lymphocyte chemoattractant *in vitro*, and co-localizes to B cell zones in lymph nodes *in vivo*. *Eur J Immunol.* 2000;30: 2808–2814.
135. Berney C, Herren S, Power CA, Gordon S, Martinez-Pomares L, Kosco-Vilbois MH. A member of the dendritic cell family that enters B cell follicles and stimulates primary antibody responses identified by a mannose receptor fusion protein. *J Exp Med.* 1999;190:851–860.
136. Leon B, Ballesteros-Tato A, Browning JL, Dunn R, Randall TD, Lund FE. Regulation of T(H)2 development by $CXCR5^+$ dendritic cells and lymphotoxin-expressing B cells. *Nat Immunol.* 2012;13:681–690.
137. S-JL W, Grouard-Vogel G, Sun W, et al. Human skin Langerhans cells are targets of dengue virus infection. *Nat Med.* 2000;6:816–820.

138. Ho L-J, Wang J-J, Shaio M-F, et al. Infection of human dendritic cells by dengue virus causes cell maturation and cytokine production. *J Immunol.* 2001;166:1499–1506.
139. Steinman RM, Granelli-Piperno A, Pope M, et al. The interaction of immunodeficiency viruses with dendritic cells. *Curr Top Microbiol Immunol.* 2003;276:1–30.
140. Ho AW, Prabhu N, Betts RJ, et al. Lung CD103+ dendritic cells efficiently transport influenza virus to the lymph node and load viral antigen onto MHC class I for presentation to CD8 T cells. *J Immunol.* 2011;187:6011–6021.
141. Sethi S, Kerksiek KM, Brocker T, Kretzschmar H. Role of the $CD8^+$ dendritic cell subset in transmission of prions. *J Virol.* 2007;81:4877–4880.
142. Cordier-Dirikoc S, Chabry J. Temporary depletion of CD11c+ dendritic cells delays lymphoinvasion after intraperitoneal scrapie infection. *J Virol.* 2008;82:8933–8936.
143. Sakhon OS, Ross B, Gusti V, Pham AJ, Vu K, lo DD. M cell-derived vesicles suggest a unique pathway for trans-epithelial antigen delivery. *Tissues Barriers.* 2015;3: e1004975.
144. Raymond CR, Aucouturier P, Mabbott NA. *In vivo* depletion of $CD11c^+$ cells impairs scrapie agent neuroinvasion from the intestine. *J Immunol.* 2007;179:7758–7766.
145. Gunn MD, Ngo VN, Ansel KM, Ekland EH, Cyster JG, Williams LT. A B-cell-homing chemokine made in lymphoid follicles activates Burkitt's lymphoma receptor-1. *Nature.* 1998;391:799–803.
146. Ansel KM, Ngo VN, Hyman PL, et al. A chemokine-driven feedback loop organizes lymphoid follicles. *Nature.* 2000;406:309–314.
147. Jang MH, Sougawa N, Tanaka T, et al. CCR7 is critically important for migration of dendritic cells in intestinal lamina propria to mesenteric lymph nodes. *J Immunol.* 2006;176:803–810.
148. Levavasseur E, Matharom P, Dorban G, et al. Experimental scrapie in '*plt*' mice: an assessment of the role of dendritic-cell migration in the pathogenesis of prion diseases. *J Gen Virol.* 2007;88:2353–2360.
149. Wathne GJ, Kissenpfennig A, Malissen B, Zurzolo C, Mabbott NA. Determining the role of mononuclear phagocytes in prion neuroinvasion from the skin. *J Leukoc Biol.* 2012;91:817–828.
150. Castro-Seoane R, Hummerich H, Sweeting T, et al. Plasmacytoid dendritic cells sequester high prion titres at early stages of prion infection. *PLoS Pathog.* 2012;8: e1002538.
151. Michel B, Meyerett-Reid C, Johnson T, et al. Incunabular immunological events in prion trafficking. *Sci Rep.* 2012;2:440.
152. Gossner A, Hunter N, Hopkins J. Role of lymph-borne cells in the early stages of scrapie agent dissemination from the skin. *Vet Immunol Immunopathol.* 2005;109:267–278.
153. Loeuillet C, Lemaire-Vielle C, Naquet P, Cesbron-Delauw M-F, Gagnon J, Cesbron J-Y. Prion replication in the hematopoietic compartment is not required for neuroinvasion in scrapie mouse model. *PLoS One.* 2010;5: e13166.
154. Carp RI, Callahan SM. In vitro interaction of scrapie agent and mouse peritoneal macrophages. *Intervirology.* 1981;16:8–13.
155. Carp RI, Callahan SM. Effect of mouse peritoneal macrophages on scrapie infectivity during extended in vitro incubation. *Intervirology.* 1982;17:201–207.
156. Maignien T, Shakweh M, Calvo P, et al. Role of gut macrophages in mice orally contaminated with scrapie or BSE. *Int J Pharm.* 2005;298:293–304.
157. Beringue V, Demoy M, Lasmezas CI, et al. Role of spleen macrophages in the clearance of scrapie agent early in pathogenesis. *J Pathol.* 2000;190:495–502.
158. Mabbott NA, Baillie JC, Hume DA, Freeman TC. Meta-analysis of co-expressed gene signatures in mouse leukocyte populations. *Immunobiology.* 2010;215:724–736.
159. Bradford BM, Sester D, Hume DA, Mabbott NA. Defining the anatomical localisation of subsets of the murine mononuclear phagocyte system using integrin alpha X (Itgax, CD11c) and colony stimulating factor 1 receptor (Csf1r, CD115)

expression fails to discriminate dendritic cells from macrophages. *Immunobiology*. 2011;216:1228–1237.
160. Hume DA, Mabbott N, Raza S, Freeman TC. Can DCs be distinguished from macrophages by molecular signatures. *Nat Immunol*. 2013;14:187–189.
161. Mok SW, Proia RL, Brinkmann V, Mabbott NA. B cell-specific S1PR1 deficiency blocks prion dissemination between secondary lymphoid organs. *J Immunol*. 2012;188:5032–5040.
162. Turner M, Gulbranson-Judge A, Quinn ME, Walters AE, MacLennan IC, Tybulewicz VLJ. Syk tyrosine kinase is required for the positive selection of immature B cells into the recirculating B cell pool. *J Exp Med*. 1997;186:2013–2021.
163. Suzuki K, Grigorova I, Phan TG, Kelly LM, Cyster JG. Visualizing B cell capture of cognate antigen from follicular dendritic cells. *J Exp Med*. 2009;206:1485–1493.
164. Andreoletti O, Litaise C, Simmons H, et al. Highly efficient prion transmission by blood transfusion. *PLoS Pathog*. 2012;8: e1002782.
165. Douet JY, Lacroux C, Litaise C, et al. Mononucleated blood cell populations display different abilities to transmit prion disease by the transfusion route. *J Virol*. 2016;90:3439–3445.
166. Edwards JC, Moore SJ, Hawthorne JA, Neale MH, Terry LA. PrP(Sc) is associated with B cells in the blood of scrapie-infected sheep. *Virology*. 2010;405:110–119.
167. Mathiason CK, Hayes-Klug J, Hays SA, et al. B cells and platelets harbour prion infectivity in the blood of deer infected with chronic wasting disease. *J Virol*. 2010;84:5097–5107.
168. Peden AH, Head MW, Ritchie DL, Bell JE, Ironside JW. Preclinical vCJD after blood transfusion in a *PRNP* codon 129 heterozygous patient. *Lancet*. 2004;354:527–529.
169. Wroe SJ, Pal S, Siddique D, et al. Clinical presentation and pre-mortem diagnosis of variant Creutzfeldt-Jakob disease associated with blood transfusion: a case report. *Lancet*. 2006;368:2061–2067.
170. Llewelyn CA, Hewitt PE, Knight RSG, et al. Possible transmission of variant Creutzfeldt-Jakob disease by blood transfusion. *Lancet*. 2004;363:417–421.
171. Phan TG, Grigorova I, Okada T, Cyster JG. Subcapsular encounter and complement-dependent transport of immune complexes by lymph node B cells. *Nat Immunol*. 2007;8:992–1000.
172. Phan TG, Green JA, Gray EE, Xu Y, Cyster JG. Immune complex relay by subcapsular sinus macrophages and noncognate B cells drives antibody affinity maturation. *Nat Immunol*. 2009;10:786–793.
173. Felten SY, Felten DL. Innervation of lymphoid tissue. In: Ader R, Felten DL, Cohen N (Eds.), Psychoneuroimmunology, 2nd ed.: Academic Press Inc., London, 1991:27–69.
174. Beekes M, Baldauf E, Diringer H. Sequential appearance and accumulation of pathognomonic markers in the central nervous system of hamsters orally infected with scrapie. *J Gen Virol*. 1996;77:1925–1934.
175. Baldauf E, Beekes M, Diringer H. Evidence for an alternative direct route of access for the scrapie agent to the brain bypassing the spinal cord. *J Gen Virol*. 1997;78: 1187–1197.
176. Beekes M, McBride PA, Baldauf E. Cerebral targeting indicates vagal spread of infection in hamsters fed with scrapie. *J Gen Virol*. 1998;79:601–607.
177. McBride PA, Beekes M. Pathological PrP is abundant in sympathetic and sensory ganglia of hamsters fed with scrapie. *Neurosci Lett*. 1999;265:135–138.
178. McBride PA, Schulz-Shaeffer WJ, Donaldson M, et al. Early spread of scrapie from the gastrointestinal tract to the central nervous system involves autonomic fibers of the splanchnic and vagus nerves. *J Virol*. 2001;75:9320–9327.
179. Glatzel M, Heppner FL, Albers KM, Aguzzi A. Sympathetic innervation of lymphoreticular organs is rate limiting for prion neuroinvasion. *Neuron*. 2001;31:25–34.

180. von Poser-Klein C, Flechsig E, Hoffmann T, et al. Alteration of B-cell subsets enhances neuroinvasion in mouse scrapie infection. *J Virol.* 2008;82:3791–3795.
181. Urayama A, Concha-Marambio L, Khan U, Bravo-Alegria J, Kharat V, Soto C. Prions efficiently cross the intestinal barrier after oral administration: study of the bioavailability, and cellular tissue distribution *in vivo*. *Sci Rep.* 2016;6:32338.
182. Sisó S, González L, Jeffrey M, Martin S, Chianini F, Steele P. Prion protein in kidneys of scrapie-infected sheep. *Vet Rec.* 2006;159:327–328.
183. Armstrong RA, Cairns NJ, Ironside JW, Lantos PL. Does the neuropathology of human patients with variant Creutzfeldt-Jakob disease reflect haematogenous spread of the disease. *Neurosci Lett.* 2003;348:37–40.
184. Gomez-Nicola D, Schetters STT, Perry VH. Differential role of CCR2 in the dynamics of microglia and perivascular macrophages during prion disease. *Glia.* 2014;62:1041–1052.
185. Gabanyi I, Muller PA, Feighery L, Oliveira TY, Costa-Pinto FA, Mucida D. Neuro-immune interactions drive tissue programming in intestinal macrophages. *Cell.* 2016;164:378–391.
186. Aucouturier P, Geissmann F, Damotte D, et al. Infected splenic dendritic cells are sufficient for prion transmission to the CNS in mouse scrapie. *J Clin Invest.* 2001;108:703–708.
187. Gousset K, Schiff E, Langevin C, et al. Prions hijack tunnelling nanotubes for intercellular spread. *Nat Cell Biol.* 2009;11:328–336.
188. Langevin C, Gousset K, Costanzo M, Richard-Le Goff O, Zurzolo C. Characterization of the role of dendritic cells in prion transfer to primary neurons. *Biochem J.* 2010;431:189–198.
189. Tanaka Y, Sadaike T, Inoshima Y, Ishiguro N. Characterisation of PrP^{Sc} transmission from immune cells to neuronal cells. *Cell Immunol.* 2012;279:145–150.
190. Zhu S, Victoria GS, Marzo L, Ghosh R, Zurzolo C. Prion aggregates transfer through tunneling nanotubes in endocytic vesicles. *Prion.* 2015;9:125–135.
191. Fevrier B, Vilette D, Archer F, et al. Cells release prions in association with exosomes. *Proc Natl Acad Sci USA.* 2004;101:9683–9688.
192. Klingeborn M, Race B, Meade-White KD, Rosenke R, Striebel JF, Chesebro B. Crucial role for prion protein membrane anchoring in the neuroinvasion and neural spread of prion infection. *J Virol.* 2011;85:1484–1494.
193. Kranich J, Krautler NJ, Falsig J, et al. Engulfment of cerebral apoptotic bodies controls the course of prion disease in a mouse strain-dependent manner. *J Exp Med.* 2010;207:2271–2281.
194. Zhan Y, Paolocelli RC, Sfronazzini F, et al. Deficient neuron-microglia signaling results in impaired functional brain connectivity and social behaviour. *Nat Neurosci.* 2014;17:400–406.
195. Prinz M, Priller J. Microglia and brain macrophages in the molecular age: from origin to neuropsychiatric disease. *Nat Rev Neurosci.* 2014;15:300–312.
196. De Lucia C, Rinchon A, Olmos-Alsonso A, et al. Microglia regulate hippocampal neurogenesis during chronic neurodegeneration. *Brain Behav Immun.* 2016;55:179–190.
197. Vincenti JE, Murphy L, Grabert K, et al. Defining the microglial response during the time course of chronic neurodegeneration. *J Virol.* 2016;90:3003–3017.
198. Boche D, Cunningham C, Gauldie J, Perry VH. Transforming growth factor-beta 1-mediated neuroprotection against excitotoxic injury in vivo. *J Cereb Blood Flow Metab.* 2003;23:1174–1182.
199. Minghetti L, Greco A, Cardone F, et al. Increased brain synthesis of prostaglandin E2 and F2-isoprostane in human and experimental transmissible spongiform encephalopathies. *J Neuropathol Exp Neurol.* 2000;59:866–871.

200. Walsh DT, Perry VH, Minghetti L. Cyclooxygenase-2 is highly expressed in microglia-like cells in a murine model of prion disease. *Glia*. 2000;29:392–396.
201. Cunningham C, Deacon R, Wells H, et al. Synaptic changes characterize early behavioural signs of the ME7 model of muring prion disease. *Eur J Neurosci*. 2003;17:2147–2155.
202. Lunnon K, Teeling JL, Tutt AL, Cragg MS, Glennie MJ, Perry VH. Systemic inflammation modulates Fc receptor expression on microglia during chronic neurodegeneration. *J Immunol*. 2011;186:7215–7224.
203. Fadok VA, Bratton DL, Konowal A, Freed PW, Westcott JY, Henson PM. Macrophages that have ingested apoptotic cells in vitro inhibit proinflammatory cytokine production through autocrine/paracrine mechanisms involving TGF-beta, PGE2, and PAF. *J Clin Invest*. 1998;101:890–898.
204. Thackray AM, McKenzie AN, Klein MA, Lauder A, Bujdoso R. Accelerated prion disease pathogenesis in the absence of interleukin-10. *J Virol*. 2004;78:13697–13707.
205. Tamguney G, Giles K, Glidden DV, et al. Genes contributing to prion pathogenesis. *J Gen Virol*. 2008;98:1777–1788.
206. Gomez-Nicola D, Fransen NL, Suzzi S, Perry VH. Regulation of microglial proliferation during chronic neurodegeneration. *J Neurosci*. 2013;33:2481–2493.
207. Mabbott NA, Bruce ME. Complement component C5 is not involved in scrapie pathogenesis. *Immunobiology*. 2004;209:545–549.
208. Striebel JF, Race B, Carroll JA, Philips GE, Chesebro B. Knockout of fractalkine receptor Cx3cr1 does not alter disease or microglial activation in prion infected mice. *J Gen Virol*. 2016;97:1481–1487.
209. Tribouillard-Tanvier D, Race B, Striebel JF, Carroll JA, Philips K, Chesebro B. Early cytokine elevation, PrPres deposition, and gliosis in mouse scrapie: no effect on disease by deletion of cytokine genes IL-12p40 and IL-12p35. *J Virol*. 2012;86:10377–10383.
210. Carroll JA, Striebel JF, Race B, Philips K, Chesebro B. Prion infection of mouse brain reveals multiple new upregulated genes involved in neuroinflammation or signal transduction. *J Virol*. 2015;89:2388–2404.
211. Gresser I, Maury C, Chandler RL. Failure to modify scrapie in mice by administration of interferon or anti-interferon globulin. *J Gen Virol*. 1983;64:1387–1389.
212. Nuvolone M, Sorce S, Schwarz P, Aguzzi A. Prion pathogenesis in the absence of NLRP3/ASC inflammasomes. *PLoS One*. 2015;10: e0117208.
213. Prinz M, Heikenwalder M, Scwarz P, Takeda K, Akira S, Aguzzi A. Prion pathogenesis in the absence of Toll-like receptor signalling. *EMBO Rep*. 2003;4:195–199.
214. Falsig J, Julius C, Margalith I, Schwarz P, Heppner FL, Agguzi A. A versatile prion replication assay in organotypic brain slices. *Nat Neurosci*. 2008;11:109–117.
215. Zhu C, Hermann US, Falsig J, et al. A neuroprotective role for microglia during prion diseaes. *J Exp Med*. 2016;213:1047–1059.
216. Alibhai J, Blanco RA, Barria MA, et al. Distribution of misfolded prion protein seeding activity alone does not predict regions of neurodegeneration. *PLoS Biol*. 2016;14: e1002579.
217. Takahashi K, Rochford CD, Neumann H. Clearance of apoptotic neurons without inflammation by microglial triggering receptor expressed on myeloid cells-2. *J Exp Med*. 2005;201:647–657.
218. Zhu C, Herrmann US, Li B, et al. Triggering receptor expressed on myeloid cells-2 is involved in prion-induced microglial activation but does not contribute to prion pathogenesis in mouse brains. *Neurobiol Aging*. 2015;36:1994–2002.
219. Erny D, Hrabe de Angelis AL, Jaitin D, et al. Host microbiota constantly control maturation and function of microglia in the CNS. *Nat Neurosci*. 2015;18:965–977.
220. Donaldson DS, Mabbott NA. The influence of the commensal and pathogenic gut microbiota on prion disease pathogenesis. *J Gen Virol*. 2016;97:1725–1738.

221. Combrinck MI, Perry VH, Cunningham C. Peripheral infection evokes exaggerated sickness behaviour in pre-clinical murine prion disease. *Neuroscience*. 2002;112:7–11.
222. Cunningham C, Wilcockson DC, Campion S, Lunnon K, Perry VH. Central and systemic endotoxin challenges exacerbate the local inflammatory response and increase neuronal death during chronic neurodegeneration. *J Neurosci*. 2005;25:9275–9284.
223. Ierna MI, Farquhar CF, Outram GW, Bruce ME. Resistance of neonatal mice to scrapie is associated with inefficient infection of the immature spleen. *J Virol*. 2006;80:474–482.
224. Turner VM, Mabbott NA. Influence of ageing on the microarchitecture of the spleen and lymph nodes. *Biogerontology*. 2017. http://dx.doi.org/10.1007/s10522-017-9707-7.
225. Kobayashi A, Donaldson DS, Erridge C, et al. The functional maturation of M cells is dramatically reduced in the Peyer's patches of aged mice. *Mucosal Immunol*. 2013;6:1027–1037.
226. Brown KL, Wathne GJ, Sales J, Bruce ME, Mabbott NA. The effects of host age on follicular dendritic cell status dramatically impair scrapie agent neuroinvasion in aged mice. *J Immunol*. 2009;183:5199–5207.
227. Mebius RE, Kraal G. Structure and function of the spleen. *Nat Rev Immunol*. 2005;5:606–616.
228. Cinamon G, Zachariah MA, Lam OM, Foss Jr FW, Cyster JG. Follicular shuttling of marginal zone B cells facilitates antigen transport. *Nat Immunol*. 2008;9:54–62.
229. Brown KL, Gossner A, Mok S, Mabbott NA. The effects of host age on the transport of complement-bound complexes to the spleen and the pathogenesis of intravenous scrapie infection. *J Virol*. 2012;86:1228–1237.
230. Turner VM, Mabbott NA. Ageing adversely affects the migration and function of marginal zone B cells. *Immunology*. 2017;151:349–362. http://dx.doi.org/10.1111/imm.12737.
231. Hunter N, Houston F, Foster J, et al. Susceptibility of young sheep to oral infection with bovine spongiform encephalopathy decreases significantly after weaning. *J Virol*. 2012;86:11856–11862.
232. Spraker TR, VerCauteren KC, Gidlewski TL, Munger RD, Walter WD, Balachandran A. Impact of age and sex of Rocky Mountain elk (*Cervus elaphus nelsoni*) on follicle counts from rectal mucosal biopsies for preclinical detection of chronic wasting disease. *J Vet Diagn Invest*. 2009;21:868–870.
233. Geremia C, Hoeting JA, Wolfe LL, et al. Age and repeated biopsy influence antemortem PrP^{CWD} testing in mule deer (*Odocoileus hemionus*) in Colorado, USA. *J Wildl Dis*. 2015;51:801–810.
234. Valleron A-J, Boelle P-Y, Will R, Cesbron J-Y. Estimation of epidemic size and incubation time based on age characteristics of vCJD in the United Kingdom. *Science*. 2001;294:1726–1728.
235. Wilesmith JW. BSE: epidemiological approaches, trials and tribulations. *Prev Vet Med*. 1993;18:33–42.
236. Dickinson AG, Fraser H, McConnell I, Outram GW. Mitogenic stimulation of the host enhances susceptibility to scrapie. *Nature*. 1978;272:54–55.
237. Bremer J, Heikenwalder M, Haybaeck J, et al. Repetitive immunization enhances the susceptibility of mice to peripherally administered prions. *PLoS One*. 2009;4: e7160.
238. Seeger H, Heikenwalder M, Zeller N, et al. Coincident scrapie infection and nephritis lead to urinary prion excretion. *Science*. 2005;310:324–326.
239. Heikenwalder M, Zeller N, Seeger H, et al. Chronic lymphocytic inflammation specifies the organ tropism of prions. *Science*. 2005;307:1107–1110.
240. Ligios C, Cancedda MG, Carta A, et al. Sheep with scrapie and mastitis transmit infectious prions through the milk. *J Virol*. 2011;85:1136–1139.
241. Heikenwalder M, Kurrer MO, Margalith I, et al. Lymphotoxin-dependent prion replication in inflammatory stromal cells of granulomas. *Immunity*. 2008;29:998–1008.

242. Maestrale C, Di Guardo G, Cancedda MG, et al. A lympho-follicular microenvironment is required for pathological prion protein deposition in chronically inflamed tissues from scrapie-affected sheep. *PLoS One*. 2013;8: e62830.
243. Sigurdson CJ, Heikenwalder M, Manco G, et al. Bacterial colitis increases susceptibility to oral prion pathogenesis. *J Infect Dis*. 2009;199:243–252.
244. Denkers ND, Telling GC, Hoover EA. Minor oral lesions facilitate transmission of chronic wasting disease. *J Virol*. 2011;85:1396–1399.
245. Crowell J, Wiley JA, Bessen RA. Lesion of the alfactory epithelium accelerates prion neuroinvasion and disease onset when prion replication is restricted to neurons. *PLoS One*. 2015;10: e0119863.
246. Brown KL, Stewart K, Bruce ME, Fraser H. Severly combined immunodeficient (SCID) mice resist infection with bovine spongiform encephalopathy. *J Gen Virol*. 1997;78:2707–2710.
247. Gill ON, Spencer Y, Richard-Loendt A, et al. Prevelent abnormal prion protein in human appendixes after bovine spongiform encephalopathy epizootic: large scale survey. *Br Med J*. 2013;f5675:347.
248. Tsukamoto T, Diringer H, Ludwig H. Absence of autoantibodies against neurofilament proteins in the sera of scarpie infected mice. *Tohoku J Exp Med*. 1985;4:483–484.
249. Clarke MC, Haig DA. Attempts to demonstrate neutralising antibodies in the sera of scrapie-infected animals. *Vet Rec*. 1966;19:647–649.
250. Sassa Y, Kataoka N, Inoshima Y, Ishiguro N. Anti-PrP antibodies detected at terminal stage of prion-affected mouse. *Cell Immunol*. 2010;263:212–218.
251. White AR, Enever P, Tayebi M, et al. Monoclonal antibodies inhibit prion replication and delay the development of prion disease. *Nature*. 2003;422:80–83.
252. Sigurdsson EM, Brown DR, Daniels M, et al. Immunization delays the onset of prion disease in mice. *Am J Pathol*. 2002;161:13–17.
253. Bachy V, Ballerini C, Gourdain P, et al. Mouse vaccination with dendritic cells loaded with prion protein peptides overcomes tolerance and delays scrapie. *J Gen Virol*. 2010;91:809–820.
254. Gourdain P, Gregoire S, Iken S, et al. Adoptive transfer of T lymphocytes sensitized against the prion protein attenuates prion invasion in scrapie-infected mice. *J Immunol*. 2009;183:6619–6628.
255. Iken S, Bachy V, Gourdain P, et al. Th2-polarized PrP-specific transgenic T-cells confer partial protection against murine scrapie. *PLoS Pathog*. 2011;7: e1002216.
256. Goñi F, Chablagoity JA, Prelli F, et al. High titres of mucosal and systemic anti-PrP antibodies abrogates oral prion infection in mucosal vaccinated mice. *Neuroscience*. 2008;153:679–686.
257. Pilon JL, Rhyan JC, Wolfe LL, et al. Immunization with a synthetic peptide vaccine fails to protect mule deer (*Odocoileus hemionus*) from chronic wasting disease. *J Wildl Dis*. 2013;49:694–698.
258. David MA, Jones DR, Tayebi M. Potential candidate camelid antibodies for the treatment of protein-misfolding diseases. *J Neuroimmunol*. 2014;272:76–85.
259. Song CH, Furuok H, Kim C-L, et al. Effect of intraventricular infusion of anti-prion protein monoclonal antibodies on disease progression in prion-infected mice. *J Gen Virol*. 2008;89:1533–1544.
260. Sonati T, Reimann RR, Falsig J, et al. The toxicity of antiprion antibodies is mediated by the flexible tail of the prion protein. *Nature*. 2013;501:102–106.
261. Reimann RR, Sonati T, Hornemann S, et al. Differential toxicity of antibodies to the prion protein. *PLoS Pathog*. 2016;12: e1005401.
262. Requena JR, Kristensson K, Korth C, et al. The priority position paper: protecting Europe's food chain from prions. *Prion*. 2016;10:165–181.

CHAPTER ELEVEN

Phenotypical Variability in Bovine Spongiform Encephalopathy: Epidemiology, Pathogenesis, and Diagnosis of Classical and Atypical Forms

Cristiano Corona, Elena Vallino Costassa, Barbara Iulini, Maria Caramelli, Elena Bozzetta, Maria Mazza, Rosanna Desiato, Giuseppe Ru, Cristina Casalone[1]

Italian Reference Centre for Animal Encephalopathies, Istituto Zooprofilattico Sperimentale del Piemonte, Liguria e Valle d'Aosta, Turin, Italy

[1]Corresponding author: e-mail address: cristina.casalone@izsto.it

Contents

Abstract

After thirty years, bovine spongiform encephalopathy (BSE) still represents the biggest crisis in the field of food safety. Initially detected in the United Kingdom in 1986, BSE spread to many other countries all over the world, involving approximately 200,000 cattle. The origin of BSE is uncertain, but epidemiological studies suggest that the source was cattle feed prepared from prion-infected animal tissues. The implementation of the drastic measures, including the ban of meat and bone meal from livestock feed and the

Progress in Molecular Biology and Translational Science, Volume 150
ISSN 1877-1173
http://dx.doi.org/10.1016/bs.pmbts.2017.06.015

removal of specified risk material from the food chain, has eventually resulted in a significant decline of the epidemic. For many years, it was believed that the disease was caused by transmission of a single prion strain. However, since 2004 two types of BSE, with distinct phenotypical characteristics, have been detected in Italy and France. These atypical types are characterized by distinct Western Blot profiles of abnormal protease-resistant prion protein, named high-type (H-BSE) and low-type (L-BSE). At present, there is no comprehensive information about the origin of the atypical BSEs (sporadic vs. acquired), and data about the pathogenesis of both atypical forms are very limited as compared to the classical type (C-BSE). This chapter will provide a well-organized overview of what is known about classical and atypical BSE. It will review information on the main epidemiological features, pathogenesis, and the criteria for routine diagnosis based on rapid tests, histological, immunohistochemical, and Western blot analysis. Furthermore, a brief overview about the most recently in vitro techniques will be also provided.

1. INTRODUCTION

Bovine spongiform encephalopathy (BSE), first diagnosed in the United Kingdom (UK) in 1986, is now classified as a potentially lethal zoonotic disease acquired *via* contaminated food. BSE spread from the UK to at least 28 other countries, mostly in Europe, with occasional cases also confirmed in Asia (Japan), the Middle East (Israel), and North America. To date, more than 112.000.000 animals have been examined in Europe, and more than 184.500 cases of BSE have been confirmed in the UK, 5.500 in Europe, and 60 in the rest of the world (Brazil, Canada, Israel, and Japan). The World Organization for Animal Health (OIE) reported only 5 cases of BSE worldwide in 2015, 3 of which for Europe, 1 for Norway, and 1 for Canada; in 2016 only 2 cases, in France and in Spain and in the current year, only one BSE case has been reported for Ireland.

While the origin of BSE remains unknown, the cause of its spread is well recognized. In fact, the use of ruminant-derived meat and bone meal (MBM) as a food source for animals is the most reliable hypothesis of its diffusion in cattle populations. MBM contains central nervous system (CNS) tissues and other tissues that are known to contain the protease-resistant, disease-associated isoform of the prion protein (PrPSc).

BSE has a long incubation period of about 2.5 to 8 years, with clinical disease usually affecting adult cattle at a peak age onset of 4 to 5 years, and all breeds being equally susceptible. BSE is characterized by altered behavior, abnormal posture changes, incoordination and difficulty in rising, decrease in milk production, and loss of body weight. Histopathologic features include neuronal and neuropil vacuolation, glial reaction, and the complete absence of inflammatory lesions.

In addition to the classical form of BSE (C-BSE), two atypical BSE prions were identified in France and in Italy in 2004.[1,2] One has higher molecular mass fragments than C-BSE and is called H-BSE; the other has a lower molecular mass and is called L-BSE, or bovine amyloidotic spongiform encephalopathy (BASE), because of the presence of PrP-positive amyloid plaques in the brain. The most likely hypothesis is that they arise spontaneously in cattle, comparable to most sporadic Creutzfeldt-Jakob diseases in humans. From an epidemiological point of view, these atypical forms are mainly detected in cattle 8 years of age or older. Most cases were identified in fallen stock and none was reported as clinically suspected, suggesting that the clinical presentation is unlike C-BSE. Since 2001, a total of 44 cases of L-type and 60 of H-type BSE have been identified in Europe according to the European Union BSE databases. The prevalence of atypical BSE cases in the rest of the world is unknown because official surveillance requirements or systematic reports are lacking.

2. EPIDEMIOLOGY

During the mid eighties routine surveillance activities carried out on the English livestock led to the identification of an emerging neurological disease affecting an increasing number of cows. In some ways the disease features were resembling scrapie, i.e., a known disease affecting small ruminants. The combination of the peculiar clinical presentation with the histopathological pattern observed at the brain level and the discovery of protein fibrils reminiscent scrapie-associated fibrils allowed the formulation of an intriguing hypothesis: a bovine spongiform encephalopathy was emerging and spreading among the cattle.[3] One year later the results from an inoculation experiment confirmed the nature of the disease[4] as a *transmissible spongiform encephalopathy*. The retrieval of archived bovine brains indicated that the first cases occurred around 1985. Based on data from nearly 200 affected herds and cases of BSE, the initial epidemic curve was consistent with that of an extended common source epidemic.[5] The emerging cases were geographically spread but there were no clear direct epidemiological relationships linking them. Most of the cases were from dairy herds of large size. All those data helped in formulating a feedborne hypothesis for the diffusion of such a spongiform disease in cattle. In particular, attention was focused on one ingredient: the inclusion in feed of infected meat and bone meal might have accounted for the transmission to cattle. A subsequent case–control study[6] provided evidence of a higher risk of disease in dairy cows herds where concentrated proprietary feed had been massively used. The hypothesis is still largely accepted and international supportive evidence has been cumulated over the years.[7–10]

To understand why meal and bone meal might have represented a new hazard, a survey of all rendering plants in operation in the UK was carried out in 1988.[11] Two major changes in the industrial processes were identified as possibly significant, i.e., a change from batch processing to continuous processing and a reduction in the use of hydrocarbon solvents to maximize the extraction of tallow: it was hypothesized that whereas the original processes through a combination of time and temperature had been able to inactivate the agent responsible of BSE, the recent changes were no longer able in preventing the infections in orally exposed cattle. Moreover the survey also revealed a geographical variation in the use of hydrocarbon solvents and in heat treatment which was negatively correlated with the incidence of BSE, as the process was abandoned entirely in England and Wales, but remained in use in Scotland where BSE was showing the lowest incidence. This finding provided the first explanation for the variation in the geographical distribution of the BSE risk.

Based on the feedborne hypothesis, an early ban (1988) of the administration of MBM to ruminants was the first effective measure applied in the UK: consistently, after peaking in 1992 (with more of 37,000 confirmed cases) the epidemic started to decline. Moreover, to further reduce the risk in 1989, a new measure was enforced, i.e., the removal of all offal from both the food and feed chains.

The spread of the disease in other European countries (initially Portugal, Ireland, and Switzerland)[12] and above all the discovery of a pathogenic relationship between BSE and a new fatal neurodegenerative disorder in humans, called variant Creutzfeldt-Jakob disease (vCJD), prompted the European Commission to commit an EU-wide risk assessment exercise to quantify the risks that Member States were facing.

In particular, the concern on the human risk of exposure to BSE *via* contaminated food was suggested after the identification in the UK of the first few cases of vCJD.[13] Besides the epidemiological observation of the emerging of a new human disease in a country that was experiencing the largest livestock epidemic, molecular and biological studies of strain-typing provided evidence that the human and animal diseases were sharing the same agent.[14–17] BSE was recognized as a zoonotic disease that was posing a serious risk to human and animal health. Risk assessment was identified as the most appropriate tool to address the urgency for the European Community to devise effective interventions aimed at the protection of public health through specific rules for the prevention, control, and eradication in cattle.

2.1 Geographical Risk Assessment

One of the main reason to seek help from risk assessment was the difficulty in obtaining a reliable epidemiological situation through routine surveillance activities: the disease was relatively rare, difficult to diagnose (only post mortem), and the farmers were not willing to report as each case was representing a stigma. In the first stages of the epidemic, the passive surveillance, based on a voluntary reporting of the cases, was the only detection strategy and resulted in a biased assessment of the spread of the disease[18] and of its risk factors. An enforced active surveillance system was made available in the majority of the European Countries only since 2001: it helped in facing the lack of an accurate data collection and in understanding better the risk factors.

Meanwhile in the second half of nineties, through its Scientific Steering Committee (SSC), the European Commission decided to designate Transmissible spongiform encephalopathies (TSE) working group formed by international experts for developing and applying the risk assessment method in establishing the risk that a country could hold undetected BSE cases in cattle population.[19]

The result was a risk assessment process entitled the Geographical Bovine Spongiform Encephalopathy Risk (GBR). The main outcome of the GBR methodology was to devise a model enabling the classification of the risk of a country to have at least one animal being infected with the BSE agent in its autochthonous bovine population. Where its presence had been already confirmed, the GBR was able to provide an indication of the level of infection and spread.[20] The GBR is based on a general model aimed to assess the likelihood of the introduction of the BSE agent in a country, its potentially recycling and amplifying or elimination.[21] It takes into account and embodies in a conceptual frame the main risk factors involved in determining the disease: the main assumption is that BSE could be initially introduced through the importation of contaminated MBM or live cattle from BSE-risk countries (the so-called external challenge); the BSE agent could potentially enter into the rendering system, and the rendering system itself could allow the amplification of the BSE agent in processing animal by-products. A country may be considered a stable system if it does not allow the BSE agent to propagate following its introduction and it is capable to prevent the import of or remove BSE infected animals before processing.

The GBR was a useful tool to provide a general frame for the epidemiology of the disease and to assess the BSE status of a country in order to support the European Commission in establishing legislation rules for international trade of cattle and their products.

2.2 Risk Factors and Effective Measures

As mentioned, in general, dairy cattle have a higher risk than beef suckler herds because of the greater use of compound feed.[6,10] Herd size has been associated to the spread of infection, even though it was not consistently across countries.[10,22–24] After the 1992 peak, the rate of decline in epidemic in the UK was not as important as expected. To identify the drivers of such a persistence of the disease within the population ad hoc studies were carried out: such studies, looking at animals from the birth cohorts before and one year after the July 1988 when the initial feed ban on feeding ruminant-derived MBM to ruminants entered into force, demonstrated the role of low-level cross-contamination of cattle feed by pig feed in BSE incidence as the epidemic evolved.[25,26] Finally, there is no evidence of animal to animal spread, but a cohort study indicated that a very low level of maternal transmission, insufficient to maintain the epidemic, may have occurred in the UK.[27]

The international legislation to effectively manage the risks associated with BSE has been devised on the basis of the identification of the mentioned risk factors. The European Regulation (EC) 999/2001 is an integrated legislative rule to manage the BSE crisis and to prevent, control, and eradicate the disease. It entered into force in 2001 and provided a strict harmonized set of rules to be implemented in all Member States. The regulation aimed at avoiding the exposure of the cattle population through potentially infected meat and bone meal also due to cross contaminations at feed mills or at farms: for the purpose an extended total feed ban preventing the administration of feed containing mammalian-derived protein to all farmed animals was enforced; the measure was combined with the removal of specified risk material from all slaughtered animals (bovines and small ruminants) to avoid the amplification of the BSE agent through the rendering process. Other prevention and control measures contained in the regulation are surveillance measures aimed to detect and eliminate BSE-infected animals.

Over time the trend of the C-BSE and, thus, the effectiveness of enforcement of the measures were monitored through the testing of a huge number of cattle (10 million cows per year on average in EU). The surveillance made clear that BSE occurred in European countries and throughout the world with more than 190,000 BSE cases in 28 countries worldwide. Most reported cases are from the UK; in other countries the epidemic peaked ten years later, in 2002 or 2003. In particular, in the UK, more than 184,000 cases of BSE have been reported and more than 3 million cows were destroyed to stop the spread of the disease. Then, following the ban on feeding ruminant-derived MBM and the total feed ban, the occurrence of BSE was dramatically reduced.

After many years, the BSE problem seems to be solved now, and since 2005 the European Union has been setting up an exit strategy from the crisis,[28] providing for the gradual easing of the measures. Today, a drastic reduction in the extent of surveillance has been enforced in EU, resulting in the restriction of the testing in most of the European Member States to the so-called at-risk groups (fallen stock and emergency slaughtered cattle): such a residual surveillance is of outmost importance as it represents an early warning tool in the case of an eventual re-emerging of the disease.

3. PATHOGENESIS OF CLASSICAL AND ATYPICAL BSE

Because the epidemiological and clinical characteristics of C-BSE make it difficult to study its pathogenesis in field conditions, the bulk of pathogenetic data are derived from experimental transmission studies. A number of knowledge gaps remains, however. How the prion agent crosses the epithelium after oral exposure to infective material is not fully understood. The most likely mechanism is *via* M-cells, which are highly phagocytic epithelial cells found within the epithelium that covers Peyer's patches (PP) and specialized in the transport of large particles and whole bacteria across the gut epithelium.[29] In prion diseases, these cells are able to transcytose the prion protein from the lumen of the gut into the epithelium. During the first 8 months post-infection (mpi), the earliest PrPSc accumulation is displayed by tingible body macrophages (TBM). TBM belong to a subset of mononuclear phagocytes found in the unique microenvironment of the germinal centers in gut-associated lymphoid tissue (GALT) of the ileocecal junction and jejunum and in the PP of the ileum.[30]

Infectivity can be found in the palatine tonsils at 6–10 mpi.[31] A peak of infectivity in the distal ileum at 12 mpi is related to the number of follicles involved and to the amount of PrPSc detectable in the follicular dendritic cells (FDC) and TBM, indicating increased clearance activity of these cells.

FDC are stromal cells residing in primary follicles and in the germinal centers of secondary and tertiary lymphoid organs. There is a second peak of PrPSc accumulation at 24 mpi, where prion protein is mainly found in the TBM and FDC of the jejunum and ileum. A third peak of PrPSc accumulation may be present, with some individual variability, between 32 and 40 mpi in the ileal PP.[32]

The enteric nervous system, and the involvement of nerve fibers of the mucosal plexus in particular, represents the way of entry of the prion into the nervous system.[33] Through the mesenteric nerves, the prion proteins accumulate in the cranial celiacomesenteric ganglion complex and then ascend to the thoracic spinal cord *via* the sympathetic nervous system (e.g., splanchnic nerves) and to the brainstem and the brain *via* the parasympathetic nervous system (e.g., vagus nerve) and nodose ganglion.

From the thoracic spinal cord, PrPSc spreads rostrally to the cranial medulla and caudally to the cauda equina.[34] From the spinal cord, PrPSc then accumulates in the dorsal root ganglia, trigeminal ganglia, and cervical ganglia.[35] The adrenal glands and sciatic nerve have also been described as positive tissues with demonstrable prion protein accumulation.[36] Between 42 and 84 mpi, PrPSc can be identified in the spindles of various muscles, including the masseter, triceps brachii, intercostal muscles, and semitendinosus.[37]

Comprehensive information about the pathogenesis of atypical BSEs in cattle is lacking. Some data are available on the peripheral distribution of H- and L-BSE agents in cattle experimentally challenged *via* the intracerebral route, but all the animals in these studies were allowed to develop clinical disease. Experimental transmission studies have demonstrated that in both H-BSE and L-BSE PrPSc accumulates in CNS tissues, peripheral ganglia and nerves, muscles (muscle spindles), adrenal glands, and retina.[38] No lymphoid tissues or gastrointestinal tissues have tested positive in atypical cases. Furthermore, a study of intraspecies transmission of a case of L-type BSE suggested that prions may propagate in the CNS and spread centrifugally by nerve pathways.[39]

4. SAMPLING

When clinical suspected cases are under investigation, the whole brain is sampled in order to go through a classical neuropathological approach to examine a representative number of cerebral areas. To allow a reliable BSE diagnosis, at least one block 0.5–1 cm in width taken from the medulla oblongata at the level of the obex[40] must be subjected to histological and immunohistochemical examination, respectively, for spongiform changes and PrPSc detection, performed on formalin fixed and paraffin-embedded tissue; furthermore immunoblot for PrPSc will be carried out on a complete coronal section taken immediately rostral or caudal to the above area.

The remaining portion of the brain tissue will be placed in 10% buffered formalin to be subjected to further neuropathological examination in case of negative results for BSE.

For monitoring purposes through active surveillance, rapid tests require a precise sampling within 1.0 cm rostral or caudal to the obex, in order to include key target sites of PrPSc accumulation. For this purpose brainstem is collected with the help of a dedicated spoon through the *foramen magnum* (Figs. 1 and 2). Sampling for rapid testing should be carried out

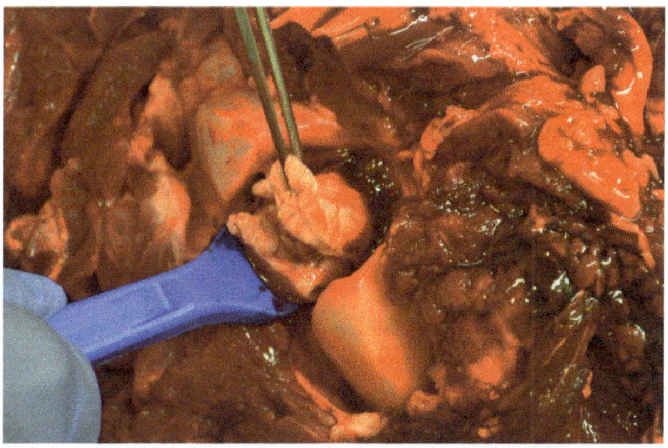

Fig. 1 Sampling of brainstem through the foramen magnum.

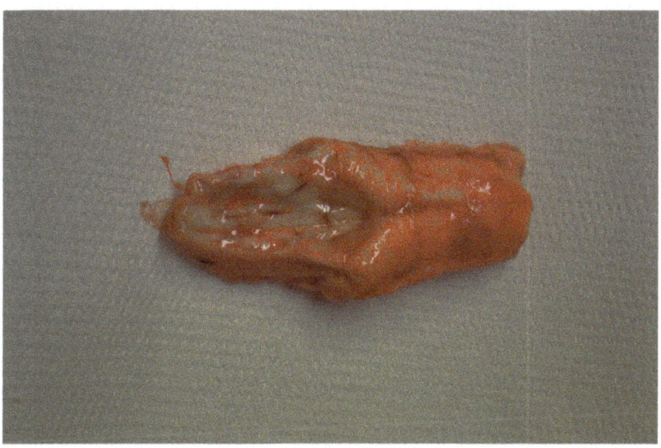

Fig. 2 Fresh brainstem at the level of the obex.

from a hemi-section of the medulla in order to preserve the contralateral one for further confirmatory tests and a possible discriminatory Western Blot (WB).

In case of autolysis of the sample, an unidentified aliquot of the medulla should be taken and tested. When a positive result is achieved, this has to be considered a valid result, while a negative test cannot be considered as a negative animal and has to be reported as appropriate (i.e., negative result in not suitable tissue).

5. RAPID TESTS

Rapid testing for BSE monitoring purposes was first introduced in the European Union in January 2001 to strengthen control programs on slaughtered animals and fallen stock. The systematic sampling of animals post-mortem in order to verify the impact of control measures and to allow the trend of the disease to be followed allowed the detection of the first BSE case in most European countries as well as of the majority of BSE cases.

Rapid tests became available for BSE active surveillance in the late 1990s and their suitability for this purpose was assessed through three evaluation exercises implemented since 1999. While in the first comparative study only the accuracy and the analytical sensitivity of the rapid tests under evaluation were assessed, robustness on poor quality negative samples and a field trial were later included in order to verify their suitability for routine activity. In 2009 the Community Reference Laboratory for TSEs assessed the analytical sensitivity of all the currently approved TSE rapid tests to determine their continued suitability for active surveillance plans.[41] Importantly, the design strategy of this study involved all tests being evaluated against the same sample set. The lowest limit of detection (LOD) of rapid tests approved for the diagnosis of classical BSE in bovines was assessed as a crucial parameter for their suitability. According to EFSA requirements for the evaluation of TSE rapid *post-mortem* tests[42] a new test will be approved for full evaluation if the LOD is better than, similar to, or no more than 2 logs poorer than the most sensitive test. This last point is considered to be a critical point for the key issue of public health and for the capacity of a test to detect BSE in cattle at early stages of incubation.

To date European Commission (EC) has assessed 19 rapid tests, 9 of which were approved for survey including a new version of the Enfer TSE test for BSE diagnosis.[43]

All currently recognized forms of BSE are detectable by these methods although an official full sensitivity and specificity evaluation for atypical forms (H- and L-types) has not been carried out. However a recent study demonstrates the suitability of BSE EU-approved rapid tests also for the detection of both atypical forms.[44]

Tests approved for active surveillance activity within the EU are listed in Annex C Chapter X of the TSE Regulation 999/2001 and subsequent amendments. To date, according to the most recent Regulation 1148/2014, the following tests can be used:

- the immunoblotting test based on a Western blotting procedure for the detection of the Proteinase K-resistant fragment PrPRes (Prionics-Check Western test)
- the sandwich immunoassay for PrPRes detection (short assay protocol) carried out following denaturation and concentration steps (Bio-Rad TeSeE SAP rapid test)
- the microplate-based immunoassay (ELISA) which detects Proteinase K-resistant PrPRes with monoclonal antibodies (Prionics-Check LIA test),
- the two-sided immunoassay using two different monoclonal antibodies directed against two epitopes presented in a highly unfolded state of bovine PrPSc (Roboscreen Beta Prion BSE EIA Test Kit).
- the immunoassay using a chemical polymer for selective PrPSc capture and a monoclonal detection antibody directed against conserved regions of the PrP molecule (IDEXX HerdChek BSE Antigen Test Kit, EIA &HerdChek BSE-Scrapie Antigen (IDEXX Laboratories)).
- the lateral-flow immunoassay using two different monoclonal antibodies to detect Proteinase K-resistant PrP fractions (Prionics Check PrioSTRIP).

All these tests, apart from the IDEXX HerdChek tests, that are based on a conformational detection technology for PrPSc using a specific aggregate capture ligand on a dextran polymer (Seprion ligand technology, Microsens Biotechnologies, London, UK), rely on the digestion of the cellular prion protein (PrPC) to allow the binding of anti-PrP antibodies to the PK resistant PrP.

The Prionics-check Western blot was the first test available for BSE surveillance. It is based on an optimized WB procedure that allows to monitor the characteristic three band pattern (non-, mono-, and di-glycosylated forms) of the protein-resistant PrPSc fragment (27–30 kD) and their position relative to the control, the top band corresponding to a protein of approximate molecular weight of 30 kD.

The other tools are represented by a lateral flow immunoassay and four semiquantitative ELISA techniques that provide qualitative results relative to a cut-off value.

Such tests provide a preliminary diagnosis from which positive or inconclusive results are subject to examinations by histology, immunohistochemistry, or WB confirmatory methods.

Whenever their first purpose is that of screening methods, for which a positive or suspect result must be followed by a confirmatory method, a combination of two rapid tests is allowed for primary screening and subsequent confirmation, according to an algorithm maintained in the TSE-LAB-NET.[45] However, a negative result according to the two rapid tests as well as a discordant one has to be further investigated through a traditional confirmatory test. Where the histopathological examination is used for that purpose, but proves to be inconclusive or negative, the tissues must be submitted to a further examination by one of the other confirmatory methods and protocols.

6. HISTOPATHOLOGICAL EXAMINATION

Histological examination is performed on formalin-fixed, paraffin-embedded brain sections stained with hematoxylin and eosin. The solitary tract nucleus (NST) and the spinal tract nucleus of the trigeminal nerve (NSTV) are the target areas at the level of the obex to be examined in order to diagnose C-BSE (Figs. 3 and 4). Vacuolation of the gray matter

Fig. 3 Section of the obex showing the target nuclei for BSE diagnosis.

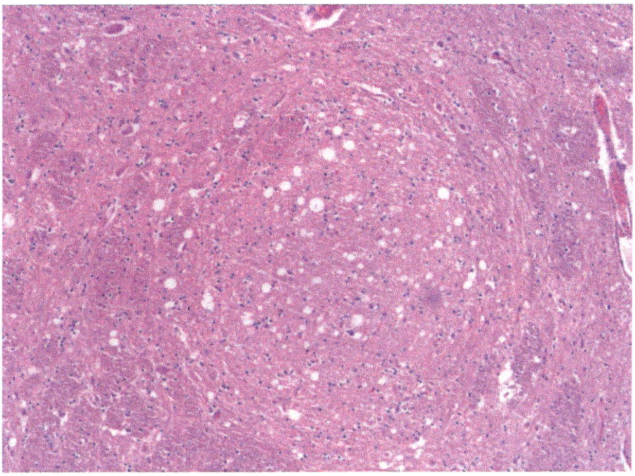

Fig. 4 H&E, nucleus of the solitary tract of C-BSE: presence of spongiosis in the neuropil (20 ×).

neuropil (spongiform change) and/or vacuolation of neurons, not necessarily associated with astrocytosis and neuronal degeneration, are the most frequent histological changes that can be observed in the CNS of C-BSE cases.[46–48] Neuronal vacuolation is especially prominent in the vestibular nucleus in the medulla oblongata at the level of the cerebellar peduncles. Other histological lesions are vacuolation in the central gray matter of the midbrain and mild spongiform changes of the neuropil at the level of the thalamus.

In natural cases of L-BSE (BASE), spongiosis is not consistently found in the brainstem, at the level of the obex or in more rostral areas. The frontal, parietal, and occipital cortices are apparently spared, and no vacuolation is detected in the olfactory bulb, piriform cortex, and hippocampus.[1] A more severe involvement of the central gray matter (periaqueductal gray) and rostral colliculus, but not the vestibular nuclear complex, has been observed in experimental cases of L-BSE. Additional brain areas, including the olfactory area, amygdala, hippocampus, and dorsal horns of spinal cord, are severely involved. The ventral and dorsal roots do not show major pathological changes.[49]

Vacuolar changes are generally observed in all the brain areas in experimental H-type BSE. Major vacuolation has been described in the thalamic nuclei and the neuropil of the central gray matter of the midbrain; however, mild vacuolation can also be present in the caudal cerebral and cerebellar cortices. Spongy changes in the vestibular and pontine nuclei are not as prominent as those seen in the other brainstem nuclei.[50]

7. IMMUNOHISTOCHEMISTRY

Immunohistochemical analysis is performed on paraffin-embedded brain tissues to highlight the presence of PrPSc accumulation. The slides are dewaxed, rehydrated, and then immersed in 98% formic acid for 25 min. After washing in distilled water, the sections are autoclaved at 121°C. Endogenous peroxidase activity is blocked in 3% hydrogen peroxide. To block nonspecific tissue antigens, the sections are incubated with horse blocking serum and then incubated with a primary monoclonal antibody directed to prion protein. After rinsing, a biotinylated secondary antibody is applied to the tissue sections, followed by the avidin-biotin peroxidase complex. Prion protein immunoreactivity is visualized using 3,3-diaminobenzidine as chromogen; the sections are then counterstained with Meyer's hematoxylin.

Several morphologic types of PrPSc deposition can be observed in the brain of C-BSE affected cattle[51] (Fig. 5):

Glial type: Radiating, branching PrPSc deposits often centered on a visible glial-type nucleus, conferring a star-like appearance, are predominantly present in the central gray matter, the cerebral lamina, and within the medial pontine nuclei in the cerebral cortex, thalamus, and obex.

Fig. 5 IHC, patterns of PrPSc of C-BSE, characterized by granular deposits and linear tracts (10×).

Granular type: Numerous small PrPSc granules are commonly found in the neuropil of the gray matter nuclei, such as the dorsal motor nucleus of vagus nerve (DMV), NST, and in the thalamic nuclei.

Intraneuronal type: Fine to coarse, sometimes confluent, granular deposits of PrPSc in the neuronal cytoplasm are often observed in the DMV, reticular formation, olivary nuclei, vestibular complex, pontine and thalamic nuclei, and hypothalamus.

Perineuronal type: Thin deposits of PrPSc are observed around individual neuronal perikarya and neuritis in the caudate and putamen nuclei of the basal ganglia and in the DMV.

Linear type: Thread-like deposits of PrPSc are seen along neuronal processes, particularly at the level of the reticular formation of the brainstem.

Coalescing type: Large granular PrPSc deposits merging to form amorphous or mesh-like masses are noted.

Intraglial type: Fine punctate PrPSc is found adjacent to glial nuclei.

A distinctive feature of natural cases of BASE is the presence of PrPSc deposition prevalently in the more rostral portions of the brain than as occurs in C-BSE. PrP-positive amyloid plaques are the hallmark characteristic of BASE: they appear as dense, unicentric, or less frequently multicentric, round structures up to 25 μm in diameter with a pale core and a dark radial periphery (Fig. 6). They are predominantly located in the thalamus, subcortical white matter, the deeper layers of the cerebral cortexes, and the

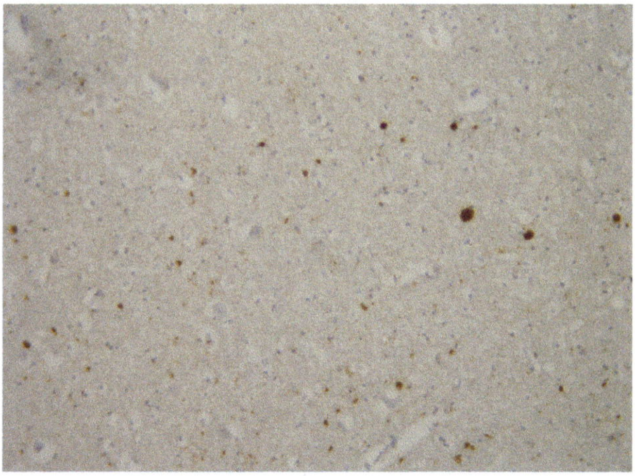

Fig. 6 IHC, patterns of PrPSc of L-type BSE (BASE), characterized by amyloid plaques (20 ×).

olfactory bulb. Other prevalent PrPSc deposition patterns are the punctate and granular types, which are mildly present in the hypoglossal and olivary nucleus and moderately present at the level of DMV nucleus, NST, NSTV, and reticular formation.[52] Glial, intraneuronal, perineuronal, and linear tracts are also frequently observed in different brain areas in BASE cases.

In experimental BASE cattle, abundant amyloid PrP-plaques are found in the subcortical white matter and deep gray nuclei, as observed in natural BASE cases. No PrP-plaques are present in the olfactory glomeruli, the cerebellum, or the spinal cord. The perineuronal pattern of PrPSc is also seen in the ventral horn neurons of the spinal cord and the dorsal root ganglion cells.[51]

In natural H-type BSE, granular, intraneuronal, linear, intraglial, and punctate PrPSc deposits in the brainstem are the most characteristic types. While mainly detected at the level of the DMNV, NST, NSTV, and in the reticular formation, a certain variability in PrPSc distribution has been described (Corona C. personal communication[53]) (Fig. 7). In experimental H-type BSE, large amounts of PrPSc are diffusely deposited throughout the CNS, including the cerebral cortex, basal ganglia, thalamus, hypothalamus, brainstem, and spinal cord. Fine or coarse particulate-type deposits in the neuropil of the gray matter throughout the brain and spinal cord are the most conspicuous type of PrPSc deposition. However, linear, perineuronal, and intraneuronal types of PrPSc staining can be observed in the cerebral cortex, basal ganglia, thalamus, and brainstem. Glial-type PrPSc deposition is

Fig. 7 IHC, pattern of PrPSc of H-type BSE, characterized by granular, intraglial, and small aggregates (20 ×).

predominantly identified in the cerebral cortex, basal ganglia, thalamus, hypothalamus, and hippocampus, and often in the cerebellar cortex, but it is not visible in the brainstem and spinal cord. Intraglial-type PrPSc deposition is highly consistent throughout the white matter of the CNS and spinal cord. Some animals show the presence of PrPSc-positive plaques scattered throughout the cerebral white matter.[50]

8. WESTERN BLOTTING

Western blotting is a widely used immunobiochemical technique for the diagnosis of prion diseases. Various different WB methods have been developed since the active surveillance system was instituted in Europe: some are used as screening tests and others to confirm the suspect cases identified by active and passive surveillance. The techniques are based on the immunodetection of the PrPSc at the level of the medulla oblongata. WB methods are very versatile since they can be applied on fresh, frozen, and autolytic tissues.[54] Comparable in diagnostic sensitivity to IHC techniques, WB remains the method of choice, along with IHC, for the confirmation of suspect BSE cases. A highly sensitive method, WB uses a large mass (2–4 g) of CNS material and several steps to concentrate PrPSc. Alternative less time-consuming and less costly methods are now employed by TSE Reference Laboratories in Europe to confirm BSE cases. Different from published methods, in-house tests for confirmatory testing must be validated; their analytical sensitivity, together with commercial tests, is continuously monitored through annual ring trials conducted by the TSE European Union Reference Laboratory (APHA, UK).

SAF-immunoblotting entails the preparation of homogenates from brainstem and digestion of the samples with proteinase K. After ultracentrifugation, the pellet is dissolved in Laemmli buffer, and an equivalent of 10 mg of wet tissue is loaded onto SDS-polyacrylamide gels. Following separation, the proteins are transferred onto a polyvinylidene difluoride (PVDF) membrane. PrPSc is detected by utilizing monoclonal antibody anti-PrP, and the presence of immunosignals is revealed by phosphatase-conjugated anti-mouse IgG developed using a chemiluminescence system and visualized on Hyperfilm ECL sheets. In positive test cases, confirmatory WB shows the presence of PrPSc characterized by an electrophoretic pattern consisting of three bands that correspond to the di-, mono-, and nonglycosylated forms, migrating at approximately 30, 25, and 19 kDa, respectively. No PrP signals are present in the bovine cases confirmed as negative

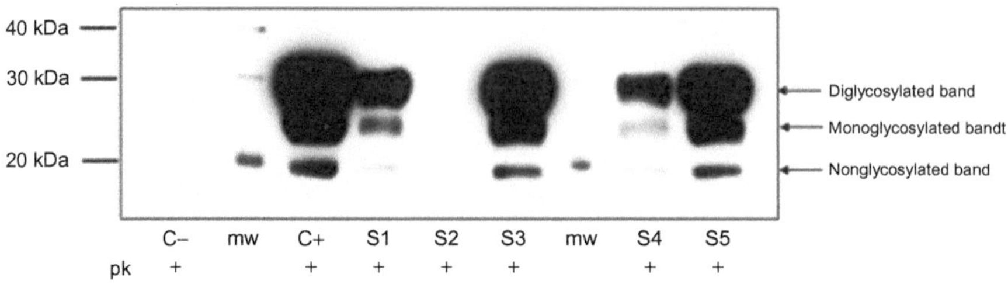

Fig. 8 Western Blotting analysis of positive and negative BSE cases. C-, negative BSE control; C+, positive BSE control; Lanes S1, S3, S4, and S5, positive BSE samples; lane S2, negative BSE sample; Mw, molecular markers. Immunodetection was performed by monoclonal antibody 6H4.

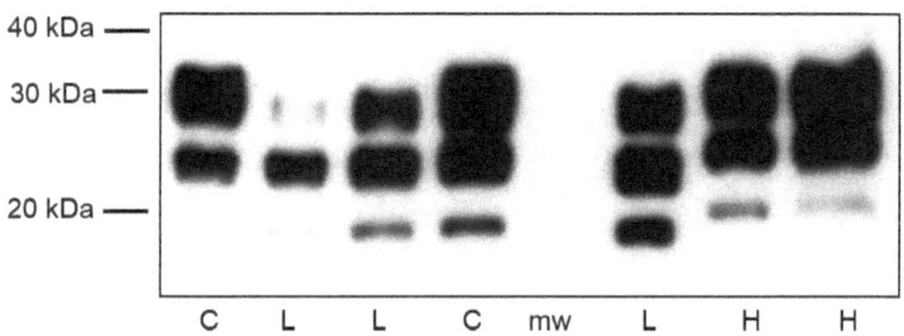

Fig. 9 Western Blotting analysis of PrPSc from classical and atypical BSE cases. C, classical BSE; L, low-type BSE; H, high-type BSE; Mw, molecular markers. Immunodetection was performed by monoclonal antibody 6H4.

since the PrPC is completely digested by proteinase K (Fig. 8). The application of immunoblotting methods is essential for evaluating the molecular features of PrPSc and so to discriminate between classical and atypical BSE isolates. The H-type is characterized by a significantly higher molecular size of the nonglycosylated PrPSc form and a conventional glycopattern, while the L-type or BASE has only a slightly lower molecular size of the nonglycosylated PrPSc form and a predominance of the monoglycosylated moiety (Fig. 9).

9. SCRAPIE-ASSOCIATED FIBRILS

Electron microscopy highlights BSE-associated fibrils, the bovine equivalent of SAF. The fibrils are composed of PrPSc and can be extracted from fresh, frozen or formalin-fixed nervous tissue by homogenization, centrifugation, and digestion with proteinase K, followed by coloring with phosphotungstic acid. Observation by electron microscopy reveals fibrils with a simple or double helix structure 100–500 nm in length.

10. IN VITRO AMPLIFICATION TECHNIQUES

A key challenge to coping with TSE diseases in humans, livestock, and wildlife is to develop sensitive and rapid high-throughput assays for the routine detection of prions and the early diagnosis of TSEs. Owing to the peculiar characteristics of prion diseases, such as the absence of a classical immune response and the absence of nucleic acids in the prion agent, many conventional methods for detecting infections by common pathogens cannot be applied. In fact, current biochemical or immunocytochemical assays have low sensitivity and may yield inconclusive results, rendering them unreliable in a clinical or preclinical setting of prion-infected hosts. In contrast, in vitro amplification techniques can be used to determine whether a tissue contains any prion seeding activity. Research in this direction has led to the development of reliable assays for the early diagnosis of prion disease in easily accessible tissues or body fluids. Both protein misfolding cyclic amplification (PMCA) and real-time quaking-induced conversion (RT-QuIC) rely on the detection of PrP structural conversion and polymerization upon the addition of PrPSc seeds contained in the infected samples. These methods are usually more sensitive than bioassays by two to three orders of magnitude, and endpoint titration can be performed in a format similar to the bioassay.[52]

10.1 Protein Misfolding Cyclic Amplification

In 2001, Soto and colleagues described a new type of in vitro prion conversion reaction called PMCA that greatly improved efficiency and sensitivity over the initial conversion reactions of the prion protein in cell-free environments (cell-free conversion assay).[55] PMCA combines cycles of sonication that result in hydrodynamic sharing of PrPSc aggregates into smaller nuclei. During incubation, the PrPSc molecules present in the nuclei imprint their abnormal conformation onto PrPC, which is subsequently incorporated into growing PrPSc aggregates. The sonication–incubation cycles are repeated to achieve exponential amplification of minute quantities of PrPSc present in a sample. In a typical PMCA reaction, brain extracts are used as a source of PrPC. The cyclic nature of the system and the possibility to refresh the substrate at each round enable the performance of as many cycles as needed to reach the amplification state needed for the detection of PrPSc in a given sample. In these conditions, the PrPSc molecules can be amplified to detectable levels by immunoblotting. PMCA is a promising platform for prion diagnosis in biological tissues or fluid samples, including blood, urine, feces, or cerebrospinal fluid, where the level of PrPSc is estimated to be in the range of picograms per mL.[56] PMCA has been successfully employed to study prion behavior, the molecular mechanism of prion replication, the cellular factors involved in prion propagation, and the still unknown aspects of prion strains and their trans'species conversion characteristics upon passage. Thus, PMCA holds promise not only as a prion detection assay but also as a tool to investigate the mechanism of prion-induced PrP conversion. Despite the advances achieved with PMCA in prion research, the technique has several pitfalls: amplification of PrPSc in apparently normal individuals, spontaneous generation of misfolded PrPSc, and possible cross-contamination. Other limitations are the time necessary to achieve optimal sensitivity and the use of brain-derived PrPSc as the amplification substrate. Clearly, there was a need to develop an accurate, high throughput diagnostic that is automated and can be easily used in a routine diagnostic laboratory.

10.2 Real-Time Quaking-Induced Conversion

To overcome the technical complexities of PMCA reactions, a new practical prion assay, quaking-induced conversion (QuIC), developed by Atarashi and colleagues[57] substitutes sonication with automated tube shaking for the conversion of recombinant PrP (rPrP) substrates. QuIC-based methods use the rPrPSc produced in bacteria as a substrate for seeded polymerization and tube shaking to break the generated polymers and provide new seeds for conversion in amplification rounds. The rPrP-QuIC method also solves the problem of using brain-derived PrPSc as the amplification substrate. Moreover, the fact that rPrPSc can be easily mutated allows investigation into the role of specific sequences or amino acids in the conversion reaction and accelerates studies on the detection of prions. One of the major refinements of the technique was the substitution of WB with real-time fluorescent color reaction based on a fluorescent amyloid-sensitive thioflavin T dye (ThT). The formation of these prion-seeded amyloid fibers is detected in real time by reading ThT fluorescence over time. In its real-time and multiwell plate format, RT-QuIC has the potential for use in high-throughput screening of samples. The method provides for a test as quantitative and sensitive as in vivo testing[52] and it has been adapted to different types of TSE. The RT-QuIC assay allows rapid and highly sensitive discrimination between prion-infected and uninfected brain tissues. Furthermore, it has been proven sensitive for detecting prions in infected tissues and fluids, including cerebrospinal fluid, saliva, nasal fluids, and blood.[57,58] The method does not detect prion infectivity in a given tissue, but rather allows for the detection of a seeding activity potentially associated with prion replication. Indeed, based upon the quantitative correlation between prion seed concentration and the lag time to the start of the conversion reaction, qRT-QuIC allows for the quantification of prion infectivity in tissues, body fluids, and excreta.[59] For quantification, the amplified PrPSc signal can be compared with that seen in endpoint titrated material run under the same conditions (such as brain homogenate from animals at the terminal stage of disease) or to PrP calibration curves. Like animal bioassays, the RT-QuIC assay can titrate the seeding activity in endpoint diluted samples.[52,57] Serial dilutions of a given sample are used as seeds, and the seeding dose (SD) giving 50% ThT-positive replicate reactions (SD50), i.e., the 50% endpoint dilution, is estimated. The SD50 is analogous to the 50% lethal dose (LD50) determined in an endpoint dilution animal bioassay. Moreover, RT-QuIC has several major advantages over animal bioassays, including practicality, high-throughput potential, rapidity, and lower cost. The quantitative aspect of qRT-QuIC suggests that it can provide reliable assessment of anti-prion therapy in vivo in order to follow the effects of therapy on the progression of prion diseases. Furthermore, since qRT-QuIC provides an ultra-sensitive method for quantifying pathological amyloid aggregate seeds, it may also be applicable to other disease-associated proteins rich in β-pleated structures that bind ThT and show seeded aggregation. Some prion strain types are known to be fairly resistant to amplification by either PMCA or RT-QuIC. However, recent studies have adapted RT-QuIC assays for the sensitive detection and discrimination of C-BSE, L-BSE, and H-BSE (Fig. 10).[60,61] Testing with the RT-QuIC assay on brain tissue from cattle affected by BSE strains showed that all three forms can be detected and distinguished with the use of particular rPrPSc substrates. RT-QuIC tests have been adapted to detect many types of prion seeding activity; however, numerous critical issues remain: the evaluation of outcomes for the antemortem diagnosis of TSE in animals of farm interest from which the least invasive and most economical biological peripheral matrix can be identified for prion testing in living subjects; the recognition of classical and atypical forms of PrPSc by a unique protocol in order to obtain a single diagnostic assay for TSE diseases; the detection of PrPSc from biological fluids by removing soluble components that could inhibit the assay; the collection of preclinical and clinical data from subjects testing PrPSc-positive in peripheral matrices to better define the distribution of peripheral TSE infectivity.

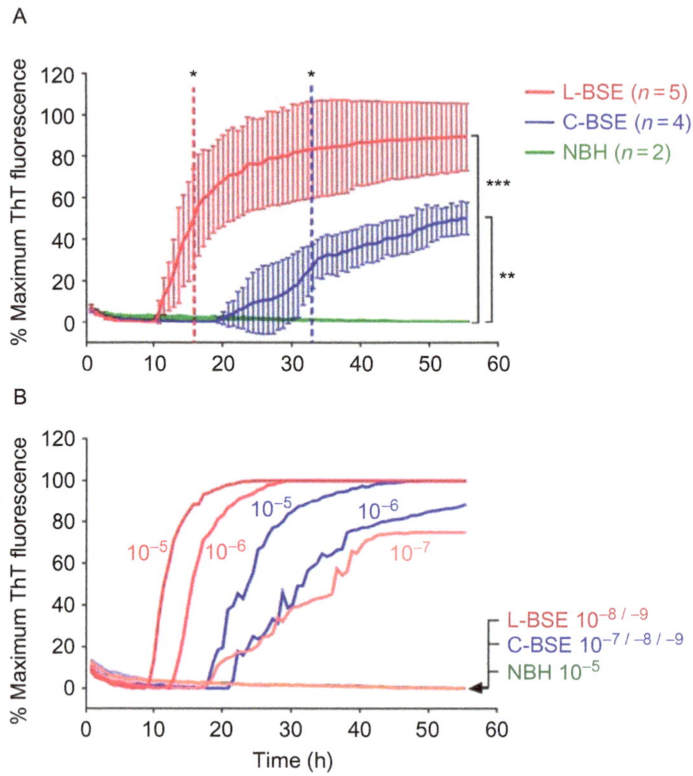

Fig. 10 *RT-QuIC sensitivity for C-BSE and L-BSE detection.* (A) L-BSE-infected (*magenta*), C-BSE-infected (*blue*), or normal negative control (NBH, *green*) 10^{-5} brain tissue dilutions were used to seed quadruplicate RT-QuIC reactions using the Ha-S rPrPSen substrate. (B) Serial dilutions (10^{-5} to 10^{-9}) of C-BSE-infected or L-BSE-infected brain tissue or a 10^{-5} dilution of uninfected brain tissue were used to seed quadruplicate RT-QuIC reactions with Ha-S rPrPSen as the substrate. The data show the average ThT fluorescence of four replicate wells. Each ThT reading is indicated as the percentage of the maximum value achievable by the plate readers as a function of reaction time. Square brackets signify the difference between the text groups and the control at the 55-h endpoint. $^{*}P < 0.05$; $^{**}P < 0.005$; $^{***}P < 0.0005$.

REFERENCES

1. Casalone C, Zanusso G, Acutis PL, et al. Identification of a second bovine amyloidotic spongiform encephalopathy: molecular similarities with sporadic Creutzfeld-Jakob disease. *Proc Natl Acad Sci U S A*. 2004;101:3065–3070.
2. Biacabe AG, Laplanche JL, Ryder S, Baron T. Distinct molecular phenotypes in bovine prion diseases. *EMBO Rep*. 2004;5(1):110–115.
3. Wells GA, Scott AC, Johnson CT, et al. A novel progressive spongiform encephalopathy in cattle. *Vet Rec*. 1987;121(18):419–420.
4. Fraser H, Bruce ME, Chree A, Mcconnell I, Wells GAH. Transmission of bovine spongiform encephalopathy to mice. *Vet Rec*. 1988;123(18):472.

5. Wilesmith JW, Wells GAH, Cranwell MP, Ryan JBM. Bovine spongiform encephalopathy—epidemiological studies. *Vet Rec.* 1988;123(25):638–644.
6. Wilesmith JW, Ryan JBM, Hueston WD. Bovine spongiform encephalopathy: case-control studies of calf feeding practices and meat and bone meal inclusion in proprietary concentrates. *Res Vet Sci.* 1992;52:325–331.
7. Wilesmith JW, Ryan JBM, Stevenson MA, et al. Temporal aspects of the epidemic of bovine spongiform encephalopathy in Great Britain: holding associated risk factors for the disease. *Vet Rec.* 2000;147:319–325.
8. La Bonnardière C, Calavas D, Abrial D, Morignat E, Ducrot C. Estimating the trend of the French BSE epidemic over six birth cohorts through the analysis of the abattoir screening in 2001 and 2002. *Vet Res.* 2004;35:299–308.
9. Ducrot C, Abrial D, Calavas D, Carpenter T. A spatio-temporal analysis of BSE cases born before and after the reinforced feed ban in France. *Vet Res.* 2005;36:839–853.
10. Ru G, Maurella C, Ponti AM, Ingravalle F, Caramelli M. Epidemiological study of the decline of BSE in Italy. *Vet Rec.* 2007;161(15):511–514.
11. Wilesmith JW, Ryan JB, Atkinson MJ. Bovine spongiform encephalopathy: epidemiological studies on the origin. *Vet Rec.* 1991;128(9):199–203.
12. Nathanson N, Wilesmith J, Griot C. Bovine spongiform encephalopathy (BSE): causes and consequences of a common source epidemic. *Am J Epidemiol.* 1997;145(11):959–969.
13. Will RG, Ironside JW, Zeidler M, et al. A new variant of Creutzfeldt-Jakob disease in the UK. *Lancet.* 1996;347(9006):921–925.
14. Bruce ME, Will RG, Ironside JW, et al. Transmissions to mice indicate that 'new variant' CJD is caused by the BSE agent. *Nature.* 1997;389(6650):498–501.
15. Collinge J, Sidle KCL, Meads J, Ironside J, Hill AF. Molecular analysis of prion strain variation and the aetiology of 'new variant' CJD. *Nature.* 1996;383:685–690.
16. Hill AF, Desbruslais M, Joiner S, et al. The same prion strain causaes vCJD and BSE. *Nature.* 1997;389:448–450.
17. Scott MR, Will R, Ironside J, et al. Compelling transgenetic evidence for transmission of bovine spongiform encephalopathy prions to humans. *Proc Natl Acad Sci U S A.* 1999;96:15137–15142.
18. Doherr MG, Oesch B, Moser M, Vandevelde M, Heim D. Targeted surveillance for bovine spongiform encephalopathy. *Invest Ophthalmol Vis Sci.* 1999;32:1492–1498.
19. Salman M, Silano V, Heim D, Kreysad J. Geographical BSE risk assessment and its impact on disease detection and dissemination. *Prev Vet Med.* 2012;105:255–264.
20. Budka H, Goossens B, Ru G. BSE and TSEs: past, present and future. *Trends Food Sci Technol.* 2008;19:S34–S39.
21. Opinion of the scientific panel on biological hazards on the revision of the geographical BSE risk assessment (GBR) methodology. *EFSA J.* 2007;463:1–3.
22. Donnelly CA, Ferguson NM, Ghani AC, Woolhouse MEJ, Watt CJ, Anderson RM. The epidemiology of BSE in cattle herds in Great Britain. I. Epidemiological processes, demography of cattle and approaches to control by culling. *Philos Trans R Soc Lond B, Biol Sci.* 1997;352:781–801.
23. Ducrot C, Arnold M, De Koeijer A, Heim D, Calavas D. Review on the epidemiology and dynamics of BSE epidemics. *Vet Res.* 2008;39:15.
24. Griffin JM, Collins JD, Nolan JP, Weavers ED. Bovine spongiform encephalopathy in the Republic of Ireland: epidemiological observations 1989-1996. *Ir Vet J.* 1997;50:593–600.
25. Wilesmith JW, Gibbs Jr CF. *Bovine Spongiform Encephalopathy: The BSE Dilemma.* Berlin, Germany: Springer Science & Business Media; 2012.
26. Stevenson MA, Morris RS, Lawson AB, Wilesmith JW, Ryan JBM, Jackson R. Area-level risks for BSE in British cattle before and after the July 1988 meat and bone meal feed ban. *Prev Vet Med.* 2005;69(1–2):129–144.

27. Wilesmith JW, Wells GA, Ryan JB, Gavier-Widen D, Simmons MM. A cohort study to examine maternally-associated risk factors for bovine spongiform encephalopathy. *Vet Rec.* 1997;141(10):239–243.
28. The TSE Roadmap COM. 322 CE, available at: http://ec.europa.eu/food/food/biosafety/bse/roadmap_en.pdf; 2005.
29. Heppner FL, Christ AD, Klein MA, et al. Transepithelial prion transport by M cells. *Nat Med.* 2001;7(9):976–977.
30. Terry LA, Marsh S, Ryder SJ, Hawkins SA, Wells GA, Spencer YI. Detection of disease-specific PrP in the distal ileum of cattle exposed orally to the agent of bovine spongiform encephalopathy. *Vet Rec.* 2003;152(13):387–392.
31. Wells GA, Spiropoulos J, Hawkins SA, Ryder SJ. Pathogenesis of experimental bovine spongiform encephalopathy: preclinical infectivity in tonsil and observations on the distribution of lingual tonsil in slaughtered cattle. *Vet Rec.* 2005;156(13):401–407.
32. Hoffmann C, Eiden M, Kaatz M, et al. BSE infectivity in jejunum, ileum and ileocaecal junction of incubating cattle. *Vet Res.* 2011;42:21.
33. Jeffrey M, González L, Espenes A, et al. Transportation of prion protein across the intestinal mucosa of scrapie-susceptible and scrapie-resistant sheep. *J Pathol.* 2006;209(1):4–14.
34. Balkema-Buschmann A, Fast C, Kaatz M, et al. Pathogenesis of classical and atypical BSE in cattle. *Prev Vet Med.* 2011;102(2):112–117.
35. Arnold ME, Ryan JB, Konold T, et al. Estimating the temporal relationship between PrPSc detection and incubation period in experimental bovine spongiform encephalopathy of cattle. *J Gen Virol.* 2007;88(pt 11):3198–3208.
36. Espinosa JC, Morales M, Castilla J, Rogers M, Torres JM. Progression of prion infectivity in asymptomatic cattle after oral bovine spongiform encephalopathy challenge. *J Gen Virol.* 2007;88(pt 4):1379–1383.
37. Okada H, Miyazawa K, Fukuda S, et al. The presence of disease-associated prion protein in skeletal muscle of cattle infected with classical bovine spongiform encephalopathy. *J Vet Med Sci.* 2014;76(1):103–107.
38. Protocol for further laboratory investigations into the distribution of infectivity of Atypical BSE. *EFSA J.* 2014;12(7):3798.
39. Iwamaru Y, Imamura M, Matsuura Y, et al. Accumulation of L-type bovine prions in peripheral nerve tissues. *Emerg Infect Dis.* 2010;16(7):1151–1154.
40. Manual of diagnostic tests and vaccines for terrestrial animal, Chapter 2.4.5: Bovine Spongiform Encephalopathy http://www.oie.int/fileadmin/Home/eng/Health_standards/tahm/2.04.05_BSE.pdf.
41. Webster K, Flowers M, Cassar C, Bayliss D. Determination of Analytical Sensitivity (Detection Limit) for Currently Approved TSE Rapid Tests. Available: http://www.efsa.europa.eu/de/scdocs/doc/1436.pdf; 2009.
42. Scientific opinion of the European food safety authority on a protocol for the evaluation of new rapid BSE post mortem tests. *EFSA J.* 2007;508:1–20.
43. European Commission: Regulation (EC) No 162/2009 of the European Parliament and of the Council of 26 February 2009 amending Annexes III and X to regulation (EC) No 999/2001 laying down rules for the prevention control and eradication of certain transmissible spongiform encephalopathies. *Off J Eur Communities.* 2009;55:11–16.
44. Meloni D, Davidse A, Langeveld JPM, et al. EU-approved rapid tests for bovine spongiform encephalopathy detect atypical forms: a study for their sensitivities. *PLoS One.* 2012;7(9), e43133.
45. http://www.tse-lab-net.eu/documents/tse-oie-guide.pdf.
46. Wells GA, Hancock RD, Cooley WA, Richards MS, Higgins RJ, David GP. Bovine spongiform encephalopathy: diagnostic significance of vacuolar changes in selected nuclei of the medulla oblongata. *Vet Rec.* 1989;125(21):521–524.

47. Wells GAH, Wilesmith JW. The neuropathology and epidemiology of bovine spongiform encephalopathy. *Brain Pathol.* 1995;5:91–103.
48. Simmons MM, Harris P, Jeffrey M, Meek SC, Blamire IW, Wells GA. BSE in Great Britain: consistency of the neurohistopathological findings in two random annual samples of clinically suspect cases. *Vet Rec.* 1996;138:175–177.
49. Lombardi G, Casalone C, D' Angelo A, et al. Intraspecies transmission of BASE induces clinical dullness and amyotrophic changes. *PLoS Pathog.* 2008;4(5), e1000075.
50. Okada H, Iwamaru Y, Imamura M, et al. Experimental H-type bovine spongiform encephalopathy characterized by plaques and glial- and stellate-type prion protein deposits. *Vet Res.* 2011;42:79.
51. Casalone C, Caramelli M, Crescio MI, Spencer YI, Simmons MM. BSE immunohistochemical patterns in the brainstem: a comparison between UK and Italian cases. *Acta Neuropathol.* 2006;111(5):444–449.
52. Wilham JM, Orrú CD, Bessen RA, et al. Rapid end-point quantitation of prion seeding activity with sensitivity comparable to bioassays. *PLoS Pathog.* 2010;6, e1001217.
53. Porcario C, Hall SM, Martucci F, et al. Evaluation of two sets of immunohistochemical and western blot confirmatory methods in the detection of typical and atypical BSE cases. *BMC Res Notes.* 2011;4:376.
54. Hayashi H, Takata M, Iwamaru Y, et al. Effect of tissue deterioration on postmortem BSE diagnosis by immunobiochemical detection of an abnormal isoform of prion protein. *J Vet Med Sci.* 2004;66(5):515–520.
55. Saborio GP, Permanne B, Soto C. Sensitive detection of pathological prion protein by cyclic amplification of protein misfolding. *Nature.* 2001;411(6839):810–813.
56. Castilla J, Saa P, Soto C. Detection of prions in blood. *Nat Med.* 2005;11(9):982–985.
57. Atarashi R, Satoh K, Sano K, et al. Ultrasensitive human prion detection in cerebrospinal fluid by real-time quaking-induced conversion. *Nat Med.* 2011;17(2):175–178.
58. Orrú CD, Bongianni M, Tonoli G, et al. A test for Creutzfeldt-Jakob disease using nasal brushings. *N Engl J Med.* 2014;371(6):519–529.
59. Henderson DM, Davenport KA, Haley NJ, Denkers ND, Mathiason CK, Hoover EA. Quantitative assessment of prion infectivity in tissues and body fluids by real-time quaking-induced conversion. *J Gen Virol.* 2015;96(pt 1):210–219.
60. Orrú CD, Favole A, Corona C, et al. Detection and discrimination of classical and atypical L-type bovine spongiform encephalopathy by real-time quaking-induced conversion. *J Clin Microbiol.* 2015;53(4):1115–1120.
61. Masujin K, Orrú CD, Miyazawa K, et al. Detection of atypical H-type bovine spongiform encephalopathy and discrimination of bovine prion strains by real-time quaking-induced conversion. *J Clin Microbiol.* 2016;54(3):676–686.

CHAPTER TWELVE

Scrapie, CWD, and Transmissible Mink Encephalopathy

Candace K. Mathiason[1]
Colorado State University, Fort Collins, CO, United States
[1]Corresponding author: e-mail address: candace.mathiason@colostate.edu

Contents

Abstract

Transmissible spongiform encephalopathies (TSEs), or prions, are neurodegenerative diseases that affect a variety of animal species, including humans. Cruetzfeldt–Jakob disease (CJD) in humans, sheep and goat scrapie, chronic wasting disease (CWD) of cervids, and transmissible mink encephalopathy (TME) of mink are classified as TSEs. According to the "protein-only" hypothesis (Prusiner, 1982),[1] prions are devoid of nucleic acids and consist of assemblies of misfolded host-encoded normal protein, the prion protein (PrP^C). Prion propagation is thought to occur by a templating mechanism during which

Progress in Molecular Biology and Translational Science, Volume 150
ISSN 1877-1173
http://dx.doi.org/10.1016/bs.pmbts.2017.07.009

PrP^C is recruited, converted to a disease-associated isoform (PrP^D), and assembled onto the growing amyloid fibril. This fibular assembly is infectious, with ability to initiate disease processes similar to other pathogenic agents. Evidence indicates that scrapie, CWD, and TME disease processes follow this rule.

1. SCRAPIE

1.1 History and Epidemiology

Scrapie, the archetype transmissible spongiform encephalopathy (TSE), is a naturally occurring prion disease of sheep, goats, and mouflon (*Ovis musimon*).[2–4] Scrapie was first reported in Europe in the 16th century (England 1732 and Germany 1759),[5–7] with references to the disease in literature from the early Chinese and Roman epochs.[8] This is likely why the disease has been referred to by so many common names, including—shakings/shakers, cuddie trot, rida (ataxia or tremor), rickets, rubbers, goggles (to stare or squint), la tremblante (trembling), traberkrankheit (trotting disease), and prurigo lumbar (itchy back).[7–10] Scrapie has the distinction of being the first prion disease evidenced to be both infectious and transmissible in natural settings.[11,12]

In the early 20th century (between 1920 and 1958), scrapie became an economical concern for sheep breeders. Most sheep breeds have been affected by scrapie, although it is more common in some breeds than others. International trade of Suffolk sheep incubating the disease lead to the introduction of scrapie in numerous countries, resulting in endemic infections in flocks around the world.[9,13–15] Progress has been achieved in the prevention of scrapie in sheep due to efficient genetic breeding programs based on eradication and control of the disease. In general, the selection plans for sheep aim to eliminate and reduce the susceptible alleles (VRQ) and to enrich the resistant allele (ARR).[16]

The codons of the prion protein (*Prnp*) that are especially significant in determining susceptibility or resistance to classical scrapie in sheep include 136, 154, and 171[17] and for atypical sheep scrapie include 141, 154, and 171.[18] *Prnp* haplotypes correlated with increased susceptibility are 136 valine (V), 154 arginine (R), and 171 glutamine (Q) (VRQ), whereas 136 alanine (A),[17,19,20] 154 histidine (H),[21] and 171 arginine (R)[22,23] are associated with resistance to natural scrapie. Codon 171 appears to have the most discernible influence, where sheep with 171 QQ are susceptible and 171RR are resistant. Goat scrapie has been less well studied. A number of polymorphisms appear to be associated with goat scrapie susceptibility and resistance.[24] While none confer complete susceptibility or resistance, K222 has been suggested as a good candidate for selective breeding programs to enhance scrapie resistance in goats.[25,26]

Natural scrapie in sheep and goats occurs in classical and atypical forms, which are distinguished on the basis of neuropathology and PrP^{Sc} glycosylation patterns on western blot.[27,28] Atypical scrapie was first reported in 1998, and has now been identified throughout Europe, mainly through active surveillance of asymptomatic sheep.[29–31] The earliest evidence for atypical scrapie in sheep dates back to 1972.[32] Atypical scrapie tends to occur in older sheep with *Prnp* genotypes considered resistant to classical scrapie (ARR/ARR).[33–35] Both forms (classical and atypical scrapie) have been demonstrated in the same naturally infected animal,[32,36,37] giving rise to questions regarding the origin of atypical scrapie. Experimental inoculation studies have suggested that the phenotype of atypical scrapie can in some instances change into that of a second strain during passage in sheep.[38] It is therefore more likely that natural TSE infections of ruminants involve mixtures of strains rather than a single strain.[32]

1.2 Clinical Signs

Classical and atypical scrapie can be distinguished by careful observation of clinical signs.[39,40] Clinical signs of classical scrapie include hunched posture, hind limb ataxia with fore limb hypermetria, scratching, and wool loss.[41,42] Whereas sheep with atypical scrapie exhibit normal posture, circling, no scratching, ataxia, visible head tremor, and no wool loss.[40] Classical scrapie is typically found among sheep between 2 and 5 years of age[42–45] and in goats >6 years of age[46] with no gender bias.

1.3 Distribution of Scrapie in Sheep Tissues

In naturally acquired scrapie, evidence suggests that the main route of entry of the infectious agent is through the gut-associated lymphoid tissues (GALTs) with subsequent spread to the central nervous system (CNS) by centripetal transport along the peripheral nervous system (PNS).[47,48] A large body of evidence indicates that scrapie infection ascends from the gut into the thoracic spinal cord and to the medulla oblongata along the splanchnic and vagus nerves. From these initial sites, infection propagates cranially and caudally within the CNS, with subsequent centrifugal spread to the sensory ganglia of the vagus and splanchnic circuits.[49] Further centrifugal transport from the brain to peripheral tissues appears to be along the PNS.[50–53] It is widely recognized that the lymphoreticular system plays an important role in the dissemination of the prion agent.[54] Infectivity and/or PrP^{Sc} have been demonstrated in peripheral tissues of natural scrapie infections, including placenta,[55,56] skeletal muscle,[52,57] pancreas,[52,58] kidney,[52,59,60] liver,[52,61] adrenal glands,[52,62,63] heart,[52] urinary bladder,[52] lung,[52] mammary gland,[52,64] and skin,[52,65] suggesting that the agent may propagate in peripheral organs.[52] The detection of PrP^{Sc} deposition in sheep has, in some cases (mammary gland and liver), been found in combination with the presence of an inflammatory response to another pathogen.[52,66]

Tissue distribution of the scrapie agent may be influenced by several factors including host genetics, dose, and strain.[47,52] For example, sheep with resistant vs susceptible *Prnp* genotypes develop natural scrapie with longer incubation periods and less or varied dissemination of PrP^{Sc} in CNS and PNS tissues.[52,67–69]

1.4 Diagnosis

Diagnosis of sheep scrapie can be achieved via a combination of clinical observation and detection of the biomarker associated with the disease, PrP^{Sc}, in lymphoid biopsy or brainstem at the level of the obex.[70] Biochemical methods for differentiation of classical and atypical scrapie are available. Atypical scrapie PrP^{Sc} is more sensitive to proteinase K (PK) digestion than classical scrapie PrP^{Sc} and has a more variable pattern on western blots, including a characteristic low molecular mass band (estimated at 7–12 kDa) not present in classical scrapie.[71] To complicate detection further, infectivity has been demonstrated in peripheral tissues of sheep with atypical scrapie devoid of PrP^{Sc},[72] suggesting that atypical scrapie may at times be difficult to detect by conventional methodologies used to surveille for scrapie.

1.5 Transmission and Pathogenesis

An increase in scrapie incidence has long been associated with the lambing season,[42] so it is thought that most animals are infected at birth or shortly thereafter[42] via contact (presumed oral consumption)[73–76] with prions in placental tissues known to contain prion infectivity[76–79] or infectious prions shed into the environment.[80] Prion infectivity has also been demonstrated in the blood,[58,81–83] milk, colostrum,[84–86] saliva,[87] and environments[45,88,89] of scrapie-infected sheep and goats. The demonstration of prion infectivity in pregnancy-related milieu (milk, colostrum, and placenta) and maternal blood suggests that maternal/vertical transmission plays an important role in early scrapie exposure. The recent demonstration of in utero scrapie transmission in sheep during clinical and preclinical stages of disease provides additional evidence to this point.[90] Scrapie transmission dynamics and efficiency would likely rival that of chronic wasting disease (CWD) in cervid species without the advent of breeding programs to select for genotypes resistant to scrapie infection.

1.6 Zoonotic Potential

Early epidemiological studies lacked support for a causal relationship between sheep scrapie and human infections,[91] and to date there is no evidence for an increased zoonotic risk in the human population. Experimental studies provide conflicting results as to the zoonotic potential of scrapie. Some studies have failed to demonstrate transmission of scrapie to non-human primates[92] and transgenic mice overexpressing the human prion protein.[93] Other studies, using natural classical and atypical ovine scrapie strains and marmoset,[94] transgenic mice,[95] or cynomolgus macaques[96] have provided reason to revisit this premise. Cynomolgus macaques, a highly relevant model for human prion disease, demonstrated susceptibility to infection by a natural classical scrapie isolate after a 10-year silent phase of disease.[96] These findings require additional studies to assess the risk natural scrapie strains pose to human public health.[97]

2. CHRONIC WASTING DISEASE

2.1 History and Epidemiology

CWD is unique among prion diseases, being the only TSE to infect free-range and captive wildlife populations. CWD has to date been identified in mule deer (*Odocoileus hemionus*), black-tailed deer (*O. hemionus*), white-tailed deer (*Odocoileus virginianus*), elk (*Cervus elaphus*), moose (*Alces alces*), and reindeer (*Rangifer tarandus*). The geographical distribution of CWD (Fig. 1) continues to expand, now including 24 states and 2 Canadian provinces of North America,[98] transport from North America to South Korea,[99,100] and most recently the first cases have been identified in Europe (Norway 2016).[101] The disease syndrome of wasting and eventual death was first observed in cervids held in captivity in Colorado and Wyoming in the late 1960s. It was initially thought to be the result of nutritional

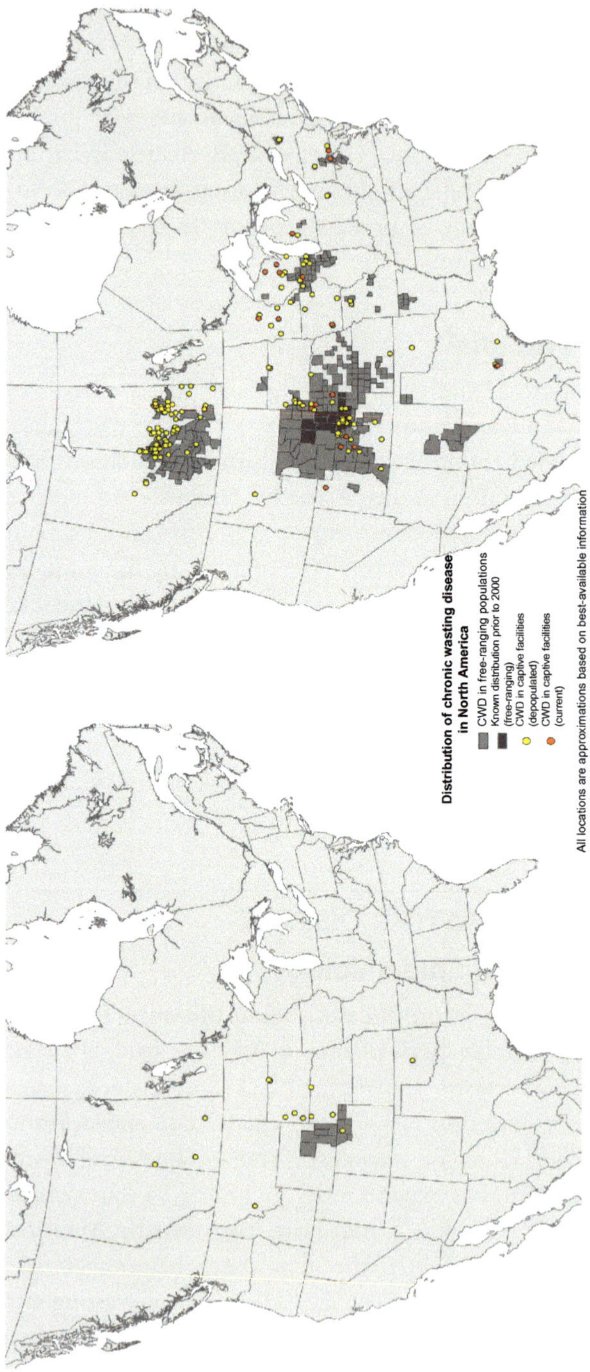

Fig. 1 Distribution of CWD in North America. *Maps provided by Bryan Richards, USGS.*

deficiency/or stress associated with their confinement. The disease was described as a spongiform encephalopathy in 1980[102] and was soon after identified in free-range elk and mule deer.[103,104] CWD is the most efficiently transmitted of the prion diseases. Free-range prevalence rates are reported to be <5% in deer and <2.5% in elk, with hot spots as high as 50% in free-range populations and >90% in captive facilities.[105] To date there is no clear evidence to explain the origin of CWD. Suggested hypotheses include the transmission of sheep scrapie to cervids, or a spontaneous conversion of the normal cervid prion protein to a misfolded confirmer that was transmissible between cervid species. Prevalence rates between males and females in free-range populations are variable dependent upon location; a slight increase in males,[106] twofold higher rates in males to females,[107] or nearly twice the infection rate in females as males.[108]

Deciphering the genetic susceptibility of cervids to CWD has been pursued. The structure of the mature prion protein is consistent across taxa, with a long, flexible N-terminal tail, three alpha-helices and a two-stranded antiparallel beta-sheet.[109] An amino acid loop connecting the beta-sheet strand to the alpha-helices is particularly rigid in cervids.[110] This rigid loop structure appears to be influential in the efficiency of PrP^{C} to PrP^{CWD} conversion[111] and may have bearing on the highly infectious nature of CWD in cervids.[110–112] Genetic analysis of CWD-affected cervid populations suggests that *Prnp* variation affects susceptibility and/or disease progression. Many of these conclusions have been based upon the proportion of affected and unaffected cervids in a population with respect to their *Prnp* genotype. Studies have evaluated disease susceptibility and progression of *Prnp* genotypes based on tissue pathogenesis, infection rates in captive animals, disease incubation period in experimentally challenged animals, and determination of genotype-specific infection and survival rates in native free-range populations. These studies in elk, mule deer, white-tailed deer, moose, and reindeer have revealed some correlations between *Prnp* sequence and CWD infection.

2.1.1 Elk

Studies in Rocky Mountain elk have found that polymorphisms at *Prnp* codon 132, encoding either a methionine or leucine may influence the susceptibility to CWD infection.[113,114] An initial report found that there was an overrepresentation of 132MM *Prnp* genotype in CWD-infected elk.[115,116] Further survey of free-ranging elk in Colorado found evidence for CWD infection in 132MM, 132ML, or 132LL in proportion to their frequency in the population.[117] Experimental oral inoculation studies in elk demonstrated modulation of incubation time dependent upon 132 expression, i.e., 132MM expressing elk exhibit the shortest time to clinical disease (23 mpi), followed by 132 ML at 40 mpi, with the longest incubation periods being associated with 132LL (60 mpi).[118,119]

2.1.2 Mule Deer and White-Tailed Deer

Mule deer carry a unique polymorphism at codon 225, serine (S), or phenylalanine (F).[120,121] While lower prevalence rates, lengthened incubation time,[122] and delayed PrP^{CWD} tissue accumulation[123] have been shown in 225SF genotype mule deer, complete resistance is not conferred. Studies of the white-tailed deer *Prnp* genotype have also been conducted. These studies have found two coding polymorphisms, Q95H and G96S, which occur at lower than expected frequencies in CWD-infected animals.[124–128] The serine substitution for glycine at residue 96 (S96G) may be associated with slower disease progression, but neither S96G nor G96G confer resistance to CWD infection.[125,129] Further investigation of the affects *Prnp* gene haplotypes/diplotypes may have on CWD resistance has been applied to a population of white-tailed deer in Wisconsin/Illinois. Results from this survey suggest that deer populations with higher frequencies of haplotype C or diplotypes AC and BC might have a reduced risk for CWD infection.[130]

2.1.3 Moose and Reindeer

Few free-range CWD-infected moose has been identified,[131] and the first natural infection of reindeer has been reported.[101] The homozygous 209MM amino acid sequence was identified in the index case of wild moose CWD, but subsequent experimental studies designed to further define CWD resistance in moose have demonstrated susceptibility in both 209MM and 209IM moose.[132] The *Prnp* genetic sequence in the first reindeer naturally infected with CWD[101] was found to be identical to that found in two experimentally infected reindeer (homozygous V (valine), G (glycine), S (serine), and V (valine) at codons 2, 129, 138, and 169, respectively).[133]

In summary, none of the cervid *Prnp* genotypes described are highly diverged from the known susceptible genotypes, suggesting that *Prnp* divergence alone is unlikely to provide complete resistance to CWD infection. Thus, a population of cervid with prolonged infectious states could contribute to environmental contamination and subsequent transmission to susceptible animals in captive and native free-range cervid populations[134,135] and may be driving the decline of some cervid species.[108]

2.2 Clinical Signs

Disease presentation in wild and captive populations is initially recognized as progressive weight loss and isolation from the herd population. As disease progresses infected cervids exhibit excessive salivation, polydipsia, polyuria, bruxism, regurgitation, ataxia, and tremors, with terminal disease presentation within 18–24 months.[136] Experimental inoculation studies have provided further observation of clinical disease progression, with similar time to terminal disease (~24 months postexposure).[137,138] The earliest signs of clinical disease in white-tailed deer inoculated with brain (oral), blood (intravenous or intraperitoneal), or saliva (oral) from CWD-infected cervids included behavioral changes (isolation, increase/decrease response to human interaction) and were seen between 12 and 20 months postinoculation. Clinical disease onset manifested primarily as perceived body muscle-mass reduction and gradual weight loss, which reached ≥20% of maximum body weight over 2–8 months. Other late-stage clinical signs included: rough hair coat due to piloerection and a body stance characterized by a low head position and wide leg stance. Changes in behavior included hyperphagia and polydipsia despite weight loss, and stereotypic movements including head tossing, repetitive and exaggerated lifting of the legs, diminished alertness, and occasional aggressive behavior in the advanced stage of disease.[137,139]

2.3 Distribution of CWD in Cervid Tissues

The distribution of CWD in cervid tissues has been investigated via microscopy (immunohistochemistry—IHC) and in vitro amplification methodology (real-time quaking-induced conversion—RT-QuIC).

2.3.1 Histopathology

The terminal brain histopathology of CWD-affected animals is largely congruent across cervid species.[101,103,140–142] Lesions within the central nervous tissue are associated with gray matter with similar severity in natural and experimental infections. Bilateral lesions including spongiosis in neuronal perikarya and neuronal processes, astrocytic hyperplasia, and hypertrophy are apparent in brain tissue of infected cervids. Florid amyloid plaques, pale fibrillar esoinophic areas of neuropil surrounded by vacuoles, are common in brain tissues.[103,142,143] Histopathology has limited practical use in assessing preclinical status of disease progression in free-range populations due to obvious inaccessibility to brain tissue biopsies or serial access to lymphoid biopsies to monitor CWD status. However, it has been extensively used to assess PrP^{CWD} accumulation and disease dynamics in experimental studies.

2.3.2 Pattern of PrP^{CWD} Tissue Accumulation

Oral exposure is the most plausible pathway by which CWD prions are introduced to deer in nature. Thus, longitudinal experimental studies conducted to emulate this presumed mode of transmission provide insight into the patterns of PrP^{CWD} tissue deposition.[123,144] Accumulation of PrP^{CWD} was detected as early as 42 dpi in follicular germinal centers of the retropharyngeal lymph nodes, Peyer's patches and ileocaecal nodes, and at 78 dpi in tonsilar tissue of orally exposed fawns.[144] A subsequent experimental study incorporating the same CWD inoculum, dose, and oral exposure in mule deer fawns provided extended analysis of PrP^{CWD} tissue deposition from 90 dpi to terminal clinical CWD disease (630–785 dpi).[123] PrP^{CWD} accumulation was observed in the GALT and ganglia of the enteric nervous system, as well as the dorsal motor nucleus of the vagus nerve (DMNV), the intermediolateral column of the spinal cord, and the vagus nerve. From there, PrP^{CWD} accumulation was detected in the rest of the brain initially affecting nuclei in the medulla, thalamus, hypothalamus, midbrain, and olfactory cortex.[123] The general progression of PrP^{CWD} accumulation was characterized by rapid and widespread involvement of lymphatic tissues followed by the tissues of the CNS and PNS, and the endocrine system of animals exhibiting terminal clinical CWD.[123] Assessment of white-tailed deer tissues harvested prior to 4 months postinfection by the RT-QuIC methodology corroborates initial prion detection in pharyngeal lymphoid tissue between 30 and 60 dpi and supports systemic dissemination prior to neuroinvasion.[145]

2.4 Diagnosis

The current United States Department of Agriculture Animal and Plant Inspection Service (USDA-APHIS) approved test for the CWD herd certification program for farmed or captive cervids is immunohistochemistry and Bio-Rad ELISA.[146] The western blot is also an official test when performed at certified laboratories. All suspect positive Bio-Rad ELISA test and suspect positive IHC test results are confirmed by the National Veterinary Services Laboratory.

Several peripheral lymphoid tissues have been evaluated for their ability to permit antemortem diagnosis of CWD. The interpretation of tests for CWD is complicated by a long incubation period, the pattern of distribution of prions throughout the body, and the influence of genetics on the progression of disease in infected animals. As a result, a truly infected animal may not have detectable prions in sampled tissue at the time of testing. In addition, the antemortem biopsy collection provides a smaller sampling of the target tissue than is evaluated postmortem, enhancing the feasibility that prion deposition may be missed. Together, these factors have implications for CWD live-animal testing that are very different than when assessing CWD status at the time of necropsy. Thus, CWD antemortem testing has important regulatory implications if used to permit the interstate movement of animals and/or release quarantines in CWD-infected or exposed herds, i.e., testing may fail to detect infected animals in the preclinical phase of disease. Additional diagnostic tools are critically needed to manage CWD.

The development of assays with sensitivity and specificity to detect prions during the protracted asymptomatic phase of disease is ongoing. Two such assays, protein misfolding cyclic amplification (PMCA)[147] and RT-QuIC,[148,149] have made considerable contribution to the understanding of CWD distribution, pathogenesis, and transmission dynamics.[134,145,150–161]

2.5 Transmission and Pathogenesis

How CWD is transmitted from one susceptible cervid to the next has been an ongoing mystery. Direct animal-to-animal contact and indirect exposure to CWD prions in the environment are thought to be the most probable modes of CWD transmission within free-range cervid populations.[162] Experimental studies have helped to define the point source of CWD infectivity that is presumed to account for horizontal transmission. Secretions and excreta (saliva, urine, and feces) from cervids infected with CWD have been shown to harbor infectivity.[139,153,163,164] Of particular concern is the finding that infectivity may be shed by infected animals throughout the protracted preclinical phase of disease that can last months to years.[164] The shedding of infectivity throughout the full disease course likely enhances exposure risk to CWD—especially as herd density and environmental prion load increase during winter pasturing. Further contribution to environmental contamination includes carcasses of infected animals, as tissues[165] including skeletal muscle,[166] antler velvet,[167] and blood[139,168] contain infectivity. CWD prions are known to retain infectivity in the environment for at least

2 years.[162] In addition, CWD infectivity can be transferred with fomites (buckets, hay, and water) establishing infection within 15–19 months post-exposure.[137] Maternal transmission may also play a role in the spread of CWD. Prions have been detected in fetal and reproductive tissues harvested from experimental[160] and naturally infected[169] cervids, and infectivity has been demonstrated within the pregnancy microenvironment of muntjac deer.[170]

2.6 Zoonotic Potential

Strong evidence of zoonotic transmission of bovine spongiform encephalopathy (BSE) to humans has led to concerns about zoonotic transmission of CWD. As noted above, CWD prions are present throughout the diseased host, including muscle, organs, blood, antler velvet, and tissues of the CNSs and PNSs. Exposure to—and consumption of—materials from CWD-infected cervids places humans at increased risk. To date there has been no known transmission of CWD to humans and the zoonotic potential is thought to be low based on experimental studies using Cynomologus macaques,[171] transgenic mice overexpressing human prion protein,[172] and in vitro assays showing low efficiency of PrP^{CWD}-directed conversion of the normal human prion protein to the aberrant disease-associated form.[173] However, squirrel monkeys are susceptible to CWD by oral inoculation[171] and recent in vitro assays showing efficiency of PrP^{CWD}-directed conversion of the normal human prion protein at the level of the protein–protein interaction suggest the barrier preventing the transmission of CWD to humans may be less robust than previously estimated.[150] Clearly, additional research is needed to further investigate the zoonotic potential of CWD.

The role reservoir species may play in CWD transmission, including enhanced zoonotic potential, has also been under investigation. Interspecies transmission of CWD to noncervid species sympatric with CWD-infected cervids, including cattle,[174,175] raccoons, coyotes, and opposums[176] has not been observed under natural conditions. Experimental studies have provided evidence that several species, including voles, mink, ferrets, cats, goats, sheep, cattle, and mice are susceptible to CWD infection when exposed via intracranial inoculation.[138,175,177–180]

There is convincing evidence that CWD strains exist.[181] Disease course, including clinical presentation, incubation period, and prion deposition vary dependent upon prion strain.[171,182] As CWD continues to emerge, infecting additional susceptible cervids as well as new cervid species,[101] the potential for strain divergence,[182,183] and interspecies events increases.[184] The distribution and prevalence of CWD strains in native cervid populations is unknown. As are factors associated with increased risk of interspecies transmission including the cervid and human genotypes that define interspecies transmission potential of CWD. The zoonotic potential of CWD, therefore, remains unknown.

3. TRANSMISSIBLE MINK ENCEPHALOPATHY

3.1 History and Epidemiology

Transmissible mink encephalopathy (TME) is an invariably fatal disease of ranched mink with similar clinical presentation and pathology as sheep scrapie.[185] It was initially identified in mink in 1947 in Wisconsin, United States, with subsequent outbreaks on five additional farms in Wisconsin in 1961.[186] TME has further been recognized on a farm in Ontario, Canada and two farms in Wisconsin that used feed from the same supplier. The most recent outbreak in North America was again in Wisconsin in 1985.[187] The disease has also been reported in Idaho, United States (1963),[185] Finland (1966),[185] East Germany (1967),[188] and the former Soviet Union (1979–early 1980s).[189]

The origin of TME is hypothesized to be the ingestion of prion-contaminated feed of ovine or bovine origin.[186,187] This hypothesis is supported by the occurrence of outbreaks on several farms using the same feed supply.[186] In the United States, tissues of cattle, mainly "downer" dairy cows, was commonly used in preparing mink feed and is now considered the likely source of the TME agent.[187] In the Soviet Union the origin of TME has been traced to feeding carcasses of scrapie-infected sheep to mink.[190] Further support for bovine or ovine-sourced encephalopathies as the origin of TME has been provided by experimental intracranial (IC) inoculation of TME into cattle[191] or sheep[192] that resulted in neurologic disease similar to BSE and scrapie, respectively. As well, the reverse, i.e., experimental IC or oral inoculation of BSE[193] or scrapie[194,195] into mink resulted in disease presentation similar to TME. Incubation times associated with clinical disease progression in these studies[196] and other cross-species transmission studies[197] have led to the suggestion that the etiology of TME is L-type BSE vs C or H-type BSE or sheep scrapie.

Male and female adult mink are affected equally with mortality rates ranging from 10% to 100%.[186] A study of the mink prion protein did not identify *Prnp* polymorphisms that influence susceptibility, incubation period, clinical course, or neuropathologic pattern.[198]

3.2 Clinical Signs

Clinical disease in affected mink manifest in 7–12 months postexposure, with either a slow progressive disease process over a period of weeks (2–7 weeks) or a rapid progressive disease course lasting ~1 week. Changes in behavior include hyperexcitability, hyeresthesia, ataxia, whole body tremors, bouts of drowsiness, and increased aggressiveness.[199]

3.3 Distribution of TME in Mink Tissues

3.3.1 Histopathology

TME histopathological changes are characterized by spongiform alterations in the neuropil of the gray matter, neuronal degeneration, and astrycytosis.[185,189,200] The most obvious changes are spongiform alterations that vary from localized to diffuse neuropil vacuolation. Neuronal degeneration is marked by dark, shrunken angular neuronal perikarya. Astrocytosis in the gray matter is prominent, with reactive astrocytes appearing as enlarged, pale naked nuclei. Amyloid plaques are not found in TME.[199,200] Severe degenerative changes are present in the thalamus, amygdala, and caudal colliculi, with more moderate lesions in the hippocampus and hypothalamus. Lesions in the brain stem are generally less intense and more variable, with limited changes in the pons, medulla oblongata, and no affect on the cerebellum.[199]

3.4 Diagnosis

Behavioral changes in mink affected with TME may be mistaken for the effects of a neurotoxin or acute infection of the CNS.[201] Upon further progression of the disease (~1 week) it is easily distinguished from other naturally occurring neurodegenerative diseases of ranched mink. Microscopic examination of brain tissue from affected mink confirms a TME diagnosis.[192]

3.5 Transmission and Pathogenesis

TME is not a naturally transmitted disease, and therefore does not sustain itself in the mink population.[192,194] Transmission is associated with foodborne contamination events.[186,187] Further study suggests the origin of TME to be L-type BSE.[197]

Experimental TME transmission can be achieved by oral, intradermal, subcutaneous, intramuscular, intraperitonal, and intracerebral inoculation.[187–189,199,202,203] Incubation times vary, dependent upon route of inoculation and titer of inoculum, ranging from 4 to 12 months—IC, 4 months; oral 7–8 months. The experimental host range includes the domestic ferret, striped skunk, pine and beech martins, raccoons, Syrian and Chinese hamsters, squirrel, rhesus and stumptail macaque monkeys, sheep, goat, and cattle, with variable incubation times post-IC inoculation between 5 (skunk) and 65 months (sheep) pi.[191,203–210]

Isolation of hamster-adapted TME strains and development of their use to investigate TSE stain adaptability have provided considerable insight into the TSE strain selection process. Two TME strains producing distinctly different clinical syndromes and brain titers in Syrian hamsters, "drowsy" (DY) and "hyper" (HY),[183,211] were isolated, enriched, and demonstrate strain-specific conversion of PrP^{C} to PrP^{RES} in vitro.[212] Further use of these distinct TSE strains has resulted in demonstration that coinfection with the two strains could result in the DY TME phenotype and PrP^{RES} conformation on first passage, but upon subsequent passages, the disease pattern converted to HY TME.[213] These findings, and subsequent investigations,[214–222] indicate that during TSE strain adaptation, there is selection of a strain-specific PrP^{RES} conformation that can determine the TSE strain phenotype.

3.6 Zoonotic Potential

As TME disease is extremely rare the zoonotic potential to humans is negligible. Yet, TME has provided insight to cross-species transmission dynamics and their potential implications to future zoonotic events of prion diseases, including the possible presence of L-BSE for many decades prior to its identification in the United States and Europe.[197]

REFERENCES

1. Prusiner SB. Novel proteinaceous infectious particles cause scrapie. *Science*. 1982;216:136–144.
2. Wood JL, Done SH. Natural scrapie in goats: neuropathology. *Vet Rec*. 1992;131:93–96.
3. Wood JL, Lund LJ, Done SH. The natural occurrence of scrapie in moufflon. *Vet Rec*. 1992;130:25–27.
4. Wood JN, Done SH, Pritchard GC, Wooldridge MJ. Natural scrapie in goats: case histories and clinical signs. *Vet Rec*. 1992;131:66–68.
5. Comber T. *Real Improvements in Agriculture. On the Principles of A. Young, Edq, Letters to Reade Peacock, Edq. and to Dr. Hunter, Physician in York, Concerning the Rickets in sheep*. 1772.

6. Gaiger H. Scrapie. *J Comp Med.* 1924;37:259–277.
7. Parry H. *Scrapie Disese in Sheep.* London: Academic Press; 1983.
8. Hornlimann B, Riesner D, Kretzschmar H, et al. *Historical Perspectives.* Berlin, New York: de Gruyter; 2007.
9. Laplanche JL, Hunter N, Shinagawa M, Williams E. *Scrapie, Chronic Wasting Disease, and Transmissible Mink Encephalopathy.* New York: Cold Spring Harbor Laboratory Press; 1999.
10. M'Fadyean NL, Disease history in sheep. *Nature.* 1914;94:222.
11. Cuille J, Chelle PI. La maladie dite tremblante du mouton est-elle inoculable? *CR Acad Sci Paris.* 1936;203:1552–1554.
12. Cuille J, Chelle PI. La tremblante du mouton est bien inoculable. *CR Acad Sci Paris.* 1938;206:78–79.
13. de Andrade CP, de Oliveira EC, Leal JS, et al. Report of outbreaks of classical scrapie in Dorper sheep and associated prion protein gene polymorphisms in affected flocks. *Tropl Anim Health Prod.* 2015;47:1203–1212.
14. Hourrigan J, Klingsporn A, Clark WW. *Slow Transmissible Diseases of the Nervous System.* vol 1 New York: Academic press; 1979.
15. Kittelberger R, Chaplin MJ, Simmons MM, et al. Atypical scrapie/Nor98 in a sheep from New Zealand. *J Vet Diagn Invest.* 2010;22:863–875.
16. Goldmann W. PrP genetics in ruminant transmissible spongiform encephalopathies. *Vet Res.* 2008;39:30.
17. Hunter N, Goldmann W, Foster JD, Cairns D, Smith G. Natural scrapie and PrP genotype: case–control studies in British sheep. *Vet Rec.* 1997;141:137–140.
18. Moum T, Olsaker I, Hopp P, et al. Polymorphisms at codons 141 and 154 in the ovine prion protein gene are associated with scrapie Nor98 cases. *J Gen Virol.* 2005;86:231–235.
19. Goldmann W, Hunter N, Foster JD, Salbaum JM, Beyreuther K, Hope J. Two alleles of a neural protein gene linked to scrapie in sheep. *Proc Natl Acad Sci USA.* 1990;87:2476–2480.
20. Hunter N, Foster JD, Benson G, Hope J. Restriction fragment length polymorphisms of the scrapie-associated fibril protein (PrP) gene and their association with susceptibility to natural scrapie in British sheep. *J Gen Virol.* 1991;72(pt 6):1287–1292.
21. Laplanche JL, Chatelain J, Westaway D, et al. PrP polymorphisms associated with natural scrapie discovered by denaturing gradient gel electrophoresis. *Genomics.* 1993;15:30–37.
22. Goldmann W, Hunter N, Smith G, Foster J, Hope J. PrP genotype and agent effects in scrapie: change in allelic interaction with different isolates of agent in sheep, a natural host of scrapie. *J Gen Virol.* 1994;75(pt 5):989–995.
23. O'Rourke KI, Holyoak GR, Clark WW, et al. PrP genotypes and experimental scrapie in orally inoculated Suffolk sheep in the United States. *J Gen Virol.* 1997;78(pt 4):975–978.
24. Acin C, Martin-Burriel I, Monleon E, et al. Prion protein gene variability in Spanish goats. Inference through susceptibility to classical scrapie strains and pathogenic distribution of peripheral PrP(sc.). *PLoS One.* 2013;8, e61118.
25. Aguilar-Calvo P, Espinosa JC, Pintado B, et al. Role of the goat K222-PrP(C) polymorphic variant in prion infection resistance. *J Virol.* 2014;88:2670–2676.
26. Curcio L, Sebastiani C, Di Lorenzo P, Lasagna E, Biagetti M. Review: a review on classical and atypical scrapie in caprine: Prion protein gene polymorphisms and their role in the disease. *Animal.* 2016;10:1585–1593.
27. Benestad SL, Sarradin P, Thu B, Schonheit J, Tranulis MA, Bratberg B. Cases of scrapie with unusual features in Norway and designation of a new type, Nor98. *Vet Rec.* 2003;153:202–208.

28. Tranulis MA, Benestad SL, Baron T, Kretzschmar H. Atypical prion diseases in humans and animals. *Top Curr Chem*. 2011;305:23–50.
29. De Bosschere H, Roels S, Benestad SL, Vanopdenbosch E. Scrapie case similar to Nor98 diagnosed in Belgium via active surveillance. *Vet Rec*. 2004;155:707–708.
30. Nentwig A, Oevermann A, Heim D, et al. Diversity in neuroanatomical distribution of abnormal prion protein in atypical scrapie. *PLoS Pathog*. 2007;3, e82.
31. Polak MP, Larska M, Langeveld JP, Buschmann A, Groschup MH, Zmudzinski JF. Diagnosis of the first cases of scrapie in Poland. *Vet J*. 2010;186:47–52.
32. Chong A, Kennedy I, Goldmann W, et al. Archival search for historical atypical scrapie in sheep reveals evidence for mixed infections. *J Gen Virol*. 2015;96:3165–3178.
33. Benestad SL, Arsac JN, Goldmann W, Noremark M. Atypical/Nor98 scrapie: properties of the agent, genetics, and epidemiology. *Vet Res*. 2008;39:19.
34. Luhken G, Buschmann A, Brandt H, Eiden M, Groschup MH, Erhardt G. Epidemiological and genetical differences between classical and atypical scrapie cases. *Vet Res*. 2007;38:65–80.
35. Luhken G, Buschmann A, Groschup MH, Erhardt G. Prion protein allele A136 H154Q171 is associated with high susceptibility to scrapie in purebred and crossbred German Merinoland sheep. *Arch Virol*. 2004;149:1571–1580.
36. Langeveld JP, Jacobs JG, Erkens JH, et al. Sheep prions with molecular properties intermediate between classical scrapie, BSE and CH1641-scrapie. *Prion*. 2014;8:296–305.
37. Mazza M, Iulini B, Vaccari G, et al. Co-existence of classical scrapie and Nor98 in a sheep from an Italian outbreak. *Res Vet Sci*. 2010;88:478–485.
38. Simmons MM, Chaplin MJ, Vickery CM, et al. Does the presence of scrapie affect the ability of current statutory discriminatory tests to detect the presence of bovine spongiform encephalopathy? *J Clin Microbiol*. 2015;53:2593–2604.
39. Konold T, Phelan L. Clinical examination protocol for the detection of scrapie. *Vet Rec*. 2014;174:257.
40. Konold T, Phelan L. Clinical examination protocol to detect atypical and classical scrapie in sheep. *J Vis Exp*. 2014;83:e51101.
41. Dickinson AG. Scrapie in sheep and goats. *Front Biol*. 1976;44:209–241.
42. Dickinson AG, Young GB, Stamp JT, Renwick CC. An analysis of natural scrapie in Suffolk sheep. *Heredity*. 1965;20:485–503.
43. Dickinson AG, Stamp J, Renwich CC. Maternal and lateral transmission of scrapie in sheep. *J Comp Pathol*. 1974;84:19–25.
44. Georgsson G, Adolfsdottir JA, Palsdottir A, Jorundsson E, Sigurdarson S, Thorgeirsdottir S. High incidence of subclinical infection of lymphoid tissues in scrapie-affected sheep flocks. *Arch Virol*. 2008;153:637–644.
45. Georgsson G, Sigurdarson S, Brown P. Infectious agent of sheep scrapie may persist in the environment for at least 16 years. *J Gen Virol*. 2006;87:3737–3740.
46. Colussi S, Vaccari G, Maurella C, et al. Histidine at codon 154 of the prion protein gene is a risk factor for Nor98 scrapie in goats. *J Gen Virol*. 2008;89:3173–3176.
47. Beekes M. Prions and prion diseases. *FEBS J*. 2007;274:575.
48. Jeffrey M, Goodsir CM, Race RE, Chesebro B. Scrapie-specific neuronal lesions are independent of neuronal PrP expression. *Ann Neurol*. 2004;55:781–792.
49. van Keulen LJ, Schreuder BE, Vromans ME, Langeveld JP, Smits MA. Pathogenesis of natural scrapie in sheep. *Arch Virol Suppl*. 2000;16:57–71.
50. Andreoletti O, Simon S, Lacroux C, et al. PrPSc accumulation in myocytes from sheep incubating natural scrapie. *Nat Med*. 2004;10:591–593.
51. Crozet C, Lezmi S, Flamant F, Samarut J, Baron T, Bencsik A. Peripheral circulation of the prion infectious agent in transgenic mice expressing the ovine prion protein gene in neurons only. *J Infect Dis*. 2007;195:997–1006.

52. Garza MC, Monzon M, Marin B, Badiola JJ, Monleon E. Distribution of peripheral PrP(Sc) in sheep with naturally acquired scrapie. *PLoS One*. 2014;9, e97768.
53. Kimberlin RH, Hall SM, Walker CA. Pathogenesis of mouse scrapie. Evidence for direct neural spread of infection to the CNS after injection of sciatic nerve. *J Neurol Sci*. 1983;61:315–325.
54. Aguzzi A, Nuvolone M, Zhu C. The immunobiology of prion diseases. *Nat Rev Immunol*. 2013;13:888–902.
55. Pattison IH. Veterinary research. *Proc R Soc Med*. 1972;65:344–345.
56. Tuo W, O'Rourke KI, Zhuang D, Cheevers WP, Spraker TR, Knowles DP. Pregnancy status and fetal prion genetics determine PrPSc accumulation in placentomes of scrapie-infected sheep. *Proc Natl Acad Sci USA*. 2002;99:6310–6315.
57. Andreoletti O, Berthon P, Marc D, et al. Early accumulation of PrPSc in gut-associated lymphoid and nervous tissues of susceptible sheep from a Romanov flock with natural scrapie. *J Gen Virol*. 2000;81:3115–3126.
58. Pattison IH, Millson G. Distribution of the scrapie agent in the tissues of experimentally inoculated goats. *J Comp Pathol*. 1962;76:233–244.
59. Ligios C, Cancedda GM, Margalith I, et al. Intraepithelial and interstitial deposition of pathological prion protein in kidneys of scrapie-affected sheep. *PLoS One*. 2007;2, e859.
60. Siso S, Jeffrey M, Steele P, et al. Occurrence and cellular localization of PrPd in kidneys of scrapie-affected sheep in the absence of inflammation. *J Pathol*. 2008;215:126–134.
61. Everest SJ, Ramsay AM, Chaplin MJ, et al. Detection and localisation of PrP(Sc) in the liver of sheep infected with scrapie and bovine spongiform encephalopathy. *PLoS One*. 2011;6, e19737.
62. Jeffrey M, Begara-McGorum I, Clark S, et al. Occurrence and distribution of infection-specific PrP in tissues of clinical scrapie cases and cull sheep from scrapie-affected farms in Shetland. *J Comp Pathol*. 2002;127:264–273.
63. McGovern G, Jeffrey M. Membrane toxicity of abnormal prion protein in adrenal chromaffin cells of scrapie infected sheep. *PLoS One*. 2013;8, e58620.
64. Ligios C, Cancedda MG, Carta A, et al. Sheep with scrapie and mastitis transmit infectious prions through the milk. *J Virol*. 2011;85:1136–1139.
65. Thomzig A, Schulz-Schaeffer W, Wrede A, et al. Accumulation of pathological prion protein PrPSc in the skin of animals with experimental and natural scrapie. *PLoS Pathog*. 2007;3, e66.
66. Ligios C, Sigurdson CJ, Santucciu C, et al. PrPSc in mammary glands of sheep affected by scrapie and mastitis. *Nat Med*. 2005;11:1137–1138.
67. Ersdal C, Ulvund MJ, Espenes A, Benestad SL, Sarradin P, Landsverk T. Mapping PrPSc propagation in experimental and natural scrapie in sheep with different PrP genotypes. *Vet Pathol*. 2005;42:258–274.
68. Gonzalez L, Dagleish MP, Martin S, et al. Diagnosis of preclinical scrapie in live sheep by the immunohistochemical examination of rectal biopsies. *Vet Rec*. 2008;162:397–403.
69. Jeffrey M, Martin S, Thomson JR, Dingwall WS, Begara-McGorum I, Gonzalez L. Onset and distribution of tissue prp accumulation in scrapie-affected suffolk sheep as demonstrated by sequential necropsies and tonsillar biopsies. *J Comp Pathol*. 2001;125:48–57.
70. Gough KC, Rees HC, Ives SE, Maddison BC. Methods for differentiating prion types in food-producing animals. *Biology*. 2015;4:785–813.
71. Le Dur A, Beringue V, Andreoletti O, et al. A newly identified type of scrapie agent can naturally infect sheep with resistant PrP genotypes. *Proc Natl Acad Sci USA*. 2005;102:16031–16036.

72. Andreoletti O, Orge L, Benestad SL, et al. Atypical/Nor98 scrapie infectivity in sheep peripheral tissues. *PLoS Pathog.* 2011;7, e1001285.
73. Hadlow WJ, Kennedy RC, Race RE. Natural infection of Suffolk sheep with scrapie virus. *J Infect Dis.* 1982;146:657–664.
74. Pattison IH. Experimental transmission of scrapie to goats and sheep by the oral route. *J Comp Pathol.* 1961;71:171–176.
75. Pattison IH, Hoare MN, Jebbett JN, Watson WA. Further observations on the production of scrapie in sheep by oral dosing with foetal membranes from scrapie-affected sheep. *Br Vet J.* 1974;130:lxv–lxvii.
76. Pattison IH, Hoare MN, Jebbett JN, Watson WA. Spread of scrapie to sheep and goats by oral dosing with foetal membranes from scrapie-affected sheep. *Vet Rec.* 1972;90:465–468.
77. Onodera T, Ikeda T, Muramatsu Y, Shinagawa M. Isolation of scrapie agent from the placenta of sheep with natural scrapie in Japan. *Microbiol Immunol.* 1993;37:311–316.
78. Race R, Jenny A, Sutton D. Scrapie infectivity and proteinase K-resistant prion protein in sheep placenta, brain, spleen, and lymph node: implications for transmission and antemortem diagnosis. *J Infect Dis.* 1998;178:949–953.
79. Schneider DA, Madsen-Bouterse SA, Zhuang D, Truscott TC, Dassanayake RP, O'Rourke KI. The placenta shed from goats with classical scrapie is infectious to goat kids and lambs. *J Gen Virol.* 2015;96:2464–2469.
80. Saunders SE, Bartelt-Hunt SL, Bartz JC. Occurrence, transmission, and zoonotic potential of chronic wasting disease. *Emerg Infect Dis.* 2012;18:369–376.
81. Clark M, Haig DA. Presence of the transmissible agent of scrapie in the serum of affected mice and rats. *Vet Rec.* 1967;80:504.
82. Dassanayake RP, Schneider DA, Herrmann-Hoesing LM, Truscott TC, Davis WC, O'Rourke KI. Cell-surface expression of PrPC and the presence of scrapie prions in the blood of goats. *J Gen Virol.* 2012;93:1127–1131.
83. Houston F, Foster JD, Chong A, Hunter N, Bostock CJ. Transmission of BSE by blood transfusion in sheep. *Lancet.* 2000;356:999–1000.
84. Konold T, Moore SJ, Bellworthy SJ, Simmons HA. Evidence of scrapie transmission via milk. *BMC Vet Res.* 2008;4:14.
85. Konold T, Moore SJ, Bellworthy SJ, et al. Evidence of effective scrapie transmission via colostrum and milk in sheep. *BMC Vet Res.* 2013;9:99.
86. Konold T, Thorne L, Simmons HA, Hawkins SA, Simmons MM, Gonzalez L. Evidence of scrapie transmission to sheep via goat milk. *BMC Vet Res.* 2016;12:208.
87. Tamguney G, Richt JA, Hamir AN, et al. Salivary prions in sheep and deer. *Prion.* 2012;6:52–61.
88. Hawkins SA, Simmons HA, Gough KC, Maddison BC. Persistence of ovine scrapie infectivity in a farm environment following cleaning and decontamination. *Vet Rec.* 2015;176:99.
89. Konold T, Hawkins SA, Thurston LC, et al. Objects in contact with classical scrapie sheep act as a reservoir for scrapie transmission. *Front Vet Sci.* 2015;2:32.
90. Spiropoulos J, Hawkins SA, Simmons MM, Bellworthy SJ. Evidence of in utero transmission of classical scrapie in sheep. *J Virol.* 2014;88:4591–4594.
91. Chatelain J, Cathala F, Brown P, Raharison S, Court L, Gajdusek DC. Epidemiologic comparisons between Creutzfeldt–Jakob disease and scrapie in France during the 12-year period 1968–1979. *J Neurol Sci.* 1981;51:329–337.
92. Gibbs Jr CJ, Gajdusek DC. Transmission and characterization of the agents of spongiform virus encephalopathies: kuru, Creutzfeldt–Jakob disease, scrapie and mink encephalopathy. *Res Publ Assoc Res Nerv Ment Dis.* 1971;49:383–410.

93. Wadsworth JD, Joiner S, Linehan JM, et al. Atypical scrapie prions from sheep and lack of disease in transgenic mice overexpressing human prion protein. *Emerg Infect Dis*. 2013;19:1731–1739.
94. Baker HF, Ridley RM, Wells GA. Experimental transmission of BSE and scrapie to the common marmoset. *Vet Rec*. 1993;132:403–406.
95. Cassard H, Torres JM, Lacroux C, et al. Evidence for zoonotic potential of ovine scrapie prions. *Nat Commun*. 2014;5:5821.
96. Comoy EE, Mikol J, Luccantoni-Freire S, et al. Transmission of scrapie prions to primate after an extended silent incubation period. *Sci Rep*. 2015;5, 11573.
97. Requena JR, Kristensson K, Korth C, et al. The priority position paper: protecting Europe's food chain from prions. *Prion*. 2016;10:165–181.
98. USGS-NWHC March 2017, posting date. Chronic Wasting Disease [Online].
99. Kim TY, Shon HJ, Joo YS, Mun UK, Kang KS, Lee YS. Additional cases of chronic wasting disease in imported deer in Korea. *J Vet Med Sci*. 2005;67:753–759.
100. Sohn HJ, Kim JH, Choi KS, et al. A case of chronic wasting disease in an elk imported to Korea from Canada. *J Vet Med Sci*. 2002;64:855–858.
101. Benestad SL, Mitchell G, Simmons M, Ytrehus B, Vikoren T. First case of chronic wasting disease in Europe in a Norwegian free-ranging reindeer. *Vet Res*. 2016;47:88.
102. Williams ES, Young S. Chronic wasting disease of captive mule deer: a spongiform encephalopathy. *J Wildl Dis*. 1980;16:89–98.
103. Williams ES, Young S. Spongiform encephalopathies in Cervidae. *Rev Sci Tech*. 1992;11:551–567.
104. Williams ES, Young S. Spongiform encephalopathy of Rocky Mountain elk. *J Wildl Dis*. 1982;18:465–471.
105. Medicine, I. S. U. C. o. V. July 2016, posting date. Chronic Wasting Disease. Iowa State University [Online].
106. Miller MW, Conner MM. Epidemiology of chronic wasting disease in free-ranging mule deer: spatial, temporal, and demographic influences on observed prevalence patterns. *J Wildl Dis*. 2005;41:275–290.
107. Samuel MD, Storm DJ. Chronic wasting disease in white-tailed deer: infection, mortality, and implications for heterogeneous transmission. *Ecology*. 2016;97:3195–3205.
108. Edmunds DR, Kauffman MJ, Schumaker BA, et al. Chronic wasting disease drives population decline of white-tailed deer. *PLoS One*. 2016;11, e0161127.
109. Aguzzi A. Unraveling prion strains with cell biology and organic chemistry. *Proc Natl Acad Sci USA*. 2008;105:11–12.
110. Sigurdson CJ, Nilsson KP, Hornemann S, et al. A molecular switch controls interspecies prion disease transmission in mice. *J Clin Invest*. 2010;120:2590–2599.
111. Kurt TD, Bett C, Fernandez-Borges N, et al. Prion transmission prevented by modifying the beta2-alpha2 loop structure of host PrPC. *J Neurosci*. 2014;34:1022–1027.
112. Gossert AD, Bonjour S, Lysek DA, Fiorito F, Wuthrich K. Prion protein NMR structures of elk and of mouse/elk hybrids. *Proc Natl Acad Sci USA*. 2005;102:646–650.
113. O'Rourke KI, Baszler TV, Miller JM, Spraker TR, Sadler-Riggleman I, Knowles DP. Monoclonal antibody F89/160.1.5 defines a conserved epitope on the ruminant prion protein. *J Clin Microbiol*. 1998;36:1750–1755.
114. Schatzl HM, Wopfner F, Gilch S, von Brunn A, Jager G. Is codon 129 of prion protein polymorphic in human beings but not in animals? *Lancet*. 1997;349:1603–1604.
115. O'Rourke KI, Besser TE, Miller MW, et al. PrP genotypes of captive and free-ranging Rocky Mountain elk (Cervus elaphus nelsoni) with chronic wasting disease. *J Gen Virol*. 1999;80(pt 10):2765–2769.
116. Spraker TR, Balachandran A, Zhuang D, O'Rourke KI. Variable patterns of distribution of PrP(CWD) in the obex and cranial lymphoid tissues of Rocky Mountain elk

(Cervus elaphus nelsoni) with subclinical chronic wasting disease. *Vet Rec.* 2004;155:295–302.
117. Perucchini M, Griffin K, Miller MW, Goldmann W. PrP genotypes of free-ranging wapiti (Cervus elaphus nelsoni) with chronic wasting disease. *J Gen Virol.* 2008;89:1324–1328.
118. Hamir AN, Gidlewski T, Spraker TR, et al. Preliminary observations of genetic susceptibility of elk (Cervus elaphus nelsoni) to chronic wasting disease by experimental oral inoculation. *J Vet Diagn Invest.* 2006;18:110–114.
119. O'Rourke KI, Spraker TR, Zhuang D, Greenlee JJ, Gidlewski TE, Hamir AN. Elk with a long incubation prion disease phenotype have a unique PrPd profile. *Neuroreport.* 2007;18:1935–1938.
120. Brayton KA, O'Rourke KI, Lyda AK, Miller MW, Knowles DP. A processed pseudogene contributes to apparent mule deer prion gene heterogeneity. *Gene.* 2004;326:167–173.
121. Heaton MP, Leymaster KA, Freking BA, et al. Prion gene sequence variation within diverse groups of U.S. sheep, beef cattle, and deer. *Mamm Genome.* 2003;14:765–777.
122. Jewell JE, Conner MM, Wolfe LL, Miller MW, Williams ES. Low frequency of PrP genotype 225SF among free-ranging mule deer (Odocoileus hemionus) with chronic wasting disease. *J Gen Virol.* 2005;86:2127–2134.
123. Fox KA, Jewell JE, Williams ES, Miller MW. Patterns of PrPCWD accumulation during the course of chronic wasting disease infection in orally inoculated mule deer (Odocoileus hemionus). *J Gen Virol.* 2006;87:3451–3461.
124. Johnson C, Johnson J, Clayton M, McKenzie D, Aiken J. Prion protein gene heterogeneity in free-ranging white-tailed deer within the chronic wasting disease affected region of Wisconsin. *J Wildl Dis.* 2003;39:576–581.
125. Johnson C, Johnson J, Vanderloo JP, Keane D, Aiken JM, McKenzie D. Prion protein polymorphisms in white-tailed deer influence susceptibility to chronic wasting disease. *J Gen Virol.* 2006;87:2109–2114.
126. Kelly AC, Mateus-Pinilla NE, Diffendorfer J, et al. Prion sequence polymorphisms and chronic wasting disease resistance in Illinois white-tailed deer (Odocoileus virginianus). *Prion.* 2008;2:28–36.
127. O'Rourke KI, Spraker TR, Hamburg LK, Besser TE, Brayton KA, Knowles DP. Polymorphisms in the prion precursor functional gene but not the pseudogene are associated with susceptibility to chronic wasting disease in white-tailed deer. *J Gen Virol.* 2004;85:1339–1346.
128. Wilson GA, Nakada SM, Bollinger TK, Pybus MJ, Merrill EH, Coltman DW. Polymorphisms at the PRNP gene influence susceptibility to chronic wasting disease in two species of deer (Odocoileus Spp.) in western Canada. *J Toxicol Environ Health A.* 2009;72:1025–1029.
129. Keane DP, Barr DJ, Bochsler PN, et al. Chronic wasting disease in a Wisconsin white-tailed deer farm. *J Vet Diagn Invest.* 2008;20:698–703.
130. Brandt AL, Kelly AC, Green ML, Shelton P, Novakofski J, Mateus-Pinilla NE. Prion protein gene sequence and chronic wasting disease susceptibility in white-tailed deer (Odocoileus virginianus). *Prion.* 2015;9:449–462.
131. Baeten LA, Powers BE, Jewell JE, Spraker TR, Miller MW. A natural case of chronic wasting disease in a free-ranging moose (Alces alces shirasi). *J Wildl Dis.* 2007;43:309–314.
132. Kreeger TJ, Montgomery DL, Jewell JE, Schultz W, Williams ES. Oral transmission of chronic wasting disease in captive Shira's moose. *J Wildl Dis.* 2006;42:640–645.
133. Mitchell GB, Sigurdson CJ, O'Rourke KI, et al. Experimental oral transmission of chronic wasting disease to reindeer (Rangifer tarandus tarandus). *PLoS One.* 2012;7, e39055.

134. Cheng YC, Hannaoui S, John TR, Dudas S, Czub S, Gilch S. Early and non-invasive detection of chronic wasting disease prions in elk feces by real-time quaking induced conversion. *PLoS One*. 2016;11, e0166187.
135. Robinson SJ, Samuel MD, O'Rourke KI, Johnson CJ. The role of genetics in chronic wasting disease of north American cervids. *Prion*. 2012;6:153–162.
136. Williams ES. Chronic wasting disease. *Vet Pathol*. 2005;42:530–549.
137. Mathiason CK, Hays SA, Powers J, et al. Infectious prions in pre-clinical deer and transmission of chronic wasting disease solely by environmental exposure. *PLoS One*. 2009;4, e5916.
138. Mathiason CK, Nalls AV, Seelig DM, et al. Susceptibility of domestic cats to chronic wasting disease. *J Virol*. 2013;87:1947–1956.
139. Mathiason CK, Powers JG, Dahmes SJ, et al. Infectious prions in the saliva and blood of deer with chronic wasting disease. *Science*. 2006;314:133–136.
140. Spraker TR, Miller MW, Williams ES, et al. Spongiform encephalopathy in free-ranging mule deer (Odocoileus hemionus), white-tailed deer (Odocoileus virginianus) and Rocky Mountain elk (Cervus elaphus nelsoni) in northcentral Colorado. *J Wildl Dis*. 1997;33:1–6.
141. Spraker TR, Zink RR, Cummings BA, Sigurdson CJ, Miller MW, O'Rourke KI. Distribution of protease-resistant prion protein and spongiform encephalopathy in free-ranging mule deer (Odocoileus hemionus) with chronic wasting disease. *Vet Pathol*. 2002;39:546–556.
142. Williams ES. Scrapie and chronic wasting disease. *Clin Lab Med*. 2003;23:139–159.
143. Liberski PP, Guiroy DC, Williams ES, Walis A, Budka H. Deposition patterns of disease-associated prion protein in captive mule deer brains with chronic wasting disease. *Acta Neuropathol*. 2001;102:496–500.
144. Sigurdson CJ, Williams ES, Miller MW, Spraker TR, O'Rourke KI, Hoover EA. Oral transmission and early lymphoid tropism of chronic wasting disease PrPres in mule deer fawns (Odocoileus hemionus). *J Gen Virol*. 1999;80(pt 10): 2757–2764.
145. Hoover CE, Davenport KA, Henderson DM, et al. Pathways of prion spread during early chronic wasting disease in deer. *J Virol*. 2017;91(10):pii: e00077-17. http://dx.doi.org/10.1128/JVI.00077-17. Print 2017 May 15. PMID: 28250130.
146. USDA-APHIS 2014, posting date. APHIS CWD program standards 2014 [Online].
147. Saborio GP, Permanne B, Soto C. Sensitive detection of pathological prion protein by cyclic amplification of protein misfolding. *Nature*. 2001;411:810–813.
148. Atarashi R, Moore RA, Sim VL, et al. Ultrasensitive detection of scrapie prion protein using seeded conversion of recombinant prion protein. *Nat Methods*. 2007;4: 645–650.
149. Colby DW, Zhang Q, Wang S, et al. Prion detection by an amyloid seeding assay. *Proc Natl Acad Sci USA*. 2007;104:20914–20919.
150. Davenport KA, Henderson DM, Bian J, Telling GC, Mathiason CK, Hoover EA. Insights into chronic wasting disease and bovine spongiform encephalopathy species barriers by use of real-time conversion. *J Virol*. 2015;89:9524–9531.
151. Denkers ND, Henderson DM, Mathiason CK, Hoover EA. Enhanced prion detection in biological samples by magnetic particle extraction and real-time quaking-induced conversion. *J Gen Virol*. 2016;97:2023–2029.
152. Elder AM, Henderson DM, Nalls AV, et al. Immediate and ongoing detection of prions in the blood of hamsters and deer following oral, nasal, or blood inoculations. *J Virol*. 2015;89:7421–7424.
153. Haley NJ, Mathiason CK, Zabel MD, Telling GC, Hoover EA. Detection of sub-clinical CWD infection in conventional test-negative deer long after oral exposure to urine and feces from CWD+ deer. *PLoS One*. 2009;4, e7990.

154. Haley NJ, Siepker C, Hoon-Hanks LL, et al. Seeded amplification of chronic wasting disease prions in nasal brushings and recto-anal mucosa-associated lymphoid tissues from elk by real-time quaking-induced conversion. *J Clin Microbiol.* 2016;54: 1117–1126.
155. Henderson DM, Davenport KA, Haley NJ, Denkers ND, Mathiason CK, Hoover EA. Quantitative assessment of prion infectivity in tissues and body fluids by real-time quaking-induced conversion. *J Gen Virol.* 2015;96:210–219.
156. Hoover CE, Davenport KA, Henderson DM, Zabel MD, Hoover EA. Endogenous brain lipids inhibit prion amyloid formation in vitro. *J Virol.* 2017;91(9):pii: e02162-16. http://dx.doi.org/10.1128/JVI.02162-16. Print 2017 May 1. PMID: 28202758.
157. John TR, Schatzl HM, Gilch S. Early detection of chronic wasting disease prions in urine of pre-symptomatic deer by real-time quaking-induced conversion assay. *Prion.* 2013;7:253–258.
158. Kurt TD, Telling GC, Zabel MD, Hoover EA. Trans-species amplification of PrP(CWD) and correlation with rigid loop 170N. *Virology.* 2009;387:235–243.
159. Meyerett C, Michel B, Pulford B, et al. In vitro strain adaptation of CWD prions by serial protein misfolding cyclic amplification. *Virology.* 2008;382:267–276.
160. Nalls AV, McNulty E, Powers J, et al. Mother to offspring transmission of chronic wasting disease in reeves' muntjac deer. *PLoS One.* 2013;8, e71844.
161. Pulford B, Spraker TR, Wyckoff AC, et al. Detection of PrPCWD in feces from naturally exposed Rocky Mountain elk (Cervus elaphus nelsoni) using protein misfolding cyclic amplification. *J Wildl Dis.* 2012;48:425–434.
162. Miller MW, Williams ES, Hobbs NT, Wolfe LL. Environmental sources of prion transmission in mule deer. *Emerg Infect Dis.* 2004;10:1003–1006.
163. Safar JG, Lessard P, Tamguney G, et al. Transmission and detection of prions in feces. *J Infect Dis.* 2008;198:81–89.
164. Tamguney G, Miller MW, Wolfe LL, et al. Asymptomatic deer excrete infectious prions in faeces. *Nature.* 2009;461:529–532.
165. Browning SR, Mason GL, Seward T, et al. Transmission of prions from mule deer and elk with chronic wasting disease to transgenic mice expressing cervid PrP. *J Virol.* 2004;78:13345–13350.
166. Angers RC, Browning SR, Seward TS, et al. Prions in skeletal muscles of deer with chronic wasting disease. *Science.* 2006;311:1117.
167. Angers RC, Seward TS, Napier D, et al. Chronic wasting disease prions in elk antler velvet. *Emerg Infect Dis.* 2009;15:696–703.
168. Mathiason CK, Hayes-Klug J, Hays SA, et al. B cells and platelets harbor prion infectivity in the blood of deer infected with chronic wasting disease. *J Virol.* 2010;84:5097–5107.
169. Selariu A, Powers JG, Nalls A, et al. In utero transmission and tissue distribution of chronic wasting disease-associated prions in free-ranging Rocky Mountain elk. *J Gen Virol.* 2015;96:3444–3455.
170. Nalls AV, McNulty E, Hoover CE, Pulscher LA, Hoover EA, Mathiason CK. Infectious prions in the pregnancy microenvironment of chronic wasting disease-infected Reeves' muntjac deer. *J Virol.* 2017;91(15):pii: e00501-17. http://dx.doi.org/10.1128/JVI.00501-17. Print 2017 Aug 1. PMID: 28539446.
171. Race B, Meade-White KD, Miller MW, et al. Susceptibilities of nonhuman primates to chronic wasting disease. *Emerg Infect Dis.* 2009;15:1366–1376.
172. Sandberg MK, Al-Doujaily H, Sigurdson CJ, et al. Chronic wasting disease prions are not transmissible to transgenic mice overexpressing human prion protein. *J Gen Virol.* 2010;91:2651–2657.

173. Barria MA, Telling GC, Gambetti P, Mastrianni JA, Soto C. Generation of a new form of human PrP(Sc) in vitro by interspecies transmission from cervid prions. *J Biol Chem.* 2011;286:7490–7495.
174. Gould DH, Voss JL, Miller MW, Bachand AM, Cummings BA, Frank AA. Survey of cattle in northeast Colorado for evidence of chronic wasting disease: geographical and high-risk targeted sample. *J Vet Diagn Invest.* 2003;15:274–277.
175. Sigurdson CJ. A prion disease of cervids: chronic wasting disease. *Vet Res.* 2008;39:41.
176. Jennelle CS, Samuel MD, Nolden CA, et al. Surveillance for transmissible spongiform encephalopathy in scavengers of white-tailed deer carcasses in the chronic wasting disease area of Wisconsin. *J Toxicol Environ Health A.* 2009;72:1018–1024.
177. Di Bari MA, Chianini F, Vaccari G, et al. The bank vole (Myodes glareolus) as a sensitive bioassay for sheep scrapie. *J Gen Virol.* 2008;89:2975–2985.
178. Di Bari MA, Nonno R, Castilla J, et al. Chronic wasting disease in bank voles: characterisation of the shortest incubation time model for prion diseases. *PLoS Pathog.* 2013;9, e1003219.
179. Heisey DM, Mickelsen NA, Schneider JR, et al. Chronic wasting disease (CWD) susceptibility of several north American rodents that are sympatric with cervid CWD epidemics. *J Virol.* 2010;84:210–215.
180. Sigurdson CJ, Mathiason CK, Perrott MR, et al. Experimental chronic wasting disease (CWD) in the ferret. *J Comp Pathol.* 2008;138:189–196.
181. Angers RC, Kang HE, Napier D, et al. Prion strain mutation determined by prion protein conformational compatibility and primary structure. *Science.* 2010;328:1154–1158.
182. Beringue V, Vilotte JL, Laude H. Prion agent diversity and species barrier. *Vet Res.* 2008;39:47.
183. Bessen RA, Marsh RF. Biochemical and physical properties of the prion protein from two strains of the transmissible mink encephalopathy agent. *J Virol.* 1992;66:2096–2101.
184. Bartz JC, Marsh RF, McKenzie DI, Aiken JM. The host range of chronic wasting disease is altered on passage in ferrets. *Virology.* 1998;251:297–301.
185. Marsh RF, Hadlow WJ. Transmissible mink encephalopathy. *Rev Sci Tech.* 1992;11:539–550.
186. Hartsough GR, Burger D. Encephalopathy of mink. I. Epizootiologic and clinical observations. *J Infect Dis.* 1965;115:387–392.
187. Marsh RF, Bessen RA, Lehmann S, Hartsough GR. Epidemiological and experimental studies on a new incident of transmissible mink encephalopathy. *J Gen Virol.* 1991;72(pt 3):589–594.
188. Hartung J, Zimmermann H, Johannsen H. Infektiose Enzephalopathie beim Nerz 1. Mitteilung: Klinisch-eqizootiologische and experimentelle Untersuchungen. *Monatsh Veterinari-med.* 1970;25:385–388.
189. Dukur II, Geller VI, Chizhov VA, Roikhel VM, Pogodina VV. Clinico-morphological study of transmissible encephalopathy of mink. *Vopr Virusol.* 1986;31:220–225.
190. Gorham JR. Viral and bacterial diseases of mink in Soviet Union. *Fur Rancher.* 1991;71:10–11.
191. Robinson MM, Hadlow WJ, Knowles DP, et al. Experimental infection of cattle with the agents of transmissible mink encephalopathy and scrapie. *J Comp Pathol.* 1995;113:241–251.
192. Hadlow WJ, Race RE, Kennedy RC. Temporal distribution of transmissible mink encephalopathy virus in mink inoculated subcutaneously. *J Virol.* 1987;61:3235–3240.
193. Robinson MM, Hadlow WJ, Huff TP, et al. Experimental infection of mink with bovine spongiform encephalopathy. *J Gen Virol.* 1994;75(pt 9):2151–2155.
194. Hanson RP, Eckroade RJ, Marsh RF, Rhein GMZ, Kanitz CL, Gustafson DP. Susceptibility of mink to sheep scrapie. *Science.* 1971;172:859–861.

195. Marsh RF, Hanson RP. *On the Origin of Transmissible Mink Enceophalopathy*. New York: Academic Press; 1979.
196. Liberski PP, Sikorska B, Guiroy D, Bessen RA. Transmissible mink encephalopathy—review of the etiology of a rare prion disease. *Folia Neuropathol*. 2009;47:195–204.
197. Comoy EE, Mikol J, Ruchoux MM, et al. Evaluation of the zoonotic potential of transmissible mink encephalopathy. *Pathogens*. 2013;2:520–532.
198. Kretzschmar HA, Neumann M, Riethmuller G, Prusiner SB. Molecular cloning of a mink prion protein gene. *J Gen Virol*. 1992;73(pt 10):2757–2761.
199. Hadlow WJ. *Portrait of Transmissible Mink Encephalopathy*. Berlin, New York: DeGuyter; 2007.
200. Eckroade RJ, Zurhein GM, Hanson RP. *Experimental Transmissible Mink Enceophalopathy: Brain Lesions and Their Sequential Development*. New York: Academic Press; 1979.
201. Loliger HC. Mink diseases. *Bull Off Int Epizoot*. 1968;70:553–566.
202. Burger D, Hartsough GR. Encephalopathy of mink. II. Experimental and natural transmission. *J Infect Dis*. 1965;115:393–399.
203. Marsh RF, Hanson RP. Physical and chemical properties of the transmissible mink encephalopathy agent. *J Virol*. 1969;3:176–180.
204. Eckroade RJ, Rhein GMZ, Marsh RF, Hanson RP. Transmissible mink encephalopathy: experimental transmission to the squirrel monkey. *Science*. 1970;169: 1088–1090.
205. Eckroade RJ, ZuRhein GM, Hanson RP. Transmissible mink encephalopathy in carnivores: clinical, light and electron microscopic studies in raccons, skunks and ferrets. *J Wildl Dis*. 1973;9:229–240.
206. Hadlow WJ, Race RE, Kennedy RC. Experimental infection of sheep and goats with transmissible mink encephalopathy virus. *Can J Vet Res*. 1987;51:135–144.
207. Hamir AN, Kehrli Jr ME, Kunkle RA, et al. Experimental interspecies transmission studies of the transmissible spongiform encephalopathies to cattle: comparison to bovine spongiform encephalopathy in cattle. *J Vet Diagn Invest*. 2011;23:407–420.
208. Hamir AN, Kunkle RA, Miller JM, Bartz JC, Richt JA. First and second cattle passage of transmissible mink encephalopathy by intracerebral inoculation. *Vet Pathol*. 2006;43:118–126.
209. Hamir AN, Miller JM, O'Rourke KI, Bartz JC, Stack MJ, Chaplin MJ. Transmission of transmissible mink encephalopathy to raccoons (Procyon lotor) by intracerebral inoculation. *J Vet Diagn Invest*. 2004;16:57–63.
210. Kimberlin RH, Cole S, Walker CA. Transmissible mink encephalopathy (TME) in Chinese hamsters: identification of two strains of TME and comparisons with scrapie. *Neuropathol Appl Neurobiol*. 1986;12:197–206.
211. Bessen RA, Marsh RF. Distinct PrP properties suggest the molecular basis of strain variation in transmissible mink encephalopathy. *J Virol*. 1994;68:7859–7868.
212. Bessen RA, Kocisko DA, Raymond GJ, Nandan S, Lansbury PT, Caughey B. Non-genetic propagation of strain-specific properties of scrapie prion protein. *Nature*. 1995;375:698–700.
213. Bartz JC, Bessen RA, McKenzie D, Marsh RF, Aiken JM. Adaptation and selection of prion protein strain conformations following interspecies transmission of transmissible mink encephalopathy. *J Virol*. 2000;74:5542–5547.
214. Ayers JI, Kincaid AE, Bartz JC. Prion strain targeting independent of strain-specific neuronal tropism. *J Virol*. 2009;83:81–87.
215. Bartz JC, Dejoia C, Tucker T, Kincaid AE, Bessen RA. Extraneural prion neuroinvasion without lymphoreticular system infection. *J Virol*. 2005;79:11858–11863.
216. Bartz JC, Kincaid AE, Bessen RA. Rapid prion neuroinvasion following tongue infection. *J Virol*. 2003;77:583–591.

217. Bartz JC, Kramer ML, Sheehan MH, et al. Prion interference is due to a reduction in strain-specific PrPSc levels. *J Virol.* 2007;81:689–697.
218. Bessen RA, Martinka S, Kelly J, Gonzalez D. Role of the lymphoreticular system in prion neuroinvasion from the oral and nasal mucosa. *J Virol.* 2009;83:6435–6445.
219. Langenfeld KA, Shikiya RA, Kincaid AE, Bartz JC. Incongruity between prion conversion and incubation period following coinfection. *J Virol.* 2016;90:5715–5723.
220. Mulcahy ER, Bessen RA. Strain-specific kinetics of prion protein formation in vitro and in vivo. *J Biol Chem.* 2004;279:1643–1649.
221. Schutt CR, Bartz JC. Prion interference with multiple prion isolates. *Prion.* 2008;2:61–63.
222. Shikiya RA, Ayers JI, Schutt CR, Kincaid AE, Bartz JC. Coinfecting prion strains compete for a limiting cellular resource. *J Virol.* 2010;84:5706–5714.

CHAPTER THIRTEEN

Infectious and Sporadic Prion Diseases

Richard Knight[1]
National CJD Research & Surveillance Unit, Centre for Clinical Brain Sciences, University of Edinburgh, Edinburgh, United Kingdom
[1]Corresponding author: e-mail address: r.knight@ed.ac.uk

Contents

Abstract

Prion diseases are progressive fatal encephalopathies characterized by a neurodegenerative pathology, the tissue deposition of abnormally folded prion protein and, in general, potential transmissibility. Creutzfeldt–Jakob disease (CJD) is the commonest

Progress in Molecular Biology and Translational Science, Volume 150
ISSN 1877-1173
http://dx.doi.org/10.1016/bs.pmbts.2017.06.010

human prion disease and occurs in three principal forms: sporadic (idiopathic), acquired (infectious), and inherited (genetic). This chapter concerns the sporadic and acquired forms. Sporadic CJD occurs worldwide and affects mainly the middle aged and elderly. There are recognized genetic risk factors—most importantly the *PRNP*-129 polymorphism. The acquired forms of CJD consist of iatrogenic CJD (accidental transmission of CJD via medical or surgical procedures) and variant CJD (vJCD) (which originated as a zoonosis via bovine spongiform encephalopathy (BSE)-contamination of human food). The main causes of iatrogenic CJD are cadaveric-derived human growth hormone treatment and dura mater surgical grafts. The *PRNP*-129 polymorphism has important effects on iatrogenic infection, including overall susceptibility and incubation period. vCJD, resulting from dietary exposure to BSE, has affected mostly the United Kingdom, followed by France. All tested cases were originally *PRNP*-129 MM, although two MV cases have been identified recently (one possible; one definite). vCJD has been secondarily transmitted via blood transfusion and a blood product. There is continuing concern over secondary transmission since there is evidence—from lymphoreticular tissue studies—of extensive subclinical infection in the UK general population, although a further recent study has caused uncertainty over the significance of the previous studies. While definitive diagnosis of CJD is pathological, recent developments in protein amplification and detection have led to significantly better clinical diagnosis.

1. INTRODUCTION

Prion diseases affect animals and humans and are classified as being sporadic (idiopathic), genetic, or infectious in nature, according to their primary causation. However, most—if not all—prion diseases are potentially infectious in certain situations (as illustrated by iatrogenic transmission of sporadic disease). This chapter deals with sporadic and infectious forms of human prion disease.

Sporadic Creutzfeldt–Jakob disease (sCJD) is the principal sporadic form of human prion disease but Variably Protease Sensitive Prionopathy has been described more recently. Iatrogenic CJD refers to the accidental transmission of sporadic or genetic disease via medical or surgical procedures. Variant CJD (vCJD) originated as a zoonotic illness, caused by BSE (bovine spongiform encephalopathy) contamination of food but secondary transmission of vCJD via blood transfusion and blood product treatment has been recognized. The most important recent advance in this area has been the development of better clinical diagnostic tests based on the underlying protein pathology.

2. SPORADIC CJD

2.1 Introduction

Sporadic CJD (sCJD) is the commonest form of human prion disease. It occurs sporadically throughout the world and affects predominantly the middle aged and elderly. Although there is a common typical clinicopathological phenotype, there are variations and some unusual presentations.

2.2 Epidemiology

It has been reported in all countries where it has been looked for. In countries with developed surveillance systems, the annual mortality rates (AMRs) are very similar being around 1–2/million/year.[1] In these countries, following the institution of surveillance, there has been a general increase in the AMR over time—thought to reflect better case ascertainment (based on increasing awareness and improved diagnostics) rather than a true increase in incidence. There is a characteristic age profile to its occurrence (Fig. 1); the increasing mortality with increasing age being typical for a neurodegenerative disease but there is a notable fall-off in incidence in the more elderly.[2] It is uncertain if this reflects a true age-related change in incidence,

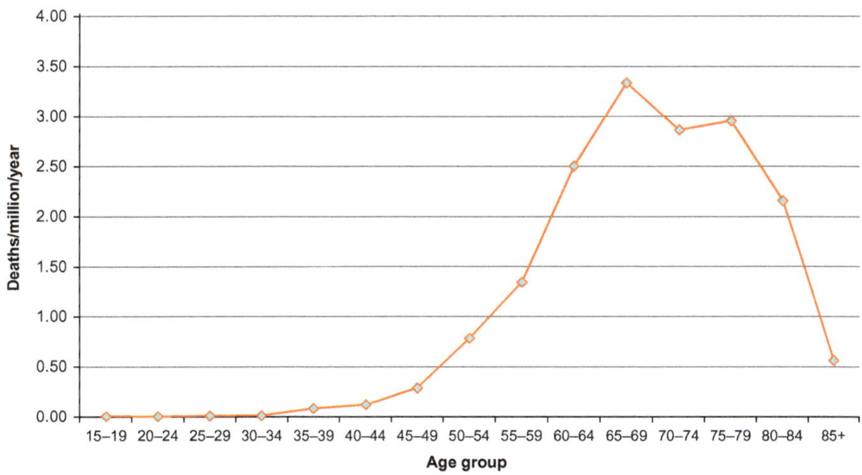

Fig. 1 The age-specific mortality of sCJD in the United Kingdom. *Graph courtesy of Ms. Janet Mackenzie, NCJDRSU, Edinburgh.*

reflects under-ascertainment in this age group, or both. Males and females are affected equally.

2.3 Cause

The cause of sCJD is unknown. Currently it is generally regarded as being a spontaneous neurodegenerative disease, probably reflecting some stochastic protein folding change or a somatic mutation of *PRNP* (the prion protein gene).[3] While there remains speculation as to whether some or even all cases arise from unidentified infection (iatrogenic or zoonotic), accumulated disease case data, and case–control studies have failed to identify any definite consistent risk factors aside from age, *PRNP* codon-129 genotype, and some other genetic variations.

There are studies implicating surgical operations as risk factors but with variable specific surgical procedures implicated.[3–6] Other studies have not shown any association with surgery and a UK study showed no association with ophthalmic surgery in particular.[7–9]

A case–control study involving 431 sCJD cases concluded "it is unlikely that a high proportion of UK sCJD cases are the result of transmission during surgery, but we cannot exclude the possibility that such transmission occurs occasionally".[10] One UK study showed no evidence of spatiotemporal case clustering although a later UK study did, consistent with some sCJD cases resulting from exposure to a common source.[11,12] de Pedro Cuesta *et al.* reviewed the methodological difficulties in studying possible risk factors such as surgery in sCJD.[13]

A European case–control study of sCJD found no definite medical, surgical, or other environmental risk factors but provided evidence that could support the role of genetic factors other than *PRNP* mutations.[14] Several studies have been undertaken to assess proposed genetic risk factors with negative results in relation to *PRNP*-117, *PRNP*-24b deletions, tau gene haplotypes, and a Cathepsin D polymorphism but positive results for the PLCXD3 gene, *GRM8*, and a polymorphism in the of a *PRNP* regulatory region.[15–19] Mead *et al.*, in their genome-wide association study of CJD, concluded that it was "likely that the *PRNP* locus contains the only strong risk factors that act universally across human prion diseases" but that available data are "consistent with several other risk loci of modest overall effects which will require further genetic association studies to provide definitive evidence."[20]

2.4 Clinical Features

The majority of cases of sCJD presents as a rapidly progressive, multifocal, encephalopathy, with dementia, cerebellar ataxia, and myoclonus as prominent features. This typical, rapidly progressive, form generally results in an akinetic mute state and death, after an illness lasting only a few weeks to a few months.[1] A significant minority of individuals present with a progressive isolated impairment before progressing on to a more general encephalopathic illness, most frequently cerebellar ataxia or visual disturbance but there are other reported isolated onsets, such as dysphasia.[21–23] There are other variations in clinical features, investigation results, age at onset, illness duration, and neuropathological characteristics. This clinicopathological heterogeneity of sCJD has been studied and correlated with the prion protein type and the *PRNP*-codon 129 genotype, resulting in a subclassification scheme (Table 1).[24]

These subclassifications have some practical diagnostic utility but the distinctions are not always clear-cut with individual cases that do not altogether fit these defined categories. The partial dependence of subclassification on

Table 1 Clinicopathological Subtypes of Sporadic CJD

sCJD Subtype 129 Genotype, Protein Type C: Cortical; T: Thalamic	**% of Cases (Approx.)**	**Some General Clinical Features** ***Note: There are also pathological features that characterize each subtype-not given here***
MM I	67	Cognitive, ataxic, visual, ~4 months duration, periodic EEG common
MV I	3	Broadly similar to MM I
VV I	3	No visual problems, ~15 months duration, no periodic EEG
MM II C	2	Progressive dementia. Duration ~16 months. No periodic EEG
MM II T	2	Usually insomnia, psychomotor hyperactivity, duration ~15 months. No periodic EEG
MV II	9	Duration ~17 months. Usually no periodic EEG
VV II	14	Duration ~15 months. Usually no periodic EEG

Data from Parchi P, Giese A, Capellari S, et al. Classification of sporadic Creutzfeldt–Jakob disease based on molecular and phenotypic analysis of 300 subjects. Ann Neurol 1999;46:224–233.

pathological prion protein type, is complicated by the fact that more than one prion protein type can be found in one brain.[25] A more complex classification, of uncertain practical clinical diagnostic utility, was devised to take this into account.[26]

The duration of survival in sCJD is linked to four factors: age at onset, sex, *PRNP*-129 genotype, and the associated abnormal prion protein type.[27]

2.5 Diagnosis

Clinical diagnosis begins with considering the possibility of sCJD, requiring awareness of a disease that even many clinical neurologists will see only rarely. In addition, the presentation of sCJD is essentially nonspecific with a potentially wide initial differential diagnosis.[28] A variety of blood tests are indicated along with cerebral imaging and CSF analysis. In sCJD, all blood tests and the cerebral CT are generally normal, with, in most cases, an acellular CSF and a normal or relatively modestly raised CSF total protein.[29] There are useful confirmatory tests, classically: the EEG (showing a characteristic periodic discharge pattern); the CSF 14-3-3 protein test; and the cerebral MRI (showing increased signal in the caudate/putamen and multifocal cortical areas).[30–32] However, while these have proved very useful (especially the MRI and CSF 14-3-3), these tests are not always sensitive (with their sensitivity varying according to the sCJD subtype) and they are, in essence, nonspecific[33] (Fig. 2).

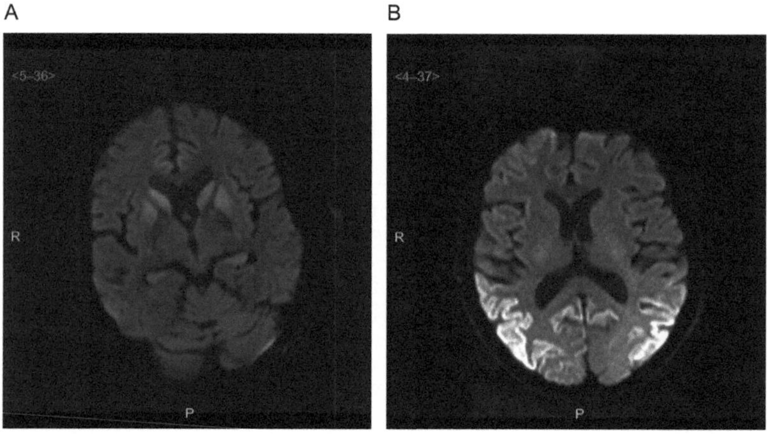

Fig. 2 The cerebral MRI (DWI) in sCJD. (A) Increased signal in the caudate/putamen in sCJD. (B) Increased signal in the cerebral cortex in sCJD. *Images courtesy of Dr. David Summers, Department of Neuroradiology, Edinburgh.*

However, tests that center on the prion protein itself and, therefore, disease-specific tests, are increasingly being employed in clinical practice. The best evaluated of these is CSF RT-QuIC (real-time quaking induced conversion) with a reported sensitivity and specificity for sCJD of 87% and 100%, respectively.[34] There are, to date, no useful blood or urine tests for sCJD although blood tests developed for vCJD have shown very occasional positive results in sCJD.[35–37] Brain biopsy remains the most definitive clinical test in life for prion disease but carries obvious risk and, if confirmatory, does not lead to any specific therapy. Arguably, it should be limited to those cases where there is a real possibility of an alternative, otherwise undiagnosed, treatable, diagnosis thereby being made. However, olfactory neurons can be obtained through nasal brushing that are then analyzable for the presence of abnormal prion protein. This has, therefore, been proposed as a test for prion disease and one that directly confirms the disease process, with disease-related prion protein being detected in neuronal tissue.[38] *PRNP* genetic analysis has two roles: the detection of pathogenic mutations (important because genetic prion disease may resemble sCJD with a family history not always being present) and the determination of the 129 polymorphisms (which may aid diagnosis, given the described subtypes of sCJD and the variation of MR findings with genotype).[33,39] The determination of the *PRNP*-219 polymorphism may be important in Japanese populations.[40]

3. VARIABLY PROTEASE SENSITIVE PRIONOPATHY

Organized surveillance of diseases, with increased interest and better diagnostic techniques might lead to the description of newly recognized clinicopathological entities as well as any newly arising ones.

What is now termed Variably Protease Sensitive Prionopathy (VPSPr) was first described in 2008 as Protease Sensitive Prionopathy. The reported initial cases were all of *PRNP*-129 VV genotype, frequently with a family history of dementia, and with brain deposition of abnormal prion protein markedly less protease resistant than found in other prion diseases. Since this first report, the phenotypic spectrum has widened and cases have been found with all three *PRNP*-129 genotypes, making an absolute distinction from sCJD more uncertain.[41–43] The available suggests that VPSPr may be difficult to distinguish clinically from some cases of sCJD and, indeed, other more common, dementias; it is, therefore, difficult to know how common it is. No associated genetic mutations have been found and its precise nosological status is unclear. It is a sporadic prion disease somewhat set apart from sCJD in relation to the overall clinical picture and the prion protein signature.[41] An experimental transmission study of UK VPSPr cases showed biological properties distinct from those of sCJD and with a limited, but not negligible, potential for infectivity.[44]

4. IATROGENIC CJD

4.1 Introduction

Iatrogenic CJD (iCJD) arises from accidental transmission of CJD via medical and surgical procedures with over 475 cases so far reported worldwide with the majority being due to cadaveric-derived human dura mater grafts or human growth hormone (hGH) treatment (Table 2).

Prion protein has been detected in sCJD pituitary glands with likely accidental inclusion of affected glands in the preparation of some pituitary hormone treatment batches.[45] While, in many cases, the evidence for infection is necessarily indirect with the precise infection source not known, prion infectivity in actual USA hGH preparations was confirmed (albeit in low amount) by experimental intracerebral inoculation of samples into squirrel monkeys.[46]

Blood-related vCJD cases are dealt with separately, under vCJD.

4.2 General Characteristics

In general, the resulting illness has no specific characteristics—being essentially like that of typical sCJD, however, hGH—cases have a rather specific clinical profile: a progressive cerebellar ataxia and lower limb dysaesthesia with other features, including cognitive impairment, occurring later in the illness.[47]

The sex and age characteristics essentially reflect the exposure and the relevant incubation period. For example, hGH was given to children and the mean age of onset of hGH-CJD in France is 24.2 years and 27.8 years in the United Kingdom.[48]

The relevant known risk factors for iCJD are the exposure to infection and, as with other human prion diseases, the *PRNP*-129 genotype.

4.3 Diagnosis

As with prion diseases in general, there are no abnormalities in routine of hematological and biochemical tests with normally acellular CSF and a total

Table 2 Iatrogenic CJD

Cause	Total Number of Cases	Countries Reporting Cases	Incubation Period Details	Comments
Human growth hormone	209	Vast majority of cases in France, United Kingdom, and United States	4–36 years	Incubation period uncertain—calculated from midpoint of hormone treatment. Risk period generally before mid-1980s
Dura mater grafts	206	Over half the cases in Japan. Others (in descending order of frequency) in France, Germany, Spain, United Kingdom, Australia, Canada, Italy, and United States. A few other countries reporting 1 or 2 cases each	16 months–23 years	Virtually all related to one manufacturer. Risk period generally up to 1987
Neurosurgical instruments	4	United Kingdom and France	12–28 months	
Brain depth electrodes	2	Switzerland	16 and 20 months	
Corneal transplants	2	United States, Germany	18 months and 30 years	See main text
Human gonadotrophin hormone	4	Australia	12–16 years	
Blood (red cells-vCJD)	3	United Kingdom	6.5–8.3 years	All related to nonleucodepleted red cells
Blood products (vCJD)	0	—	—	

protein level that is either normal or nonspecifically modestly high. A positive 14-3-3, periodic EEG discharges and sCJD-like MR changes may be seen but there no data on the utility of CSF RT-QuIC. *PRNP* analysis is important to exclude pathogenic mutations/genetic prion disease. The diagnosis rests essentially on suspecting CJD in someone with an identifiable relevant risk exposure.

4.4 Human Growth Hormone (hGH) CJD

Treatment with cadaveric-derived hGH was first associated with iCJD in 1985, leading to its ban in many countries and the alternative use of biosynthetic growth hormone. Exact figures are not available but a review in 2000 listed 226, and a 2017 paper over 240, cases worldwide, with the majority of cases in France and the United Kingdom.[49,50]

Rates of iCJD in hGH-treated individuals vary with country (highest in France at around 6%–10% and lowest in the United States at around 0.4%–1%).[49] In France, a cohort effect was seen with all their cases occurring in a specific time period; in the United States, no cases have been seen in those treated after 1977, when an important change in hormone production was introduced; no cohort effect has been seen in the United Kingdom and new cases have been reported up to 2016.[47,49,50] Very long incubation periods have, therefore, been seen—the longest being 42 years.[49]

The *PRNP*-129 polymorphism has important effects on susceptibility and incubation period in hGH cases. There are theoretical reasons for considering that transmission is more likely between humans with the same *PRNP*-129 genotype, however, the genotype of the infecting donors cannot be known. In French hGH cases, 129 homozygotes had a shorter incubation period than heterozygotes.[48] In the United Kingdom cases, there has been a change in genotype distribution over the years: initially only 4% of cases were MM and most were VV, but in later cases, the VV% has declined and the MM% increased.[47] In 2003, when MM up only 4% of UK hGh cases, MM made up 62% of French cases; this interesting difference suggests the infecting donor genotype might have been different in the two countries.[48]

One interesting development in this disease has been the reported finding of the amyloid beta (Aβ) protein (associated with Alzheimer's disease) in the brains and blood vessels of a proportion of hGH-CJD cases, as well as in some cases treated with cadaveric-derived hormone but who died from non-CJD causes.[50] This suggests that Aβ, via peripheral administration, may have a seeding effect in the CNS although the significance of this is uncertain.

4.5 Human Gonadotrophin CJD

Four cases of CJD (with incubation periods from 12 to 14 years) related to another cadaveric-derived pituitary hormone (human gonadotrophin, for infertility) have been reported—all related to treatment in Australia. The relevant treatment period was 1964–85, with around 1447 individuals being treated. In many other countries, this hormone was prepared from urinary rather than pituitary sources.[51,52]

4.6 Human Dura Mater CJD

In 2012, there were 228 cases reported worldwide due to human cadaveric-derived dura mater grafts, with 142 of these being in Japan, where surgical usage of such grafts was particularly high.[49] These grafts have been used generally in neurosurgical repair of dural defects and so involve direct CNS exposure—this is reflected in the relatively shorter incubation period than for the hGH-CJD cases (range 1–25 years, mean 11.8 in Japanese cases).[53] The *PRNP*-129 polymorphism is important, with homozygosity being overrepresented in dura mater cases, although this observation is affected by the fact that most cases have occurred in Japan, where the normal population is largely MM.[54] The frequency of infection in dura mater cases is more uncertain (due to poor data on the true frequency of usage of such grafts) but is calculated to vary from around 0.03% in Japan.[49] Nearly all cases have resulted from one product (Lyodura) produced in Germany before 1987.[49] The site of a dura mater graft might be expected to influence the presenting features of resulting illness and while this has been described in some individual case reports, there is no consistent overall evidence for this.[55,56]

Study of the Japanese cases has defined two clinic-pathological phenotypes: plaque type and nonplaque type dura mater CJD; the latter being characterized by prion plaques in the brain, relatively slow illness progression, and no or late EEG periodic discharges while the former being like sCJD.[53]

4.7 Other Causes

Instances of transmission have been reported relating to surgery, depth EEG electrodes, and corneal transplantation (Table 2).[56] Given the potential transmissibility of prion diseases, it is notable how few such instances have been identified, despite organized disease surveillance in many countries.

Although reports sometimes detail three instances of corneal transplantation, however in only one instance did the illness develop in the recipient both at a reasonable time following transplant and from a donor with pathologically proven CJD.[57] In a second case, CJD was pathologically confirmed in donor and recipient but with a rather long interval of 30 years between transplant and recipient illness.[58] In the third case, there was no confirmation of CJD in the donor.[59]

As noted above, a UK study found no evidence that ophthalmic surgery is a risk factor for CJD.[9]

5. VARIANT CJD

5.1 Introduction

vCJD arose primarily as a zoonosis attributed to dietary contamination with infectious material from BSE affected cattle. Secondary, human-to-human transmission has occurred via blood transfusion and the therapeutic use of blood product (factor VIII).

5.2 Epidemiology

vCJD, initially termed "new variant CJD," was first reported in the United Kingdom in 1996.[60] Subsequently, cases have been identified in a number of countries but particularly the United Kingdom and France. The current world figures are given in Table 3.

The number of intrinsic BSE cases in a country is not the only relevant risk factor; visitors to the United Kingdom being potentially exposed to risk during their stay and there were exports of potentially contaminated material from the United Kingdom to other countries—as has been discussed in relation to the French vCJD cases.[61,62]

vCJD has affected a younger age group than is typical for sCJD. In the United Kingdom, for 178 definite or probable vCJD cases, the median age at onset was 26 years (range: 12–74) (sCJD: 67 years), and the median age at death was 28 years (sCJD: 68 years) [NCJDRSU website/annual report]. The reported 15 French cases in were also young, although slightly older than the UK cases with a median age at onset of 37 years (range: 18–57).[62] However, it is important to note that vCJD has affected older people (74 years old being the oldest UK person affected) and other forms of

Table 3 Variant CJD: Reported Cases

Country[a]	Number	Notes
United Kingdom	178	Includes three blood transfusion cases. 1 case was *PRNP*-129 MV
France	27	1: UKR
Spain	5	
Rep. of Ireland	4	
United States	4	2: UKR
Italy	3	
Netherlands	3	
Portugal	2	
Canada	2	1: UKR
Saudi Arabia	1	
Japan	1	
Taiwan	1	1: UKR

[a]By convention, cases are attributed to their country of residence at time of disease. This is *not* necessarily the country in which infection took place. UKR means that those cases were known to have had at least 6 months residence in the United Kingdom between 1980 and 1996 (the agreed UK dietary BSE risk period). However, some other cases are considered to have been exposed in a country other than their country of attribution.
Definite and probable cases, as defined by accepted surveillance criteria, as at 25.04.17.

prion disease may affect younger people (such as atypical sCJD and genetic prion disease). Within the United Kingdom, there was a differential incidence of vCJD cases between the north and the south.[63] There is no definite sex difference in vCJD cases.

5.3 Cause

The geographical and temporal facts concerning BSE and vCJD are strong epidemiological arguments for a causal connection between the two but there is additional experimental evidence. This includes: the neuropathological features of the illness in Macaques inoculated intracerebrally with BSE; the identical Western blot migration patterns of protein from vCJD and BSE; the similar incubation periods and brain lesion profiles in vCJD and BSE brain inoculated mice and the similar transmission characteristics in wild-type mice inoculated with vCJD brain/lymphoreticular tissue and BSE, are similar.[64–67]

Human food BSE contamination is the highly probable initial cause of vCJD and a case–control study provided supportive evidence.[68] It particularly notable that vCJD has never affected a UK individual born after 1989, the year in which important UK human food protection measures were enacted.

For non-UK cases, there is a more convincing link to the level of importation of potentially contaminated beef products from the United Kingdom between 1980 and 1996, than to indigenous cases of BSE.[61] In France, this importation peaked in 1994, explaining the five-year time difference (2000 and 2005) between the peaks of United Kingdom and French vCJD cases.[62,69] The most likely relevant specific food items are uncertain but MRM (mechanically recovered meat) is considered one important factor. A predictive study based on the level of consumption of beef MRM and head meat estimated the number of vCJD cases in the United Kingdom at 362 (168 for the 1940–69 birth cohort and 194 for the post-1969 birth cohort).[70] A comparable study predicted 33 cases in France (12 for the 1940–69 birth cohort and 21 for the post-1969 birth cohort).[69]

There are estimates that potentially very large amounts of BSE material entered human food in the United Kingdom but there have been relatively few cases of vCJD.[71] Fortunately, there appears to be a high interspecies BSE transmission barrier between cattle and man with experimental evidence to support this.[72] As with human prion diseases in general, the *PRNP*-129 polymorphism is an important factor. Up to 2016, all definite and probable vCJD cases (as define by the agreed surveillance criteria) with genetic testing, were of 129-MM genotype, although one publication reported an MV individual classified as possible vCJD.[73] In 2016, a definite vCJD case of 129 MV genotype was reported.[74] The 129 polymorphism has potential effects on overall susceptibility and incubation period; relevant experimental studies have provided evidence for both and also supported the notion that subclinical infection may have two components: a preclinical period before disease is evident and also a permanent subclinical phase (either a permanent failure for disease to occur or an incubation period longer than the natural lifespan of the infected individual).[72] With infection establishing itself first in lymphoreticular tissue (without detectable clinical manifestations) and neurological involvement occurring later, it was suggested that surveys of surgically removed tissues such as tonsil and appendix could be analyzed for abnormal prion protein to gain an estimate of the number of subclinical cases in the population.[75–78] Several UK studies have been undertaken in the United Kingdom, all using strictly anonymized specimens. The three biggest studies have involved mostly appendices. The first study found three positive appendices on 12,674 specimens from the birth cohort in which most vCJD cases have arisen (1961–85) leading to an estimation of prevalence of healthy carriers of 237 per million or 1 per 4000 but with very wide 95% confidence intervals.[77] A second, larger, study found 16 positive samples on 32,441 from two different birth cohorts, 1961–81 and 1941–60. The global estimated prevalence was 493 per million (95% confidence interval 282–801) or 1 per 2000 (95% confidence interval 1/3500–1/1250).[79] Although these studies were predicated on the assumption that abnormal prion protein in lymphoreticular tissue indicates BSE/vCJD infection, it was thought important to confirm this. A further UK population study was, therefore, undertaken, comparing three time/birth cohort periods: before likely dietary BSE exposure, during the accepted BSE period, and after establishment of the full UK human dietary protection measures. The full results are not yet published but positive appendices were found in all three time periods, with no statistically significant difference between them.[80] This must call into question either the current assumptions made about human population BSE exposure or the acceptance that a positive result indicates BSE/vCJD infection.

The *PRNP*-129 genotype was determined for two out of three patients of the first study (all valine–valine) and for all the patients of the second study (eight methionine–methionine, four methionine–valine, and four valine–valine), indicating that the BSE infection may affect individuals of any 129-genotype and, indirectly, that 129-genotype may affect incubation period or susceptibility to clinical disease. Based on the available data and various assumptions, Garske and Ghani have modeled the likely future course of vCJD in the United Kingdom.[81]

5.4 Secondary Transmission

The primary lymphoreticular involvement in vCJD, with evidence for subclinical infection, caused concern that secondary human-to-human transmission could occur through non-CNS surgery or even via human blood use. To date, no surgical transmission has been identified and a study of dental surgery in particular showed no evidence of a link with vCJD.[68,82] However, there have been four reported instances of transmission of vCJD infection via red blood cell transfusion-3 resulting in clinical vCJD (all *PRNP*-129 MM with the fourth being an instance of asymptomatic infection in the recipient (*PRNP*-129 MV), as evidenced by lymphoreticular abnormal prion protein deposition.[83–87] There is also one reported probable instance of asymptomatic infection in the recipient of a blood product (factor VIII) manufactured from plasma of UK donors.[88] Animal experiments in sheep with BSE have provided evidence for potential infectivity in all blood components and even after leucoreduction.[89,90]

With dietary contamination being the most likely cause, two possible explanations emerge for the relatively young age of vCJD cases: (1) possibly different dietary habits in the young led to greater exposure and (2) there is age-related biologically susceptibility to infection and disease. In relation to the second possibility, the young have a greater amount of gut-associated lymphoreticular tissue and a higher incidence of potentially relevant infections (such as tonsillitis) that might increase susceptibility to the BSE agent in diet.[91–93]

5.5 Genetic Factors

Apart from *PRNP*-129, other possible genetic risk factors have been studied. In a genome-wide association study, Mead *et al.* identified additional candidate loci that they felt merited further investigation.[20] Other studies have found evidence to implicate a Cathepsin D polymorphism, MTMR7 and PLCXD3.[94–96]

5.6 Clinical Features

vCJD is significantly slower in progression than typical sCJD. The median duration of illness in 178 definite and probable UK cases was 14 months (range: 6–114) and 16 months (range: 8–24) in 15 reported French cases. In the United Kingdom, the median duration of sCJD is 4 months (range: 1–74).[97] Like all prion diseases, with no presently effective treatment, progression to death is the inevitable course.

Psychiatric features frequently dominate the earlier part of the illness and may be the sole presenting feature.[98] Characteristic specifically neurological features include painful sensory symptoms (limb pain, cold feelings, dysesthesiae, parasthesiae, and numbness), cerebellar ataxia, cognitive impairment, and involuntary movements (including tremor, chorea, or dystonia but especially myoclonus).[99–101]

Although over half of cases develop at least one neurological feature within 2 months of illness onset, this is often a fairly nonspecific problem, also it may be difficult to recognize or interpret in the context of a significant psychiatric presentation and clear neurological features may not appear until 4–6 months of illnesses.[98]

It is not surprising that a serious progressive brain disease can produce, at some stage of the illness, a wide variety of other clinical abnormalities such as extensor plantar responses and primitive reflexes, but these are generally nonspecific features. Occulomotor problems may occur, sometimes associated with complaints of diplopia (NCJDRSU, personal data). Notably, certain specific problems are uncommon: epileptic seizures have been documented in only 8% and typical extrapyramidal features in only around 4% of UK cases (NCJDRSU, personal data).

5.7 Diagnosis

vCJD typically presents in a nonspecific way, with a potentially very wide differential diagnosis including primary psychiatric illnesses. Diagnosis starts with awareness of the possibility and vCJD is an exceedingly rare illness (at the time of writing, there are no known living cases in the world) with an important geographical distribution, making it a particularly unlikely diagnosis in many countries. Leaving aside possible nonprion diseases, the important differential diagnosis of vCJD importantly includes genetic prion disease (where there may be no known family history) and atypical sCJD. *PRNP* analysis allows codon 129 genotyping; currently, an MV or VV genotype might make the diagnosis less likely, but certainly cannot exclude it. Additionally, atypical sCJD is an important differential diagnosis of vCJD, and atypical sCJD features are more likely in MV and VV genotype individuals.

Presently, the only absolute confirmatory test is neuropathological examination, requiring either a brain biopsy or an autopsy. In relation to the former, the same comments can be made as were made above for sCJD.

However, there is no reason to believe that the examination of olfactory neurons via nasal brushing in vCJD would not have the diagnostic utility that has been reported in sCJD.[38,102]

In addition, because of lymphoreticular involvement in vCJD, tonsil biopsy has been used to obtain a tissue confirmation of the presence of relevant abnormal prion protein and, as such, is arguably a highly specific clinical test and a definitely positive result has not been seen in any clinical condition other than vCJD.[103] However, it is procedure that has potential morbidity and a negative result cannot totally exclude vCJD. Since lymphoreticular involvement precedes clinical illness in BSE/vCJD infection, a positive tonsil biopsy could, conceivably, reflect preclinical infection rather than relevant illness, a possibility supported by the finding of positive results in UK general population studies described above. However, if a tonsil biopsy is being undertaken in an appropriate clinical illness context, a positive result is highly likely to be diagnostically relevant.

CSF examination is an important part of the diagnostic process as various inflammatory CNS diseases may present in a similar fashion to vCJD. As in sCJD, the routine CSF examination is either normal or only nonspecifically abnormal, usually with no pleocytosis, and, at most, only a relatively modest total protein elevation. CSF 14-3-3 testing is less sensitive for vCJD than sCJD, being positive in only around 50% of cases.[104] Raised CSF tau, especially phosphorylated tau, is more sensitive than 14-3-3, for vCJD but elevations may be seen in some other neurodegenerative diseases, reducing its specificity.[105]

Interestingly and notably, CSF RT-QuIC, has proven consistently negative in cases of vCJD [34] (A.J.E. Green, personal communication). It is, therefore, not positively diagnostically useful but is potentially helpful in differentiating cases of sporadic and variant disease. There is a report of a PMCA-based urine test being sensitive and specific for vCJD; this is an interesting result but relates to a small number of specimens tested in a single study.[106]

Given the present failure to develop reliable blood tests for sCJD, it is notable that significant progress in this regard has been made with vCJD. Edgeworth *et al.* reported a test called the "Direct Detection Assay" that was studied further with a reported sensitivity of around 71% for clinical vCJD and a specificity of 100% for CJD-although with occasional positive results in sCJD.[35,36] More recently, two significantly more sensitive tests—both PMCA based—have been reported.[37,107] Concha-Marambio *et al.* reported a 100% sensitivity and 100% specificity on testing 14 vCJD specimens and 153 controls (including sCJD specimens).[107] Bougard *et al.* reported a 100% sensitivity and a 100% diagnostic specificity on testing 18 vCJD samples, 67 sCJD samples and 238 non-CJD controls. However, one of the 67 sCJD samples was positive. They also reported positive results in samples taken presymptomatically from two vCJD cases, i.e., they confirmed positivity in preclinical as well as clinical vCJD.[37]

Nontissue-based clinical confirmatory investigation has heavily relied on cerebral MR imaging which, in the majority of cases, shows characteristic high signal in the posterior thalamus (the so-called "pulvinar sign"); the dorsomedial thalamic nuclei and the periaqueductal gray matter may also be involved.[31,108] High signal may be seen also in the caudate and putamen (areas characteristically involved in sCJD), but the "pulvinar sign" consists of signal intensity that is *higher* in the posterior thalamus than in these anterior basal ganglia areas. However, some cases of vCJD may show MR changes that are truly characteristic of sCJD.[74] Cortical signal change—common in sCJD—is not usually seen in vCJD and cerebral atrophy is uncommon.[108]

Different MR sequences have different sensitivities for vCJD, in descending order: DWI > FLAIR > PD > T2 > T1. In a study of 86 neuropathologically confirmed cases of vCJD, 9% of T1 images, 81% of PD images, and 100% of FLAIR images showed the characteristic pulvinar sign—though the authors state they have seen a probable case without this feature on FLAIR (Fig. 3).[108]

The reported MR data relate to routine clinical practice, not formal, designed, studies, so it is difficult to comment on the evolution of MR changes during the illness course. The pulvinar sign may be absent in some cases even late in the illness, but it is worth repeating the scan if the initial images do not show it.[108]

The EEG may be normal in the early stages of vCJD; with disease progression, nonspecific loss of normal background rhythms and excess slow-wave activity is seen. The typical periodic discharges found in many cases of sCJD either do not develop in vCJD or are seen rarely and only in very late, preterminal, stages as described in two case reports.[109,110]

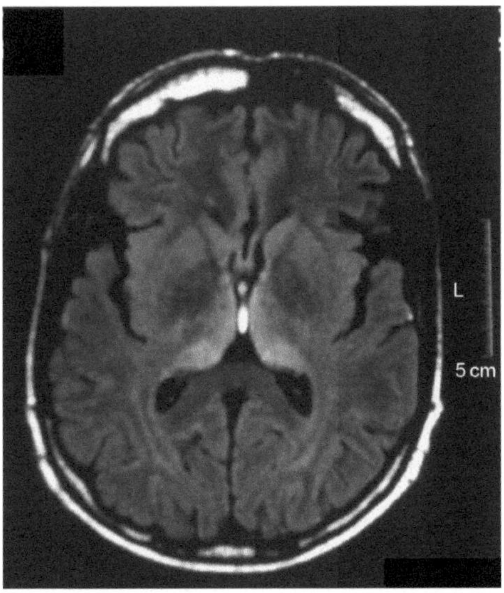

Fig. 3 The cerebral MRI (FLAIR) in vCJD. Increased signal in the posterior thalamus in vCJD—the "Pulvinar Sign." *Image courtesy of Dr. David Summers, Department of Neuroradiology, Edinburgh.*

Diagnostic criteria for vCJD are used in international surveillance and research to classify cases into definite, probable, and possible categories.[97]

These criteria have been reviewed and validated in one study showing that, in terms of sensitivity, 83% of 106 definite cases would be classifiable as probable vCJD (with six cases classified as possible vCJD). The specificity was found to 100% with no autopsy/cerebral biopsy-proven non-vCJD cases being classified as probable vCJD in life.[97]

The described clinical phenotype (including investigation results) is based on 129-MM cases an, conceivably, it might be different in non-MM129 genotypes. The single reported definite case of MV-vCJD showed some clinical phenotypic variation from the commonest MM picture but was within the general clinical envelope of previously described vCJD cases.[74] The other reported case of MV genotype (only designated as possible vCJD) was not clinically distinct from other vCJD cases.[73]

REFERENCES

1. Ladogana A, Puopolo M, Croes EA, et al. Mortality from Creutzfeldt–Jakob disease and related disorders in Europe, Australia and Canada. *Neurology*. 2005;64(9):1586–1591.
2. Knight R. Creutzfeldt–Jakob disease: a rare cause of dementia in elderly persons. *Clin Infect Dis*. 2006;43(3):340–346.
3. Collins S, Law MG, Fletcher A, Boyd A, Kaldor J, Masters CL. Surgical treatment and risk of sporadic Creutzfeldt–Jakob disease: a case–control study. *Lancet*. 1999;353(9154): 693–697.
4. Ward HJ, Everington D, Croes EA, et al. Sporadic Creutzfeldt–Jakob disease and surgery: a case–control study using community controls. *Neurology*. 2002;59(4):543–548.
5. Mahillo-Fernandez I, de Pedro-Cuesta J, Bleda MJ, et al. Surgery and risk of sporadic Creutzfeldt–Jakob disease in Denmark and Sweden: registry-based case–control studies. *Neuroepidemiology*. 2008;31(4):229–240.
6. de Pedro-Cuesta J, Mahillo-Fernández I, Rábano A, et al. Nosocomial transmission of sporadic Creutzfeldt–Jakob disease: results from a risk-based assessment of surgical interventions. *JNNP*. 2011;82(2):204–212.
7. Harries-Jones R, Knight R, Will RG, Cousens S, Smith PG, Matthews WB. Creutzfeldt–Jakob disease in England and Wales, 1980–1984: a case–control study of potential risk factors Harries–Jones et al. *JNNP*. 1988;51:1113–1119.
8. Hamaguchi T, Noguchi-Shinohara M, Nozaki I, et al. Medical procedures and risk for sporadic Creutzfeldt–Jakob disease, Japan, 1999-2008. *Emerg Infect Dis*. 2009;15(2): 265–271.
9. Sanchez-Juan P, Ward HJ, De Silva R, Knight RS, Will RG. Ophthalmic surgery and Creutzfeldt–Jakob disease. *Br J Ophtalmol*. 2004;88(4):446–449.
10. Ward HJ, Everington D, Cousens SN, et al. Risk factors for sporadic Creutzfeldt–Jakob disease. *Ann Neurol*. 2008;63(3):347–354.
11. Cousens S, Harries-Jones R, Knight R, Will RG, Smith PG, Matthews WB. Geographical distribution of cases of Creutzfeldt–Jakob disease in England and Wales 1970-84. *JNNP*. 1990;53:459–465.

12. Linsell L, Cousens SN, Smith PG, et al. A case–control study of sporadic Creutzfeldt–Jakob disease in the United Kingdom: analysis of clustering. *Neurology*. 2004;63(11):2077–2083.
13. de Pedro CJ, Ruiz Tovar M, Ward H, et al. Sensitivity to biases of case–control studies on medical procedures, particularly surgery and blood transfusion, and risk of Creutzfeldt–Jakob disease. *Neuroepidemiology*. 2012;39(1):1–18.
14. van Duijn CM, Delasnerie-Lauprêtre N, Masullo C, et al. Case–control study of risk factors of Creutzfeldt–Jakob disease in Europe during 1993-95. European Union (EU) collaborative study Group of Creutzfeldt–Jakob disease (CJD). *Lancet*. 1998;351(9109): 1081–1085.
15. Bishop MT, Sanchez-Juan P, Knight RS. Splice site SNPs of phospholipase PLCXD3 are significantly associated with variant and sporadic Creutzfeldt–Jakob disease. *BMC Med Genet*. 2013;14(1):91.
16. Bishop MT, Pennington C, Heath CA, Will RG, Knight RSG. *PRNP* variation in UK sporadic and variant Creutzfeldt Jakob disease highlights genetic risk factors and a novel non-synonymous polymorphism. *BMC Med Genet*. 2009;10:146. http://dx.doi.org/10.1186/1471-2350-10-146.
17. Sanchez-Juan P, et al. A genome wide association study links glutamate receptor pathway to sporadic Creutzfeldt–Jakob disease risk. *PLoS One*. 2015;10:e0123654. http://dx.doi.org/10.1371.
18. Kovacs GG, Sanchez-Juan P, Strobel T, et al. Cathepsin D (C224T) polymorphism and modulation of Creutzfeldt–Jakob disease phenotype. *Alzheimer Dis Assoc Disord*. 2010;24(1):104–107.
19. Sanchez-Juan P, Bishop MT, Croes EA, et al. A polymorphism in the regulatory region of PRNP is associated with increased risk of sporadic Creutzfeldt–Jakob disease. *BMC Med Genet*. 2011;12:73.
20. Mead S, Uphill J, Beck J, et al. Genome-wide association study in multiple human prion diseases suggests genetic risk factors additional to *PRNP*. *J Hum Mol Genet*. 2012;21(8):1897–1906.
21. Cooper SA, Murray KL, Heath CA, Will RG. Knight RSG sporadic Creutzfeldt–Jakob disease with cerebellar ataxia at onset in the UK. *J Neurol Neurosurg Psychiatry*. 2006;77:1273–1275.
22. Cooper SA, Murray KL, Heath CA, Will RG. Knight RSG isolated visual symptoms at onset in sporadic Creutzfeldt–Jakob disease: the clinical phenotype of the "Heidenhain variant" *Br J Ophthalmol*. 2005;89:1341–1342.
23. Salwa El Tawil S, Chohan G, Mackenzie J, et al. Isolated language impairment as the primary presentation of sporadic Creutzfeldt Jakob disease. *Acta Neurol Scand*. 2017;135(3):316–323.
24. Parchi P, Giese A, Capellari S, et al. Classification of sporadic Creutzfeldt–Jakob disease based on molecular and phenotypic analysis of 300 subjects. *Ann Neurol*. 1999;46: 224–233.
25. Puoti G, Giaccone G, Rossi G, Canciani B, Bugiani O, Tagliavini F. Sporadic Creutzfeldt–Jakob disease: co-occurrence of different types of PrP(Sc) in the same brain. *Neurology*. 1999;53:2173–2176.
26. Parchi P, Strammiello R, Notari S, et al. Incidence and spectrum of sporadic Creutzfeldt–Jakob disease variants with mixed phenotype and co-occurrence of PrPSc types: an updated classification. *Acta Neuropathol*. 2009;118:659–671.
27. Pocchiari M, Puopolo M, Croes EA, et al. Predictors of survival in sporadic Creutzfeldt–Jakob disease and other human transmissible spongiform encephalopathies. *Brain*. 2004;127(10):2348–2359.
28. Mead S, Rudge P. CJD mimics and chameleons. *Pract Neurol*. 2017;17:113–121.

29. Green A, Sanchez-Juan P, Ladogana A, et al. CSF analysis in patients with sporadic CJD and other transmissible spongiform encephalopathies. *Eur J Neurol.* 2007;14(2): 121–124.
30. Chohan G, Pennington C, Mackenzie JM, et al. The role of cerebrospinal fluid 14-3-3 and other proteins in the diagnosis of sporadic Creutzfeldt–Jakob disease in the UK: a 10-year review. *J Neurol Neurosurg Psychiatry.* 2010;81:1243.
31. Collie DA, Sellar RJ, Zeidler M, Colchester A, Knight RSG, Will RG. MRI of Creutzfeldt–Jakob disease: imaging features and recommended MRI protocol. *Clin Radiol.* 2001;56:726–739.
32. Zerr I, Pocchiari M, Collins S, et al. Analysis of EEG and CSF 14-3-3 proteins as aids to the diagnosis of Creutzfeldt–Jakob disease. *Neurology.* 2000;55(6):811–815.
33. Meissner B, Kallenberg K, Sanchez-Juan P, et al. MRI lesion profiles in sporadic Creutzfeldt–Jakob disease. *Neurology.* 2009;72:1994–2001.
34. McGuire L, Peden AH, Orru CD, et al. Real time quaking-induced conversion analysis of cerebrospinal fluid in sporadic Creutzfeldt–Jakob disease. *Ann Neurol.* 2012;72:278–285.
35. Edgeworth JA, Farmer M, Sicilia A, et al. Detection of prion infection in variant Creutzfeldt–Jakob disease: a blood-based assay. *Lancet.* 2011;377:487–493.
36. Jackson Graham S, Burk-Rafel J, Edgeworth JA, et al. Population screening for variant CJD using a novel blood test. Diagnostic accuracy and feasibility study. *JAMA Neurol.* 2014;7(4):421–428.
37. Bougard D, Brandel JP, Belondrade M, et al. Prion Detection in the Plasma of Presymptomatic Blood Donors and Symptomatic Subjects with Variant Creutzfeldt–Jakob disease. *Sci Transl Med.* 2016;8:37. 0ra182.
38. Orru Christina D, Matilde B, Giovanni T, et al. A test for Creutzfeldt–Jakob disease using nasal brushings. *N Engl J Med.* 2014;371:519–529.
39. Kovacs GG, Puopolo M, Ladogana A, et al. Genetic prion disease: the EUROCJD experience. *Hum Genet.* 2005;118:166–174.
40. Shibuya S, Higuchi J, Shin RW, Tateishi J, Kitamoto T. Codon 219 Lys allele of *PRNP* is not found in sporadic Creutzfeldt–Jakob disease. *Ann Neurol.* 1998;43(6):826–828.
41. Gambetti P, Puoti G, Zou WG. Variably protease-sensitive prionopathy: a novel disease of the prion protein. *J Mol Neurosci.* 2011;45:422–424.
42. Zou WQ, Puoti G, Xiao X, Yuan J, Qing L, Cali I. Variably protease-sensitive prionopathy: a new sporadic disease of the prion protein. *Ann Neurol.* 2010;68: 162–172.
43. Head MW, Yull HM, Ritchie DL, et al. Variably protease-sensitive prionopathy in the UK: a retrospective review 1991–2008. *Brain.* 2013;136:1102–1115.
44. Diack AB, Ritchie DL, Peden AH, et al. Variably protease-sensitive prionopathy, a unique prion variant with inefficient transmission properties. *Emerg Infect Dis.* 2014;20(12):1969–1979.
45. Peden AH, Ritchie DL, Uddin HP, et al. Abnormal prion protein in the pituitary in sporadic and variant Creutzfeldt–Jakob disease. *J Gen Virol.* 2007;88: 1068–1072.
46. Gibbs Jr CJ, Asher DM, Brown PW, Fradkin JE, Gajdusek DC. Creutzfeldt–Jakob disease infectivity of growth hormone derived from human pituitary glands. *N Engl J Med.* 1993;328:358–359.
47. Rudge P, Jaunmuktane Z, Adlard P, et al. Iatrogenic CJD due to pituitary-derived hormone with genetically determined incubation times of up to 40 years. *Brain.* 2015;138:3386–3399.
48. Brandel JP, Preece M, Brown P, et al. Distribution of codon 129 genotype in human growth hormone-treated CJD patients in France and the UK. *Lancet.* 2003;362: 128–130.

49. Brown P, Brandel J-P, Sato T, et al. Iatrogenic Creutzfeldt–Jakob disease, final assessment. *Emerg Infect Dis*. 2012;18:901–907.
50. Riitchie Diane L, Adlard P, Peden A, et al. Amyloid-β accumulation in the CNS in human growth hormone recipients in the UK. *Acta Neuropathol*. 2017; 134(2):221–240. http://dx.doi.org/10.1007/s00401-017-1703-0. [Epub ahead of print].
51. Cochius JI, Burns RJ, Blumbergs PC, Mack K, Aldeman CP. Creutzfeldt–Jakob disease in a recipient of human pituitary-derived gonadotrophin. *Aust N Z J Med*. 1990;20: 592–593.
52. Cochius JI, Hyman N, Esiri MM. Creutzfeldt–Jakob disease in a recipient of human pituitary-derived gonadotrophion: a second case. *J Neurol Neurosurg Psychiatry*. 1992;55:1094–1095.
53. Yamada M, et al. Dura mater graft-associated Creutzfeldt–Jakob disease in Japan: Clinicopathological and molecular characterization of the two distinct subtypes. *Neuropathology*. 2009;29:609.
54. Nurmi MH, Bishop M, Strain L, et al. The normal population distribution of *PRNP* codon 129 polymorphism. *Acta Scand Neurol*. 2003;108:374–378.
55. Heath CA, Barker RA, Esmonde TFG, et al. Dura mater-associated Creutzfeldt–Jakob disease: experience from surveillance in the UK. *J Neurol Neurosurg Psychiatry*. 2006;77:880–882.
56. Brown P, Preece M, Brandel J-P, et al. Iatrogenic Creutzfeldt–Jakob disease at the millennium. *Neurology*. 2000;55(8):1075–1081.
57. Duffy P, Wolf J, Collins G, DeVoe AG, Streeten B, Cowen D. Possible person to person transmission of Creutzfeldt–Jakob disease. *N Engl J Med*. 1974;290: 692–693.
58. Heckmann JG, Lang CJG, Petruch F, et al. Transmission of Creutzfeldt–Jakob disease via a corneal transplant. *J Neurol Neurosurg Psychiatry*. 1997;63:388–390.
59. Uchiyama S, Ishida C, Yago S, Kurumaya H, Kitamoto T. An autopsy case of Creutzfeldt–Jakob disease associated with corneal transplantation [in Japanese]. *Dementia*. 1994;8:466–473.
60. Will RG, Ironside JW, Zeidler M, et al. A new variant of Creutzfeldt–Jakob disease in the UK. *Lancet*. 1996;347:921–925.
61. Sanchez-Juan P, Cousens SN, Will RG, van Duijn CM. Source of variant Creutzfeldt–Jakob disease outside United Kingdom. *Emerg Infect Dis*. 2007;13:1166–1169.
62. Brandel JP, Heath CA, Head MW, et al. Variant Creutzfeldt–Jakob disease in France and the United Kingdom: evidence for the same agent strain. *Ann Neurol*. 2009;65: 249–256.
63. Cousens S, Smith PG, Ward H, et al. Geographical distribution of variant Creutzfeldt–Jakob disease in great Britain, 1994-2000. *Lancet*. 2001;357(9261):1002–1007.
64. Bruce ME, Will RG, Ironside JW, et al. Transmissions to mice indicate that 'new variant' CJD is caused by the BSE agent. *Nature*. 1997;389:498–501.
65. Hill AF, Desbruslais M, Joiner S, et al. The same prion strain causes vCJD and BSE. *Nature*. 1997;389:448–450.
66. Lasmézas CI, Deslys JP, Demaimay R, et al. BSE transmission to macaques. *Nature*. 1996;381:743–744.
67. Ritchie DL, Boyle A, McConnell I, Head MW, Ironside JW, Bruce ME. Transmissions of variant Creutzfeldt–Jakob disease from brain and lymphoreticular tissue show uniform and conserved bovine spongiform encephalopathy-related phenotypic properties on primary and secondary passage in wild-type mice. *J Gen Virol*. 2009;90:3075–3082.
68. Ward HJ, Everington D, Cousens SN, et al. Risk factors for variant Creutzfeldt–Jakob disease: a case–control study. *Ann Neurol*. 2006;59:111–120.

69. Chadeau-Hyam M, Alpérovitch A. Risk of variant Creutzfeldt–Jakob disease in France. *Int J Epidemiol.* 2005;34:46–52.
70. Cooper JD, Bird SM. Predicting incidence of variant Creutzfeldt–Jakob disease from UK dietary exposure to bovine spongiform encephalopathy for the 1940 to 1969 and post-1969 birth cohorts. *Int J Epidemiol.* 2003;32:784–791.
71. Valleron AJ, Boelle PY, Will R, Cesbron JY. Estimation of epidemic size and incubation time based on age characteristics of vCJD in the United Kingdom. *Science.* 2001;294:1726–1728.
72. Bishop MT, Hart P, Aitchison L, et al. Predicting susceptibility and incubation time of human-to-human transmission of vCJD. *Lancet Neurol.* 2006;5:393–398.
73. Kaski D, Mead S, Hyare H, et al. Variant CJD in an individual heterozygous for PRNP codon 129. *Lancet.* 2009;374:2128.
74. Mok t, Jaunmuktane Z, Joiner S, et al. Variant Creutzfeldt–Jakob Disease in a Patient with Heterozygosity at *PRNP* Codon 129. *NEJM.* 2017;376:292–294.
75. Brandel JP, Heath CA, Head MW, et al. Detection of variant Creutzfeldt–Jakob disease infectivity in extraneural tissues. *Lancet.* 2001;358:208–209.
76. Hilton DA, Fathers E, Edwards P, Ironside JW, Zajicek J. Prion immunoreactivity in appendix before clinical onset of variant Creutzfeldt–Jakob disease. *Lancet.* 1998;352:703–704.
77. Hilton DA, Ghani AC, Conyers L, et al. Prevalence of lymphoreticular prion protein accumulation in UK tissue samples. *J Pathol.* 2004;203:733–739.
78. Head MW, Ritchie D, Smith N, et al. Peripheral tissue involvement in sporadic, iatrogenic, and variant Creutzfeldt–Jakob disease: an immunohistochemical, quantitative, and biochemical study. *Am J Pathol.* 2004;164:143–153.
79. Gill ON, Spencer Y, Richard-Loendt A, et al. Prevalent abnormal prion protein in human appendixes after bovine spongiform encephalopathy epizootic: large scale survey. *BMJ.* 2013;347:f5675.
80. Public Health England. Health protection report. In: *Summary Results of the Third National Survey of Abnormal Prion Prevalence in Archived Appendix Specimens. Infection Report.* 2016;10(26).
81. Garske T, Ghani AC. Uncertainty in the tail of the variant Creutzfeldt–Jakob disease epidemic in the UK. *PLoS One.* 2010;5(12):e15626.
82. Everington D, Smith AJ, Ward HJT, Letters S, Will RG, Bagg J. Dental treatment and risk of variant CJD—a case control study. *Br Dent J.* 2007;202(8):E19 [discussion 470-1].
83. Peden AH, Head MW, Ritchie DL, Bell JE, Ironside JW. Preclinical vCJD after blood transfusion in a PRNP codon 129 heterozygous patient. *Lancet.* 2004;364:527–529.
84. Llewelyn CA, Hewitt PE, Knight RS, et al. Possible transmission of variant Creutzfeldt–Jakob disease by blood transfusion. *Lancet.* 2004;363:417–421.
85. Wroe SJ, Pal S, Siddique D, et al. Clinical presentation and pre-mortem diagnosis of variant Creutzfeldt–Jakob disease associated with blood transfusion: a case report. *Lancet.* 2006;368(9552):2061–2067.
86. Urwin PJ, Mackenzie JM, Llewelyn CA, Will RG, Hewitt PE. Creutzfeldt–Jakob disease and blood transfusion: updated results of the UK transfusion medicine epidemiology review study. *Vox Sang.* 2016;110:310–316.
87. Health Protection Agency. Fourth case of transfusion-associated variant-CJD. *Health Prot Rep.* 2007;1(3).
88. Peden A, McCardle L, Head MW, et al. Variant CJD infection in the spleen of a neurologically asymptomatic UK adult patient with haemophilia. *Haemophilia.* 2010;16:296–304.
89. Houston F, Foster JD, Chong A, Hunter N, Bostock CJ. Transmission of BSE by blood transfusion in sheep. *Lancet.* 2000;356:999–1000.

90. McCutcheon S, Alejo Blanco AR, Houston EF, et al. All clinically-relevant blood components transmit prion disease following a single blood transfusion: a sheep model of vCJD. *PLoS One*. 2011;6(8):e23169.
91. St R, Hunter N, Matthews L, et al. Comparative evidence for a link between Peyer's patch development and susceptibility to transmissible spongiform encephalopathies. *BMC Infect Dis*. 2006;6:5.
92. Brown KL, Wathne GJ, Sales J, Bruce ME, Mabbott NA. The effects of host age on follicular dendritic cell status dramatically impair scrapie agent neuroinvasion in aged mice. *J Immunol*. 2009;183(8):5199–5207.
93. Donaldson DS, Sehgal A, Rios D, Williams IR, Mabbott NA. Increased abundance of M cells in the gut epithelium dramatically enhances oral prion disease susceptibility. *PLoS Pathog*. 2016;12(12):e1006075.
94. Bishop MT, Kovacs GJ, Sanchez-Juan P, Knight RSG. Cathepsin D SNP associated with increased risk of variant Creutzfeldt–Jakob disease. *BMC Med Genet*. 2008;9:31. http://dx.doi.org/10.1186/1471-2350-9-31.
95. Sanchez-Juan P, Bishop MT, Aulchenko YS, et al. Genome-wide study links MTMR7 gene to variant Creutzfeldt–Jakob risk. *Neurobiol Aging*. 2012;33:1487.e21–1487.e28.
96. Bishop MT, Sanchez-Juan P, Knight RS. Splice site SNPs of phospholipase PLCXD3 are significantly associated with variant and sporadic Creutzfeldt–Jakob disease. *BMC Med Genet*. 2013;14(1):91.
97. Heath CA, Cooper SA, Murray K, et al. Diagnosing variant Creutzfeldt–Jakob disease: a retrospective analysis of the first 150 cases in the UK. *JNNP*. 2011;82(6):646–651.
98. Spencer Michael D, Knight Richard SG, Will RG. First hundred cases of variant Creutzfeldt–Jakob disease: retrospective case note of early psychiatric and neurological features. *BMJ*. 2002;342:1479–1482.
99. Zeidler M, Stewart GE, Barraclough CR, et al. New variant Creutzfeldt–Jakob disease: neurological features and diagnostic tests. *Lancet*. 1997;350(9082):903–907.
100. McLeod M-A, Stewart GE, Zeidler M, Will R, Knight R. Sensory features of variant Creutzfeldt–Jakob disease. *J Neurol*. 2002;249(6):706–711.
101. Heath CA, Cooper SA, Murray K, et al. Validation of diagnostic criteria for variant CJD. *Ann Neurol*. 2010;67(6):761–770.
102. Bongianni M, Orru C, Groveman BR, et al. Diagnosis of human prion disease using real-time quaking-induced conversion testing of olfactory mucosa and cerebrospinal fluid samples. *JAMA Neurol*. 2017;74(2):155–162.
103. Hill AF, Butterworth RJ, Joiner S, et al. Investigation of variant Creutzfeldt–Jakob disease and other human prion diseases with tonsil biopsy samples. *Lancet*. 1999;353:183–189.
104. Green AJE, Thompson EJ, Stewart GE, et al. Use of 14-3-3 and other brain-specific proteins in CSF in the diagnosis of variant Creutzfeldt–Jakob disease. *JNNP*. 2001;70:744–748.
105. Green AJE, Andrews MM, Bishop MT, et al. Elevated phosphorylated tau pT-181 in a possible PRNP codon 129 MV vCJD case. *JNNP*. 2010;81(12):1408–1409.
106. Moda F, Gambetti P, Notari S, et al. Prions in the urine of patients with variant Creutzfeldt–Jakob disease. *NEJM*. 2014;371(6):530–539.
107. Concha-Marambio L, Pritzkow S, Moda F, et al. Highly sensitive and specific detection of prionsin blood of variant Creutzfeldt–Jakob disease patients. *Sci Transl Med*. 2016;8:370ra183.
108. Collie DA, Summers DM, Sellar RJ, et al. Diagnosing variant Creutzfeldt–Jakob disease with the Pulvinar sign: MR imaging findings in 86 neuropathologically confirmed cases. *Am J Neuroradiol*. 2003;24:1560–1569.

109. Masahito Yamada on behalf of the Variant CJD Working Group. The first Japanese case of variant Creutzfeldt–Jakob disease showing periodic electroencephalogram. *Lancet*. 2006;367:874.
110. Binelli S, Agazzi P, Giaccone G, et al. Periodic electroencephalogram complexes in a patient with variant Creutzfeldt–Jakob disease. *Ann Neurol*. 2006;59:423–427.

FURTHER READING

111. Hewitt PE, Llewelyn CA, Mackenzie J, et al. Creutzfeldt–Jakob disease and blood transfusion: results of the UK transfusion medicine epidemiology review study. *Vox Sang*. 2006;91:221–230.

CHAPTER FOURTEEN

Neuropathology of Human Prion Diseases

Diane L. Ritchie, James W. Ironside[1]
National CJD Research & Surveillance Unit, Centre for Clinical Brain Sciences, School of Clinical Sciences, University of Edinburgh, Western General Hospital, Edinburgh, United Kingdom
[1]Corresponding author: e-mail address: james.ironside@ed.ac.uk

Contents

Abstract

The human prion diseases comprise sporadic, genetic, and acquired disorders. These are rare conditions with a heterogeneous clinicopathologic phenotype, which can make diagnosis challenging. A combined clinical, genetic, neuropathologic and biochemical approach to diagnosis is therefore essential. Since prion infectivity is the highest in tissues from the central nervous system, special laboratory precautions are required for the

Progress in Molecular Biology and Translational Science, Volume 150
ISSN 1877-1173
http://dx.doi.org/10.1016/bs.pmbts.2017.06.011

safe handling of these tissues. Neuropathologic assessment is generally performed following autopsy, when the fixed brain should be adequately sampled and studied by conventional stains and immunohistochemistry for the abnormal form of the prion protein. Frozen brain tissue is also required for DNA extraction for prion protein gene sequencing and for Western blot analysis of protease-resistant prion protein. The microscopic assessment of the nature and degree of spongiform change, neuronal loss, gliosis, and abnormal prion protein deposition in the brain can be used to determine the major categories of human prion disease. This information can be combined with clinical, genetic data, and biochemical data to allow an accurate diagnosis of a human prion disease and facilitates subclassification into recognized disease subtypes, for example in sporadic Creutzfeldt–Jakob disease. The spectrum of human prion diseases continues to expand and neuropathology will play a key role in the recognition and understanding of any further novel entities or disease variants that may emerge in the future.

1. INTRODUCTION

The common human neurodegenerative diseases are associated with the accumulation of abnormal disease-associated proteins in the brain, for example β-amyloid and hyperphosphorylated tau in Alzheimer's disease and α-synuclein in Parkinson's disease. Prion diseases are rare neurodegenerative disorders that are associated with the accumulation of an abnormal isoform of the prion protein (PrP^{Sc}), which is derived by misfolding of the normal cellular isoform of the prion protein (PrP^{C}).[1] Prion diseases differ from other human neurodegenerative disorders since they are transmissible between individuals and occur in sporadic, genetic, and acquired forms (Table 1).[2] The transmissible agent is unconventional and differs from bacteria and viruses in its physical, chemical, and biological properties. The major, if not the sole, component of the agent is PrP^{Sc}, which acts as an infectious misfolded amyloidogenic protein (or prion) as proposed in the prion hypothesis.[1] The prion hypothesis offers a framework within which to classify and investigate human prion diseases (Table 1). Human prion diseases are characterized neuropathologically by spongiform change in affected gray matter regions, accompanied by variable neuronal loss, reactive gliosis, and the accumulation of PrP^{Sc} in the brain, as occurs in Creutzfeldt–Jakob disease (CJD).[2]

2. CLASSIFICATION OF HUMAN PRION DISEASES

Human prion diseases have previously been classified according to their clinicopathologic phenotype (for example, CJD, Gerstmann–Sträussler–Scheinker syndrome (GSS), or fatal familial insomnia (FFI)) and etiology

Table 1 Human Prion Diseases Classified According to Etiology

Etiology	Disease
1. Idiopathic (sporadic)	Sporadic Creutzfeldt–Jakob disease (CJD)
	Sporadic fatal insomnia
	Variably protease-sensitive prionopathy
2. Genetic (inherited)	Genetic CJD
	Gerstmann–Sträussler–Scheinker disease
	Prion protein cerebral amyloid angiopathy
	Fatal familial insomnia
3. Acquired (infectious)	Kuru
	Iatrogenic CJD
	Variant CJD

(sporadic, familial, and acquired), but current classification is increasingly dependent on both genetic and molecular criteria, specifically the presence of pathogenic mutations, insertions, deletions, and polymorphisms in the prion protein gene (*PRNP*) and the biochemical analysis of the isoforms of disease-associated prion protein in the brain, combined with the results of analysis of the neuropathologic lesions in the brain.[2,3]

2.1 Sporadic CJD

Sporadic CJD (sCJD) is the commonest form of human prion disease, occurring most frequently in the seventh decade of life, with an incidence of 1–2 per million population per annum.[2] Cases of sCJD appear to arise spontaneously, perhaps due to a somatic mutation in the *PRNP* or a consequence of spontaneous misfolding of PrP^{C}, resulting in the generation of PrP^{Sc}. The commonest presenting clinical feature is rapidly progressive dementia, accompanied by other neurologic signs and symptoms, including myoclonus, ataxia, and visual problems. Death usually occurs within 6 months of the disease onset. Electroencephalography (EEG) and cerebrospinal fluid (CSF) 14-3-3 protein analysis are helpful investigations in the clinical diagnosis of sCJD (see below); the detection of PrP^{Sc} in the CSF by the real-time quaking-induced conversion assay (RT-QuIC) has recently proven to be of high specificity and sensitivity as a diagnostic tool.[4] Magnetic resonance imaging (MRI) of the brain in sCJD characteristically shows signal

Table 2 *PRNP* Codon 129 Polymorphisms in Normal Populations and in Prion Diseases

	***PRNP* Codon 129 Polymorphisms (%)**		
	MM	**MM**	**MV**
Normal population[a]	39	11	50
Sporadic CJD[b]	60	20	20
Variant CJD[c]	99		1

[a]Data pooled from five studies (http://www.cjd.ed.ac.uk/sites/default/files/report24.pdf).
[b]UK deaths from sporadic CJD 1996–2014 for whom genetic data are available ($n=856$) (http://www.cjd.ed.ac.uk/sites/default/files/report24.pdf).
[c]National CJD Research & Surveillance Unit 2017 (http://www.cjd.ed.ac.uk/sites/default/files/worldfigs_0.pdf).
CJD: Creutzfeldt–Jakob disease; M: methionine; V: valine.

hyperintensities in the cortical ribbon and in the basal ganglia on diffusion-weighted imaging.[2]

However, sCJD is heterogeneous in terms of clinicopathologic phenotype; individual cases broadly conform to a range of subtypes which are classified according to the *PRNP* codon 129 polymorphism (Table 2) and the isoform (type) of the protease-resistant prion protein (PrP^{res}) in the brain, and are associated with characteristic neuropathologic features (see below).[3]

2.2 Variably Protease-Sensitive Prionopathy

Variably protease-sensitive prionopathy (VPSPr) is the most recently identified human prion disease.[5] VPSPr appears to be a sporadic disease of unknown etiology. Patients do not have any risk factors for acquired human prion diseases, nor have there been any *PRNP* mutations associated with VPSPr. Clinical diagnostic criteria for VPSPr have not yet been defined; patients are generally elderly and present with a range of neurological abnormalities including movement disorders, ataxia, and cognitive decline. The duration of the clinical illness is longer than for sCJD. A diagnosis of VPSPr is difficult to make in life and many of the reported cases have been diagnosed after autopsy based on the neuropathological findings and brain PrP biochemistry (see below). Retrospective studies have suggested that VPSPr is a rare disorder, but its rather poorly defined clinical features suggest that it may be underascertained and hence no accurate figures for its incidence and prevalence are available.[6]

2.3 Genetic Prion Diseases

Genetic prion diseases are inherited diseases that occur as autosomal dominant disorders with high penetrance. Genetic prion diseases are estimated to represent approximately 10%–15% of all forms of human prion disease.[2] Not all cases have a family history of prion disease, but some patients give a family history of other neurological disorders, possibly as a result of misdiagnosis of prion diseases in the past. Cases without a family history may occur as a consequence of spontaneous mutation. Over 40 different pathogenic *PRNP* mutations (missense, insertion, deletion, and amber mutations) have been identified to date.[2] The age at onset of clinical symptoms is remarkably variable, even within affected families.

A wide range of clinicopathologic phenotypes have been reported, including genetic CJD, GSS, and FFI. The *PRNP* polymorphisms at codons 129 and 219 on both the mutant and nonmutated alleles have a substantial influence on the clinicopathologic phenotype.

- Genetic CJD closely resembles sCJD, in its clinical and pathologic features; *PRNP* sequencing in cases of apparent sCJD with no family history may reveal occasional cases with a *PRNP* mutation association with genetic CJD, the commonest of which is the E200K mutation.
- GSS presents clinically as a progressive spinocerebellar syndrome with variable neurologic signs and symptoms including pyramidal signs and progressive cognitive decline, but dementia are not always present. Some affected families have previously been given a diagnosis of other forms of hereditary ataxia, such as spinocerebellar ataxia. The P102L mutation was the first mutation to be associated with GSS; an increasing number of point mutations also associated with GSS have subsequently been identified.[2]
- FFI is characterized clinically by sleep disturbance and dysautonomia, with other variable neurologic signs and symptoms including myoclonus, cognitive abnormalities, and pyramidal signs. FFI was recognized clinically for many years before the underlying *PRNP* abnormality was identified: the *PRNP* D178N–129M haplotype.[2]
- The insertional, deletion, and amber *PRNP* mutations are associated with variable clinical phenotypes, some of which resemble sCJD while others resemble GSS; in the past, patients with these mutations have been misclassified clinically as Alzheimer's disease or Huntingdon's disease.
- A novel clinicopathologic phenotype of genetic human prion disease has recently been reported, characterized by diarrhea, autonomic neuropathy,

and a systemic PrP amyloidosis. A previously unreported *PRNP* Y163X truncation mutation was found in affected family members.[7]

2.4 Iatrogenic CJD

Iatrogenic CJD (iCJD) was first reported in 1974 in a patient who had received a corneal transplant from a donor with sCJD. Over 450 cases of iCJD have subsequently been reported, with occasional cases representing transmission by contaminated neurosurgical instruments and intracerebral electrodes. However, the majority of cases have occurred in recipients of human dura mater grafts human or pituitary growth hormone. Incubation periods vary from around 2–4 years in cases associated with neurosurgery or intracerebral electrodes to over 3 decades in some human pituitary hormone recipients. Iatrogenic CJD has a variable clinicopathologic phenotype; some patients have clinical features similar to those in sCJD, whereas progressive cerebellar signs and symptoms with other focal neurologic abnormalities and late dementia are common in human pituitary hormone recipients who develop iCJD.[2]

2.5 Variant CJD

Variant CJD (vCJD) was first reported in 1996 as a novel prion disease that was linked to dietary exposure to the bovine spongiform encephalopathy (BSE) agent the United Kingdom. Over 230 cases of vCJD have since been identified worldwide, with 178 cases occurring in the United Kingdom.[8] Three cases of vCJD have been transmitted by blood transfusion in the United Kingdom. Although the number of new vCJD cases worldwide is declining, the number of future vCJD cases is uncertain; evidence of asymptomatic vCJD infection has been found in around 1 in 2000 of UK appendix specimens tested.[9]

Variant CJD has a younger age of onset than sCJD, with most cases occurring in the third decade of life. In contrast, the clinical illness is longer (mean 14 months) and usually begins with psychiatric and/or sensory manifestations followed by ataxia, pyramidal and extrapyramidal signs and cognitive impairment. Cranial MRI shows hyperintensity in the posterior thalamus (the "pulvinar sign") on T2-weighted and fluid-attenuated inversion recovery (FLAIR) images.[2] All confirmed cases of vCJD worldwide were *PRNP* codon 129 methionine homozygotes, but in 2016 a case of definite vCJD with a *PRNP* codon 129 heterozygote (methionine/valine) genotype was identified,[10] raising the possibility of further similar cases in the future (Table 2).

2.6 Kuru

Kuru occurred as an epidemic among the Fore tribe of Papua New Guinea in the 1950s as a neurologic disorder associated with ritualistic cannibalism.[2] Kuru was responsible for many deaths in the women and children of the tribe. The incidence of kuru declined after this ritual was discouraged. The final patients with kuru died in the early 21st century, with incubation periods extending over 4 decades. Kuru is now extinct and will therefore not be considered at any greater length here.

2.7 Diagnostic Criteria for Human Prion Diseases

Human prion diseases benefit from internationally agreed diagnostic criteria that refer to clinical features, duration of illness, results of clinical investigations, *PRNP* analysis, and neuropathology to allow suspected cases to be classified as "possible," "probable," or "definite."[11] A "definite" diagnosis requires neuropathological confirmation.

Useful clinical investigations include:

1. EEG: periodic triphasic complexes in 2/3 of sCJD cases.
2. MRI of the brain: increased signal in T2 sequences in the basal ganglia and cortex in sCJD; increased signal in T2/FLAIR sequences in the posterior thalamus and midbrain in vCJD (the "pulvinar sign").
3. CSF protein analysis: elevated levels of 14.3.3 and S100 proteins in sCJD; normal 14.3.3 levels with elevated phospho-tau levels in vCJD.[2]
4. RT-QuIC positivity for PrP^{Sc} in CSF or other tissues.[4,12]

   For more detailed information, please see: "Diagnostic criteria for human prion disease."[11]

3. THE AUTOPSY IN HUMAN PRION DISEASES

3.1 When to Suspect a Prion Disease Prior to Autopsy

The differential clinical diagnosis of a progressive neurodegenerative disorder not infrequently includes sCJD and other human prion diseases. The likelihood of a prion disease under these circumstances can be assessed by reference to the diagnostic criteria for human prion diseases.[11] However, sCJD is not a uniform disorder and not all cases show the typical changes described in the investigations above; furthermore, some patients may have a prolonged duration of illness, while others have atypical presentations, such as sudden stroke-like features.

Genetic prion diseases usually, but not always, have a family history of a neurodegenerative disorder. Although this may have been identified as genetic CJD, GSS of FFI, the disease occurring in earlier affected family members may have been misdiagnosed, particularly if *PRNP* sequencing had not been performed. Phenotypic variability also occurs in genetic prion diseases, even within a single affected family.

The clinical history should also be reviewed for other potential risk factors for a human prion disease, such as a previous history of treatment with human growth hormone, or previous neurosurgery that includes a dura mater graft. Not all the relevant clinical details for these and other risk factors may be available before an autopsy is to be performed, in which case a precautious approach (see below) is required in view of the uncertainty concerning potential exposure to human prions.

3.2 Autopsy Considerations

Special equipment is required when performing an autopsy on a case of suspected human prion disease, preferably including a dedicated "high risk" autopsy room with trained mortuary staff, personal protective equipment, and single use-disposable instruments. If a dedicated autopsy room is not available, the autopsy may proceed in the mortuary as long as no other autopsies are being performed at the same time and care is taken to minimize any potential contamination of the mortuary environment.

The autopsy should be performed according to WHO guidelines for personal protection and the cleaning and decontamination of reusable instruments and the mortuary environment[13]; if potential CJD autopsies are performed frequently, consideration should be given for establishing and using a dedicated set of instruments that are not used in general autopsies. In addition to retaining fixed and frozen brain tissue in accordance with the consent given for the autopsy (see below), sampling of other tissues can also be of diagnostic value, particularly the lymphoreticular system (spleen, tonsil, and lymph nodes) in suspected cases of vCJD.

4. NEUROPATHOLOGY

4.1 Macroscopic Pathology and Tissue Sampling

Macroscopic examination of the brain in cases of prion disease shows no specific abnormalities, with age-related changes present. Cerebral and cerebellar atrophy may be present in cases with a long duration of illness, when the term "panencephalopathic CJD" may be appropriate. Interestingly, the hippocampi in such cases usually do not appear to be proportionally atrophied. Cerebellar atrophy may present in some cases of sCJD (particularly the VV2 subtype) and in other groups of human prion disease with significant cerebellar involvement, including GSS, vCJD, and iCJD in human growth hormone recipients.

Extensive sampling of the brain for microscopy is recommended in cases of suspected prion disease, as the nature and distribution of the lesions can be variable in each disease entity, and between cases of the same disease. Blocks should be taken from all regions of the cerebral cortex, the hippocampus, basal ganglia, thalamus, cerebellum, and brain stem. Formic acid pretreatment of fixed tissue blocks prior to processing and embedding in paraffin wax is advised to significantly reduce potential infectivity, including blocks of any non-CNS tissues sampled.[13]

4.2 Brain Biopsies

Brain biopsies in patients with suspected prion disease are not performed as a routine investigation, since the neurosurgical instruments that make contact with brain tissue have to be destroyed after use if the diagnosis is confirmed, to eliminate the possibility of subsequent iatrogenic transmission following reuse.[13] Routine hospital processing and decontamination of instruments do not remove prion infectivity, which binds avidly to steel surfaces. However, occasionally a brain biopsy may be required to exclude the possibility of an alternative treatable condition (such as cerebral vasculitis) in the differential diagnosis. In this situation, it is important to obtain both fixed and frozen tissue from the biopsy specimen to allow both neuropathological and biochemical investigations to aid diagnosis; the fixed tissue can be pretreated with formic acid prior to embedding to reduce prion infectivity.[13]

4.3 Microscopy

4.3.1 Microscopic Pathology

The characteristic neuropathological features of human prion diseases comprise spongiform change in affected gray matter regions, with reactive astrocytosis and microglial proliferation, accumulation of PrP^{Sc} in a variety of patterns, and amyloid plaque formation in some cases.[2] These features vary in their nature, distribution, and severity between different disease entities and are also highly variable within a single brain, reinforcing the need for adequate sampling of the brain for histological diagnosis.

4.3.2 Spongiform Change

Spongiform change is characterized by numerous rounded vacuoles in the gray matter that can vary from 2 to 20 μm in diameter. These occur in variable numbers and can range from small discrete vacuoles (microvacuolar spongiform change) to large coalescent vacuoles (confluent spongiform change) that give a sponge-like appearance (Fig. 1). It is important not to confuse tissue fixation and processing artifacts such as retraction artifact with

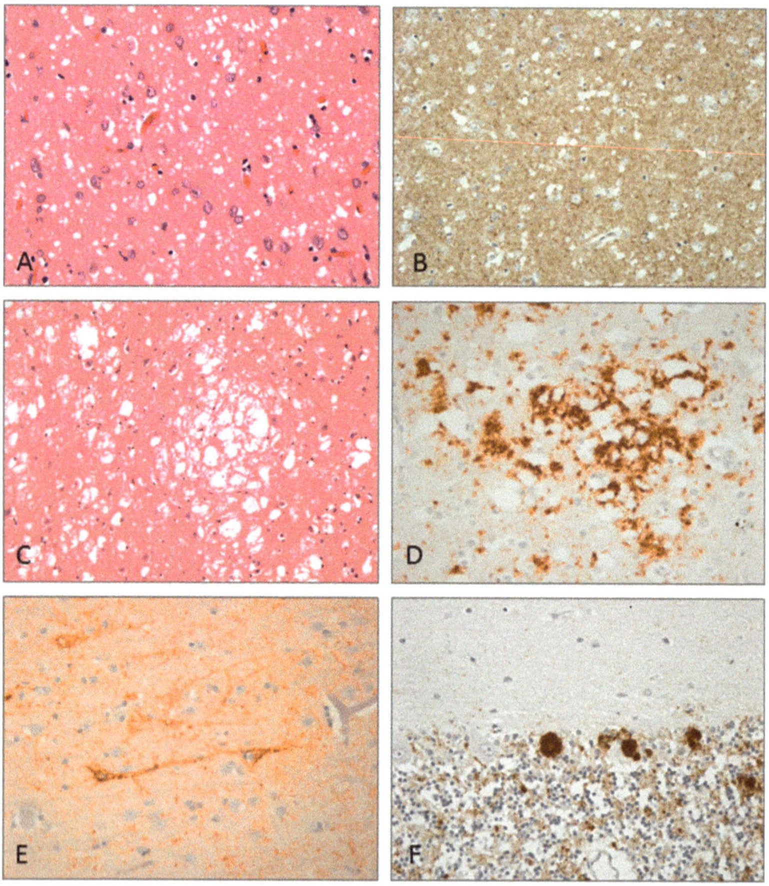

Fig. 1 (A) Microvacuolar spongiform change in the cerebral cortex in the MM/MV1 subtype of sCJD (hematoxylin and eosin). (B) Granular/synaptic PrP accumulation in the MM/MV1 subtype of sCJD (12F10 antibody). (C) Confluent spongiform change in the MM2 (cortical) subtype of sCJD (hematoxylin and eosin). (D) Perivacuolar accumulation of abnormal PrP around a region of confluent spongiform change (12F10 antibody). (E) Perineuronal accumulation of abnormal PrP in the VV2 subtype of sCJD (12F10 antibody). (F) Intense labeling of a kuru-type plaque in the cerebellum in the MV2 subtype of sCJD (12F10 antibody).

spongiform change. The rounded vacuoles of spongiform change should also be distinguished from gray matter vacuolation in a range of unrelated conditions, including cerebral edema and metabolic or toxic encephalopathies. In these conditions, vacuolation is usually also present in the white matter, unlike in human prion diseases. The vacuolation in metabolic encephalopathies and toxic disorders is usually accompanied by other histological abnormalities that allow a distinction from human prion diseases. However, other neurodegenerative disorders can exhibit focal gray matter vacuolation that is indistinguishable from spongiform change, particularly in the temporal cortex in cases of Lewy body dementia. Immunohistochemistry for disease-associated prion protein in such cases (and in the other nonprion disorders mentioned above) is negative, and a full histological assessment will allow an appropriate diagnosis to be made.

Status spongiosis is the neuropathological term used when severe cerebral cortical gray matter vacuolation is accompanied by extensive neuronal loss, gliosis, and collapse of the cortical architecture. Status spongiosis is characteristically present in panencephalopathic sCJD, but it is a nonspecific finding that can occur in other disorders associated with severe cortical neuronal loss, including frontotemporal lobar degeneration and severe hypoxic brain damage.

4.3.3 Amyloid Plaques

A variety of amyloid plaques with different morphologies has been identified in some sporadic, genetic, and acquired human prion diseases (Figs. 1 and 2). These plaques are composed of PrP^{Sc} arranged in dense β-sheet aggregates, which differ in composition and morphology from the amyloid plaques in other neurodegenerative disorders such as Alzheimer's disease and are most frequently present in the cerebellar cortex. Tinctorial stains for amyloid plaques such as the Congo red stain can be used to demonstrate these plaques, but they are most easily visualized using immunohistochemistry for disease-associated prion protein. A range of distinct plaque morphologies have been described:

- Kuru-type plaques, composed of a solid core and radiating fibrils, were first described in kuru, but are also present as a diagnostic feature in the MV2 subtype of sCJD,[3] and in some iCJD cases, including dura mater graft and human pituitary growth hormone recipients.[2] They occur most frequently in the cerebellar cortex (granular layer and molecular layer), but can occasionally be found in other gray matter regions, including the cerebral cortex and basal ganglia.

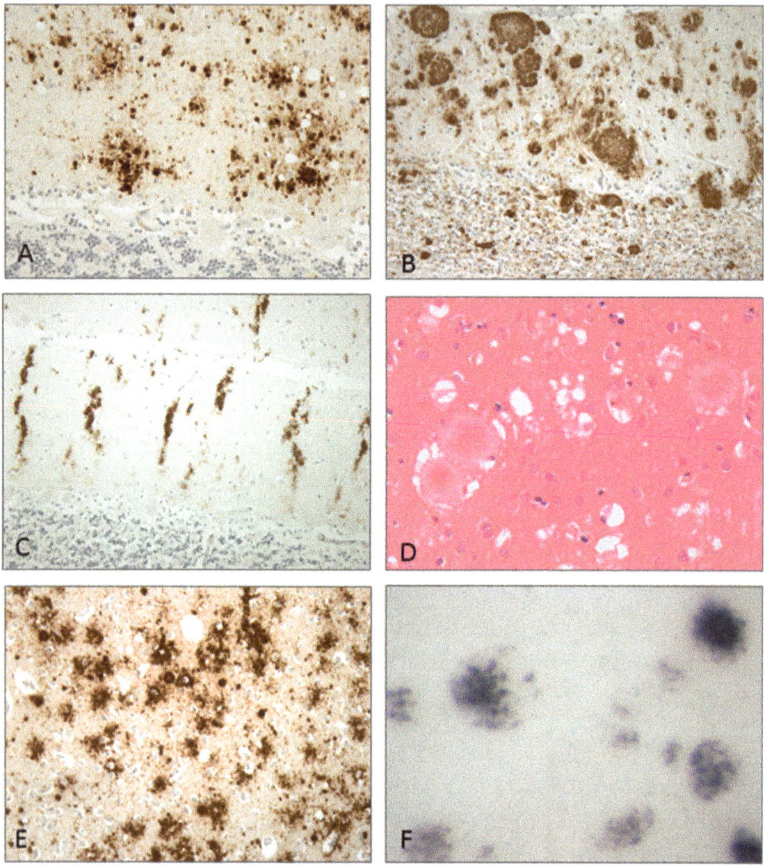

Fig. 2 (A) Numerous microplaques in the cerebellar cortex in VPSPr (*PRNP* codon 129 VV) (12F10 antibody). (B). Multicentric amyloid plaques in the cerebellar cortex in GSS (*PRNP* codon 129 P102L) (12F10 antibody). (C) Unusual linear deposits of abnormal PrP in the cerebellar flat cortex in genetic CJD with four octapeptide repeat inserts in the *PRNP* (12F10 antibody). (D). Florid plaques in the cerebral cortex in vCJD surrounded by a "halo" of spongiform change (hematoxylin and eosin). (E) PrP immunohistochemistry shows widespread accumulation of PrP^{Sc} in florid plaques and in multiple smaller deposits in vCJD (12F10 antibody). (F) A paraffin-embedded tissue blot demonstrates protease-resistant prion protein accumulation in the germinal centers of the spleen in vCJD (3F4 antibody).

- Microplaques, which are a diagnostic feature of VPSPr and occur most frequently in the molecular layer of the cerebellar cortex.[5] These small lesions may also be present in other gray matter regions, including the cerebral cortex, basal ganglia, and thalamus.

- Multicentric plaques, which are large irregular lesions with multiple cores that give a multicentric appearance on both tinctorial stains and immunohistochemistry for PrP. They are diagnostic of GSS and can be found most frequently in the cerebellar cortex, also occasionally in the cerebral cortex and basal ganglia. Multicentric plaques are associated with several different *PRNP* mutations, the most common of which is the P102L mutation.[2]
- Florid plaques, which are composed of a central eosinophilic amyloid core with radiating peripheral fibrils surrounded by a corona of vacuolation. They are a diagnostic feature of vCJD in both *PRNP* codon 129 MM and MV genotypes and are present in large numbers in the cerebral cortex and cerebellar cortex.[2]

Immunohistochemistry for prion protein is essential for the neuropathological diagnosis of human prion diseases and is described in further detail below.

4.3.4 Sporadic CJD

Sporadic CJD has a variable clinical and pathological phenotype, which has been found to broadly correspond to a subclassification based on the results of *PRNP* sequencing and Western blot analysis for PrP^{res}.[3] The combinations of the *PRNP* codon 129 polymorphism (Table 2) and the PrP^{res} isoform (Fig. 3) allows for six combinations (MM1, MV1, MM2, MV2, VV1, and VV2) that appear to be associated with certain neuropathologic features, often accompanied by suggestive clinical findings[3] (Fig. 1). There is some overlap of the features between these groups; for example, the MM1 and MV1 subgroups share common clinical and neuropathologic features and are often combined into a single subgroup, MM/MV1. In contrast, the MM2 subgroup is heterogeneous in terms of both clinical and pathological features and is usually subdivided into the MM2 cortical (MM2C) and MM2 thalamic (MM2T) subgroups. The MM2T subgroup corresponds to a phenotype that is closely resembles the genetic prion disease FFI and is now often referred to as sporadic fatal insomnia.[2,3]

The diagnostic neuropathologic features in the sCJD subgroups defined in this classification are summarized in Table 3. However, not all cases of sCJD fit easily into this classification, since some cases have atypical clinical and pathologic features. Also, Western blot analysis of the brain can show combinations of types 1A and 2A PrP^{res} either in different areas of the same

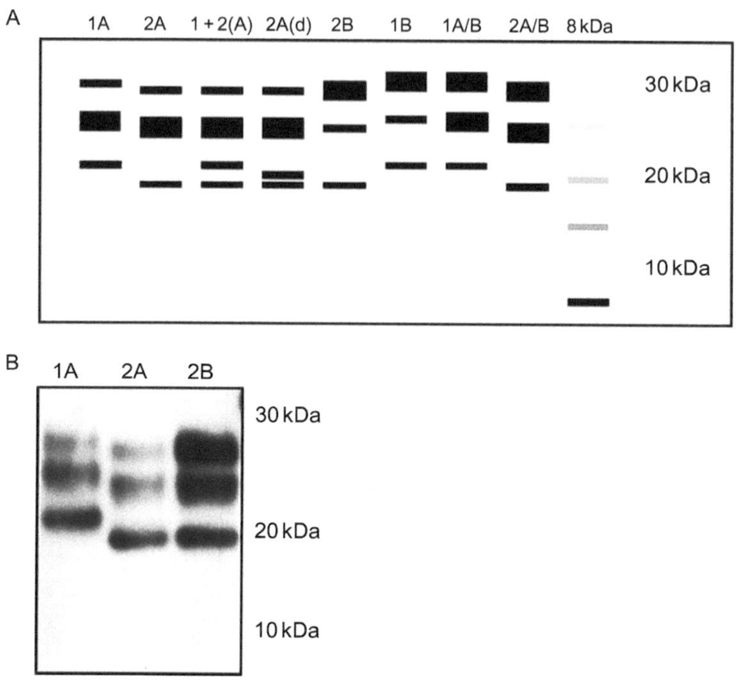

Fig. 3 (A) A diagrammatic representation of the main protease-resistant prion protein (PrP^{res}) types found in the human prion disease brain as determined by proteinase K digestion and Western blot analysis. The classification of the banding pattern has two components, one numerical depending on the migration of the bands and the other alphabetical depending on their relative abundance. The pattern is termed type 1 if the nonglycosylated (*bottom*) band is ~21 kDa, type 2 if the nonglycosylated band is ~19 kDa or type 1+2 if both bands are found. Type 2 can occur as a doublet and is termed type 2(d). The pattern is given the suffix A if the *middle* or *bottom* (mono- or nonglycosylated) bands predominate, B if the top (diglycosylated) band predominates or A/B if the glycosylated bands (*middle* and *top*) predominate at the expense of the nonglycosylated (*bottom*) band. A pattern dominated by a low molecular mass unglycosylated band is termed 8 kDa. The faint ladder of bands that sometimes accompanies the 8 kDa band is shown in *gray*. Types 1A, 2A, 2A(d), and 1+2(A) are characteristic of sporadic and iatrogenic CJD. Type 2B is associated with variant CJD. Types 1B, 1A/B, and 2A/B are often found in genetic CJD, GSS, and FFI, respectively. The 8 kDa pattern is characteristic of some cases of GSS and of VPSPr. (B) shows a representative Western blot image of sCJD MM1 subtype (1A), sCJD VV2 subtype (2A), and vCJD (2B).

brain, or even in the same brain region in sCJD (Fig. 3). Although an updated classification system taking some of these findings into account has been proposed by Parchi *et al.*[14] this has not yet been universally adopted.

Table 3 Sporadic CJD Subtypes Classified by *PRNP* Codon 129 Genotype and PrP^{res} Type[a]

Subtype	Median Age at Onset (Years)	Median Duration of Illness (Months)	Frequency (%)	Neuropathologic Diagnostic Features	PrP Deposition Patterns
MM1 MV1	66 73	3 5	57 6	Widespread microvacuolar spongiform change in cerebral cortex and cerebellum	Granular, synaptic
MM2 cortical	52	17	7	Confluent spongiform change in cerebral cortex	Perivacuolar
MM2 thalamic (sporadic fatal insomnia)	53[b]	16[b]	<1	Thalamic neuronal loss and gliosis in anterior nuclei. Limited spongiform change	Granular, synaptic
VV1	53	10	2	Microvacuolar spongiform change: Cerebral cortex (temporal lobe)	Synaptic
VV2	66	6	14	Linear microvacuolar/confluent spongiform change in cerebral cortex layer 3; severe cerebellar neuronal loss and gliosis	Perineuronal, synaptic, plaque-like
MV2	65	11	14	Kuru-type plaques in cerebellar cortex Widespread microvacuolar ± confluent spongiform change	Kuru plaques, synaptic, plaque-like

[a]Cases with mixed PrP^{Sc} subtypes are not included.
[b]Data represent a single case.
Data from the UK National Creutzfeldt–Jakob Disease Research & Surveillance Unit (www.cjd.ed.ac.uk/).

4.3.5 VPSPr

VPSPr is the most recently identified human prion disease, which (as its name suggests) is defined not by its clinical or pathologic features, but by the findings on Western blot examination for PrP^{res} in the brain[5] (Fig. 3). In contrast to sCJD, VPSPr occurs most commonly in the *PRNP* codon 129VV genotype, with the codon 129 MM genotype being the least frequent. VPSPr is characterized neuropathologically by the presence of spongiform change with intermediately sized vacuoles and amyloid microplaques, which occur most commonly in the cerebellum[6] (Fig. 2). The neuropathology of VPSPr is influenced by the *PRNP* codon 129 genotype, with the MM genotype usually containing more plaques in the cerebrum.[6] The PrP^{Sc} in the brain in VPSPr is relatively poorly resistant to proteinase K digestion, resulting in the appearance of characteristic, N- and C-terminally truncated ~8 kDa PrP^{res} bands in Western blots, sometimes accompanied by a faint ladder of bands extending into the 18–30 kDa range[5] (Fig. 3).

4.3.6 Genetic CJD

Microscopic examination of the brain shows features similar to sCJD, usually corresponding to the MM/MV1 subtype, with microvacuolar spongiform change and granular/synaptic accumulation of disease-associated PrP on immunohistochemistry. The microscopic features can vary depending on the associated *PRNP* mutation.[2] The commonest *PRNP* mutation to be associated with genetic CJD is the E200K mutation, which has a worldwide distribution. The mutant allele is usually *PRNP* codon 129-M. Western blot analysis of brain PrP^{res} characteristically shows type 1B or 1A/B type (Fig. 3).

4.3.7 GSS

Multicentric PrP amyloid plaques in the cerebellar cortex are the diagnostic neuropathological feature of GSS (Fig. 2). Spongiform change is often absent in the cerebellum and is variable in its distribution in other brain regions. Some of the associated *PRNP* mutations exhibit particular neuropathologic features, such as neurofibrillary tangles in the cerebral cortex in cases with the *PRNP* F198S mutation. Western blot analysis of the brain PrP^{res} in GSS demonstrates a range of findings, including types 1B, 1A/B, or the 8 kDa pattern, the latter usually found in brain regions containing amyloid plaques (Fig. 3).

4.3.8 FFI

FFI is characterized by severe thalamic gliosis and neuronal loss in the anterior thalamic nuclei and occasionally in other thalamic and hypothalamic nuclei. Spongiform change and disease-associated PrP accumulation in the thalamus are difficult to detect, but may be present in the entorhinal cortex or cerebellum. FFI is associated with the *PRNP* D178N-129M haplotype. The *PRNP* D178N mutation is also found in some patients with genetic CJD, but in a D178N–129V haplotype.[2] Western blot analysis in FFI shows a brain PrP^{res} type 2A/B (Fig. 3).

4.3.9 Insertional, Deletion, and Amber PRNP Mutations

This group of genetic prion disease exhibits a range of neuropathologic features that vary according to the associated genetic abnormality. Some of the smaller insertional mutations have features resembling sCJD, but in cases with four octapeptide repeat inserts or more PrP immunohistochemistry can demonstrate characteristic linear deposits in the cerebellar molecular layer (Fig. 2). In some of the rare amber mutations disease-associated PrP accumulates in the walls of intracerebral blood vessels, resulting in a PrP amyloid angiopathy.[2] Western blot analysis of brain PrP^{res} in patients with insertional mutations usually shows PrP^{res} types resembling sCJD type 1A or 2A (Fig. 3).

4.3.10 Variant CJD

Large numbers of florid plaques within the cerebral cortex and the cerebellum are a diagnostic neuropathologic feature of vCJD (Fig. 2). Other characteristic features include extensive disease-associated PrP accumulation as small cluster plaques and diffuse deposits in the brain on immunohistochemistry or paraffin-embedded tissue (PET) blot analysis (Fig. 2). Spongiform change is widespread, particularly in the caudate nucleus and putamen. The thalamus shows severe neuronal loss and gliosis in the pulvinar, corresponding to the characteristic MRI abnormalities. Western blot analysis shows a uniform brain PrP^{res} type 2B (Fig. 3).

Disease-associated PrP is also present outside the CNS in follicular dendritic cells within germinal centers in lymphoreticular tissues including the tonsil, lymph nodes, and spleen[15] (Fig. 2). Western blot analysis of these tissues also demonstrates PrP^{res} type 2B (Fig. 3). Tonsil biopsy has been employed as a diagnostic test for some patients with suspected vCJD, using both immunohistochemistry for disease-associated PrP and Western blot analysis for PrP^{res}.

4.3.11 Iatrogenic CJD

The neuropathology of iCJD in general resembles that of sCJD, but can vary depending on the source of infection and the patient's *PRNP* codon 129 subtype.[2] Kuru-type plaques can occur in the cerebellum in some human growth hormone and dura mater graft recipients, not only in patients with a *PRNP* codon 129 MV genotype, but also in patients with a *PRNP* codon 129 MM genotype. This finding may be influenced by the strain of the transmissible prion agent involved. Western blot analysis shows a range of brain PrP^{res} isoforms resembling those found in sCJD, but *PRNP* codon 129 MM iCJD cases with kuru-type plaques in the cerebellum may contain a PrP^{res} type that is intermediate in size between type 1 and type 2, either on its own or as a doublet (Fig. 3).

Recent studies of iatrogenic CJD in human growth hormone and human dura mater graft recipients have found evidence of increased Aβ deposition in the brain, both as diffuse and neuritic plaques and as amyloid angiopathy.[16] Since low levels of Aβ can be present in both the pituitary gland and dura mater, these findings raise the possibility that iatrogenic amyloid seeding of Aβ has occurred in the brains of these patients by a prion-like mechanism.

5. METHODS FOR THE DETECTION OF PrP^{Sc} AND PrP^{res} IN TISSUES

5.1 Immunohistochemistry for PrP

Immunohistochemistry for PrP detection is essential in the neuropathologic diagnosis of human prion diseases.[2] Disease-associated prion protein can accumulate in the brain in a variety of patterns, ranging from widespread synaptic and perineuronal accumulations to perivacuolar, granular, and plaque-like accumulation. The recognition of these different patterns of accumulation is required both for diagnosis and for the identification of the different subtypes of sCJD[3] (Table 3; Fig. 1).

Most of the available antibodies to PrP do not distinguish between the normal cellular form of the protein, PrP^C, and the disease-associated isoform, PrP^{Sc}, so it is necessary to introduce pretreatment steps to optimize the labeling of the disease-associated isoform and to minimize the labeling of the normal isoform.[17] The pretreatment protocols generally include steps to denature PrP^C in the tissue sections before the antibodies are applied, for example by autoclaving the sections, or the use of proteinase K to digest PrP^C in the tissue sections.[17] The available anti-PrP antibodies recognize a range of different epitopes on the protein. Consequently, the use of these different antibodies may produce differential labeling in tissue sections, as observed recently in VPSPr, where the use of four different antibodies resulted in differential labeling of the microplaques in the brain.[6] Immunohistochemistry can also be used to detect disease-associated prion protein in non-CNS tissues, particularly in the germinal centers of lymphoid tissues and autonomic ganglia in vCJD[15] (Fig. 2).

5.2 Paraffin-Embedded Tissue Blot for PrP^{res}

In order to improve the sensitivity and specificity in the detection of protease-resistant PrP (PrP^{res}) in tissue sections, a method known as the paraffin-embedded tissue (PET) blot has been modified for use on human tissues[18] (Fig. 2). This method involves cutting and deparaffinizing the section, mounting it on a nitrocellulose membrane and performing a longer digestion with proteinase K than is possible on glass-mounted slides, which helps denature PrP^C more completely before immunostaining.[18] The PET blot sections are visualized on a dissecting microscope, since the membrane is not fully translucent. This method has proven to be very helpful in cases where the interpretation of PrP immunohistochemistry has been problematic, such as in small brain biopsy samples, or where only very limited amounts of disease-associated prion protein are present, in a tissue section for example in non-CNS tissues in vCJD[18] (Fig. 2).

5.3 Western Blot for PrP^{res}

The identification and classification of PrP^{Sc} in the brain by Western blot analysis is essential for the diagnosis and classification of human prion diseases.[2] This makes it essential to store at least one frozen tissue sample from each biopsy or autopsy cases of suspected prion disease; the frozen tissue can also be used for DNA extraction and *PRNP* sequencing or codon 129 analysis if required.

Western blotting is used to detect PrP^{Sc} in CNS and non-CNS tissues, and to characterize or type the protease-resistant fragments that are generated on the basis of their molecular weight and glycosylation status (Fig. 3). Several different Western blot protocols and nomenclatures for typing have been published, but the Parchi et al. classification is the widest use.[3] In addition to being diagnostically informative, PrP^{res} typing can also be used to explore the relationships between PrP^{Sc} biochemistry, neuropathology, and the clinical disease phenotype in human prion diseases (Fig. 3).

5.4 Recent Laboratory Techniques for PrP^{Sc} Detection in Human Tissues

A range of other techniques have been developed recently for the detection of PrP^{Sc} in human prion disease tissues. While all are used as research tools, some are now being used diagnostically, representing good examples of translational research.

- Real-time quaking-induced conversion (RT-QuIC). This in vitro PrP^{Sc} amplification method is now in use in a highly sensitive and specific CSF test for sCJD.[4] It has also been used to detect PrP^{Sc} in nasal brushings containing olfactory epithelium from patients with sCJD.[12]
- Protein misfolding cyclical amplification. This in vitro amplification technique[19] can specifically detect PrP^{Sc} in CNS and non-CNS tissues in vCJD. More recently, it has been used to detect PrP^{Sc} in the urine of patients with vCJD (but not in sCJD patients)[20] and in the blood in vCJD,[21] raising the possibility of its use in a blood-based assay for vCJD.

6. CONCLUSIONS

The diagnosis of human prion diseases in a Neuropathology Laboratory now requires histological, immunohistochemical, biochemical, and genetic investigations. These established methods are being supplemented by a range of recent techniques that enhance the sensitivity and specificity of PrP^{Sc} detection. However, the established neuropathological techniques are still required for the full characterization of human prion diseases, the recognition of disease subtypes, and the identification of novel prion diseases in the 21st century.

ACKNOWLEDGMENTS

We thank Dr. Mark Head and Ms. Helen Yull for the images and legends for Fig. 3, and Ms. Jan MacKenzie for data on UK human pion diseases. The National CJD Research & Surveillance Unit is funded by the Department of Health and the Scottish Government (PR-ST-0614-00008). The views expressed in this publication are those of the authors and not necessarily those of the Department of Health or the Scottish Government. The Edinburgh Brain Bank is supported by the Medical Research Council (MRC G0900580).

REFERENCES

1. Prusiner SB. Novel proteinaceous infectious particles cause scrapie. *Science.* 1982;216:136–144.
2. Head MW, Ironside JW, Ghetti B, Jeffrey M, Piccardo P, Will RG. Prion diseases. In: Love S, Budka H, Ironside JW, Perry A, eds. *Greenfield's Neuropathology.* 9th ed. Boca Raton, FL: CRC Press; 2015:1016–1086.

3. Parchi P, Giese A, Capellari S, et al. Classification of sporadic Creutzfeldt–Jakob disease based on molecular and phenotypic analysis of 300 subjects. *Ann Neurol.* 1999;46:224–233.
4. McGuire LI, Poleggi A, Poggolini I, et al. Cerebrospinal fluid real-time quaking-induced conversion is a robust and reliable test for sporadic Creutzfeldt–Jakob disease: an international study. *Ann Neurol.* 2016;80:160–165.
5. Gambetti P, Dong Z, Yuan J, et al. A novel human disease with abnormal prion protein sensitive to protease. *Ann Neurol.* 2008;63:697–708.
6. Head MW, Yull HM, Ritchie DL, et al. Variably protease-sensitive prionopathy in the UK: a retrospective review 1991–2008. *Brain.* 2013;136:1102–1115.
7. Mead S, Gandhi S, Beck J, et al. A novel prion disease associated with diarrhea and autonomic neuropathy. *N Engl J Med.* 2013;4:1904–1914.
8. National Creutzfeldt–Jakob Disease Research & Surveillance Unit. Variant CJD Cases Worldwide. http://www.cjd.ed.ac.uk/sites/default/files/worldfigs.pdf. Accessed 17.01.17 2017.
9. Gill ON, Spencer Y, Richard-Loendt A, et al. Prevalent abnormal prion protein in human appendixes after bovine spongiform encephalopathy epizootic: large scale survey. *BMJ.* 2013;347:f5675. http://dx.doi.org/10.1136/bmj.f5675.
10. Mok T, Jaunmuktane Z, Joiner S, et al. Variant Creutzfeldt-Jakob disease in a patient with heterozygosity at PRNP codon 129. *N Engl J Med.* 2017;376:292–294.
11. National Creutzfeldt–Jakob Disease Research & Surveillance Unit. Diagnostic Criteria for Human Prion Disease. http://www.cjd.ed.ac.uk/sites/default/files/criteria_0.pdf. Accessed 17.01.17 2017.
12. Zanusso G, Monaco S, Pocchiari M, Caughey B. Advanced tests for early and accurate diagnosis of Creutzfeldt–Jakob disease. *Nat Rev Neurol.* 2016;12:325–333.
13. World Health Organisation. WHO Infection Control Guidelines for Transmissible Spongiform Encephalopathies. WHO/CDS/CSR/APH/200.3 http://www.who.int/csr/resources/publications/bse/whocdscsraph2003.pdf. Accessed 17.01.17 2017.
14. Parchi P, Strammiello R, Notari S, et al. Incidence and spectrum of sporadic Creutzfeldt–Jakob disease variants with mixed phenotype and co-occurrence of PrP^{Sc} types: an updated classification. *Acta Neuropathol.* 2009;118:659–671.
15. Head MW, Ritchie D, Smith N, et al. Peripheral tissue involvement in sporadic, iatrogenic and variant Creutzfeldt–Jakob disease: an immunohistochemical, quantitative and biochemical study. *Am J Pathol.* 2004;164:143–153.
16. Ritchie DL, Adlard P, Peden AH, et al. Amyloid-ß accumulation in the CNS in human growth hormone recipients in the UK. *Acta Neuropathol.* 2017;134:221–240.
17. Kovacs GG, Head MW, Hegyi I, et al. Immunohistochemistry for the prion protein: comparison of different monoclonal antibodies in human prion disease subtypes. *Brain Pathol.* 2002;12:1–11.
18. Ritchie DL, Head MW, Ironside JW. Advances in the detection of prion protein in peripheral tissues of variant Creutzfeldt–Jakob disease patients using paraffin-embedded tissue blotting. *Neuropathol Appl Neurobiol.* 2004;30:360–368.
19. Saborio GP, Permanne B, Soto C. Sensitive detection of pathologic prion protein by cyclic amplification of protein misfolding. *Nature.* 2001;411:810–813.
20. Moda F, Gambetti P, Notari S, et al. Prions in the urine of patients with variant Creutzfeldt–Jakob disease. *N Engl J Med.* 2014;371:530–539.
21. Concha-Marambio L, Pritzkow S, Moda F, et al. Detection of prions in blood from patients with variant Creutzfeldt–Jakob disease. *Sci Transl Med.* 2016;8(370):370ra183. http://dx.doi.org/10.1126/scitranslmed.aaf6188.

CHAPTER FIFTEEN

The Structure of the Infectious Prion Protein and Its Propagation

Jesús R. Requena*, Holger Wille†,1

*CIMUS Biomedical Research Institute, University of Santiago de Compostela-IDIS, Santiago de Compostela, Spain

†Centre for Prions and Protein Folding Diseases, University of Alberta, Edmonton, Alberta, Canada

[1]Corresponding author: e-mail address: wille@ualberta.ca

Contents

Abstract

The prion diseases, which include Creutzfeldt–Jakob disease in humans, chronic wasting disease in cervids (i.e., deer, elk, moose, and reindeer), bovine spongiform encephalopathy in cattle, as well as sheep and goat scrapie, are caused by the conversion of the cellular prion protein (PrP^{C}) into a disease-causing conformer (PrP^{Sc}). PrP^{C} is a regular, GPI-anchored protein that is expressed on the cell surface of neurons and many other cell types. The structure of PrP^{C} is well studied, based on analyses of recombinant PrP, which is thought to mimic the structure of native PrP^{C}. The mature protein contains an N-terminal, unfolded domain and a C-terminal, globular domain that consists of three α-helices and only a small, two-stranded β-sheet. In contrast, PrP^{Sc} was found to contain predominantly β-structure and to aggregate into a variety of quaternary structures, such as oligomers, amorphous aggregates, amyloid fibrils, and two-dimensional crystals. The tendency of PrP^{Sc} to aggregate into these diverse forms is also responsible for our incomplete knowledge about its molecular structure. Nevertheless, the repeating nature of the more regular PrP^{Sc} aggregates has provided informative

Progress in Molecular Biology and Translational Science, Volume 150
ISSN 1877-1173
http://dx.doi.org/10.1016/bs.pmbts.2017.06.009

insights into the structure of the infectious conformer, albeit at limited resolution. These data established a four-rung β-solenoid architecture as the main element of its structure. Moreover, the four-rung β-solenoid architecture provides a molecular framework for an autocatalytic propagation mechanism, which could explain the conversion of PrP^C into PrP^{Sc}.

1. INTRODUCTION

PrP^{Sc} was the first prion, i.e., infectious protein, which was identified.[1] It was isolated as the major component of the infectious proteinaceous material present in the brains of Syrian hamsters infected with scrapie.[2] Several other prions have subsequently been identified in yeasts and filamentous fungi[3] and a number of mammalian brain proteins associated with neurodegenerative diseases, such as misfolded Aβ, microtubule-associated protein tau, or α-synuclein, are strong prion candidates.[4,5] However, PrP^{Sc} continues to be the quintessential and at the same time the most enigmatic prion. As discussed with detail in other chapters of this book, PrP^{Sc} is the causal agent of the transmissible spongiform encephalopathies, a group of fatal neurodegenerative diseases that include Creutzfeldt–Jakob disease (CJD) and Kuru, both affecting humans,[6] bovine spongiform encephalopathy (BSE), scrapie, which affects sheep and goats, and chronic wasting disease, which affects cervids.[6,7] In all these cases, PrP^{Sc} has the potential to transmit not only within the brain of an affected individual, but also among individuals of the same or even different species. This transmission can occur by an oral route[6,7] or iatrogenically.[7] Of note, PrP^{Sc} is the only known prion accepted to date to have caused local disease outbreaks (Kuru, variant CJD) and larger epizootics (BSE, CWD).

What is the mechanism underlying PrP^{Sc} infectivity? Classic infectious agents base their reproduction and propagation on duplication of nucleic acids, whose propagation is in turn based on the reproduction of their primary structure. In contrast, propagation of prions involves reproduction of their secondary, tertiary, and quaternary structures, i.e., their conformation.[5,6] PrP^{Sc} coerces PrP^C, a glycosylphosphatidylinositol-anchored (GPI-anchored) membrane protein with the same primary but different secondary, tertiary, and quaternary structures, to adopt the PrP^{Sc} conformation. To understand how this transformation is forced onto PrP^C it is essential to first know the structure of PrP^{Sc}. This chapter presents a comprehensive review of what we currently know about the structure of PrP^{Sc} and how it encodes a possible mechanism of its own replication, and hence, of PrP^{Sc} propagation.

2. INFECTIOUS PRION PROTEIN STRUCTURE

2.1 Spectroscopic Techniques

Fourier-transform infrared (FTIR) spectroscopy and circular dichroism (CD) spectroscopy have demonstrated that PrP^{Sc} and its N-terminally truncated variant, PrP 27–30, have a high β-sheet content.[8–10] For a long time, the FTIR data were interpreted to imply that PrP^{Sc} and PrP 27–30 retained some α-helical structure from the original PrP^{C} fold, possibly the two C-terminal α-helices. However, more recent studies have shown that the ~1660 cm^{-1} band in the FTIR spectra that had been attributed to α-helical conformation, is also present in the spectrum of recombinant PrP amyloid fibrils[11] known to have a parallel in-register β-structure and to be completely devoid of α-helical structure.[12] Consequently, the ~1660 cm^{-1} FTIR band has been reassigned to turns and coils.[11,13] Thus, the current consensus is that PrP 27–30 contains predominantly β-sheet, presumably well in excess of 50% (reviewed in Ref.[14]).

2.2 Chemical Probes

Chemical probes have been successfully used to obtain structural information on PrP^{Sc}. The amino group-specific, bifunctional cross-linker bis-(sulfosuccinimidyl)-suberate (BS^3) was used to cross-link 263K Syrian hamster (SHa) PrP 27–30. Mass spectrometry-based analysis, of tryptic peptides obtained after complete denaturation of cross-linked PrP 27–30, identified cross-linking of two amino-terminal Gly_{90} residues, resulting in a Gly_{90}-Gly_{90} covalent cross-link. This means that the amino-termini of successive PrP units in the PrP 27–30 stack are within 1.14 nm, the distance of the BS^3 spacer arm.[15] At the time, this distance constraint was interpreted to limit the maximum number of β-solenoid rungs in a PrP^{Sc} molecule to three at the most, as the amino termini in a four-rung β-solenoid would lie ~1.92 nm apart, if stacked head-to-tail. Allowing for some flexibility, it was reasoned theoretically that the ~1.44 nm separating the amino termini of two three-rung β-solenoids might be within the cross-linking capability of BS^3. However, a more parsimonious explanation of these results would entail a head-to-head stacking of two four-rung β-solenoid structures, whose amino termini would be placed at a distance of 0.48 nm. In fact, a head-to-head stacking of PrP^{Sc} units has been suggested based on three-dimensional (3D) reconstructions of individual PrP^{Sc} amyloid fibrils that were analyzed via electron cryomicroscopy.[16]

In an alternative approach, surface labeling of 263K SHa PrP 27–30 with the tyrosine-specific reagent tetranitromethane (TNM) showed that the C-terminal region of PrP^{C} undergoes a substantial structural rearrangement during conversion to PrP^{Sc}, attested by reduced reactivities of Tyr_{225} and Tyr_{226}.[17] This finding contradicted the hypothesis suggesting that the C-terminal α-helices might be conserved in PrP^{Sc}.

2.3 Limited Proteolysis

Limited proteolysis using proteinase K (PK) has been used to probe the structure of PrP^{Sc}. The underlying rationale is that PK will readily cleave flexible and accessible random coil stretches that connect individual β-strands, which, together, make up the structure of PrP^{Sc}, as discussed in Section 2.1.[18–20]

After PK digestion, the resulting peptides can be analyzed and characterized by epitope analysis and mass spectrometry. Analyses of hamster-adapted scrapie 263K and Drowsy (Dy) PrP^{Sc}[19] revealed mainly "classic" PK cleavages at positions 86/101. However, additional PK cleavages at residues 117, 119, 135, 139, 142, and 154 were observed in both strains. The presence of asparagine-linked sugars and the GPI-anchor impeded the analysis of the C-terminal portion of PrP^{Sc}. A similar analysis of GPI-anchorless PrP^{Sc}, in which the lack of the GPI anchor and an almost complete lack of glycosylation permitted the mass spectrometry-based analyses of the entire PrP molecule, revealed cleavage sites at residues 81, 85, 89, 116, 118, 133, 134, 141, 152, 153, 162, 169, and 179, respectively.[20] These results suggest that GPI-anchorless PrP^{Sc}, 263K SHa PrP^{Sc}, and Dy SHa PrP^{Sc} share a common molecular architecture. Interestingly, no cleavage sites were detected beyond position 179, indicating that the C-terminal stretch of GPI-anchorless PrP^{Sc} (and by extrapolation, PrP^{Sc} in general) is particularly resistant to PK. Other studies have reported similar, minor cleavage sites in a variety of PrP^{Sc} strains (PrP^{Sc} strains, fully described in other chapters of this book, are conformational variants of PrP^{Sc} with distinct biological, biochemical, and structural properties). Zou *et al.*[21] described PK-resistant, PrP^{Sc}-derived peptides from human CJD samples spanning from residues 154/156 and 162/167 to the C-terminus. Zanusso *et al.*[22] described two additional, amino-terminally truncated PrP^{Sc} peptides (MW of 16/17 kDa) from human CJD samples, analogous to the GPI-anchorless PrP^{Sc} peptides Gly_{141}–Ser_{232} and Met_{133}–Ser_{232}/Ser_{134}–Ser_{232}. These results were combined to draft a tentative map of the β-strands and random coil loops that define PrP^{Sc}.[23]

2.4 Diffraction Approaches

X-ray crystallography is routinely used to solve the structures of proteins and other macromolecules, often to atomic resolution. As a well-developed structural biology method, it includes a large variety of individual techniques that can be applied to solve a structure. Nevertheless, X-ray crystallography has stringent demands, as it requires the protein to form well-ordered, 3D crystals, which is nearly impossible to achieve with aggregation prone proteins, such as PrP^{Sc}, except with small amyloidogenic peptides.[24]

In contrast, small-angle X-ray scattering (SAXS) provides a measure of the analyte size in suspension without the need for sample orientation. In fact, the random orientation of the protein aggregates in solution allows calculation of the overall aspect ratio of the aggregate. Therefore, a sufficiently dispersed sample can provide molecular or protein aggregate dimensions via the radius of gyration.[25] A sample of purified Syrian hamster PrP 27–30 prion fibers was analyzed by SAXS at the Elettra synchrotron (Trieste, Italy). Extensive clumping of the PrP 27–30 fibers created a polydisperse state, however, the high-quality SAXS pattern obtained could be fitted to a model based on infinitely long cylinders with a log–normal intensity distribution, a hard-sphere structure factor, and a general *Porod* term accounting for the effect of lateral aggregation of the fibers.[25] The diameter calculated for the cylinders from this fit was 11.0 ± 0.2 nm. This measurement offered an estimation of the diameter of PrP^{Sc} amyloid fibers in suspension, independent from errors or distortions introduced by the negative stain procedure used on carbon film-adsorbed fibers in TEM-based fiber diameter measurements.[26,27] On the other hand, this diameter agreed well the previously obtained TEM-based measurements, and corresponds to a maximum diameter of approximately 5.5 nm for each of the protofilaments that make up the fibers.

Among the diffraction approaches, X-ray fiber diffraction has contributed the most informative insights concerning the structure of PrP^{Sc}.[27] For these analyses, PrP^{Sc} and PrP 27–30 were purified from the brains of infected mice or SHa, and oriented samples of the resulting amyloid fibrils were analyzed using synchrotron-based X-ray sources. The X-ray fiber diffraction patterns from all prion strains that were analyzed displayed three orders of meridional diffraction at 0.96, 0.64, and 0.48 nm, indicative of a 1.92 nm repeating unit along the fibril axis.[27] In addition, the diffraction patterns lacked a strong equatorial diffraction at ~1.0 nm, which itself is characteristic of generic, stacked β-sheet amyloid structures found in, for example, parallel in-register β-sheet amyloids.[27,28] Together, the fiber diffraction data revealed that the structure of the infectious prion consists of a β-helical or β-solenoidal architecture containing four-rungs of β-structure ($4 \times 0.48\ nm^2 = 1.92\ nm$).[27]

2.5 Imaging

A variety of techniques can be used to visualize the structure of infectious prions in their various aggregation states. Depending on the underlying physical principles, the type of information that can be obtained and the achievable resolution vary considerably.

2.5.1 Scanning Probe Microscopy

Scanning probe microscopy (SPM), also known under the somewhat misleading term "atomic force microscopy" or "AFM," can visualize individual prion particles adsorbed onto a flat surface such as freshly cleaved mica. A suitably sharp probe can then be used to scan the surface to detect any adhering protein aggregates, thereby producing a 2D scanning profile that includes not only the protein aggregate size in *X* and *Y* dimensions, but also height information (*Z* dimension). In fact, the size of the scanning probe (i.e., the size of the tip) convolutes the measurements in the *X* and *Y* dimensions, rendering them less accurate than other imaging techniques (e.g., electron microscopy). In contrast, the height measurements of adsorbed protein aggregates are less affected by the tip size and thus provide reliable height measurements that are otherwise difficult to obtain, for example via electron microscopy.

Recombinant prion protein (recPrP) amyloids have been investigated via SPM in many studies due to the general lack of biosafety concerns, the ease of preparing recPrP amyloid samples, and the simplicity of visualizing protein aggregates via SPM (e.g., Refs.[29,30]). SPM was used to describe different polymorphs of recPrP amyloid that were found within a single growth condition[29] or even within individual amyloid fibrils.[30]

Attempts to visualize PrP^{Sc} by means of SPM were first performed on prion-infected N2a cells, also known as ScN2a cells.[31] Here large fibrillar features were observed on the surface of infected ScN2a cells, but not on uninfected N2a control cells. The size of these surface features was well beyond the expected size of individual amyloid fibrils. Therefore, the molecular origin of these cell surface structures remained uncertain, but it is conceivable that the bundling of PrP^{Sc} amyloid fibrils may have produced these fibrillar surface features.

More detailed information on the structure of PrP^{Sc} amyloid fibrils was obtained via SPM from purified preparations of 22L and RML prions, with or without GPI anchor.[26] Here, the height and helical periodicity of individual amyloid fibrils were determined and compared with measurements obtained via negative stain electron microscopy (compare with Section 2.5.2). Consistently, the protofilament diameter (3.1–3.5 nm), as determined by negative stain electron microscopy, was narrower than the SPM values for the protofilament height (5.5–5.6 nm). This difference could be explained by the penetration of the fibril periphery by the heavy metal ions of the negative stain, giving the protofilaments a narrower appearance in the resulting electron micrographs. Alternatively, these differences could also be attributed, in part, to the tip convolution, which would result in larger SPM values for the protofilament height.

More recently, SPM was used to compare the morphologies of in vitro generated recPrP amyloid fibrils that were generated either without additives or with preparations that were treated identically but included palmitoyloleoylphosphatidylglycerol (POPG) and RNA in the reaction mixture.[32] In absence of POPG and RNA, regular amyloid fibrils were generated, but these preparations lacked prion infectivity. In contrast, the samples that included POPG and RNA were generally amorphous in structure, but were found to contain a high prion titer. In any case, the SPM measurements were not of sufficient resolution to explain the observed differences in infectivity, which will require higher-resolution techniques.

2.5.2 Negative Stain Electron Microscopy

Negative stain electron microscopy was the technique that allowed the first visualization of infectious prions as "prion rods,"[33] not long after their initial description.[1] Earlier accounts of "abnormal fibrillar structures," termed "scrapie-associated fibrils" or "SAFs," in synaptosomal preparations from brains of scrapie-sick mice and hamsters were thought to represent "β-pleated protein fibrils" of uncertain relevance with regard to scrapie infectivity.[34]

Initially, it was thought that the formation of prion rods depended on both the use of detergents and limited proteolysis of PrP^{Sc}, which results in the N-terminally truncated PrP 27–30.[35] Only later did it become evident that the formation of PrP^{Sc} amyloid fibrils is independent of any proteolytic cleavage and requires merely the extraction of PrP^{Sc} from its native lipid environment either by use of detergents or other binding agents, such as polyoxometalates.[36] Some prion strain–host combinations, for example

Sc237 prions in Syrian hamsters, result in the widespread deposition of PrP^{Sc} amyloid plaques in scrapie-infected brain tissue,[37] while other combinations, such as RML prions in FVB mice, remain completely free of any detectable PrP^{Sc} amyloid deposits.[38,39] In the latter case, PrP^{Sc} amyloid fibrils form readily in vitro upon detergent extraction,[36] emphasizing the fact that the absence or presence of amyloid properties is independent from prion infectivity.[40]

As mentioned earlier, negative stain electron microscopy is commonly used to measure the fibril diameters of amyloid fibrils such as PrP^{Sc} or PrP 27–30 amyloid. Depending on the prion strain and host species, the absence or presence of the GPI anchor, and other experimental details, the average protofilament diameters ranged between 3 and 6 nm.[26,27] Based on the presence of heavy metal contrasting agents, the high intensity illumination conditions, and the often long exposures, no high-resolution structural detail can be obtained from such negative stain electron micrographs (compare with Section 2.4).

The discovery of 2D crystals composed of PrP 27–30 allowed the use of multistep image processing routines to extract low-resolution structural information on this conformer.[41] Moreover, the fact that a truncated version of PrP 27–30—termed $PrP^{Sc}106$, which lacked the N-terminal residues 23–89 and carried an internal deletion covering residues 141–176 (mouse numbering),[42] formed isomorphous 2D crystals, allowed a difference mapping approach to locate the molecular differences between PrP 27–30 and $PrP^{Sc}106$.[41,43,44] The resulting data were used to restrain molecular models for the structure of PrP^{Sc} and to develop the concept of a parallel β-helix or β-solenoid as the key element of the infectious conformer.[41,43] Nevertheless, the use of negative stain electron microscopy limited the amount of structural data that could be obtained from these 2D crystals.

2.5.3 Electron Cryomicroscopy

To overcome the limitations of negative stain electron microscopy, higher-resolution structural data can be obtained, in principal, through the use of cryo-low dose electron microscopy approaches. For example, by collecting electron micrographs of frozen hydrated PrP^{Sc} amyloid fibrils at different viewing angles it is possible to reconstruct 3D tomograms containing individual amyloid fibrils and small bundles thereof.[16,45] However, the dose fractionation that is necessary to collect the different views limits the resolution that can be achieved with electron tomography. Subtomogram averaging allows to visualize the helical nature of individual fibrils, but little structural detail was gained beyond their overall fibril topology.

However, higher-resolution data on the structure of individual amyloid fibrils can be obtained from 3D reconstructions that take advantage of the helical symmetry within these fibrils (Fig. 1). For a GPI-anchorless form of PrP^{Sc}, cryo-low dose electron micrographs were collected, and images from individual PrP^{Sc} amyloid fibrils were used to generate 3D reconstructions that revealed important details about the molecular architecture of the protein subunits.[16] Fourier-transform analyses of unprocessed images from individual PrP^{Sc} amyloid fibrils routinely detected signals at 0.48 nm, corresponding to the cross-β signals that were seen with X-ray fiber diffraction (vide supra). 3D reconstructions of isolated PrP^{Sc} amyloid fibrils were

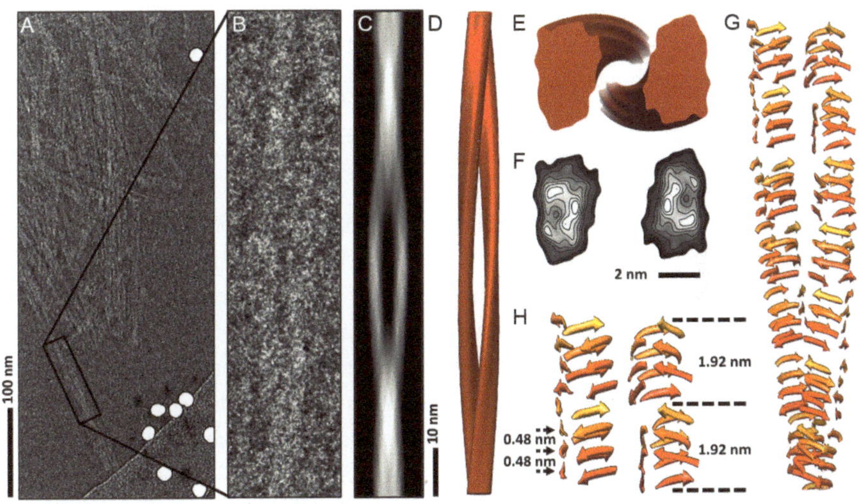

Fig. 1 Electron cryomicroscopy analysis of infectious prion protein amyloid fibrils. (A) Section of a cryo-electron micrograph showing GPI-anchorless prion fibrils. A single isolated and twisted fibril used for the 3D reconstruction is enclosed by a *black box*. (B) Close-up view of the isolated prion fibril. (C) Reprojected image of the 3D fibril map for comparison with the unprocessed image (B). (D) 3D reconstruction of the GPI-anchorless prion fibril. (E) Cross-section of the reconstructed fibril showing two distinct protofilaments. (F) Contoured density maps of the cross-section with lines contoured at increasing levels of 0.125σ. (G) Cartoon depicting the proposed configuration of the polypeptide chains in the prion fibril. Please note that this is NOT an atomistic model. (H) Close-up view of the possible ß-sheet stacking in a four-rung ß-solenoid architecture for illustration purposes only. *Different colors* represent different ß-solenoid rungs. Characteristic distances of the four-rung ß-solenoid architecture are labeled. *Figure and figure legend taken from Zweckstetter M, Requena JR, and Wille H. Elucidating the structure of an infectious protein.* PLoS Pathog. *2017;13:e1006229.*

able to resolve two intertwined protofilaments for each of the fibrils (Fig. 1). The 3D reconstructions provided sufficient detail to calculate the molecular volume and thereby the height of individual PrP^{Sc} molecules, indicating a height of ~1.8 nm.[16] This measurement compared well with the repeating unit size that was determined by X-ray fiber diffraction as 1.92 nm.[27]

An alternative image processing approach, which is known as "single particle averaging," averaged and classified hundreds of images of short fibril sections. This technique showed repetitive features along the protofilaments with an average height of ~2 nm.[16] Here too, Fourier-transform analyses indicated the presence of the characteristic 0.48 nm signal associated with highly organized β-strands oriented perpendicular to the fibril axis. Moreover, Fourier-transform analyses also detected additional signals at ~4 nm, indicating additional repetitive structures along the fibril axis. This ~4 nm signal is consistent with a dimeric arrangement of the protein,[16] similar to what was also observed in other amyloid fibrils.[46,47] Such an arrangement might be head-to-head/tail-to-tail.

2.6 NMR Spectroscopy

Analysis of PrP^{Sc} via NMR spectroscopy depends on the availability of bona fide, truly infectious, recombinant PrP^{Sc}, as it requires high titer, ^{13}C- and ^{15}N-labeled samples. In this respect, while there were a number of early reports on the preparation of samples with at least some infectivity, it was not until the seminal report by Wang *et al.*[48] that structural studies of recombinant PrP^{Sc} became feasible. Even then, limited yields continue to be an issue for NMR-based studies, which require substantial amounts of structurally homogenous samples. The latter requirement is a serious limitation for all structural biology approaches, particularly given the ability of PrP to adopt a variety of infectious conformers, which are commonly referred to as "strains."[49] In any case, the aggregated nature of the PrP^{Sc} samples requires the use of solid-state NMR (ssNMR) techniques, to compensate for the lack of molecular tumbling.

Müller *et al.* reported on ssNMR studies of infectious PrP amyloid fibers.[50] These fibers were prepared by incubation of recombinant, full-length sheep PrP (residues 23–230) with denaturants under agitation, the same conditions that have been extensively used to prepare "classic" recPrP amyloid fibers devoid of infectivity (but vide infra).[12,27,51] However, the amyloid samples prepared by Müller *et al.* exhibited a modest degree of infectivity: 3 out of 12 challenged animals developed disease symptoms with long incubation times. "Classic" recPrP amyloids consist of stacks of flat, one-layer PrP units that are arranged in-register,[12,52] exhibiting a structure that is different from that of bona fide PrP^{Sc}.[27] Unexpectedly, such recPrP amyloid fibers sometimes propagate in the brains of experimental animals after intracerebral inoculation, eventually generating PrP^{Sc} and becoming pathogenic.[51] This phenomenon has been explained by a process termed "deformed templating."[53] Müller *et al.* obtained ssNMR spectra of good quality from their samples. Although only a few individual signals could be unequivocally assigned, a rich set of information could be collected from the spectra by a variety of ingenious approaches. These data allowed Müller *et al.* to conclude that the C-terminal region of their amyloid sample, from position ~155 to the C-terminus constitutes a β-sheet-rich core. The same data also indicated that the N-terminal stretch, from the amino-terminus to position ~114, is flexible, and the central residue ~115–155 region would have the characteristics of α-helical structure.[50] This architecture coincides with that proposed for classic recPrP amyloids,[12,52] except for the central α-helical region, suggested to be unfolded in recPrP amyloids. Clearly, the fibers prepared and described by Müller *et al.* are reminiscent of classic in-register β-sheet stacks, and therefore different from PrP^{Sc}.[27] Nevertheless, this study paves the way for ssNMR analyses of recombinant PrP^{Sc}, once sufficient quantities of bona fide, PrP^{Sc}-like infectious material becomes available.[48]

While it is clearly not PrP^{Sc}, an infectious, recombinant PrP 23–144 amyloid generated in vitro deserves some consideration here.[54] The Y145Stop PrP variant associates with a human prion disease characterized by deposition of amyloid plaques in the brain.[55] Human, mouse, and SHa PrP 23–144 readily form amyloid in vitro, with ssNMR spectra featuring narrow lines, which allowed the assignment of individual amino acid residues.[56–60] Furthermore, these studies demonstrated that the PrP 23–144 amyloid forms in-register stacks whose β-core spans residues ~112–140. Interestingly, secondary structure prediction algorithms identify three separate β-strands within this core, connected by short loops.[60] This strongly suggests that the cross-section of the PrP 23–144 stack is not a simple hairpin, but likely has a richer, more complex shape.[61] The interest of these studies was boosted by the recent discovery that this recombinant amyloid is fully infectious, with attack rates of 100%. Intriguingly, a very efficient deformed templating takes place upon successive passages in vivo, as attested by changes in the pattern of PK-resistant PrP fragments in the brain of infected animals.[54] The complete structure of the infectious recombinant PrP 23–144 amyloid should provide a clearer understanding of the deformed templating phenomenon and is likely to provide interesting insights on the structure of PrP^{Sc} itself.

In summary, ssNMR studies of recombinant, infectious PrP^{Sc} were encouraged by the success of the fungal prion Het-s,[62] whose general architecture was solved by of ssNMR. Interestingly, the Het-s prion contains a two-rung β-solenoid, which has considerable resemblance to the structure of PrP^{Sc}.

2.7 Molecular Modeling

In absence of comprehensive experimental data on the structure of PrP^{Sc}, a large number of molecular models were proposed starting in the mid-1990s (reviewed in Ref.[14]). The earliest models were based on sparse experimental data (FTIR and CD spectroscopies) and subsequently served as tools to interpret various other data on the structure of PrP^{Sc}. In the intervening years, additional experimental data constrained the structure of PrP^{Sc} to an increasing degree (vide supra), and, thus, resulted in more refined molecular models. Moreover, the attempts to refute or support specific molecular models helped to sharpen the arguments and interpretations of experimental data with respect to the molecular features of the models in question.

Since a detailed discussion of the proposed molecular models was published earlier,[14] we will not revisit the arguments in favor or in opposition for each of these models. Nevertheless, the experimental evidence for a four-rung β-solenoid as the main feature of the PrP^{Sc} structure (vide supra) supports renewed modeling efforts to answer open questions about the molecular structure of PrP^{Sc} and to refine future experimental work.

3. PROPAGATION OF INFECTIOUS PRIONS

The fact that the architecture of PrP^{Sc} consists of a four-rung ß-solenoid allows us to derive several corollaries with regard to propagation of the infectious fold. First, it is obvious that the templating mechanism that controls propagation of PrP^{Sc} in vivo cannot consist of in-register stacking of identical residues in successive layers of protein molecules, as proposed by the parallel in-register ß-sheet model.[63] Instead, templating based on a four-rung ß-solenoid architecture must necessarily involve predominantly the upper- and lowermost ß-solenoid rungs. The "unpaired" ß-strands in these two rungs are inherently aggregation prone, and they can propagate their hydrogen-bonding pattern into any amyloidogenic peptide they encounter.[64] In fact, the edge strands of native proteins that contain a ß-solenoid are capped by loops or α-helices to block unregulated ß-sheet propagation. Furthermore, elimination of such structures through protein-engineering results in edge-to-edge-driven oligomerization of the "decapped" ß-solenoids.[65] Therefore, the upper- and lowermost ß-solenoid rungs in PrP^{Sc} can template an incoming, unfolded PrP molecule, and mold it into an additional ß-solenoid rung. Once this supplementary ß-solenoid rung is formed, it offers a fresh, "sticky" surface that can continue templating the remaining, unfolded "tail" of the incoming PrP molecule, until a second rung is generated. This process will be repeated two more times until the entire incoming PrP molecule has been converted into four newly formed rungs of ß-solenoid structure. Afterward, the templating can continue on the next, unfolded PrP molecule, ad infinitum.

The way in which templating occurs must necessarily be based on either a head-to-tail or a head-to-head/tail-to-tail orientation. In the former case, templating of ß-sheets would involve heterotypic contacts between different parts of the molecule. In the latter case, the same protein stretches will come into homotypic contact. Cryo-EM analyses of GPI-anchorless PrP^{Sc} fibrils provided evidence for a vertical, dimeric arrangement; specifically, a ~4 nm signal in Fourier-transform analyses individual fibril segments[16] suggested a head-to-head and tail-to-tail arrangement along the fibril axis. However, higher resolution data are required to distinguish between these alternative dimer options with certainty.

The question how this propagation mechanism can reproduce prions from different strains with relatively high fidelity poses an interesting puzzle. Ultimately, a high-resolution structure of PrP^{Sc} should allow to answer this and related questions. Furthermore, the molecular details of the propagation mechanism must also be able to explain the species barrier that prevents transmission of prions between different species (however imperfectly). Similarly, the details of the templating mechanism must also explain, why other proteins are not coerced to adopt this fold and are not converted into β-solenoid versions of themselves.

4. FUTURE APPROACHES TO INVESTIGATE THE STRUCTURE OF PRIONS

What can we expect in the near future from ongoing studies of the structure of PrP^{Sc}? Can we expect a "*eureka* moment" when the structure of PrP^{Sc} will be suddenly unveiled at an atomic resolution? Most likely, not. Instead, it is very probable that over the next few years we will witness a continued, incremental accrual of structural restraints that will, step by step, increase the details and clarity with which we can perceive the basic architecture of this conformer. Information will come from all the analytical techniques mentioned, but in our opinion two of them are called to play the most critical roles: cryo-EM and NMR spectroscopy.

Cryo-EM is on a continuous path of technical improvement; recent advances such as the CMOS direct electron detectors, electron counting/movie mode, and phase plates have boosted the sensitivity and performance of this technique to heretofore unprecedented limits. As described earlier, we have taken advantage of some of these advances to elucidate the basic architecture of PrP^{Sc} as a four-rung β-solenoid.[16] In the coming years, we will use these techniques to study the structures of various prion strains from animal and human sources, hoping that one prion strain/host combination will provide particularly high-resolution results. Moreover, we will be expanding our sample repertoire to also include various infectious recombinant PrP^{Sc} preparations, which may allow us to compare and integrate our results with those obtained via ssNMR.

As also mentioned, the availability of bona fide, infectious recombinant PrP^{Sc} has opened the possibility to solve its structure using ssNMR, a powerful technique that has proven to be decisive to solve the structure of the Het-s prion.[62] Owing to its needs and the inherent difficulties of working with aggregated protein species, this approach may not reach atomic resolution, but it should provide enough resolution to fully understand the fold of PrP^{Sc}. In this respect, it is important to acknowledge that ssNMR requires substantial amounts (i.e., milligram quantities) of alternatively labeled versions of the protein under investigation. Therefore, obtaining sufficient amounts of isotopically labeled, recombinant PrP constitutes a bottleneck. However, recent and unpublished advances in optimization and scale-up of recombinant PrP^{Sc} propagation in vitro may soon allow the generation of the required amounts of isotopically labeled, infectious material (Joaquín Castilla, personal communication). Once this preparative goal is achieved, the results that were obtained with Het-s[62] illustrate the general feasibility of this approach. It should be noted, however, that Het-s ssNMR spectra contained strikingly narrow resonance lines: down to 0.2 ppm in ^{13}C–^{13}C homonuclear DREAM correlation spectra from uniformly ^{15}N- and ^{13}C-labeled samples.[66] These resonance lines are comparable to those of microcrystalline proteins, indicating a highly ordered atomic structure for part of the fibrils. This is likely to be a consequence of the fact that the Het-s prion has a cellular function, and therefore its structure has been optimized by evolutionary selection.[66] In contrast, the ssNMR spectra of PrP^{Sc} are likely to be much more heterogeneous and will probably contain much broader signals. This is illustrated by the feasibility of a nearly complete sequence-specific chemical shift assignment of Het-s spectra,[66] which was not possible for partially infectious recombinant PrP amyloid fibers.[50] However, despite these potential complications, the identification of key structural features, such as a definitive localization of secondary structure elements should be achievable.

5. CONCLUDING REMARKS

A number of recent breakthroughs, including electron cryomicroscopy studies that took advantage of images of unprecedented quality, and of the opportunities afforded by the repetitive structure of PrP^{Sc} fibers, have allowed deciphering the basic architecture of mammalian prions. In turn, these findings provide, for the first time, an initial understanding of the propagation mechanism of mammalian prions. The molecular forces responsible for templating and that define this process—hydrogen-bonding, charge interactions, aromatic stacking, and steric constraints—are fundamentally similar to those operating during DNA replication. However, they lack its exquisite precision and the complex proofreading mechanisms that provide the flexibility of nucleic acid replication. The higher complexity of the PrP^{Sc} structure, as compared to that of nucleic acids, will require particular efforts to achieve a complete understanding of key phenomena associated with prion propagation, such as the transmission barriers between different species and the existence of strains. To fully understand these phenomena, it will be necessary to identify the specific structural differences between β-solenoids corresponding to different PrP^{Sc} strains. Such differences will likely include the specific topography of the main propagative β-solenoid rungs (i.e., the upper- and lowermost rungs). We strongly believe that this knowledge lies within our reach and should be obtainable within the next few years.

ACKNOWLEDGMENTS

The authors would like to acknowledge support from European Commission grant FP7 222887 "Priority," Spanish Ministry of Education grant BFU2006-04588/BMC, and Spanish Ministry of Economy grant BFU2013-48436-C2-1-P (all to J.R.R.) and grants from the Alberta Prion Research Institute (APRI 201600012 and APRI 201600029) and the Alberta Livestock & Meat Agency (ALMA 2016A001R) (all to H.W.).

REFERENCES

1. Prusiner SB. Novel proteinaceous infectious particles cause scrapie. *Science*. 1982;216:136–144.
2. Bolton DC, McKinley MP, Prusiner SB. Identification of a protein that purifies with the scrapie prion. *Science*. 1982;218:1309–1311.
3. Wickner RB, Taylor KL, Edskes HK, Maddelein ML, Moriyama H, Roberts BT. Prions of yeast as heritable amyloidoses. *J Struct Biol*. 2000;130:310–322.
4. Jaunmuktane Z, Mead S, Ellis M, et al. Evidence for human transmission of amyloid-β pathology and cerebral amyloid angiopathy. *Nature*. 2015;525:247–250.
5. Prusiner SB. A unifying role for prions in neurodegenerative diseases. *Science*. 2012;336:1511–1513.
6. Prusiner SB. Prions. *Proc Natl Acad Sci USA*. 1998;95:13363–13383.
7. Aguzzi A, Calella AM. Prions: protein aggregation and infectious diseases. *Physiol Rev*. 2009;89:1105–1152.
8. Caughey BW, Dong A, Bhat KS, Ernst D, Hayes SF, Caughey WS. Secondary structure analysis of the scrapie-associated protein PrP 27–30 in water by infrared spectroscopy. *Biochemistry*. 1991;30:7672–7680.
9. Pan KM, Baldwin M, Nguyen J, et al. Conversion of alpha-helices into beta-sheets features in the formation of the scrapie prion proteins. *Proc Natl Acad Sci USA*. 1993;90:10962–10966.
10. Safar J, Roller PP, Gajdusek DC, Gibbs Jr CJ. Conformational transitions, dissociation, and unfolding of scrapie amyloid (prion) protein. *J Biol Chem*. 1993;268:20276–20284.
11. Smirnovas V, Baron GS, Offerdahl DK, Raymond GJ, Caughey B, Surewicz WK. Structural organization of brain-derived mammalian prions examined by hydrogen-deuterium exchange. *Nat Struct Mol Biol*. 2011;18:504–506.
12. Cobb NJ, Sönnichsen FD, McHaourab H, Surewicz WK. Molecular architecture of human prion protein amyloid: a parallel, in-register beta-structure. *Proc Natl Acad Sci USA*. 2007;104:18946–18951.
13. Baron GS, Hughson AG, Raymond GJ, et al. Effect of glycans and the glycophosphatidylinositol anchor on strain dependent conformations of scrapie prion protein: improved purifications and infrared spectra. *Biochemistry*. 2011;50:4479–4490.
14. Requena JR, Wille H. The structure of the infectious prion protein: experimental data and molecular models. *Prion*. 2014;8:60–66.
15. Onisko B, Guitián Fernández E, Louro Freire M, et al. Probing PrPSc structure using chemical cross-linking and mass spectrometry: evidence of the proximity of Gly90 amino termini in the PrP 27–30 aggregate. *Biochemistry*. 2005;44:10100–10109.
16. Vázquez-Fernández E, Vos MR, Afanasyev P, et al. The structural architecture of an infectious mammalian prion using electron cryomicroscopy. *PLoS Pathog*. 2016;12: e1005835.
17. Gong B, Ramos A, Vázquez-Fernández E, et al. Probing structural differences between PrPC and PrPSc by surface nitration and acetylation: evidence of conformational change in the C-terminus. *Biochemistry*. 2011;50:4963–4972.
18. Hubbard SJ. The structural aspects of limited proteolysis of native proteins. *Biochim Biophys Acta*. 1998;1382:191–206.
19. Sajnani G, Pastrana MA, Dynin I, Onisko B, Requena JR. Scrapie prion protein structural constraints obtained by limited proteolysis and mass spectrometry. *J Mol Biol*. 2008;382:88–98.
20. Vázquez-Fernández E, Alonso J, Pastrana MA, et al. Structural organization of mammalian prions as probed by limited proteolysis. *PLoS One*. 2012;7: e50111.
21. Zou WQ, Capellari S, Parchi P, Sy MS, Gambetti P, Chen SG. Identification of novel proteinase K-resistant C-terminal fragments of PrP in Creutzfeldt–Jakob disease. *J Biol Chem*. 2003;278:40429–40436.

22. Zanusso G, Farinazzo A, Prelli F, et al. Identification of distinct N-terminal truncated forms of prion protein in different Creutzfeldt–Jakob disease subtypes. *J Biol Chem*. 2004;279:38936–38942.
23. Silva CJ, Vázquez-Fernández E, Onisko B, Requena JR. Proteinase K and the structure of PrPSc: the good, the bad and the ugly. *Virus Res*. 2015;207:120–126.
24. Sawaya MR, Sambashivan S, Nelson R, et al. Atomic structures of amyloid cross-beta spines reveal varied steric zippers. *Nature*. 2007;447:453–457.
25. Amenitsch H, Benetti F, Ramos A, Legname G, Requena JR. SAXS structural studies of PrPSc reveals ~11 diameter of basic double intertwined fibrils. *Prion*. 2013;7:496–500.
26. Sim VL, Caughey B. Ultrastructures and strain comparison of under-glycosylated scrapie prion fibrils. *Neurobiol Aging*. 2009;30:2031–2042.
27. Wille H, Bian W, McDonald M, et al. Natural and synthetic prion structure from X-ray fiber diffraction. *Proc Natl Acad Sci USA*. 2009;106:16990–16995.
28. Wan W, Wille H, Stöhr J, et al. Structural studies of truncated forms of the prion protein PrP. *Biophys J*. 2015;108:1548–1554.
29. Anderson M, Bocharova OV, Makarava N, Breydo L, Salnikov VV, Baskakov IV. Polymorphism and ultrastructural organization of prion protein amyloid fibrils: an insight from high resolution atomic force microscopy. *J Mol Biol*. 2006;358:580–596.
30. Makarava N, Ostapchenko VG, Savtchenko R, Baskakov IV. Conformational switching within individual amyloid fibrils. *J Biol Chem*. 2009;284:14386–14395.
31. Wegmann S, Miesbauer M, Winklhofer KF, Tatzelt J, Muller DJ. Observing fibrillar assemblies on scrapie-infected cells. *Pflügers Arch*. 2008;456:83–93.
32. Piro JR, Wang F, Walsh DJ, Rees JR, Ma J, Supattapone S. Seeding specificity and ultrastructural characteristics of infectious recombinant prions. *Biochemistry*. 2011;50:7111–7116.
33. Prusiner SB, McKinley MP, Bowman KA, et al. Scrapie prions aggregate to form amyloid-like birefringent rods. *Cell*. 1983;35:349–358.
34. Merz PA, Somerville RA, Wisniewski HM, Iqbal K. Abnormal fibrils from scrapie-infected brain. *Acta Neuropathol*. 1981;54:63–74.
35. McKinley MP, Meyer RK, Kenaga L, et al. Scrapie prion rod formation in vitro requires both detergent extraction and limited proteolysis. *J Virol*. 1991;65:1340–1351.
36. Levine DJ, Stöhr J, Falese LE, et al. Mechanism of scrapie prion precipitation with phosphotungstate anions. *ACS Chem Biol*. 2015;10:1269–1277.
37. DeArmond SJ, McKinley MP, Barry RA, Braunfeld MB, McColloch JR, Prusiner SB. Identification of prion amyloid filaments in scrapie-infected brain. *Cell*. 1985;41:221–235.
38. Godsave SF, Wille H, Kujala P, et al. Cryo-immunogold electron microscopy for prions: toward identification of a conversion site. *J Neurosci*. 2008;28:12489–12499.
39. Godsave SF, Wille H, Pierson J, Prusiner SB, Peters PJ. Plasma membrane invaginations containing clusters of full-length PrP^{Sc} are an early form of prion-associated neuropathology *in vivo*. *Neurobiol Aging*. 2013;34:1621–1631.
40. Wille H, Zhang G-F, Baldwin MA, Cohen FE, Prusiner SB. Separation of scrapie prion infectivity from PrP amyloid polymers. *J Mol Biol*. 1996;259:608–621.
41. Wille H, Michelitsch MD, Guenebaut V, et al. Structural studies of the scrapie prion protein by electron crystallography. *Proc Natl Acad Sci USA*. 2002;99:3563–3568.
42. Supattapone S, Bosque P, Muramoto T, et al. Prion protein of 106 residues creates an artificial transmission barrier for prion replication in transgenic mice. *Cell*. 1999;96:869–878.
43. Govaerts C, Wille H, Prusiner SB, Cohen FE. Evidence for assembly of prions with left-handed beta-helices into trimers. *Proc Natl Acad Sci USA*. 2004;101:8342–8347.
44. Wille H, Govaerts C, Borovinskiy A, et al. Electron crystallography of the scrapie prion protein complexed with heavy metals. *Arch Biochem Biophys*. 2007;467:239–248.

45. Terry C, Wenborn A, Gros N, et al. *Ex vivo* mammalian prions are formed of paired double helical prion protein fibrils. *Open Biol.* 2016;6: 160035.
46. Mizuno N, Baxa U, Steven AC. Structural dependence of HET-s amyloid fibril infectivity assessed by cryoelectron microscopy. *Proc Natl Acad Sci USA.* 2011;108:3252–3257.
47. White HE, Hodgkinson JL, Jahn TR, et al. Globular tetramers of beta(2)-microglobulin assemble into elaborate amyloid fibrils. *J Mol Biol.* 2009;389:48–57.
48. Wang F, Wang X, Yuan CG, Ma J. Generating a prion with bacterially expressed recombinant prion protein. *Science.* 2010;327:1132–1135.
49. Safar J, Wille H, Itri V, et al. Eight prion strains have PrP(Sc) molecules with different conformations. *Nat Med.* 1998;4:1157–1165.
50. Müller H, Brener O, Andreoletti O, et al. Progress towards structural understanding of infectious sheep PrP-amyloid. *Prion.* 2015;8:344–358.
51. Baskakov IV, Legname G, Prusiner SB, Cohen FE. Folding of prion protein to its native α-helical conformation is under kinetic control. *J Biol Chem.* 2001;276:19687–19690.
52. Tycko R, Savtchenko R, Ostapchenko VG, Makarava N, Baskakov IV. The α-helical C-terminal domain of full-length recombinant PrP converts to an in-register parallel β-sheet structure in PrP fibrils: evidence from solid state nuclear magnetic resonance. *Biochemistry.* 2010;49:9488–9497.
53. Makarava N, Kovacs GG, Savtchenko R, et al. Genesis of mammalian prions: from non-infectious amyloid fibrils to a transmissible prion disease. *PLoS Pathog.* 2011;7: e1002419.
54. Choi JK, Cali I, Surewicz K, Kong Q, Gambetti P, Surewicz WK. Amyloid fibrils from the N-terminal prion protein fragment are infectious. *Proc Natl Acad Sci USA.* 2016;113:13851–13856.
55. Kitamoto T, Iizuka R, Tateishi J. An amber mutation of prion protein in Gerstmann–Sträussler syndrome with mutant PrP plaques. *Biochem Biophys Res Commun.* 1993;192:525–531.
56. Helmus JJ, Surewicz K, Nadaud PS, Surewicz WK, Jaroniec CP. Molecular conformation and dynamics of the Y145Stop variant of human prion protein in amyloid fibrils. *Proc Natl Acad Sci USA.* 2008;105:6284–6289.
57. Helmus JJ, Surewicz K, Surewicz WK, Jaroniec CP. Conformational flexibility of Y145Stop human prion protein amyloid fibrils probed by solid-state nuclear magnetic resonance spectroscopy. *J Am Chem Soc.* 2010;132:2393–2403.
58. Helmus JJ, Surewicz K, Apostol MI, Surewicz WK, Jaroniec CP. Intermolecular alignment in Y145Stop human prion protein amyloid fibrils probed by solid-state NMR spectroscopy. *J Am Chem Soc.* 2011;133:13934–13937.
59. Jones EM, Wu B, Surewicz K, et al. Structural polymorphism in amyloids: new insights from studies with Y145Stop prion protein fibrils. *J Biol Chem.* 2011;286:42777–42784.
60. Theint T, Nadaud PS, Surewicz K, Surewicz WK, Jaroniec CP. 13C and 15N chemical shift assignments of mammalian Y145Stop prion protein amyloid fibrils. *Biomol NMR Assign.* 2017;11:75–80.
61. Skora L, Zweckstetter M. Determination of amyloid core structure using chemical shifts. *Protein Sci.* 2012;21:1948–1953.
62. Wasmer C, Lange A, Van Melckebeke H, Siemer AB, Riek R, Meier BH. Amyloid fibrils of the HET-s(218–289) prion form a beta solenoid with a triangular hydrophobic core. *Science.* 2008;319:1523–1526.
63. Groveman BR, Dolan MA, Taubner LM, Kraus A, Wickner RB, Caughey B. Parallel in-register intermolecular β-sheet architectures for prion-seeded prion protein (PrP) amyloids. *J Biol Chem.* 2014;289:24129–24142.
64. Richardson JS, Richardson DC. Natural beta-sheet proteins use negative design to avoid edge-to-edge aggregation. *Proc Natl Acad Sci USA.* 2002;99:2754–2759.

65. Bryan AW, Starner-Kreinbrink JL, Hosur R, Clark PL, Berger B. Structure-based prediction reveals capping motifs that inhibit β-helix aggregation. *Proc Natl Acad Sci USA*. 2011;108:11099–11104.
66. Ritter C, Maddelein ML, Siemer AB, et al. Correlation of structural elements and infectivity of the HET-s prion. *Nature*. 2005;435:844–888.

CHAPTER SIXTEEN

Protein Misfolding Cyclic Amplification of Infectious Prions

Fabio Moda[1]
IRCCS Foundation Carlo Besta Neurological Institute, Milan, Italy
[1]Corresponding author: e-mail address: fabio.moda@istituto-besta.it

Contents

Abstract

Transmissible spongiform encephalopathies, or prion diseases, are a group of incurable disorders caused by the accumulation of an abnormally folded prion protein (PrP^{Sc}) in the brain. According to the "protein-only" hypothesis, PrP^{Sc} is the infectious agent able to propagate the disease by acting as a template for the conversion of the correctly folded prion protein (PrP^{C}) into the pathological isoform. Recently, the mechanism of PrP^{C} conversion has been mimicked in vitro using an innovative technique named protein misfolding cyclic amplification (PMCA). This technology represents a great tool for studying diverse aspects of prion biology in the field of basic research and diagnosis. Moreover, PMCA can be expanded for the study of the misfolding process associated to other neurodegenerative diseases, including Alzheimer's disease, Parkinson's disease, and frontotemporal lobar degeneration.

1. INTRODUCTION

Transmissible spongiform encephalopathies (TSEs) are a group of devastating and incurable disorders which affect humans and animals. Human prion diseases can be sporadic, genetic, or acquired by infection.[1] Sporadic Creutzfeldt–Jakob disease (sCJD) represents the most common form of human TSEs.[2] Animal prion diseases include scrapie in sheep,[3]

Progress in Molecular Biology and Translational Science, Volume 150
ISSN 1877-1173
http://dx.doi.org/10.1016/bs.pmbts.2017.06.016

chronic wasting disease (CWD) in mule deer and elk,[4] and bovine spongiform encephalopathy (BSE) in cattle.[5] This latter is also known as "mad cow" disease and can be transmitted to humans causing the variant form of CJD (vCJD).[6] Although phenotypically different, all these diseases are caused by a proteinaceous infectious agent, known as prion (PrP^{Sc}), which is a conformationally modified isoform of the normal prion protein (PrP^{C}).[7]

PrP^{Sc} is able to transmit the disease by converting PrP^{C} into the pathological isoform.[8] Unfortunately, there are no chemical differences that enable discrimination between these two proteins.[9] However, the conversion causes an increase in the amount of β-sheet (at the expenses of the α-helices) in the secondary structure of the protein.[10] Following these structural changes PrP^{Sc} became insoluble in mild detergent[11] and partially resistant to proteinase K (PK) digestion. On the contrary, PrP^{C} is soluble in detergent and completely degraded by PK. These features are exploited to distinguish between PrP^{C} and PrP^{Sc}.[11,12] Currently, PrP^{Sc} is considered the only validated biomarker for TSEs and its postmortem detection within the brain is required for a definitive diagnosis of prion disease.[13–15]

In 2001, Claudio Soto invented an innovative technique named protein misfolding cyclic amplification (PMCA),[16] which is able to reproduce the phenomenon of prion replication in vitro, in an accelerated manner. Basically, PMCA consists of an incubation phase and a sonication phase to which samples containing minute amount of PrP^{Sc} are subjected. Brains collected from healthy animals (wild-type or transgenic models) are homogenized (BH) and used as sources of PrP^{C}. During the incubation phase, PrP^{Sc} induces PrP^{C} conversion and aggregation. The sonication phase disrupts these aggregates of PrP^{Sc} into small polymers which act as seeds able to induce further conversion of PrP^{C}. Both phases are repeated in a cyclic manner (Fig. 1). At the end of the reaction, an aliquot of the amplified sample is diluted into freshly prepared BH and subsequently exposed to additional cycles of PMCA (also referred to as *rounds* of amplification). The amplification increases the amount of PrP^{Sc} molecules to a level that can be detected with the current analytic techniques (e.g., Western blot).[17–19] An important improvement of PMCA efficiency was reached with the addition of Teflon beads to the reaction.[20] Notably, PMCA is able to amplify many prion strains with high fidelity thus suggesting that the molecular memory of each strain is completely retained in the conformation of PrP^{Sc}.[21]

For this reason, PMCA represents a great tool for studying (i) prion biology, (ii) the mechanism of prion transmission, (iii) the species barrier, (iv) the

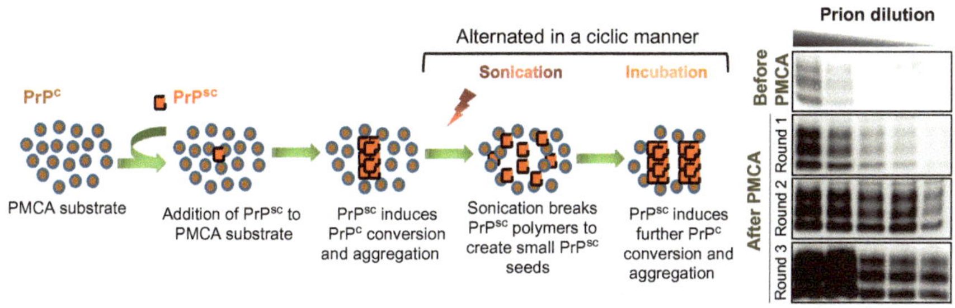

Fig. 1 Schematic representation of PMCA. Minute amounts of infectious prions (PrP^{Sc}) are incubated with an excess of normal prion protein (PrP^{C}). During the incubation phase PrP^{Sc} recruits and converts PrP^{C} into a pathological isoform. Newly generated polymers of PrP^{Sc} grow by incorporating converted PrP^{C}. During the sonication phase, PrP^{Sc} polymers are fragmented into smaller PrP^{Sc} species that can recruit and convert other molecules of PrP^{C}. Thereby, cyclic alternation of sonication and incubation phases enable exponential amplification of PrP^{Sc} that can be finally detected by Western blot.

strain phenomena, (v) the efficacy of decontaminating procedure, and (vi) the efficacy of therapeutic compounds to interfere with prion propagation in vitro. Considering its high sensitivity, this technique has been recently used as supportive test in the field of prion diagnosis.

2. PMCA APPLICATIONS IN BASIC RESEARCH

One of the most intriguing aspects of prions is their ability to be transmissible between individuals of the same or different species.[22] This feature, of course, imposes serious concerns for public health. The efficiency of transmission seems to depend upon different factors, including the *Prnp* sequence homology between the host PrP^{C} and the infectious PrP^{Sc}.[23] Thus, minimal differences could efficiently reduce the transmission efficiency and give rise to the so-called "species barrier."[24] The species barrier reduces the efficacy (in terms of attack rate and incubation time) of prions transmission between different species but is less efficient when transmissions occur within the same species.[25] To study the zoonotic potential of the most common animal prion strains, including BSE, CWD, and scrapie, transgenic mice expressing human PrP^{C} (TgHu) were intracerebrally inoculated with these prion strains. [6,26–29] Such type of experiments demonstrated that BSE and scrapie were infectious to TgHu mice, while CWD failed to induce any prion pathology.[28,30] To date, real cases of BSE transmission to humans were reported (causing vCJD) and animal bioassay demonstrated that both BSE and vCJD are caused by the same prion strain.[6] On the contrary, no cases of scrapie or CWD transmission to humans were described so far, for example after the consumption of scrapie or CWD infected food. Unfortunately, animal bioassays are time consuming and expensive because it is necessary to wait for long time until clinical signs of the disease can be observed. Considering the ability of PMCA to amplify prion strains in a rapid way while maintaining their infectious and biochemical properties, the zoonotic potential of these animal prion strains (BSE, CWD, and scrapie) has been assessed by means of PMCA using brain of TgHu mice as substrate of reaction.[19,31,32] These results confirmed the extraordinary ability of PMCA to faithfully reflect the phenomenon of the species barrier observed in vitro[33–37]. Moreover, after prolonged cycles of PMCA, the species barrier could be overcome.[33] This is not surprising since the species barrier can be overcome also in vivo, when prions are serially transmitted within the same species after a cross-species transmission.[38] In this case, prions are able to gradually adapt to a new host and the species barrier is reduced thus increasing the attack rate and the severity of the disease in infected animals. For example, a significant species barrier is known to limit the transmission of prions between hamsters and mice.[39] PMCA experiments performed by mixing PrP^{Sc} from RML-infected mice and PrP^{C} from healthy hamsters generated a hamster's PrP^{Sc} which slowly adapted to the new species and was finally infectious to hamsters.[33] Therefore, PMCA represents a great tool for reproducing the phenomenon of prion strain adaptation which occurs in vivo and causes the generation of new strains characterized by unique biochemical and infectious features.[40–44] Another phenomenon that can arise (at very low frequency) after prolonged cycles of amplification without the addition of infectious PrP^{Sc} in the reaction, is the spontaneous (or de novo) generation of prions, which mimic the spontaneous generation of PrP^{Sc} which occurs in vivo in sporadic forms of prion disease.[45,46]

Finally, PMCA has been applied to study the molecular mechanism of prion replication with the aim of understanding the role of different cellular components, such as metals, polyanions, and RNA to sustain or interfere with the process of PrP^{C} conversion.[47–50] In this regard, Ma and colleagues were able to generate bona fide prions using recombinant PrP with the addition of RNA and lipids as catalyzers, thus providing a strong evidence in favor of the protein-only hypothesis.[51]

All these observations indicate that PMCA faithfully recapitulates all the features of the species barrier in a cell-free system and that new prion strains can be generated, adapted, and stabilized upon crossing species barrier in vitro.[33,52] For example, PMCA experiments demonstrated that, although not transmissible to TgHu mice, after serial cycles of amplification and adaptation, CWD-PrP^{Sc} is able to convert human PrP^{C} and produce PrP^{Sc} with different biochemical properties.[53,54] Therefore, the real ability of CWD to infect humans still need to be demonstrated.

Transmission of prions between humans (causing iatrogenic CJD—iCJD) occurred as a consequence of using inappropriate techniques to sterilize instruments that were previously used to manipulate sCJD prion-infected samples.[55] Additionally, corneal transplants,[56] dura mater grafts,[57] stereotactic EEG recordings,[58] and treatments with human-derived pituitary growth hormones[59] were also able to efficiently transmit CJD between humans. Sometimes, it is difficult to determine whether a CJD case might have occurred spontaneously or it has been iatrogenically transmitted. Recent observations indicate that PMCA is able to efficiently amplify trace amount of PrP^{Sc} collected from brains of iCJD cases, especially those characterized by the presence of PrP plaques.[60] Therefore, this feature might enable discrimination between sporadic and iatrogenic cases of CJD. Indeed, the efficiency of iCJD amplification is comparable to that of vCJD, likely because both types of PrP^{Sc} are acquired (with respect to sCJD that instead develops spontaneously) and characterized by the presence of amyloid plaques. The efficiency of PMCA to detect vCJD–PrP^{Sc} is extremely high and well documented.[61,62] In particular, PMCA enabled detection of minute amount of PrP^{Sc} circulating in blood of presymptomatic and symptomatic patients with vCJD[63–65] (see Section 3 for details).[66,67] Therefore, PMCA analysis of blood samples collected from donors that were potentially exposed to BSE might help to identify subjects at risk of developing vCJD and minimize the possibility of iatrogenic transmission of the disease between humans.

3. APPLICATIONS OF PMCA AS SUPPORTIVE TEST IN PRION DIAGNOSIS

Beyond the great applications of PMCA to study different aspects of prion biology, this technique is extremely sensitive and can detect very low levels of PrP^{Sc} present in different tissues of humans and animals affected by prion diseases. To date, PrP^{Sc} is the only disease-specific marker for prion diseases[13,68,69] and definitive diagnosis can only be made postmortem by neuropathological analysis aimed at demonstrating the presence of spongiform alterations and PrP^{Sc} accumulation in the brain.[13] Minute amounts of PrP^{Sc} are present in peripheral tissues (e.g., lymphoid tissues, cerebrospinal fluid, urine, and blood)[69–73] but, in many cases, the current analytic techniques (such as Western blot analysis) are not sensitive enough to detect them. With the help of PMCA, PrP^{Sc} can be amplified to a detectable level. For diagnostic purposes, urine and blood represent the best candidates to be used for setting up a rapid and noninvasive PMCA-based diagnostic test for different types of prion diseases. Through PMCA, PrP^{Sc} was detected in samples of urine or blood from prion-infected hamsters, mice, sheep, and cervids.[72–77] Recently, we and others have demonstrated that urine and blood collected in the symptomatic or presymptomatic stage of vCJD contain trace amount of PrP^{Sc} that can be efficiently amplified by means of PMCA.[66,67,78] These results confirm that undetectable level of prions circulate in blood of subjects silently incubating vCJD who may never develop clinical symptoms but remaining still able to transmit the disease to other subjects.[79] This represents a serious concern for public health, especially for patients who need periodical transfusions of blood. On the contrary, PMCA analysis of blood and urine collected from patients with sporadic CJD failed to demonstrate the presence of infectious PrP^{Sc}.[66,78] These data might indicate that (i) blood and urine of sCJD patients do not contain infectious prions or (ii) that the amplification of sCJD requires experimental settings different to those used to amplify vCJD. With the discovery that the bank vole's prion protein (BVPrP) is able to sustain the amplification of a multitude of prion strains belonging to different species,[80–83] PMCA was adapted using the bank vole brain homogenates as substrate of reaction. Such modifications enabled the amplification of particular human prion strains that were reluctant to amplify with the previous conditions (for example, using brain homogenates of TgHu mice), such as the sporadic CJD.[84] By taking advantage of this discovery, we have further improved PMCA for an efficient detection of PrP^{Sc} contained in the olfactory mucosa of patients with a genetic form of prion disease, known as fatal familial insomnia.[85] Taken together, these findings indicate that PMCA represents a great tool for an efficient, specific, and rapid detection of a variety of prions present in peripheral tissues, and can be used to set up an early and noninvasive diagnostic test for these devastating diseases. Early diagnosis of human prion diseases is of fundamental importance for improving therapeutic perspectives, as treatments should start in the early stage of the disease before severe and irreversible brain damages occur. As for humans, PMCA has been also successfully applied for the detection of different animal prion strains, including scrapie in hamsters,[86] mice,[77] and sheep,[87] BSE in cattle,[86,88] and CWD in cervids.[76]

4. PMCA APPLICATIONS IN DRUG SCREENING

Central to prion pathogenesis is the process of conversion of PrP^{C} to PrP^{Sc}. This process is self-propagating and PrP^{Sc} acts as a conformational template which recruits PrP^{C} and induces its misfolding.[89,90] Therefore, inhibition of this process is one of the most important targets for TSE therapies. All therapeutic approaches to prion diseases are generally assessed in cells and in a wide range of susceptible animals (wild-type, transgenic, and other more complex models). Unfortunately, in vivo experiments are long, time consuming and expensive. In the past, almost 50 different drugs were tested but all of them failed to induce any evident beneficial effect in prion-infected animals.[91] Moreover, the discovery that prions can undergo to a process of adaptation[38,40] suggests that, after the administration of specific antiprion compounds, resistant strains might arise. For this reason, considering the ability of PMCA to amplify prions with high fidelity while mimicking different aspects which occur in vivo (e.g., the species barrier and the phenomenon of prion adaptation), it represents an important and simple tool for performing a rapid and high-throughput screening of different compounds to evaluate their efficacy to interfere with the process of prion conversion. Fig. 2 shows an

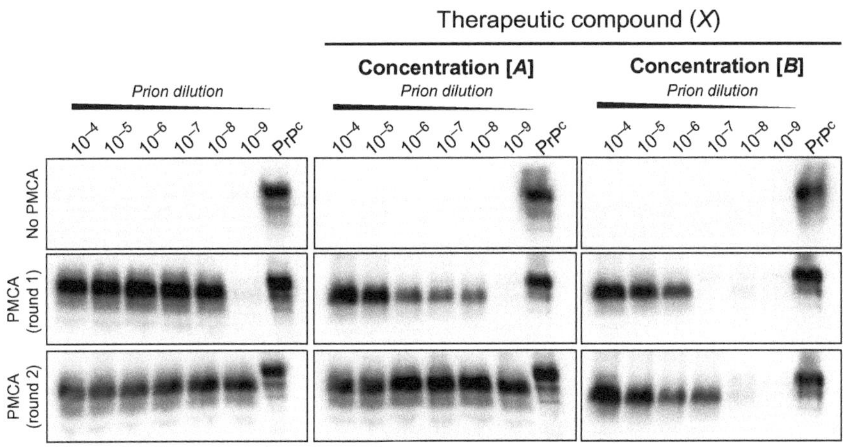

Fig. 2 Examples of PMCA application in drug screening. Two concentrations ([*A*] and [*B*]) of compound *X* are tested simultaneously in order to assess the most efficient one to interfere with the process of prion conversion. Different dilutions of PrP^{Sc} are prepared (ranging from 10^{-4} to 10^{-9}) and two rounds of amplification are performed. As easily appreciable, concentration [*B*] is more efficient to partially interfere with prion conversion if compared to concentration [*A*] and the control (without compound) where no differences in amplification efficiency were detected. PrP^{C} refers to normal prion protein.

example of PMCA application to study the inhibitory effects of specific compounds. Two concentrations (*A* and *B*) of the compound *X* are evaluated simultaneously. This compound is not able to completely block prion replication, however, concentration *B* efficiently reduced the magnitude of amplification if compared to controls (untreated with compound *X*) and samples treated with concentration *A*.

5. PMCA APPLICATIONS TO EVALUATE PRION INACTIVATION

Conventional sterilization methods are not able to inactivate prions.[92] The efficacy of specific reagents or procedures (for example autoclaving) to inactivate prions, are generally evaluated in vivo. In particular, susceptible animals are intracerebrally inoculated with decontaminated samples in order to assess the presence of residual infectivity. Unfortunately, this type of experiments is expensive and results can be collected after a long period of time. Therefore, PMCA represents a good alternative for a rapid assessment of the efficacy of specific sterilization procedures (chemical or physical) to completely remove prion infectivity.[93] This technique can detect smaller order of magnitude of prions than animal bioassay and has been applied to study the efficacy of different prion inactivation procedures.[94–97] Additionally, through quantitative PMCA, it is possible to estimate the concentration of infectious prions present in the sample before and, eventually, after treatment with inefficient decontaminants.[98]

6. PMCA AND PRION-LIKE DISORDERS

Compelling evidences suggest that, among neurodegenerative diseases, Alzheimer's disease (AD), frontotemporal dementia, Parkinson's disease (PD), Huntington disease, and amyotrophic lateral sclerosis might possess prion-like properties.[99,100] All these diseases are characterized by the accumulation of misfolded proteins in the brain which aggregate following the seeding nucleation model typical of prion diseases.[101,102] For instance, aggregates of Aβ and tau are found in the brain of patients with AD, while aggregates of α-synuclein are found in the brain of patients with PD. These proteins adopt a β-sheet-rich conformation and aggregate in a very efficient manner.[103] Extensive in vitro and in vivo studies demonstrated that the misfolding of these proteins can be transmitted in a prion-like manner.[104–109] Indeed, inoculation of pure or synthetic Aβ aggregates in transgenic animals of AD accelerates the progression of the disease thus demonstrating their ability to self-propagate and hence they might be considered as prions.[110] Similarly, inoculations of α-synuclein[111,112] and tau[113] in susceptible animals accelerated the development of the disease. Therefore, giving the similarities between prion diseases and other neurodegenerative disorders, PMCA can be extended for the amplification of the abnormal folding of these proteins as well. Very recent data indicate that optimized PMCA was able to efficiently and sensitively amplify the misfolding of α-synuclein[114] and tau[115] proteins. This feature might be exploited to (i) set up an early diagnostic test able to detect trace amount of specific pathological proteins circulating in peripheral tissues of patients in the early stage of dementia and (ii) perform a fast and high-throughput screening of compounds able to interfere with the process of protein aggregation.

REFERENCES

1. Gambetti P, Kong Q, Zou W, Parchi P, Chen SG. Sporadic and familial CJD: classification and characterisation. *Br Med Bull.* 2003;66:213–239.
2. Parchi P, Giese A, Capellari S, et al. Classification of sporadic Creutzfeldt–Jakob disease based on molecular and phenotypic analysis of 300 subjects. *Ann Neurol* 1999;46(2):224–233.
3. Jeffrey M, Gonzalez L. Classical sheep transmissible spongiform encephalopathies: pathogenesis, pathological phenotypes and clinical disease. *Neuropathol Appl Neurobiol* 2007;33(4):373–394.
4. Sigurdson CJ. A prion disease of cervids: chronic wasting disease. *Vet Res* 2008;39(4):41.
5. Novakofski J, Brewer MS, Mateus-Pinilla N, Killefer J, McCusker RH. Prion biology relevant to bovine spongiform encephalopathy. *J Anim Sci* 2005;83(6):1455–1476.
6. Hill AF, Desbruslais M, Joiner S, et al. The same prion strain causes vCJD and BSE. *Nature* 1997;389(6650):448–450526.
7. Prusiner SB. Novel proteinaceous infectious particles cause scrapie. *Science* 1982;216(4542):136–144.
8. Soto C, Saborio GP. Prions: disease propagation and disease therapy by conformational transmission. *Trends Mol Med* 2001;7(3):109–114.
9. Stahl N, Baldwin MA, Teplow DB, et al. Structural studies of the scrapie prion protein using mass spectrometry and amino acid sequencing. *Biochemistry* 1993;32(8): 1991–2002.
10. Pan KM, Baldwin M, Nguyen J, et al. Conversion of alpha-helices into beta-sheets features in the formation of the scrapie prion proteins. *Proc Natl Acad Sci USA.* 1993;90(23):10962–10966.
11. Bolton DC, McKinley MP, Prusiner SB. Identification of a protein that purifies with the scrapie prion. *Science.* 1982;218(4579):1309–1311.
12. McKinley MP, Bolton DC, Prusiner SB. A protease-resistant protein is a structural component of the scrapie prion. *Cell.* 1983;35(1):57–62.

13. Soto C. Diagnosing prion diseases: needs, challenges and hopes. *Nat Rev Microbiol.* 2004;2(10):809–819.
14. Safar JG, Geschwind MD, Deering C, et al. Diagnosis of human prion disease. *Proc Natl Acad Sci USA.* 2005;102(9):3501–3506.
15. Soto C. Prion hypothesis: the end of the controversy? *Trends Biochem Sci.* 2011;36(3):151–158.
16. Saborio GP, Permanne B, Soto C. Sensitive detection of pathological prion protein by cyclic amplification of protein misfolding. *Nature.* 2001;411(6839):810–813.
17. Saa P, Castilla J, Soto C. Ultra-efficient replication of infectious prions by automated protein misfolding cyclic amplification. *J Biol Chem.* 2006;281(46): 35245–35252.
18. Bieschke J, Weber P, Sarafoff N, Beekes M, Giese A, Kretzschmar H. Autocatalytic self-propagation of misfolded prion protein. *Proc Natl Acad Sci USA.* 2004;101(33): 12207–12211.
19. Castilla J, Saa P, Hetz C, Soto C. In vitro generation of infectious scrapie prions. *Cell.* 2005;121(2):195–206.
20. Gonzalez-Montalban N, Makarava N, Ostapchenko VG, et al. Highly efficient protein misfolding cyclic amplification. *PLoS Pathog.* 2011;7(2), e1001277.
21. Shikiya RA, Bartz JC. In vitro generation of high-titer prions. *J Virol.* 2011;85(24):13439–13442.
22. Collinge J. Mammalian prions and their wider relevance in neurodegenerative diseases. *Nature.* 2016;539(7628):217–226.
23. Hill AF, Joiner S, Linehan J, Desbruslais M, Lantos PL, Collinge J. Species-barrier-independent prion replication in apparently resistant species. *Proc Natl Acad Sci USA.* 2000;97(18):10248–10253.
24. Collinge J. Prion diseases of humans and animals: their causes and molecular basis. *Annu Rev Neurosci.* 2001;24:519–550.
25. Hill AF, Collinge J. Prion strains and species barriers. *Contrib Microbiol.* 2004;11:33–49.
26. Asante EA, Linehan JM, Desbruslais M, et al. BSE prions propagate as either variant CJD-like or sporadic CJD-like prion strains in transgenic mice expressing human prion protein. *EMBO J.* 2002;21(23):6358–6366.
27. Beringue V, Herzog L, Reine F, et al. Transmission of atypical bovine prions to mice transgenic for human prion protein. *Emerg Infect Dis.* 2008;14(12): 1898–1901.
28. Cassard H, Torres JM, Lacroux C, et al. Evidence for zoonotic potential of ovine scrapie prions. *Nat Commun.* 2014;5:5821.
29. Padilla D, Beringue V, Espinosa JC, et al. Sheep and goat BSE propagate more efficiently than cattle BSE in human PrP transgenic mice. *PLoS Pathog.* 2011;7(3), e1001319.
30. Sandberg MK, Al-Doujaily H, Sigurdson CJ, et al. Chronic wasting disease prions are not transmissible to transgenic mice overexpressing human prion protein. *J Gen Virol.* 2010;91(pt 10):2651–2657.
31. Levavasseur E, Privat N, Martin JC, et al. Molecular modeling of prion transmission to humans. *Virus.* 2014;6(10):3766–3777.
32. Weber P, Giese A, Piening N, et al. Generation of genuine prion infectivity by serial PMCA. *Vet Microbiol.* 2007;123(4):346–357.
33. Castilla J, Gonzalez-Romero D, Saa P, Morales R, De Castro J, Soto C. Crossing the species barrier by PrP(Sc) replication in vitro generates unique infectious prions. *Cell.* 2008;134(5):757–768.
34. Green KM, Castilla J, Seward TS, et al. Accelerated high fidelity prion amplification within and across prion species barriers. *PLoS Pathog.* 2008;4(8), e1000139.

35. Meyerett C, Michel B, Pulford B, et al. In vitro strain adaptation of CWD prions by serial protein misfolding cyclic amplification. *Virology*. 2008;382(2):267–276.
36. Bossers A, Belt P, Raymond GJ, Caughey B, de Vries R, Smits MA. Scrapie susceptibility-linked polymorphisms modulate the in vitro conversion of sheep prion protein to protease-resistant forms. *Proc Natl Acad Sci USA*. 1997;94(10): 4931–4936.
37. Raymond GJ, Hope J, Kocisko DA, et al. Molecular assessment of the potential transmissibilities of BSE and scrapie to humans. *Nature*. 1997;388(6639):285–288.
38. Baskakov IV. The many shades of prion strain adaptation. *Prion*. 2014;8(2).
39. Kimberlin RH, Walker CA. Evidence that the transmission of one source of scrapie agent to hamsters involves separation of agent strains from a mixture. *J Gen Virol*. 1978;39(3):487–496.
40. Li J, Browning S, Mahal SP, Oelschlegel AM, Weissmann C. Darwinian evolution of prions in cell culture. *Science*. 2010;327(5967):869–872.
41. Ghaemmaghami S, Ahn M, Lessard P, et al. Continuous quinacrine treatment results in the formation of drug-resistant prions. *PLoS Pathog*. 2009;5(11), e1000673.
42. Makarava N, Savtchenko R, Baskakov IV. Selective amplification of classical and atypical prions using modified protein misfolding cyclic amplification. *J Biol Chem*. 2013;288(1):33–41.
43. Li J, Mahal SP, Demczyk CA, Weissmann C. Mutability of prions. *EMBO Rep*. 2011;12(12):1243–1250.
44. Gonzalez-Montalban N, Baskakov IV. Assessment of strain-specific PrP(Sc) elongation rates revealed a transformation of PrP(Sc) properties during protein misfolding cyclic amplification. *PLoS One*. 2012;7(7), e41210.
45. Deleault NR, Harris BT, Rees JR, Supattapone S. Formation of native prions from minimal components in vitro. *Proc Natl Acad Sci USA*. 2007;104(23):9741–9746.
46. Barria MA, Mukherjee A, Gonzalez-Romero D, Morales R, Soto C. De novo generation of infectious prions in vitro produces a new disease phenotype. *PLoS Pathog*. 2009;5(5), e1000421.
47. Lucassen R, Nishina K, Supattapone S. In vitro amplification of protease-resistant prion protein requires free sulfhydryl groups. *Biochemistry*. 2003;42(14):4127–4135.
48. Deleault NR, Lucassen RW, Supattapone S. RNA molecules stimulate prion protein conversion. *Nature*. 2003;425(6959):717–720.
49. Nishina K, Jenks S, Supattapone S. Ionic strength and transition metals control PrPSc protease resistance and conversion-inducing activity. *J Biol Chem*. 2004;279(39): 40788–40794.
50. Orem NR, Geoghegan JC, Deleault NR, Kascsak R, Supattapone S. Copper (II) ions potently inhibit purified PrPres amplification. *J Neurochem*. 2006;96(5): 1409–1415.
51. Wang F, Wang X, Yuan CG, Ma J. Generating a prion with bacterially expressed recombinant prion protein. *Science*. 2010;327(5969):1132–1135.
52. Barria MA, Gonzalez-Romero D, Soto C. Cyclic amplification of prion protein misfolding. *Methods Mol Biol*. 2012;849:199–212.
53. Barria MA, Telling GC, Gambetti P, Mastrianni JA, Soto C. Generation of a new form of human PrP(Sc) in vitro by interspecies transmission from cervid prions. *J Biol Chem*. 2011;286(9):7490–7495.
54. Johnson CJ, Aiken JM, McKenzie D, Samuel MD, Pedersen JA. Highly efficient amplification of chronic wasting disease agent by protein misfolding cyclic amplification with beads (PMCAb). *PLoS One*. 2012;7(4), e35383.
55. Williams D. New thoughts on CJD, surgical instruments and disease transmission. *Med Device Technol*. 2006;17(1):7–8.

56. Duffy P, Wolf J, Collins G, DeVoe AG, Streeten B, Cowen D. Letter: possible person-to-person transmission of Creutzfeldt–Jakob disease. *N Engl J Med*. 1974;290(12): 692–693.
57. Centers for Disease Control and Prevention. Creutzfeldt–Jakob disease associated with cadaveric dura mater grafts—Japan, January 1979–May 1996. *MMWR Morb Mortal Wkly Rep*. 1997;46(45):1066–1069.
58. Bernoulli C, Siegfried J, Baumgartner G, et al. Danger of accidental person-to-person transmission of Creutzfeldt–Jakob disease by surgery. *Lancet*. 1977;1(8009):478–479.
59. Gibbs Jr CJ, Joy A, Heffner R, et al. Clinical and pathological features and laboratory confirmation of Creutzfeldt–Jakob disease in a recipient of pituitary-derived human growth hormone. *N Engl J Med*. 1985;313(12):734–738.
60. Xiao X, Yuan J, Qing L, et al. Comparative study of prions in iatrogenic and sporadic Creutzfeldt–Jakob disease. *J Clin Cell Immunol*. 2014;5(4), pii: 240.
61. Jones M, Peden AH, Prowse CV, et al. In vitro amplification and detection of variant Creutzfeldt–Jakob disease PrPSc. *J Pathol*. 2007;213(1):21–26.
62. Lacroux C, Comoy E, Moudjou M, et al. Preclinical detection of variant CJD and BSE prions in blood. *PLoS Pathog*. 2014;10(6), e1004202.
63. Davidson LR, Llewelyn CA, Mackenzie JM, Hewitt PE, Will RG. Variant CJD and blood transfusion: are there additional cases? *Vox Sang*. 2014;107(3):220–225.
64. Wallis JP. Variant CJD and blood transfusion. *Vox Sang*. 2015;108(4):432–433.
65. Llewelyn CA, Hewitt PE, Knight RS, et al. Possible transmission of variant Creutzfeldt–Jakob disease by blood transfusion. *Lancet*. 2004;363(9407):417–421.
66. Concha-Marambio L, Pritzkow S, Moda F, et al. Detection of prions in blood from patients with variant Creutzfeldt–Jakob disease. *Sci Transl Med*. 2016;8(370): 370ra183.
67. Bougard D, Brandel JP, Belondrade M, et al. Detection of prions in the plasma of presymptomatic and symptomatic patients with variant Creutzfeldt–Jakob disease. *Sci Transl Med*. 2016;8(370):370ra182.
68. Prusiner SB. Prions. *Proc Natl Acad Sci USA*. 1998;95(23):13363–13383.
69. Brown P, Cervenakova L, Diringer H. Blood infectivity and the prospects for a diagnostic screening test in Creutzfeldt–Jakob disease. *J Lab Clin Med*. 2001;137(1):5–13.
70. Aguzzi A. Prion diseases, blood and the immune system: concerns and reality. *Haematologica*. 2000;85(1):3–10.
71. Wadsworth JD, Joiner S, Hill AF, et al. Tissue distribution of protease resistant prion protein in variant Creutzfeldt–Jakob disease using a highly sensitive immunoblotting assay. *Lancet*. 2001;358(9277):171–180.
72. Gonzalez-Romero D, Barria MA, Leon P, Morales R, Soto C. Detection of infectious prions in urine. *FEBS Lett*. 2008;582(21 – 22):3161–3166.
73. Castilla J, Saa P, Soto C. Detection of prions in blood. *Nat Med*. 2005;11(9):982–985.
74. Saa P, Castilla J, Soto C. Presymptomatic detection of prions in blood. *Science*. 2006;313(5783):92–94.
75. Thorne L, Terry LA. In vitro amplification of PrPSc derived from the brain and blood of sheep infected with scrapie. *J Gen Virol*. 2008;89(pt 12):3177–3184.
76. Haley NJ, Seelig DM, Zabel MD, Telling GC, Hoover EA. Detection of CWD prions in urine and saliva of deer by transgenic mouse bioassay. *PLoS One*. 2009;4(3), e4848.
77. Tattum MH, Jones S, Pal S, Collinge J, Jackson GS. Discrimination between prion-infected and normal blood samples by protein misfolding cyclic amplification. *Transfusion*. 2010;50(5):996–1002.
78. Moda F, Gambetti P, Notari S, et al. Prions in the urine of patients with variant Creutzfeldt–Jakob disease. *N Engl J Med*. 2014;371(6):530–539.
79. Bishop MT, Hart P, Aitchison L, et al. Predicting susceptibility and incubation time of human-to-human transmission of vCJD. *Lancet Neurol*. 2006;5(5):393–398.

80. Nonno R, Di Bari MA, Cardone F, et al. Efficient transmission and characterization of Creutzfeldt–Jakob disease strains in bank voles. *PLoS Pathog*. 2006;2(2), e12.
81. Chandler RL, Turfrey BA. Inoculation of voles, Chinese hamsters, gerbils and guinea-pigs with scrapie brain material. *Res Vet Sci*. 1972;13(3):219–224.
82. Agrimi U, Nonno R, Dell'Omo G, et al. Prion protein amino acid determinants of differential susceptibility and molecular feature of prion strains in mice and voles. *PLoS Pathog*. 2008;4(7), e1000113.
83. Di Bari MA, Nonno R, Castilla J, et al. Chronic wasting disease in bank voles: characterisation of the shortest incubation time model for prion diseases. *PLoS Pathog*. 2013;9(3), e1003219.
84. Cosseddu GM, Nonno R, Vaccari G, et al. Ultra-efficient PrP(Sc) amplification highlights potentialities and pitfalls of PMCA technology. *PLoS Pathog*. 2011;7(11), e1002370.
85. Redaelli V, Bistaffa E, Zanusso G, et al. Detection of prion seeding activity in the olfactory mucosa of patients with fatal familial insomnia. *Sci Rep*. 2017;7:46269.
86. Soto C, Anderes L, Suardi S, et al. Pre-symptomatic detection of prions by cyclic amplification of protein misfolding. *FEBS Lett*. 2005;579(3):638–642.
87. Thorne L, Holder T, Ramsay A, et al. In vitro amplification of ovine prions from scrapie-infected sheep from Great Britain reveals distinct patterns of propagation. *BMC Vet Res*. 2012;8:223.
88. Murayama Y, Yoshioka M, Masujin K, et al. Sulfated dextrans enhance in vitro amplification of bovine spongiform encephalopathy PrP(Sc) and enable ultrasensitive detection of bovine PrP(Sc). *PLoS One*. 2010;5(10).
89. Prusiner SB. Molecular biology of prion diseases. *Science*. 1991;252(5012):1515–1522.
90. Prusiner SB. Molecular biology and pathogenesis of prion diseases. *Trends Biochem Sci*. 1996;21(12):482–487.
91. Fernandez-Borges N, Elezgarai SR, Erana H, Castilla J. Animal models for testing anti-prion drugs. *Curr Top Med Chem*. 2013;13(19):2504–2521.
92. Taylor DM. Resistance of transmissible spongiform encephalopathy agents to decontamination. *Contrib Microbiol*. 2001;7:58–67.
93. Murayama Y, Yoshioka M, Horii H, et al. Protein misfolding cyclic amplification as a rapid test for assessment of prion inactivation. *Biochem Biophys Res Commun*. 2006;348(2):758–762.
94. Pritzkow S, Wagenfuhr K, Daus ML, et al. Quantitative detection and biological propagation of scrapie seeding activity in vitro facilitate use of prions as model pathogens for disinfection. *PLoS One*. 2011;6(5), e20384.
95. Morales R, Buytaert-Hoefen KA, Gonzalez-Romero D, et al. Reduction of prion infectivity in packed red blood cells. *Biochem Biophys Res Commun*. 2008;377(2):373–378.
96. Saunders SE, Bartz JC, Vercauteren KC, Bartelt-Hunt SL. An enzymatic treatment of soil-bound prions effectively inhibits replication. *Appl Environ Microbiol*. 2011;77(13):4313–4317.
97. Ding N, Neumann NF, Price LM, et al. Inactivation of template-directed misfolding of infectious prion protein by ozone. *Appl Environ Microbiol*. 2012;78(3):613–620.
98. Chen B, Morales R, Barria MA, Soto C. Estimating prion concentration in fluids and tissues by quantitative PMCA. *Nat Methods*. 2010;7(7):519–520.
99. Brundin P, Melki R, Kopito R. Prion-like transmission of protein aggregates in neurodegenerative diseases. *Nat Rev Mol Cell Biol*. 2010;11(4):301–307.
100. Frost B, Diamond MI. Prion-like mechanisms in neurodegenerative diseases. *Nat Rev Neurosci*. 2010;11(3):155–159.
101. Soto C, Estrada L, Castilla J. Amyloids, prions and the inherent infectious nature of misfolded protein aggregates. *Trends Biochem Sci*. 2006;31(3):150–155.

102. Soto C. Unfolding the role of protein misfolding in neurodegenerative diseases. *Nat Rev Neurosci.* 2003;4(1):49–60.
103. Prusiner SB. Biology and genetics of prions causing neurodegeneration. *Annu Rev Genet.* 2013;47:601–623.
104. Meyer-Luehmann M, Coomaraswamy J, Bolmont T, et al. Exogenous induction of cerebral beta-amyloidogenesis is governed by agent and host. *Science.* 2006;313(5794):1781–1784.
105. Clavaguera F, Bolmont T, Crowther RA, et al. Transmission and spreading of tauopathy in transgenic mouse brain. *Nat Cell Biol.* 2009;11(7):909–913.
106. Frost B, Jacks RL, Diamond MI. Propagation of tau misfolding from the outside to the inside of a cell. *J Biol Chem.* 2009;284(19):12845–12852.
107. Munch C, O'Brien J, Bertolotti A. Prion-like propagation of mutant superoxide dismutase-1 misfolding in neuronal cells. *Proc Natl Acad Sci USA.* 2011;108(9):3548–3553.
108. Pearce MM, Spartz EJ, Hong W, Luo L, Kopito RR. Prion-like transmission of neuronal huntingtin aggregates to phagocytic glia in the drosophila brain. *Nat Commun.* 2015;6:6768.
109. Mougenot AL, Nicot S, Bencsik A, et al. Prion-like acceleration of a synucleinopathy in a transgenic mouse model. *Neurobiol Aging.* 2012;33(9):2225–2228.
110. Stohr J, Watts JC, Mensinger ZL, et al. Purified and synthetic Alzheimer's amyloid beta (Abeta) prions. *Proc Natl Acad Sci USA.* 2012;109(27):11025–11030.
111. Bernis ME, Babila JT, Breid S, Wusten KA, Wullner U, Tamguney G. Prion-like propagation of human brain-derived alpha-synuclein in transgenic mice expressing human wild-type alpha-synuclein. *Acta Neuropathol Commun.* 2015;3:75.
112. Luk KC, Kehm V, Carroll J, et al. Pathological alpha-synuclein transmission initiates Parkinson-like neurodegeneration in nontransgenic mice. *Science.* 2012;338(6109):949–953.
113. Sanders DW, Kaufman SK, DeVos SL, et al. Distinct tau prion strains propagate in cells and mice and define different tauopathies. *Neuron.* 2014;82(6):1271–1288.
114. Herva ME, Zibaee S, Fraser G, Barker RA, Goedert M, Spillantini MG. Anti-amyloid compounds inhibit alpha-synuclein aggregation induced by protein misfolding cyclic amplification (PMCA). *J Biol Chem.* 2014;289(17):11897–11905.
115. Meyer V, Dinkel PD, Rickman Hager E, Margittai M. Amplification of tau fibrils from minute quantities of seeds. *Biochemistry.* 2014;53(36):5804–5809.

CHAPTER SEVENTEEN

Amplified Detection of Prions and Other Amyloids by RT-QuIC in Diagnostics and the Evaluation of Therapeutics and Disinfectants

Byron Caughey[1], Christina D. Orru, Bradley R. Groveman, Andrew G. Hughson, Matteo Manca, Lynne D. Raymond, Gregory J. Raymond, Brent Race, Eri Saijo, Allison Kraus
Laboratory of Persistent Viral Diseases, Rocky Mountain Laboratories, National Institute for Allergy and Infectious Diseases, National Institutes of Health, Hamilton, MT, United States
[1]Corresponding author: e-mail address: bcaughey@nih.gov

Contents

Abstract

Among the most sensitive, specific and practical of methods for detecting prions are the real-time quaking-induced conversion (RT-QuIC) assays. These assays exploit the fundamental self-propagating activity of prions to amplify the presence of prion seeds by as much as a trillion-fold. The reactions can detect most of the known mammalian prion diseases, often with sensitivities greater than those of animal bioassays. RT-QuIC assays

Progress in Molecular Biology and Translational Science, Volume 150
ISSN 1877-1173
http://dx.doi.org/10.1016/bs.pmbts.2017.06.003

are performed in multiwell plates with fluorescence detection and have now reached the sensitivity and practicality required for routine prion disease diagnostics. Some key strains of prions within particular host species, e.g., humans, cattle, and sheep, can be discriminated by comparison of RT-QuIC responses with different recombinant prion protein substrates. The most thoroughly validated diagnostic application of RT-QuIC is in the diagnosis of sporadic Creutzfeldt–Jakob disease (sCJD) using cerebrospinal fluid. Diagnostic sensitivities as high as 96% can be achieved in less than 24 h with specificities of 98%–100%. The ability, if needed, to also test nasal swab samples can increase the RT-QuIC sensitivity for sCJD to virtually 100%. In addition to diagnostic applications, RT-QuIC has also been used in the testing of prion disinfectants and potential therapeutics. Mechanistically related assays are also now being developed for other protein misfolding diseases.

1. INTRODUCTION

The fact that prions are composed of host molecules and lack agent-specific nucleic genomes has long impeded the detection of prion infections by common serological or PCR methods that are used to detect conventional pathogens. Consequently, guidelines for the definitive diagnosis of prion diseases have required detection of abnormal prion protein (PrP) in neurological tissue, which in most cases, is only collected postmortem. In humans, antemortem diagnoses of suspected or probable sporadic Creutzfeldt–Jakob disease (sCJD) can be made using a confluence of clinical signs, MRI imaging, electroencephalography, and analyses of nonspecific biomarkers in cerebrospinal fluid (CSF), but the accuracy of such determinations has been suboptimal. Accordingly, many labs have worked toward development of ultrasensitive tests for prion-associated forms of PrP, e.g., PrP^{Sc}, including prion seed-amplification reactions such as protein misfolding cyclic amplification (PMCA),[1,2] the amyloid seeding assay,[3] quaking-induced conversion (QuIC),[4] and real-time QuIC (RT-QuIC)[5,6] (reviewed in Refs. 7,8). In this context, "seed" is used to mean PrP oligomers or multimers that are capable of growing by incorporating more PrP monomers.[9–11]

RT-QuIC assays were originated by Atarashi and colleagues for human applications[6] and Wilham and colleagues for animal applications.[5] They have become among the more practical and widely applied amplification assays for prion disease diagnostics (reviewed in Refs. 8,12–14). In RT-QuIC reactions, test samples that may or may not contain prion (PrP^{Sc}) seeds are mixed with a large stoichiometric excess of recombinant

PrP in its normal monomeric, nonamyloid, proteinase K-sensitive form ($rPrP^{Sen}$, i.e., the "substrate") (Fig. 1). Any prion seeds in the test sample can then recruit $rPrP^{Sen}$ substrate molecules as they grow into larger amyloid fibrils that are then detected by an amyloid-sensitive fluorescent dye, such as thioflavin T. This seeded polymerization mechanism can be extraordinarily efficient, allowing billion- to trillion-fold seed amplifications. As a result attograms of PrP^{Sc}, representing approximately 10–40 particles, can be detected.[5,16,17] Large studies by multiple groups have confirmed the antemortem diagnostic accuracy of RT-QuIC for human CJD.[7,13,17–27] There have also been many studies describing applications of RT-QuIC to the detection and diagnosis of many animal prion diseases. Recent developments include applications of RT-QuIC to drug screening,[28,29] the monitoring of therapeutic trials,[30] and the identification and characterization of new prion disinfectants.[31] Step-wise protocols and practical notes on

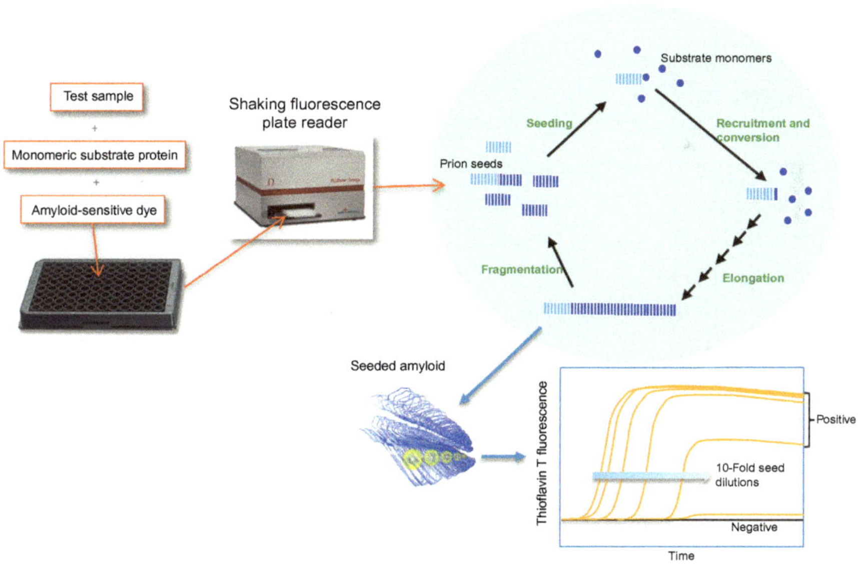

Fig. 1 RT-QuIC seeded polymerization assay format. Prion or prion-like multimeric protein seeds in a test sample are mixed with a vast excess of monomeric substrate molecules in 96-well plates. The seeds incorporate the monomers and induce their conformational conversion into amyloid fibrils that enhance the fluorescence of the dye thioflavin T (ThT) as the reaction progresses in "real time." With dilution of the seed sample, the lag phase prior to increase in fluorescence tends to increase and the maximum fluorescence decreases. The rPrP amyloid fibril model shown here is a hypothetical computational model based on a parallel in-register intermolecular beta sheet architecture.[15]

performing RT-QuIC assays will be published elsewhere (Or ru *et al.*, Meth Mol Biol, in press) and are available upon request. Here we provide an update on recent developments in the use of RT-QuIC for these various applications.

2. DIAGNOSTIC APPLICATIONS OF RT-QuIC

2.1 Human CJD Diagnosis Using CSF

CSF is usually collected from patients with rapidly progressive dementia or other clinical signs that are suggestive of human prion disease. Analyses of 14-3-3 and tau protein levels in CSF have been helpful in prion disease diagnosis, but elevated levels of these biomarkers of neuronal damage are not specific to prion diseases. In contrast, RT-QuIC measures what appear to be the etiologic agents and positive responses have proven to be almost 100% specific for prion disease. Atarashi and colleagues first reported of the diagnostic utility of RT-QuIC of CSF specimens for sCJD.[6] Since then, multiple studies have demonstrated diagnostic sensitivities of 77%–89% and specificities of 98%–100% for CJD using various first-generation RT-QuIC reaction conditions.[7,18–20,23,24,27,32] This specificity is higher than those obtained from measurements of 14-3-3 and tau. Helpful practical tips for analyzing human CSF specimens by RT-QuIC have been provided by Cramm and colleagues.[24] Overall, these studies attest to the utility, specificity, and robust transferability of the RT-QuIC technology to many laboratories for antemortem diagnosis for sCJD.

2.2 Second-Generation RT-QuIC Assay for CJD Diagnosis Using CSF

The fact that a significant percentage of sCJD cases (11%–23%, depending upon the study) were missed by the first-generation RT-QuIC assays indicated room for improvement. This prompted the development of an improved RT-QuIC assay for CSF (IQ-CSF), which is more sensitive and rapid.[20] In contrast to the 2.5–5 days required for the first-generation CSF RT-QuIC analyses, the IQ-CSF assay usually identifies positive CSF samples within 4–14 h with better analytical sensitivity. Applications of IQ-CSF to blinded panels of sCJD cases and controls by one group of investigators have cumulatively obtained positive RT-QuIC assays from 152 of 161 CJD cases but not from 103 non-CJD patients, giving 94% (CI: 90%–97%) diagnostic sensitivity and 100% specificity (CI:

97%–100%).[20,26] Another large independent study that included both retrospective and prospective cohorts of a total of 272 neuropathologically confirmed cases and controls gave IQ-CSF sensitivities of 92%–95% and specificities of 98.5%–100%.[17] These results indicate that IQ-CSF RT-QuIC should enhance the speed and accuracy of antemortem CJD diagnosis.

2.3 Nasal Brushings as Diagnostic Specimens

Despite enhancements in the diagnostic accuracy and practicality provided by IQ-CSF testing, it remained important to further improve the sensitivity and to have other types of accessible diagnostic specimens that could be tested in the event that CSF is unavailable or RT-QuIC-negative despite other clinical signs of prion disease. One such specimen is the olfactory mucosa (OM) which Dr. Gianluigi Zanusso has shown can be collected by gently brushing the upper nasal cavity.[21,22] Earlier studies by Zanusso and colleagues had demonstrated the detection of pathologic PrP in olfactory neuroepithelium by immunological methods.[33] Moreover, Bessen and colleagues had detected RT-QuIC prion-seeding activity in lavages from scrapie-infected hamsters.[5,34,35] When nasal brushings from human CJD cases ($n=43$) and non-CJD controls ($n=43$) were tested, a diagnostic sensitivity of $\geq$97% and specificity of 100% were obtained (cumulative data from two reports).[21,22]

In these initial studies, OM samples were collected using disposable cytobrushes that might cause mild discomfort to the patient or recovery of blood cells. However, softer, less abrasive, flexible flocked swabs can provide gentler sampling without reducing the recovery of prion-seeding activity or the overall RT-QuIC diagnostic sensitivity.[25] When OM, CSF, or both were analyzed by RT-QuIC, the overall diagnostic sensitivity for sCJD patients ($n=61$) improved to 100% (CI: 93%–100%) while maintaining 100% specificity with a total of 66 non-CJD controls. Importantly, negative RT-QuIC results were even obtained for 16 patients who were initially suspected of having CJD, but were ultimately given other nonprion disease diagnoses upon neuropathological examination. Thus, RT-QuIC analyses of CSF and/or OM specimens have allowed decisive discrimination of sCJD and non-CJD patients.

With regard to genetic prion diseases of humans, Bongianni and colleagues have shown that RT-QuIC analysis of eight symptomatic cases with genetic mutations causing CJD or Gerstmann–Straussler–Scheinker syndrome (GSS) gave 75% sensitivity.[25] In addition, Redaelli and colleagues have recently demonstrated that OM specimens collected from two patients with the genetic prion disease fatal familial insomnia have prion-seeding activity that is detectable by both RT-QuIC and PMCA.[36]

2.4 Bovine Spongiform Encephalopathies

Classical (C-BSE) and atypical (H-BSE and L-BSE) forms of BSE can be detected and discriminated in brain tissue using RT-QuIC assays with different substrates.[37,38] Although RT-QuIC seeding activity can often be detected in the CSF of cattle infected with H-BSE, effective antemortem diagnosis of the various forms of BSE awaits the identification of more accessible diagnostic specimens with adequate seeding activity.

2.5 Sheep and Goat Scrapie

As is the case with BSE in cattle, both classical and atypical (Nor98) strains of scrapie have been identified in sheep. Both types of scrapie can be detected using RT-QuIC assays with different $rPrP^{Sen}$ substrates.[5,39] For example, classical scrapie has been detected down to 10^{-8} dilutions of scrapie brain tissue using a hamster-sheep (Ha-S) chimeric $rPrP^{Sen}$ substrate.[39] However, this substrate was not able to detect atypical Nor98 scrapie. In contrast, a bank vole $rPrP^{Sen}$ substrate detected both classical and Nor98 scrapie. By comparing the RT-QuIC responses using these two $rPrP^{Sen}$ substrates, one can discriminate these scrapie strains.

RT-QuIC assays have also been recently adapted to the detection of scrapie in goats.[40] In this application, RT-QuIC is at least 10^4-fold more sensitive than the ELISA and Western blot assays for the detection of scrapie seeding activity in brain tissue.

2.6 Cervid Chronic Wasting Disease

A number of further applications of RT-QuIC for the detection of prion-seeding activity in chronic wasting disease (CWD)-infected cervids have been described. These include the detection of prions in brain tissue,[5] urine, feces,[41] saliva,[42] blood,[43] CSF,[44] lymph nodes,[45] nasal brushings, and RAMALT,[46,47] in some cases even in preclinical animals. RT-QuIC tests of these samples should be useful in diagnostics and surveillance, as well as in basic studies of CWD biology and transmission.

2.7 Virtually Universal Detection of Prions With a Single Substrate

As noted earlier many different $rPrP^{Sen}$ substrate sequences and reaction conditions have been developed in efforts to improve the detection of various mammalian prions. However, it is also possible to sensitively detect most, if not all, known prions of mammals using a single $rPrP^{Sen}$ substrate derived from bank voles.[39] Indeed, this assay has been applied successfully to 28 different prion diseases from humans, cattle, sheep, cervids, and rodents, including several that had previously been undetectable by RT-QuIC or PMCA. Moreover, comparison of the relative RT-QuIC amplification kinetics using bank vole $rPrP^{Sen}$, and other $rPrP^{Sen}$ molecules, has allowed discrimination of different strains of BSE,[37] scrapie,[39] and Creutzfeldt–Jakob disease.[39] Western blot analysis of protease-resistant RT-QuIC conversion products generated using bank vole $rPrP^{Sen}$ has also aided the discrimination of classical and atypical Nor98 scrapie in sheep[39] and H- and L-type BSE in cattle.[37]

3. QUANTITATION OF RELATIVE SEED CONCENTRATIONS

Both end-point dilution and lag phase analyses have been used to compare relative amounts of seeding activity in samples using RT-QuIC assays.[5,20,21,34,48,49] With end-point dilution analysis serial dilutions of the sample are tested in replicate wells. The dilution (or sample dose) that gives positive RT-QuIC reactions in 50% of the replicate reactions (i.e., the SD_{50}) is estimated, often with the aid of Spearman–Kärber analysis.[50] The SD_{50} is analogous to the 50% lethal dose (LD_{50}) in animal bioassays. The concentration of SD_{50} units per amount of the original specimen can then be calculated as a measure of the relative concentrations of prion seeds. Gray and colleagues have provided further helpful statistical considerations in discriminating prion disease-positive and -negative samples when seed concentrations are near the end point.[51]

Applications of end-point dilution RT-QuIC have allowed quantitation of prion-seeding activity even in tissue samples with PrP^{Sc} that is undetectable by more conventional means such as immunohistochemistry or Western blotting. This is presumably because of RT-QuIC's greater sensitivity and/or ability to detect less aggregated or PK-resistant forms of PrP^{Sc}.[49,52] This ability has aided the characterization of the regional specificity of cellular responses and resulting pathological implications of PrP^{Sc} forms that were previously invisible.[52]

Lag phases in RT-QuIC reactions can also provide quantitative comparisons of seeding activities based on observations of an inverse relationship between the prion seed concentration in the sample.[5,53] The use of this correlation has been formalized by comparison of experimental prion-seeded polymerization lag phases to standard curves[51,54,55] which reduces the number of dilutions of a given test sample that are required for quantitation. This advantage is partially offset, however, by the need to include additional reactions to establish a standard curve for each experiment. Lag phases for a given seed sample can vary between sample types (matrix effects), experiments, and plate readers. Thus, the inclusion of standard curves on each plate, using dilutions of the same sample type, is advisable. Nonetheless, under properly controlled conditions these kinetics-based approaches to RT-QuIC quantitation can be useful for comparing closely similar sample types.

4. THERAPEUTIC APPLICATIONS OF RT-QuIC

There are several ways that RT-QuIC assays might be helpful in the development of prion disease therapeutics.[8,12] First, the demonstrated antemortem diagnostic accuracy of RT-QuIC should improve cohort selection in therapeutic trials. Second, because RT-QuIC measures a disease-specific and etiological biomarker, it could be used to monitor the efficacy of therapeutics that target prion/PrP^{Sc} accumulation. In this type of application, the ability to collect serial diagnostic specimens as noninvasively as possible would be important. Nasal swab samples from the OM are attractive in this regard because they can be collected fairly simply and painlessly. Another potential advantage is that the natural turnover of olfactory sensory neurons in the OM might accentuate the effects of prion/PrP^{Sc} inhibitors on steady-state levels of PrP^{Sc}. Conversely, the same natural turnover may cause difficulty in obtaining an accurate readout of the prion-seeding activity in the rest of the CNS. Future testing will determine the implications of this neuronal turnover. Another potential disadvantage of nasal swab sampling is inconsistency in the amounts of olfactory sensory neurons recovered, but this issue might be overcome by reference to a neuronal marker such as olfactory marker protein.

Finally, RT-QuIC can be used directly or indirectly in the evaluation of effects and mechanisms of inhibitors of PrP^{Sc} formation in vitro.[28–30,56]

Direct applications would involve adding test compounds to prion-seeded RT-QuIC reactions. An advantage of RT-QuIC for such purposes is the possibility of testing against many different types of prions and $rPrP^{Sen}$ substrates. A disadvantage is that RT-QuIC recapitulates only part of the refolding process in PrP^{Sc} formation. Nonetheless, it should still be useful in the prescreening of chemical stabilizers of PrP^{C} (i.e., molecular chaperones) as well as compounds that block the seeding surfaces on PrP^{Sc} aggregates. Another potential confounding factor in screening for inhibitors by adding them to RT-QuIC reactions is that certain compounds may quench the amyloid-enhanced fluorescence of thioflavin T. Thus, apparent "hits" that appear to inhibit prion-seeded RT-QuIC reactions should be evaluated secondarily by other means. For example, inhibition of the formation of PK-resistant RT-QuIC products could be confirmed by Western blotting. Of course, the need for confirmatory tests is not peculiar to RT-QuIC testing, because all types of in vitro drug screens require multitiered in vitro and in vivo confirmations of compound efficacy before a drug candidate is validated.

As for indirect in vitro uses of RT-QuIC for drug screening, Kondru and colleagues have shown that the assay can be used to assess effects of test compounds on PrP^{Sc} accumulation in scrapie-infected organotypic brain slice cultures.[29]

5. EVALUATING PRION DISINFECTANTS USING RT-QuIC

RT-QuIC has also been helpful in the testing of antiprion and antiamyloid disinfectants.[31,57] Prion infectivity and RT-QuIC positivity are strongly correlated in ex vivo samples, and RT-QuIC is at least as sensitive analytically as animal bioassays.[5,6,20,49,53] This implies that if a treatment eliminates RT-QuIC seeding activity, it will likely also eliminate prion infectivity. This was shown to be true for a newly identified disinfectant, hypochlorous acid, as well as the established disinfectants, hypochlorite bleach, and concentrated sodium hydroxide.[31] However, one antiprion agent, Environ LpH™, eliminated prion infectivity (as measured by animal bioassay) but not RT-QuIC seeding activity. Given that synthetic recombinant PrP amyloids can have RT-QuIC seeding activity but no apparent infectivity in animals, this is not surprising. Fully infectious PrP^{Sc} comprises only a subset of PrP particles with RT-QuIC seeding activity, so treatments may neutralize PrP^{Sc} infectivity without proportionally affecting all types of PrP seeding activity. Thus, in applying RT-QuIC assays to other antiprion treatments, infectious tissue-derived prions, rather than noninfectious synthetic amyloid seeds, should be used as test specimens. Hits from such screens against PrP^{Sc} are likely to be effective against infectivity; however, some treatments that neutralize infectivity might be missed if they do not also eliminate all noninfectious seeds as well.

6. TECHNICAL CONSIDERATIONS IN RT-QuIC APPLICATIONS

Prior work has identified experimental factors that can be influential in the optimization of RT-QuIC assays[58]: (1) The kinetics of RT-QuIC reactions can sometimes be accelerated by the use of N-terminally truncated $rPrP^{Sen}$ substrates (e.g., Syrian hamster PrP^{Sen}, residues 90–231) while maintaining a clear kinetic distinction between prion-seeded and mock-seeded reactions. The result can be much shorter assay times without loss of sensitivity. (2) Raising temperatures, shaking speeds, or shaking periods can also accelerate RT-QuIC reactions without compromising specificity. (3) Under our usual range of reaction conditions, pH 7.4 has promoted more rapid RT-QuIC reactions than more acidic pHs. (4) Small variations in the low SDS concentrations that we prefer to use can substantially affect assay performance and must be carefully controlled. (5) Finally, although "real-time" measurements of the thioflavin T fluorescence in single all-purpose instruments that can shake, incubate, and read fluorescence are undoubtedly helpful in the development of new RT-QuIC assays, this capability may not always be necessary. Once routine assay conditions and appropriate cutoff times for a give sample type have been established, RT-QuIC throughput and cost effectiveness can be greatly improved by using one device for the shaking and incubation and another for fluorescence measurements. The shaking/incubation phase can be performed in a multiplate thermoshaker (without expensive fluorescence optics) for a chosen reaction time (s). At the end the plates can be transferred to a separate fluorescence plate reader (without expensive shaking and incubation capabilities) for rapid fluorescence readings.[58] This system allows for the time-consuming reactions to be performed on multiple relatively low-cost thermoshakers, with the fluorescence measurements being taken on a single-plate reader. We estimate that this strategy can reduce the overall instrument costs by several fold per plate. Collectively, such refinements can dramatically reduce the costs and improve the performance of RT-QuIC assays, which are now often performed in a matter of hours rather than days.

7. ADAPTATION OF RT-QuIC PLATFORM TO OTHER PROTEIN FOLDING DISEASES

Following the success of RT-QuIC assays in detecting prions and diagnosing prion diseases, we and others have started to develop similar assays for other protein misfolding diseases. There is a critical need for clear early antemortem diagnostic tests for many neurodegenerative protein misfolding diseases. Many of the causative protein aggregates of these diseases are fibrillar and have prion-like seeding or self-propagating activities in vitro and in vivo. Recently, assays exploiting the amplification power of seeded polymerization mechanisms for diagnostic purposes have begun to appear for synucleinopathies and tauopathies. For example, Fairfoul and colleagues reported the first α-synuclein RT-QuIC assay which detected α-synuclein seeds in the CSF of patients with dementia with Lewy bodies and Parkinson's disease with sensitivities of 92% and 95%, respectively, with 100% specificity with respect to Alzheimer's and other controls.[59] Shahnawaz and colleagues have developed a related assay called αSyn-PMCA that, also using CSF specimens, identified Parkinson's disease cases with a sensitivity of 88.5% and specificity of 96.9%.[60] We have also developed an α-synuclein-based RT-QuIC that was used to show the inhibition of fibrillar α-synuclein seeding activity by the antiprion disinfectant hypochlorous acid.[31] In the same study, we developed a tau RT-QuIC assay to demonstrate hypochlorous acid inhibition of aggregated tau seeds formation. With further refinements, the tau RT-QuIC assay allowed ultrasensitive and specific detection of the tau seeds associated with a tauopathy called Pick disease.[61] Tau seeds were detected in CSF specimens from Pick disease cases ($n=8$), and not in non-Pick's controls ($n=21$), suggesting a potential basis for Pick disease diagnosis using CSF. However, in this study, the Pick CSF samples were collected postmortem so it remains to be determined if this prototypic tau RT-QuIC test is useful for antemortem diagnosis. Given that there are many different tauopathies and synucleinopathies, we hope that these prototypic α-synuclein and tau seed-amplification assays can be adapted to these other fatal neurodegenerative diseases to aid in antemortem diagnostics as well as the development and application of therapeutics.

ACKNOWLEDGMENTS

This work was supported in part by the Intramural Research Program of the NIAID.

REFERENCES

1. Castilla J, Saa P, Morales R, Abid K, Maundrell K, Soto C. Protein misfolding cyclic amplification for diagnosis and prion propagation studies. *Methods Enzymol.* 2006;412:3–21.
2. Atarashi R, Moore RA, Sim VL, et al. Ultrasensitive detection of scrapie prion protein using seeded conversion of recombinant prion protein. *Nat Methods.* 2007;4(8):645–650.
3. Colby DW, Zhang Q, Wang S, et al. Prion detection by an amyloid seeding assay. *Proc Natl Acad Sci USA.* 2007;104(52):20914–20919.
4. Atarashi R, Wilham JM, Christensen L, et al. Simplified ultrasensitive prion detection by recombinant PrP conversion with shaking. *Nat Methods.* 2008;5(3):211–212.
5. Wilham JM, Orrú CD, Bessen RA, et al. Rapid end-point quantitation of prion seeding activity with sensitivity comparable to bioassays. *PLoS Pathog.* 2010;6(12):e1001217.
6. Atarashi R, Satoh K, Sano K, et al. Ultrasensitive human prion detection in cerebrospinal fluid by real-time quaking-induced conversion. *Nat Med.* 2011;17(2):175–178.
7. McGuire LI, Peden AH, Orru CD, et al. RT-QuIC analysis of cerebrospinal fluid in sporadic Creutzfeldt-Jakob disease. *Ann Neurol.* 2012;72(2):278–285.
8. Zanusso G, Monaco S, Pocchiari M, Caughey B. Advanced tests for early and accurate diagnosis of Creutzfeldt-Jakob disease. *Nat Rev Neurol.* 2016;12(6):325–333.
9. Lansbury Jr PT, Caughey B. The chemistry of scrapie infection: Implications of the 'ice 9' metaphor. *Chem Biol.* 1995;2:1–5.
10. Kocisko DA, Come JH, Priola SA, et al. Cell-free formation of protease-resistant prion protein. *Nature.* 1994;370:471–474.
11. Jarrett JT, Lansbury Jr PT. Seeding "one-dimensional crystallization" of amyloid: a pathogenic mechanism in Alzheimer's disease and scrapie? *Cell.* 1993;73:1055–1058.
12. Schmitz M, Cramm M, Llorens F, et al. The real-time quaking-induced conversion assay for detection of human prion disease and study of other protein misfolding diseases. *Nat Protoc.* 2016;11(11):2233–2242.
13. Atarashi R, Sano K, Satoh K, Nishida N. Real-time quaking-induced conversion: a highly sensitive assay for prion detection. *Prion.* 2011;5(3):150–153.
14. Orru CD, Wilham JM, Vascellari S, Hughson AG, Caughey B. New generation QuIC assays for prion seeding activity. *Prion.* 2012;6(2):147–152.
15. Groveman BR, Dolan MA, Taubner LM, Kraus A, Wickner RB, Caughey B. Parallel in-register intermolecular beta-sheet architectures for prion-seeded prion protein (PrP) amyloids. *J Biol Chem.* 2014;289(35):24129–24142.
16. Orru CD, Wilham JM, Raymond LD, et al. Prion disease blood test using immunoprecipitation and improved quaking-induced conversion. *mBio.* 2011;2(3):e00078–00011.
17. Foutz A, Appleby BS, Hamlin C, et al. Diagnostic and prognostic value of human prion detection in cerebrospinal fluid. *Ann Neurol.* 2017;81(1):79–92.
18. Sano K, Satoh K, Atarashi R, et al. Early detection of abnormal prion protein in genetic human prion diseases now possible using real-time QUIC assay. *PLoS One.* 2013;8(1):e54915.
19. McGuire LI, Poleggi A, Poggiolini I, et al. Cerebrospinal fluid real-time quaking-induced conversion is a robust and reliable test for sporadic Creutzfeldt-Jakob disease: an international study. *Ann Neurol.* 2016;80(1):160–165.
20. Orru CD, Groveman BR, Hughson AG, Zanusso G, Coulthart MB, Caughey B. Rapid and sensitive RT-QuIC detection of human Creutzfeldt-Jakob disease using cerebrospinal fluid. *MBio.* 2015;6(1).
21. Orru CD, Bongianni M, Tonoli G, et al. A test for Creutzfeldt-Jakob disease using nasal brushings. *N Engl J Med.* 2014;371(6):519–529.
22. Zanusso G, Bongianni M, Caughey B. A test for Creutzfeldt-Jakob disease using nasal brushings. *N Engl J Med.* 2014;371(19):1842–1843.
23. Cramm M, Schmitz M, Karch A, et al. Characteristic CSF prion seeding efficiency in humans with prion diseases. *Mol Neurobiol.* 2015;51(1):396–405.

24. Cramm M, Schmitz M, Karch A, et al. Stability and reproducibility underscore utility of RT-QuIC for diagnosis of Creutzfeldt-Jakob disease. *Mol Neurobiol.* 2016;53: 1896–1904.
25. Bongianni M, Orrù CD, Groveman BR, et al. Diagnosis of human prion disease using real-time quaking-induced conversion testing of olfactory mucosa and cerebrospinal fluid samples. *JAMA Neurol.* 2017;74(2):1–8.
26. Groveman BR, Orru CD, Hughson AG, et al. Extended and direct evaluation of RT-QuIC assays for Creutzfeldt-Jakob disease diagnosis. *Ann Clin Transl Neurol.* 2017;4(2):139–144.
27. Lattanzio F, Abu-Rumeileh S, Franceschini A, et al. Prion-specific and surrogate CSF biomarkers in Creutzfeldt-Jakob disease: diagnostic accuracy in relation to molecular subtypes and analysis of neuropathological correlates of p-tau and Abeta42 levels. *Acta Neuropathol.* 2017;133(4):559–578.
28. Ferreira NC, Marques IA, Conceicao WA, et al. Anti-prion activity of a panel of aromatic chemical compounds: in vitro and in silico approaches. *PLoS One.* 2014;9(1):e84531.
29. Kondru N, Manne S, Greenlee J, et al. Integrated organotypic slice cultures and RT-QuIC (OSCAR) assay: implications for translational discovery in protein misfolding diseases. *Sci Rep.* 2017;7:43155.
30. Schmitz M, Cramm M, Llorens F, et al. Application of an in vitro-amplification assay as a novel pre-screening test for compounds inhibiting the aggregation of prion protein scrapie. *Sci Rep.* 2016;6:28711.
31. Hughson AG, Race B, Kraus A, et al. Inactivation of prions and amyloid seeds with hypochlorous acid. *PLoS Pathog.* 2016;12(9):e1005914.
32. Park JH, Choi YG, Lee YJ, et al. Real-time quaking-induced conversion analysis for the diagnosis of sporadic Creutzfeldt-Jakob disease in Korea. *J Clin Neurol.* 2016;12(1):101–106.
33. Zanusso G, Ferrari S, Cardone F, et al. Detection of pathologic prion protein in the olfactory epithelium in sporadic Creutzfeldt-Jakob disease. *N Engl J Med.* 2003;348(8):711–719.
34. Bessen RA, Wilham JM, Lowe D, et al. Accelerated shedding of prions following damage to the olfactory epithelium. *J Virol.* 2011;86:1777–1788.
35. Bessen RA, Shearin H, Martinka S, et al. Prion shedding from olfactory neurons into nasal secretions. *PLoS Pathog.* 2010;6:e1000837.
36. Redaelli V, Bistaffa E, Zanusso G, et al. Detection of prion seeding activity in the olfactory mucosa of patients with fatal familial insomnia. *Sci Rep.* 2017;7:46269.
37. Masujin K, Orru CD, Miyazawa K, et al. Detection of atypical H-type bovine spongiform encephalopathy and discrimination of bovine prion strains by real-time quaking-induced conversion. *J Clin Microbiol.* 2016;54(3):676–686.
38. Orru CD, Favole A, Corona C, et al. Detection and discrimination of classical and atypical L-type bovine spongiform encephalopathy by real-time quaking-induced conversion. *J Clin Microbiol.* 2015;53(4):1115–1120.
39. Orru CD, Groveman BR, Raymond LD, et al. Bank vole prion protein as an apparently universal substrate for RT-QuIC-based detection and discrimination of prion strains. *PLoS Pathog.* 2015;11(6):e1004983.
40. Dassanayake RP, Orru CD, Hughson AG, et al. Sensitive and specific detection of classical scrapie prions in the brains of goats by real-time quaking-induced conversion. *J Gen Virol.* 2016;97(3):803–812.
41. John TR, Schatzl HM, Gilch S. Early detection of chronic wasting disease prions in urine of pre-symptomatic deer by real-time quaking-induced conversion assay. *Prion.* 2013;7(3):253–258.
42. Henderson DM, Manca M, Haley NJ, et al. Rapid antemortem detection of CWD prions in deer saliva. *PLoS One.* 2013;8(9):e74377.

43. Elder AM, Henderson DM, Nalls AV, et al. In vitro detection of prionemia in TSE-infected cervids and hamsters. *PLoS One*. 2013;8(11):e80203.
44. Haley NJ, Van de Motter A, Carver S, et al. Prion-seeding activity in cerebrospinal fluid of deer with chronic wasting disease. *PLoS One*. 2013;8(11):e81488.
45. Haley NJ, Carver S, Hoon-Hanks LL, et al. Detection of chronic wasting disease in the lymph nodes of free-ranging cervids by real-time quaking-induced conversion. *J Clin Microbiol*. 2014;52(9):3237–3243.
46. Haley NJ, Siepker C, Hoon-Hanks LL, et al. Seeded amplification of chronic wasting disease prions in nasal brushings and recto-anal mucosa associated lymphoid tissues from elk by real time quaking-induced conversion. *J Clin Microbiol*. 2016;54:1117–1126.
47. Haley NJ, Siepker C, Walter WD, et al. Antemortem detection of chronic wasting disease prions in nasal brush collections and rectal biopsy specimens from white-tailed deer by real-time quaking-induced conversion. *J Clin Microbiol*. 2016;54(4): 1108–1116.
48. Orru CD, Hughson AG, Race B, Raymond GJ, Caughey B. Time course of prion seeding activity in cerebrospinal fluid of scrapie-infected hamsters after intratongue and intracerebral inoculations. *J Clin Microbiol*. 2012;50:1464–1466.
49. Vascellari S, Orru CD, Hughson AG, et al. Prion seeding activities of mouse scrapie strains with divergent PrPSc protease sensitivities and amyloid plaque content using RT-QuIC and eQuIC. *PLoS One*. 2012;7(11):e48969.
50. Dougherty RM. Animal virus titration techniques. In: RJC H, ed. *Techniques in Experimental Virology*. New York: Academic Press, Inc.; 1964:183–186.
51. Gray JG, Graham C, Dudas S, Paxman E, Vuong B, Czub S. Defining and assessing analytical performance criteria for transmissible spongiform encephalopathy-detecting amyloid seeding assays. *J Mol Diagn*. 2016;18(3):454–467.
52. Alibhai J, Blanco RA, Barria MA, et al. Distribution of misfolded prion protein seeding activity alone does not predict regions of neurodegeneration. *PLoS Biol*. 2016;14(11): e1002579.
53. Peden AH, McGuire LI, Appleford NE, et al. Sensitive and specific detection of sporadic Creutzfeldt-Jakob disease brain prion protein using real-time quaking induced conversion. *J Gen Virol*. 2012;93:438–449.
54. Shi S, Mitteregger-Kretzschmar G, Giese A, Kretzschmar HA. Establishing quantitative real-time quaking-induced conversion (qRT-QuIC) for highly sensitive detection and quantification of PrPSc in prion-infected tissues. *Acta Neuropathol Commun*. 2013;1(1):44.
55. Henderson DM, Davenport KA, Haley NJ, Denkers ND, Mathiason CK, Hoover EA. Quantitative assessment of prion infectivity in tissues and body fluids by real-time quaking-induced conversion. *J Gen Virol*. 2015;96(pt 1):210–219.
56. Vieira TC, Cordeiro Y, Caughey B, Silva JL. Heparin binding confers prion stability and impairs its aggregation. *FASEB J*. 2014;28(6):2667–2676.
57. Mori T, Atarashi R, Furukawa K, et al. A direct assessment of human prion adhered to steel wire using real-time quaking-induced conversion. *Sci Rep*. 2016;6:24993.
58. Orru CD, Hughson AG, Groveman BR, et al. Factors that improve RT-QuIC detection of prion seeding activity. *Viruses*. 2016;8(5):140.
59. Fairfoul G, McGuire LI, Pal S, et al. Alpha-synuclein RT-QuIC in the CSF of patients with alpha-synucleinopathies. *Ann Clin Transl Neurol*. 2016;3(10):812–818.
60. Shahnawaz M, Tokuda T, Waragai M, et al. Development of a biochemical diagnosis of Parkinson disease by detection of alpha-synuclein misfolded aggregates in cerebrospinal fluid. *JAMA Neurol*. 2017;74(2):163–172.
61. Saijo E, Ghetti B, Zanusso G, et al. Ultrasensitive and selective detection of 3-repeat tau seeding activity in Pick disease brain and cerebrospinal fluid. *Acta Neuropathol*. 2017;133(5):751–765.

CHAPTER EIGHTEEN

Biochemical Characterization of Prions

Michele Fiorini[1], Matilde Bongianni, Salvatore Monaco, Gianluigi Zanusso
University of Verona, Verona, Italy
[1]Corresponding author: e-mail address: michele.fiorini@univr.it

Contents

Abstract

Prion disease or transmissible spongiform encephalopathies are characterized by the presence of the abnormal form of the prion protein (PrP^{Sc}). The pathological and transmissible properties of PrP^{Sc} are enciphered in its secondary and tertiary structures. Since it's well established that different strains of prions are linked to different conformations

Progress in Molecular Biology and Translational Science, Volume 150
ISSN 1877-1173
http://dx.doi.org/10.1016/bs.pmbts.2017.06.012

of PrP^{Sc}, biochemical characterization of prions seems a preliminary but reliable approach to detect, analyze, and compare prion strains.

Experimental biochemical procedures might be helpful in distinguishing PrP^{Sc} physicochemical properties and include resistance to proteinase K (PK) digestion, insolubility in nonionic detergents, PK-resistance under denaturing conditions and sedimentation properties in sucrose gradients. This biochemical approach has been extensively applied in human prion disorders and subsequently expanded for PrP^{Sc} characterization in animals.

In particular, in sporadic Creutzfedlt–Jakob disease (sCJD) PrP^{Sc} is characterized by two main glycotypes conventionally named Type 1 and Type 2, based on the apparent gel migration at 21 and 19 kDa of the PrP^{Sc} PK-resistant fragment. An additional PrP^{Sc} type was identified in sCJD characterized by an unglycosylated dominant glycoform pattern and in 2010 a variably protease-sensitive prionopathy (VPSPr) was reported showing a PrP^{Sc} with an electrophoretic ladder like pattern. Additionally, the presence of PrP^{Sc} truncated fragments completes the electrophoretic characterization of different prion strains.

By two-dimensional (2D) electrophoretic analysis additional PrP^{Sc} pattern was identified, since this procedure provides information about the isoelectric point and the different peptides length related to PK cleavage, as well as to glycosylation extent or GPI anchor presence.

We here provide and extensive review on PrP^{Sc} biochemical analysis in human and animal prion disorders. Further, we show that PrP^{Sc} glycotypes observed in CJD share similarities with PrP^{Sc} in bovine spongiform encephalopathy forms (BSE).

1. INTRODUCTION

Prion diseases are fatal neurodegenerative disorders affecting humans and animals with a sporadic, inherited, or infectious etiology.[1,2] The common feature of these disorders is the accumulation of a pathologic prion protein (PrP^{Sc}) in the nervous tissue. The crucial event in the pathogenesis of the disease is the conversion of the normal host-encoded cellular prion protein (PrP^C), to a misfolded pathogenic isoform named PrP^{Sc}.[3] In fact, PrP^C and PrP^{Sc} share the same primary structure but, while PrP^C secondary structure is rich in α-helices, PrP^{Sc} is characterized by increased β-sheet content and decreased α-helices. As consequence of conformational change PrP^{Sc} acquires tertiary and quaternary different feature leading to new physicochemical properties: PrP^{Sc} is insoluble in nondenaturing detergents and partially resistant to proteolysis. PrP^{Sc} is also infective since it can propagate abnormal conformation to normal PrP^C in a self-propagating mechanism.[4]

Mature PrP^C is a 209 aminoacids glycoprotein, with two potential N-linked glycosylation sites at codon 181 and 197, anchored to the plasma membrane by a glycosylphosphatidylinositol molecule. PrP^C N-terminal is characterized by an unstructured domain, composed by four copies of an octapeptide sequence (PHGGGWGQ) between residues 40 and 91, while C-terminal region is composed by two short β-sheet structures of four residues each and three α-helical regions.[4] The octarepeat sequence is a Cu^{2+} binding motif.

Western blot analysis of prion protein is characterized by three bands corresponding to the three glycoforms: di-, mono-, and unglycosylated; after proteinase-K (PK) treatment electrophoretic pattern of the three bands shifts to a lower apparent molecular weight due to the cleavage of the N-terminal PK-sensitive sequence of the protein. PrP^{Sc} PK-resistant glycoforms are also named PrP27–30[5] (Fig. 1). For simplicity, along this chapter we use PrP^{Sc} to define pathological PrP, including PrP27–30 and the truncated fragments.

Since the same prion protein can be the cause of several different disease phenotypes, variations in prions were referred to as strains. Prion strains are characterized by experimental transmission and distinguished according to differences in incubation time and regional distribution of cerebral lesions. Biochemically it is well established that these different disease phenotypes are associated with distinct PrP^{Sc} conformations, reflected in the pattern of PK-induced proteolytic cleavage of PrP^{Sc}.[6–8]

Early evidence of the connection between prion conformation and strain features was shown by molecular analysis of transmissible mink encephalopathy by Bessen and Marsh.[9] They studied the two known strains, called

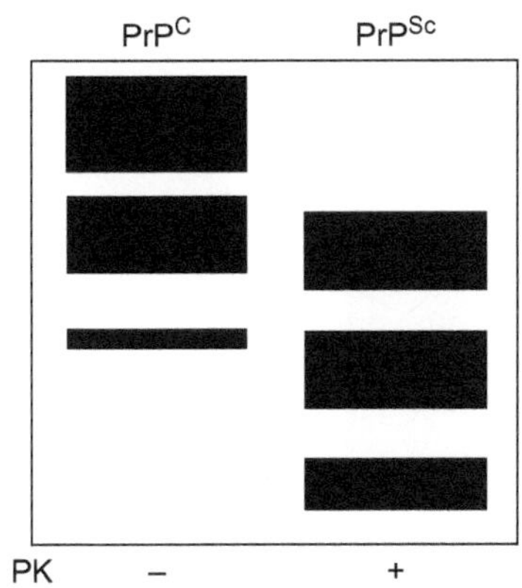

Fig. 1 Schematic representation of human western blot profile of PrP before (–) and after (+) PK treatment.

hyper (HY) and drowsy (DY), showing different clinical signs, incubation period, brain titer, brain lesion profile, and pathogenicity after transmission to laboratory animals. In particular, they found that the degradation rate of PrP^{TME} by PK was peculiar for each strain and correlated with inactivation of the TME titer; after PK digestion and sequence determination, HY PrP^{TME} was at least 10 aminoacids longer than DY PrP^{TME} at the N-terminal PK-resistant end of the protein. Finally differences in biochemical and physical properties were easily resumed by western blot after PK treatment showing two peculiar electrophoretic pattern in terms of molecular weight: HY with a core fragment at about 21 kDa and DY at about 19 kDa[9] (Fig. 2).

In human sporadic CJD (sCJD) six molecular phenotypes are defined, considering the molecular weight of the PK-resistant core fragment (Type 1 at 21 kDa and Type 2 at 19 kDA) and the polymorphism at codon 129 (MM, MV, and VV), reflecting the clinicopathological features of the disease.[8]

Western blot analysis shows peculiar migration pattern associated to different prion strains: evidence for PK cleavage at different sites in the N-terminal region of PrP^{Sc} both in animal and human prion diseases. In addition, based on the occupancy of the N-glycosylation sites or the ratio of the three PK-resistant PrP^{Sc} glycoforms (diglycosylated, monoglycosylated, and unglycosylated), further biochemical types of PrP^{Sc} can be classified.[8] In the last few years, we introduced two-dimensional (2D) analysis of prions representing an important step forward in biochemical characterization of prion strain. In particular, 2D analysis allows to recognize specific biochemical fingerprints of different prion strains otherwise undetectable by conventional western blot analysis.[5]

Additional biochemical procedures can help in differentiating prion strains such as testing resistance to denaturing conditions or investigating the size and the solubility properties of aggregates.[10,11]

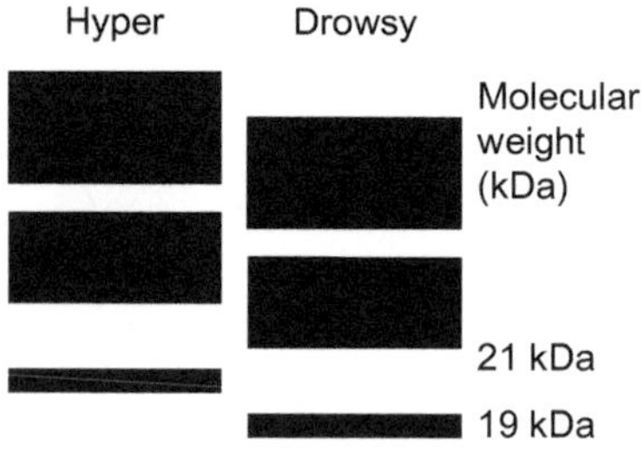

Fig. 2 Schematic representation of transmissible mink encephalopathy western blot profile drowsy and hyper molecular types.

Aim of this chapter is to describe how biochemical characterization of prions results of crucial importance in finding and discriminating strains, not only in describing prion strains but also in large-scale epidemiological studies and surveillance.

2. 1D ANALYSIS OF MOLECULAR STRAINS

2.1 Human Prion Disorders

2.1.1 Sporadic Forms

In human prion disorders, several different PrP^{Sc} types have been described based on the electrophoretic migration and the glycosylation profile of PrP27–30. In human sCJD, PrP^{Sc} separates as two main glycotypes: PrP^{Sc} Type 1 and PrP^{Sc} Type 2A with an apparent gel migration of the unglycosylated isoform at 21 and 19 kDa, respectively. Types 1 and 2A show similar glycosylation profile[9] (Figs. 3 and 4).

In 2007, a novel PrP^{Sc} glycotype was reported, mainly characterized by the unglycosylated isoform and slightly by the monoglycosylated isoform, but completely lacking the diglycosylated one; this new strain was named Type U. For this prion strain genetic or inherited origins were excluded[12] (Figs. 3 and 4).

In 2010, an additional sporadic form of prion disease has been described, called variably protease-sensitive prionopathy due to relative sensitivity to protease digestion of disease associated PrP^{Sc} (VPSPr). VPSPr is characterized by a peculiar ladder-like electrophoretic profile of the PK-resistant PrP fragments. The electrophoretic profile after western blot was detected

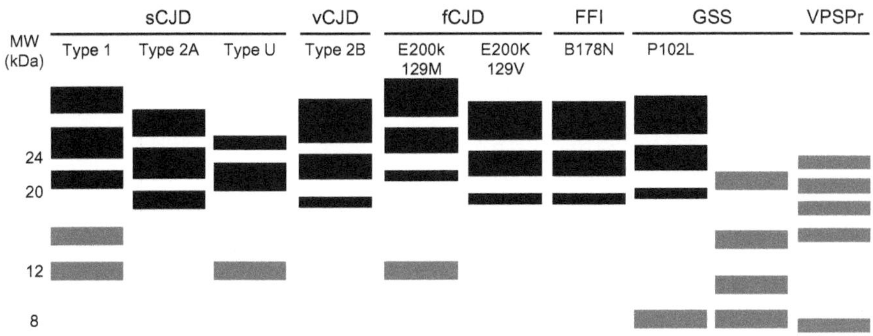

Fig. 3 Schematic representation of human western blot profile of all CJD molecular types.

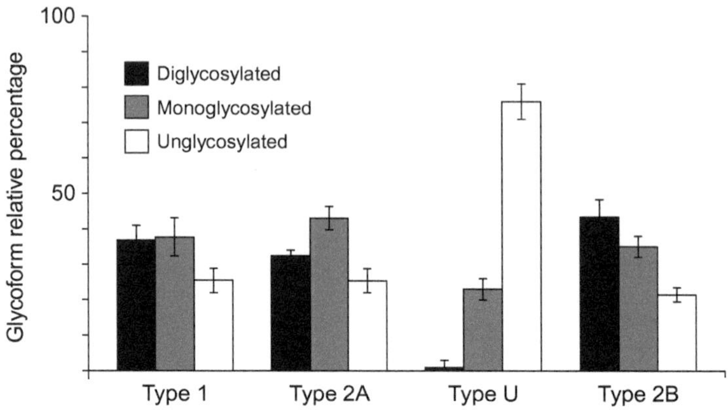

Fig. 4 Relative glycoform profile of four sCJD molecular strains.

exclusively by the use of unique anti-PrP antibody (clone 1E4) and consisted of five major bands migrating at approximately 26, 23, 20, 17, and 7 kDa[13] (Fig. 3).

2.1.2 Genetic Forms

Genetic prion disorders include familial Creutzfeldt–Jakob Disease (fCJD), fatal familial insomnia (FFI), and Gerstmann–Sträussler–Scheinker syndrome (GSS), each form characterized by distinct disease phenotypes, neuropathological hallmarks, and pattern of PrP deposition, respectively. Western blot analysis in fCJD and FFI shows the typical triplet of PrP27–30 differentially PK cleaved at residues ~82 or at ~97 with variable pattern of glycosylation.[14,15] In GSS, the PK-digested western blot pattern is defined by the enrichment of constitutive internal fragments of 8 and 11 kDa. P102L GSS is an exception presenting hybrid behavior between GSS and other genetic forms: its peculiar pattern is composed by both PrP27–30 and the 8 kDa internal fragment[7,16] (Fig. 3).

2.1.3 Infectious Forms

Early known cases of iatrogenic CJD (iCJD), except cannibalistic kuru, have been caused by medical procedures such as implantation of dura mater grafts and treatment with growth hormone derived from cadaveric pituitary glands obtained from patients with CJD. Less frequent incidences of human prion disease have resulted from iatrogenic transmission of CJD during corneal transplantation, contaminated electroencephalographic (EEG) electrode implantation, and surgical operations using contaminated instruments or apparatus. In iCJD, PrP27–30 is almost identical to that of sCJD in terms of molecular weight and glycoform profile.[17] In contrast, variant CJD (vCJD),[18] followed by bovine spongiform encephalopathy (BSE) exposure, shows a PrP^{Sc} with an unglycosylated band of 19 kDa (comigrating with Type 2A) and a diglycosylated dominant pattern, similar to PrP27–30 observed in cattle with BSE.[19]

2.2 Classical and Atypical BSE Forms

BSE forms have always been considered defined by a single strain characterized by a Western blot profile with a diglycosylated pattern and a 20 kDa core PK-resistant fragment. In 2003, Casalone *et al.* discovered a new disease phenotype in two Italian old cows showing a cerebral PrP amyloidosis, sparing the brainstem. Western blot pattern showed a lower molecular weight of the prion core fragment and a PrP^{Sc} monoglycosylated dominant pattern distinct from that observed in classical BSE.[20] Contemporary, other BSE cases where reported in France showing a PrP^{Sc} with glycosylation profile similar to that of classical BSE but with a higher molecular weight.[21] These new atypical molecular types where named L-Type BSE or BASE (bovine amyloidotic spongiform encephalopathy) and H-Type BSE on the basis of the lower or higher migration of PrP^{Sc} at western blot analysis compared to classical BSE. These forms of BSE associated with distinct molecular profiles reflect different clinicopathological features, indicating that each biochemical signature is associated with a distinct form of BSE (Fig. 5).

2.3 Scrapie

Natural scrapie is characterized by the presence of multiple circulating prion strains. However, molecular analysis of PrP^{Sc} is mainly characterized by a biochemical pattern, resembling to classical BSE in terms of molecular weight (20 kDa core fragment) and glycoform profile (diglycosylated)[22] (Fig. 5).

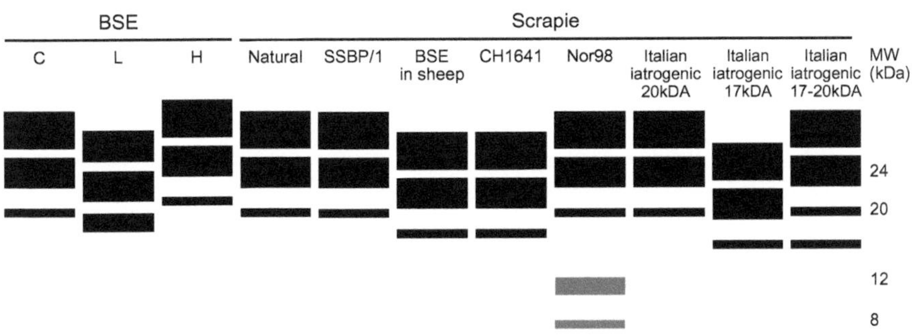

Fig. 5 Schematic representation of human western blot profile of sporadic, genetic, and acquired CJD molecular types.

A second form of scrapie was described in 1998, as an unprecedented form, named Nor98, from the year of its description. This form of scrapie was defined "atypical" based on several criteria: clinical presentation, molecular characteristics of PrP^{Sc}, distribution of PrP^{Sc} within infected sheep, genotypes of affected sheep and epidemiology. In particular, from a molecular view, Nor98 associate PrP^{Sc} was characterized by a faster migration of PrP27–30, by a smaller N-terminally truncated fragment of approximately 12/14 kDa and by an internal fragment of about 7 kDa[23] (Fig. 5).

In 2002, we described a molecular typing of PrP^{Sc} in sheep and goats belonging to a single flock accidentally exposed to an infected vaccination against *Mycoplasma agalactiae* in 1997 and 1998. A group of animals showed a PrP^{Sc} with a 20 kDa core fragment with a diglycosylated dominant pattern while other group showed the cooccurrence of two PrP^{Sc} types one, overlapping that observed in the first group and a second PrP^{Sc} with a highly glycosylated pattern and a PrP^{Sc} protease-resistant backbone of 17 kDa. Additional PrP^{Sc} conformational studies indicated that the 20 kDa isoform showed physicochemical properties indistinguishable from PrP^{Sc} in sheep with natural field scrapie[24] (Fig. 5).

However, in UK several experimental scrapie strains were isolated and used to compare novel emerging field prion strains. In particular, SSBP/1 was obtained from a pool of brain homogenates of three different sheeps suffering from scrapie and CH1641, isolated in 1999 from a natural scrapie.[25] Since PrP^{Sc} fragment size observed in the brain tissue of sheep experimentally infected with BSE was identical to that of CH 1641 it was hypothesized the BSE strain of agent might circulate in sheep,[26] however neuropathological pattern and transmission studies in rodents indicated that represented two distinct strains.[27,28] The same conclusion was raised by SSBP/1 scrapie strain, which appeared to be very close from that found in natural scrapie cases.[29]

3. MOLECULAR ANALYSIS OF PRION STRAINS BY 2D-PAGE ANALYSIS

2D-PAGE analysis is an electrophoretic analysis that couples two principles of proteins electrophoretic separation, in particular the isoelectric point and the molecular mass. This analysis gives the possibility to investigate posttranslational modifications of proteins such as the extent of glycosylation, phosphorylation, the presence of GPI anchor and differences in peptides sequence, based on their polarity. In particular, following limited PK proteolysis and glycans removal by PNG-ase treatment, unglycosylated PrP peptides separate as trains of spots showing PrP peptides with distinct N-ragged ends. Additional C-terminal PrP^{Sc} truncated fragments can be revealed, providing additional information on PrP^{Sc} molecular signature.

3.1 2D-PAGE Analysis of PrP^{Sc} in sCJD Molecular Subtypes

In sCJD, we have identified by 2D-PAGE analysis two major patterns of C-terminal protease-resistant PrP fragments among different molecular phenotypes. In all sCJD cases with a Type 1 PrP^{Sc} we detected the presence of C-terminal fragments (CTFs) of 16–17 and 12–14 kDa. This CTFs pattern was also observed in subjects with MM at codon 129 with a Type 2 PrP^{Sc}. Conversely, VV and MV patients with a Type 2 PrP^{Sc} have CTFs migrating at 17.5–18 kDa. Taken together, these findings indicate that the mechanism involved in the formation of CTFs is not influenced by codon 129 and by PrP^{Sc} glycotype. Finally, the biochemical patterns of PrP^{Sc} following 2D-PAGE analysis resulting from the combination of PrP^{Sc} glycotype and PrP27–30 lower truncated PrP species, defined three distinct PrP^{Sc} fingerprints in sCJD; a first group with Type 1 PrP^{Sc}, regardless codon 129, a second group of MM2 subjects, and a third group of MV2/VV2 cases.[8] Interestingly, this biochemical distinction had a correlation with CJD strains obtained following transmission in transgenic humanized mice[30] (Fig. 6).

3.2 Molecular Signature of PrP^{Sc} in P102L GSS Mutation

PrP^{Sc} molecular signature in GSS cases carrying a P102L mutation is characterized by a train of spots with an isoelectric point between 6.3 and 8.3 similar to that observed in PrP^{Sc} Type 2 in sCJD (Fig. 7A and B) but with CTFs pattern similar to Type 1 PrP^{Sc} (compare Figs. 6 and 7B). The GSS specific 8 kDa fragment, corresponding to ~82–146 aminoacidic sequence,

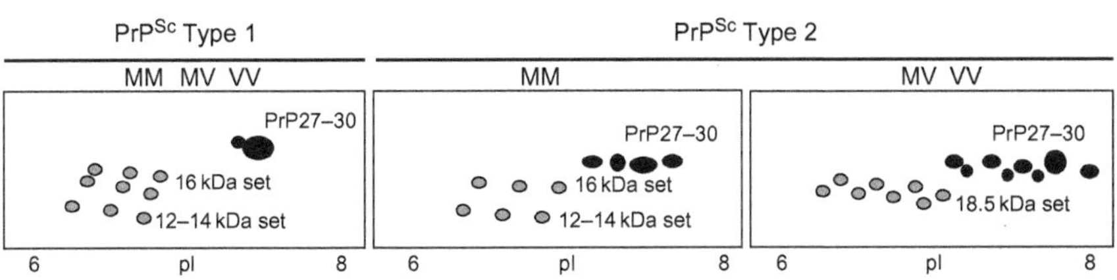

Fig. 6 Schematic 2D pattern of PrP27–30 and C-terminal PrP fragments in sCJD subtypes.

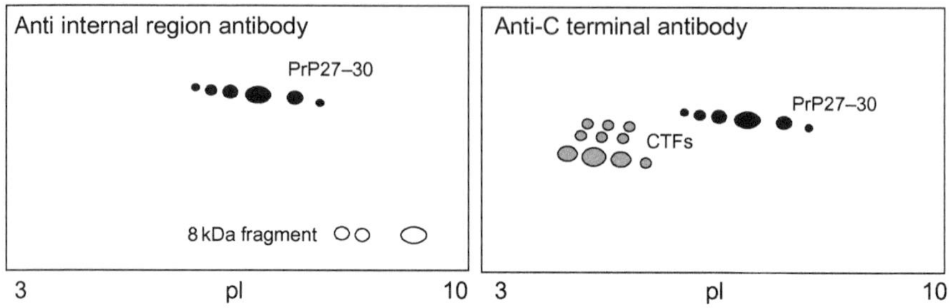

Fig. 7 2D pattern of PrP27–30, internal 8 kDa fragment, and CTFs in a GSS P102L case.

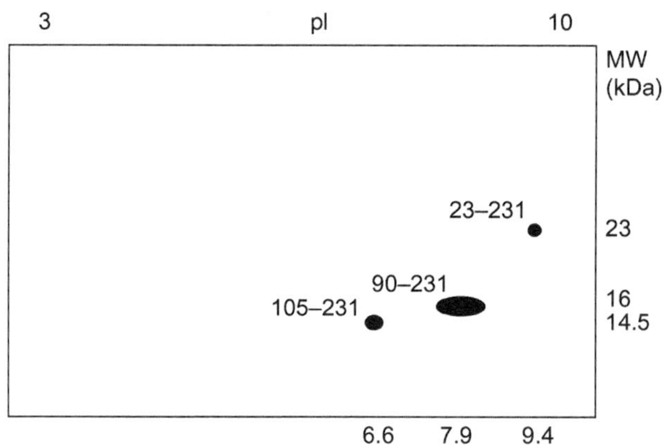

Fig. 8 Scheme of 2D mapping of synthetic PrP peptides.

can be identified by western blot with 3F4 antibody (PrP epitope between 109 and 112) (Fig. 7A) and shows a measured isoelectric point around 9. Since this fragment lacks posttranslational modifications, as expected, the 2D-PAGE migration of PrP^{Sc} corresponds to the pI of the PrP peptide sharing the same sequence at 9.0.

3.3 The Influence of GPI Anchor in PrP^{Sc} Migration in sCJD

Recombinant prion protein peptides with known aminoacidic sequence were used to determine molecular coordinates in the 2D map in terms of molecular weight and isoelectric point. Since 2D-migration of the synthetic peptides exactly correspond to theoretical coordinates, it confirms the reliability of 2D-PAGE analysis (Fig. 8 and Table 1). To establish the pI and molecular weight contributions of GPI anchor, we compared PrP migration by 2D-PAGE in wild-type mice and GPI-anchorless mice showing that the

Table 1 Theoretical Molecular Weight and Isoelectric Point of Recombinant PrP Synthetic Peptides

PrP Peptides	Molecular Mass (kDa)	Isoelectric Point
23–231	22.9	9.39
90–231	16.2	7.95
105–231	14.7	6.56

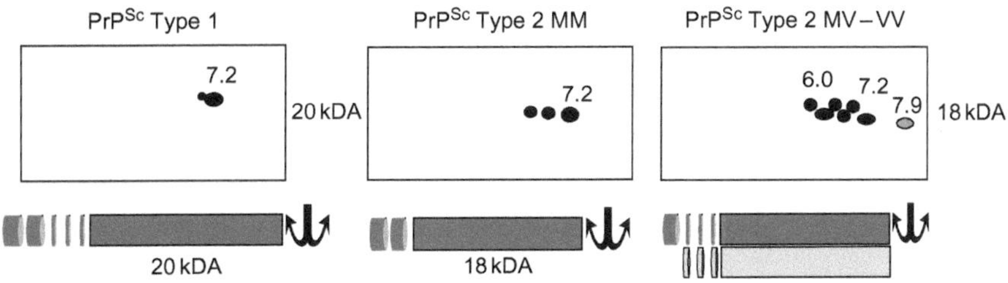

Fig. 9 Schematic representation of anchored and anchorless PK-resistant PrP core fragments in sCJD subtypes.

GPI anchor improves molecular weight of about 2 kDa and shifts measured pI to the basic pole of about 1 pH unit.[7]

Therefore we extended 2D-PAGE analysis to all sCJD molecular Type 1 and 2 subtypes showing that PrP^{Sc} migrated one unit of pH more acidic than expected from the theoretical pH of PrP peptides indicating that both PrP^{Sc} are GPI anchored. However, an extra isoform at pI about eight was detected in MV-2/VV-2 subtypes indicating that only in these forms PrP^{Sc} might undergo to GPI-anchor cleavage (Fig. 9).

3.4 Cooccurrence of PrP^{Sc} Types in sCJD

Cooccurrence of PrP^{Sc} Types 1 and 2 have been previously reported within a single individual, evidenced by the copresence of two bands at 19 and 21 kDa by western blot[31] (Fig. 10A). By 2D analysis it is possible to distinguish PrP^{Sc} types based on the different pI, as reported in Fig. 10B.

3.5 Molecular Signatures of PrP^{Sc} in BSE Forms

2D-PAGE analysis of PrP^{Sc} in classical and atypical BSEs shows distinct biochemical signatures in the three disease phenotypes. As mentioned earlier, in H-Type BSE PrP^{Sc} is characterized by a slower migration compare to the other forms and by a consistent presence of truncated fragments at

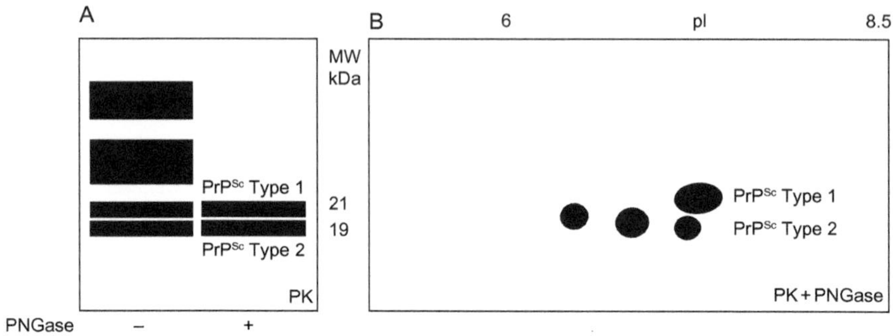

Fig. 10 Representation of human western blot profile of PrP^{Sc} cooccurrence in sCJD before and after PNGase treatment (A); 2D pattern of PrP^{Sc} cooccurrence in sCJD before PK and PNGase treatment (B).

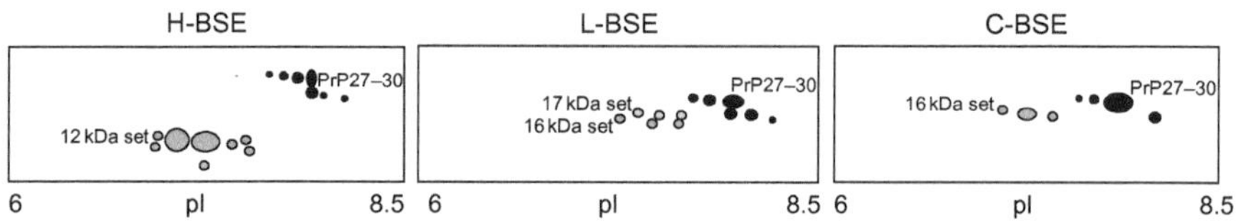

Fig. 11 Schematic 2D pattern of PrP27–30 and C-terminal PrP fragments in BSE subtypes.

12 kDa. In contrast, in C-T and L-Type, except the distinct migration of PrP^{Sc}, additional truncated fragments were not observed (Fig. 11).

3.6 PrP^{Sc} GPI-Anchored and -Anchorless Forms in BSE Forms

A schematic 2D pattern of deglycosylated PrP^{Sc} is reported in Fig. 12.

PrP^{Sc} core fragments are characterized by two sets of spots shifted of about 1.0 unit of pI and 2.0 kDa. Since 2D-PAGE migration of the lower basic spots correspond to PrP peptides theoretical coordinates, these correspond to GPI-anchorless forms, while the higher molecular weight spots are GPI anchored.

2D-PAGE analysis indicates that classical BSE shows only a single PrP^{Sc} anchorless isoform, while L-BSE and H-BSE atypical forms show anchorless isoforms.

3.7 Cooccurrence of PrP^{Sc} Types in Scrapie

2D-PAGE analysis of brain tissue from scrapie cases following iatrogenic exposure in Italy in 2002, clearly shows the overlap of 2D-PAGE pattern of PrP^{Sc}, as reported in Fig. 13. Each single PrP^{Sc} isoform of 17 kDa strain is distinctly identified and separated from the 20 kDa strain.[24]

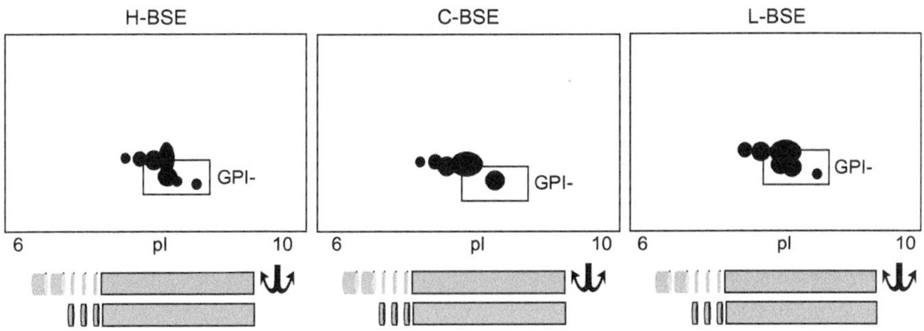

Fig. 12 Schematic representation of anchored and anchorless PK-resistant PrP core fragments in BSE molecular strains.

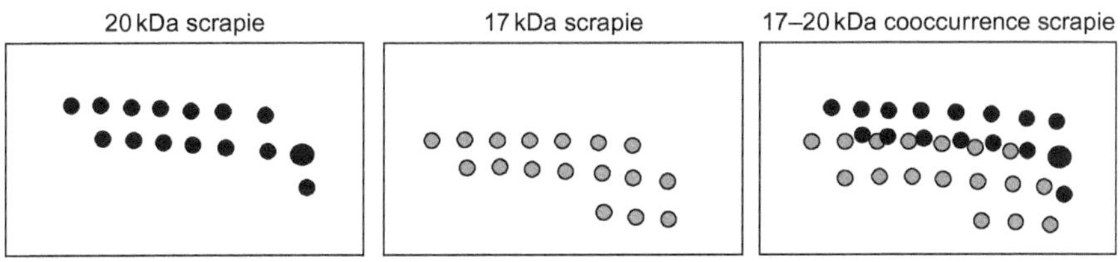

Fig. 13 Schematic 2D pattern of iatrogenic Italian scrapie molecular types described by Zanusso *et al.* (2003).

4. BIOCHEMICAL ASSAYS TO CHARACTERIZE AND DISTINGUISH PRION STRAINS

Since the chemicophysical properties of PrP^{Sc} are related to distinct conformations, by testing conformational stability to denaturing agents or sedimentation properties and size of aggregates we might be able to discriminate different prion strains.

4.1 Velocity Sedimentation in Sucrose Step Gradients

Ultracentrifugation in Sucrose Step Gradients coupled to the ionic surfactant sarkosyl determines the sizes of PrP aggregates based on solubility properties and consequently to prion strains conformational properties.[32]

4.1.1 Size Aggregates in sCJD

PrP^{Sc} shows three distinct patterns of sedimentation, before and after PK treatment (Fig. 14). In MV1 sCJD subtype, PrP is composed by soluble forms of small size and by large insoluble aggregates, before and after PK

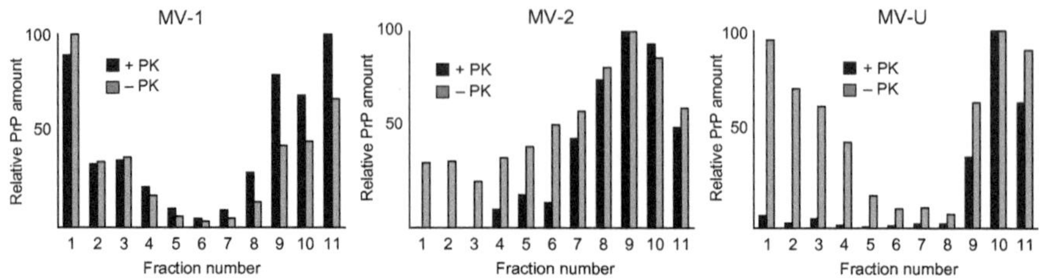

Fig. 14 Fractionation of sCJD PrP aggregates. Brain homogenates from frontal cortexes of MV-1 and MV-2 and MV-U were sedimented in a 10% (fraction 1) to 60% (fraction 11) sucrose gradient. After sedimentation, half samples were digested with PK. Relative cellular (*gray*) and PK-resistant (*black*) percentage of fraction.

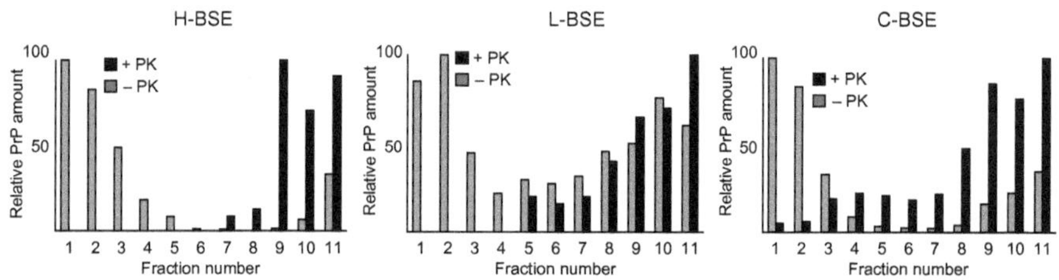

Fig. 15 Fractionation of BSE PrP aggregates. Brain homogenates from frontal cortexes of H-BSE and L-BS and C-BSE were sedimented in a 10% (fraction 1) to 60% (fraction 11) sucrose gradient. After sedimentation, half samples were digested with PK. Relative cellular (*gray*) and PK-resistant (*black*) percentage of fraction distribution.

treatment. In MV2 sCJD subtype, PrP is widely distributed all over the gradient, while after PK digestion PrP^{Sc} is detected mainly at the bottom of the gradient as insoluble aggregates of large size. Disease-associated PrP in MV-U sCJD subtype shows a sedimentation pattern sharing properties with both PrP^{Sc} Types 1 and 2.

4.1.2 PrP Sedimentation Aggregates in Different Forms of BSE

BSE forms molecular subtypes of PrP show a sedimentation pattern characterized by both small soluble and large insoluble aggregates (Fig. 15). In L-Type BSE, the relative amount of insoluble forms is consistently higher than other BSEs. After PK treatment, all BSE brain samples show the predominant presence of insoluble species except in C-Type BSE which shows also an additional small amount of soluble forms (Fig. 15).

It's interesting to note that from the comparison between human and cattle sedimentation patterns a similarity between MV-2 and L-BSE can be observed.

4.2 Conformational Stability in Increasing Concentration of GdnHCl

Conformational stability can be tested by PK digestion after incubation in increasing concentration of chaotropic agents: each PrP^{Sc} conformer can maintain its properties and its PK resistance below a guanidine concentration limit, then can be completely digested.[10] The strength of conformational state is related to the GdnHCl concentration exposure needed to denature the protein and giving the possibility to PK to digest it completely.

4.2.1 Conformational Stability Assay in sCJD

In sCJD, Types 1 and 2, PrP^{Sc} are more resistant to guanidine denaturation than Type U: an exposure to 2 M guanidine is enough to disrupt conformational resistance to PK degradation in Type U, while in Types 1 and 2, PrP^{Sc} digestion is observed at 2.5 M guanidine concentration. The optical density of prion electrophoretic bands after incubation in increasing guanidine concentrations and PK treatment is representative of the distinct kinetics of PrP proteolysis (Fig. 16). Type U western blot shows the presence of two conformers after incubation in 1.5 M guanidine and PK digestion, indicating that two molecular strains are hidden in the 3D conformation.[33] Inoculation experiments in bank voles of the same MV-U case showed it contains two different strains with different incubation time and different lesion profile associated with two different molecular electrophoretic patterns after transmission.[33]

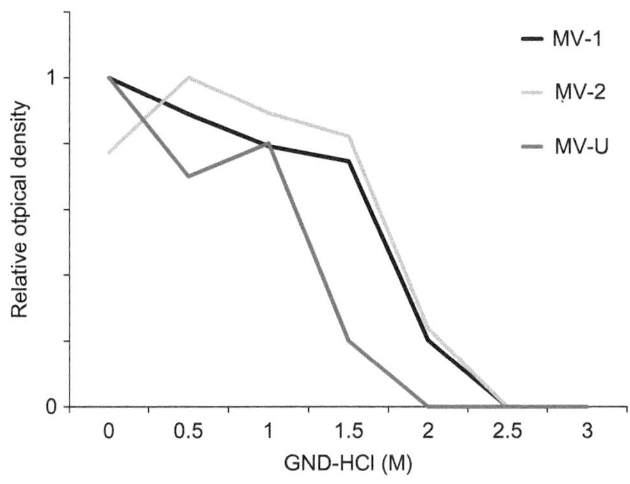

Fig. 16 Relative quantification of denaturation transitions for the MV-1, MV-2, and MV-U types.

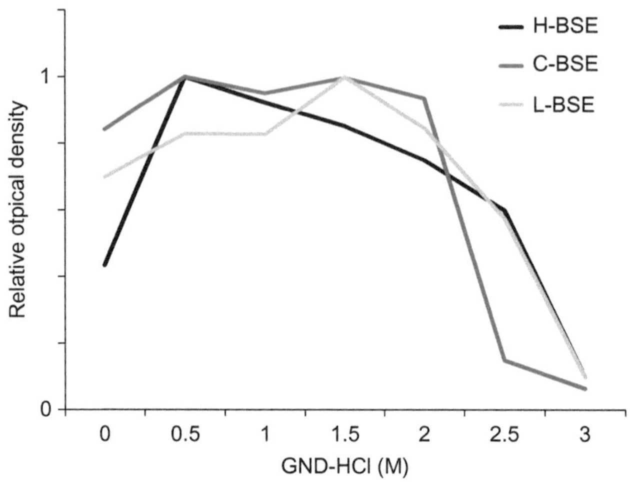

Fig. 17 Relative quantification of denaturation transitions for the H-BSE, C-BSE, and L-BSE.

4.2.2 Conformational Stability Assay in BSEs

Classical BSE can be distinguished from atypical forms of BSE since its conformational stability is maintained until 2 M guanidine concentration, while atypical forms show a higher resistance to 2.5 M guanidine. However, all BSE forms show a lesser extent of PrP^{Sc} with proteolysis resistance forms until 3 M guanidine (Fig. 17).

5. MOLECULAR AND CHEMICOPHYSICAL SIMILARITIES BETWEEN HUMAN AND CATTLE TRANSMISSIBLE SPONGIFORM ENCEPHALOPATHIES FORMS

Several similarities between human and cattle PrP^{Sc} forms can be found at molecular level in terms of both electrophoretic pattern of separation following 1D- and/or 2D-PAGE analysis, and chemicophysical properties of different prion strains conformation.

5.1 Comparison of Human and Cattle Molecular PrPSc by 1D Analysis

A comparative analysis between PrP^{Sc} types in human CJD indicates that PrP^{Sc} Type 1, PrP^{Sc} Type 2A, MV at codon 129, and PrP^{Sc} Type 2B, associated with vCJD, show biochemical similarities to H-Type, L-Type, and C-Type BSEs in terms of molecular weight or glycoform ratio. PrP^{Sc} Type 1 and H-Type BSE share the same molecular weight; PrP^{Sc} Type 2A and L-Type BSE show similar molecular weight and glycosylation profile;

finally, PrP^{Sc} Type 2B shows a similar diglycosylated profile with that observed in classical BSE (compare Figs. 3 and 5). However, the link between CJD and BSE has been extensively studied by several transmission experiments in rodents and transgenic mice.[34–36]

5.2 Comparison of Human and Cattle Molecular Signature by 2D Analysis of PrP^{Sc}

2D-PAGE analysis confirms qualitative similarities between human and cattle PrP^{Sc} molecular types, described by 1D. The molecular signature of PrP^{Sc}, following 2D-PAGE analysis, of H-BSE and Type 1 sCJD shows an abundance of CTFs, as specific finding. Type 2A sCJD and L-BSE show the same pattern of the CTFs at 18 kDa. C-BSE and vCJD share the same CTFs pattern of the upper set of spots at 16 kDa, while the lower one at 12–14 kDa is present only in vCJD (compare Figs. 6 and 11).

6. CONCLUSIONS

In this chapter, we have described several applications of molecular analysis for prions characterization, showing a panel of different biochemical approaches able to detect molecular differences and similarities among prion conformers.

In particular, we have shown by both 1D and 2D electrophoretic separations PrP^{Sc} patterns in human and animal prion disorders based on electrophoretic migration of PrP27–30 and on CTFs.

By 2D analysis, we improved the identification of molecular prion strains showing distinct molecular signatures reflecting clinicopathological phenotypes.

Furthermore, we enhanced similarities between human and cattle prions in terms of molecular and physicochemical properties, giving, in some cases, possible clues about similar origin of the strain.

Finally, such a rapid and informative biochemical analysis should be very helpful for large-scale epidemiological studies and surveillance in finding and discriminating strains (sCJD, vCJD, atypical BSEs).

REFERENCES

1. Prusiner SB. Prions. *Proc Natl Acad Sci USA*. 1998;95(23):13363–13383.
2. Zanusso G, Monaco S, Pocchiari M, Caughey B. Advanced tests for early and accurate diagnosis of Creutzfeldt–Jakob disease. *Nat Rev Neurol*. 2016;6:325–333.
3. Baskakov IV, Breydo L. Converting the prion protein: what makes the protein infectious. *Biochim Biophys Acta*. 2007;1772(6):692–703.

4. Pan KM, Baldwin M, Nguyen J, et al. Conversion of alpha-helices into beta-sheets features in the formation of the scrapie prion proteins. *Proc Natl Acad Sci USA*. 1993;90(23):10962–10966.
5. Parchi P, Zou W, Wang W, et al. Genetic influence on the structural variations of the abnormal prion protein. *Proc Natl Acad Sci USA*. 2000;97(18):10168–10172.
6. Piccardo P, Dlouhy SR, Lievens PM, et al. Phenotypic variability of Gerstmann–Sträussler–Scheinker disease is associated with prion protein heterogeneity. *J Neuropathol Exp Neurol*. 1998;57(10):979–988.
7. Zanusso G, Fiorini M, Ferrari S, et al. Gerstmann–Sträussler–Scheinker disease and "anchorless prion protein" mice share prion conformational properties diverging from sporadic Creutzfeldt–Jakob disease. *J Biol Chem*. 2014;289(8):4870–4881.
8. Zanusso G, Farinazzo A, Prelli F, et al. Identification of distinct N-terminal truncated forms of prion protein in different Creutzfeldt–Jakob disease subtypes. *J Biol Chem*. 2004;279(37):38936–38942.
9. Bessen RA, Marsh RF. Distinct PrP properties suggest the molecular basis of strain variation in transmissible mink encephalopathy. *J Virol*. 1994;68(12):7859–7868.
10. Safar J, Wille H, Itri V, et al. Eight prion strains have PrP(Sc) molecules with different conformations. *Nat Med*. 1998;4(10):1157–1165.
11. Parchi P, Giese A, Capellari S, et al. Classification of sporadic Creutzfeldt–Jakob disease based on molecular and phenotypic analysis of 300 subjects. *Ann Neurol*. 1999;46(2):224–233.
12. Zanusso G, Polo A, Farinazzo A, et al. Novel prion protein conformation and glycotype in Creutzfeldt–Jakob disease. *Arch Neurol*. 2007;64(4):595–599.
13. Zou WQ, Puoti G, Xiao X, et al. Variably protease-sensitive prionopathy: a new sporadic disease of the prion protein. *Ann Neurol*. 2010;68(2):162–172.
14. Monari L, Chen SG, Brown P, et al. Fatal familial insomnia and familial Creutzfeldt–Jakob disease: different prion proteins determined by a DNA polymorphism. *Proc Natl Acad Sci USA*. 1994;91(7):2839–2842.
15. Hainfellner JA, Parchi P, Kitamoto T, Jarius C, Gambetti P, Budka H. A novel phenotype in familial Creutzfeldt–Jakob disease: prion protein gene E200K mutation coupled with valine at codon 129 and type 2 protease-resistant prion protein. *Ann Neurol*. 1999;45(6):812–816.
16. Parchi P, Petersen RB, Chen SG, et al. Molecular pathology of fatal familial insomnia. *Brain Pathol*. 1998;8(3):539–548.
17. Will RG. Acquired prion disease: iatrogenic CJD, variant CJD, kuru. *Br Med Bull*. 2003;66:255–265.
18. Will RG, Ironside JW, Zeidler M, et al. A new variant of Creutzfeldt–Jakob disease in the UK. *Lancet*. 1996;347(9006):921–925.
19. Hill AF, Desbruslais M, Joiner S, et al. The same prion strain causes vCJD and BSE. *Nature*. 1997;389(6650):448–450. 526.
20. Casalone C, Zanusso G, Acutis P, et al. Identification of a second bovine amyloidotic spongiform encephalopathy: molecular similarities with sporadic Creutzfeldt–Jakob disease. *Proc Natl Acad Sci USA*. 2004;101(9):3065–3070.
21. Polak MP, Zmudzinski JF, Jacobs JG, Langeveld JP. Atypical status of bovine spongiform encephalopathy in Poland: a molecular typing study. *Arch Virol*. 2008;153(1):69–79.
22. Baron TG, Madec JY, Calavas D, Richard Y, Barillet F. Comparison of French natural scrapie isolates with bovine spongiform encephalopathy and experimental scrapie infected sheep. *Neurosci Lett*. 2000;284(3):175–178.
23. Benestad SL, Sarradin P, Thu B, et al. Cases of scrapie with unusual features in Norway and designation of a new type, Nor98. *Vet Rec*. 2003;153:202–208.
24. Zanusso G, Casalone C, Acutis P, et al. Molecular analysis of iatrogenic scrapie in Italy. *J Gen Virol*. 2003;84(pt 4):1047–1052.

25. Hope J, Wood SC, Birkett CR, Chong A, Bruce ME, Cairns D, Goldmann W, Hunter N, Bostock CJ. Molecular analysis of ovine prion protein identifies similarities between BSE and an experimental isolate of natural scrapie, CH1641. J Gen Virol. 1999; 80 (pt 1):1-4. Erratum in: *J Gen Virol* 2000; 81(Pt 3):851.
26. Stack MJ, Chaplin MJ, Clark J. Differentiation of prion protein glycoforms from naturally occurring sheep scrapie, sheep-passaged scrapie strains (CH1641 and SSBP1), bovine spongiform encephalopathy (BSE) cases and Romney and Cheviot breed sheep experimentally inoculated with BSE using two monoclonal antibodies. *Acta Neuropathol.* 2002;104(3):279–286.
27. Jeffrey M, González L, Chong A, et al. Ovine infection with the agents of scrapie (CH1641 isolate) and bovine spongiform encephalopathy: immunochemical similarities can be resolved by immunohistochemistry. *J Comp Pathol.* 2006;134(1):17–29.
28. Baron T, Biacabe AG. Molecular behaviors of "CH1641-like" sheep scrapie isolates in ovine transgenic mice (TgOvPrP4). *J Virol.* 2007;81(13):7230–7237.
29. Martucci F, Acutis P, Mazza M, et al. Detection of typical and atypical bovine spongiform encephalopathy and scrapie prion strains by prion protein motif-grafted antibodies. *J Gen Virol.* 2009;90(pt 4):1048–1053.
30. Bishop MT, Will RG, Manson JC. Defining sporadic Creutzfeldt–Jakob disease strains and their transmission properties. *Proc Natl Acad Sci USA.* 2010;107(26):12005–12010.
31. Puoti G, Giaccone G, Rossi G, Canciani B, Bugiani O, Tagliavini F. Sporadic Creutzfeldt–Jakob disease: co-occurrence of different types of PrP(Sc) in the same brain. *Neurology.* 1999;53(9):2173–2176.
32. Tzaban S, Friedlander G, Schonberger O, et al. Protease-sensitive scrapie prion protein in aggregates of heterogeneous sizes. *Biochemistry.* 2002;41(42):12868–12875.
33. Galeno R, Di Bari MA, Nonno R, et al. Prion strain characterization of a novel subtype of Creutzfeldt–Jakob disease. *J Virol.* 2017;91(11):1–12.
34. Bruce ME, Will RG, Ironside JW, et al. Transmissions to mice indicate that 'new variant' CJD is caused by the BSE agent. *Nature.* 1997;389(6650):498–501.
35. Lombardi G, Casalone C, D' Angelo A, et al. Intraspecies transmission of BASE induces clinical dullness and amyotrophic changes. *PLoS Pathog.* 2008;4(5): e1000075.
36. European Food Safety Authority (EFSA), European Centre for Disease Prevention and Control (ECDC). Joint Scientific Opinion on any possible epidemiological or molecular association between TSEs in animals and humans. *EFSA J.* 2011;9(1):1945 [1111 pp.].

FURTHER READING

37. Yull HM, Ritchie DL, Langeveld JP, et al. Detection of type 1 prion protein in variant Creutzfeldt–Jakob disease. *Am J Pathol.* 2006;168(1):151–157.
38. Peretz D, Scott MR, Groth D, et al. Strain-specified relative conformational stability of the scrapie prion protein. *Protein Sci.* 2001;10(4):854–863.

CHAPTER NINETEEN

Omics of Prion Diseases

Silvia Vanni[1]
Scuola Internazionale Superiore di Studi Avanzati (SISSA), Trieste, Italy
[1]Corresponding author: e-mail address: svanni@sissa.it

Contents

Abstract

Prion diseases are unique neurodegenerative pathologies that can occur with sporadic, genetic, and acquired etiologies. Human and animal prion diseases can be recapitulated in laboratory animals with good reproducibility providing highly controlled models for studying molecular mechanisms of neurodegeneration. In this chapter the overall area of *omics* research in prion diseases is described. The term *omics* includes all fields of studies that employ a comprehensive, unbiased, and high-throughput approach to areas of research such as functional genomics, transcriptomics, and proteomics. These kind of approaches can be extremely helpful in identifying disease susceptibility factors and pathways that are dysregulated upon the onset and the progression of the disease. Herein, the most important research about the various forms of prion pathologies in human and in models of prion diseases in animals is presented and discussed.

1. INTRODUCTION

Genomics, intended as the field investigating the genome of a given organism through DNA analyses, represents a crucial area of study within

Progress in Molecular Biology and Translational Science, Volume 150
ISSN 1877-1173
http://dx.doi.org/10.1016/bs.pmbts.2017.05.004

the biological sciences. Neurodegenerative disorders, and in particular prion diseases, are characterized by complexity in terms of symptoms and pathogenesis whose mechanisms are difficult to decipher using with traditional approaches focusing on limited portions of specific pathways.[1] To date, an in-depth knowledge of the molecular processes that lead to neurodegeneration onset and progression in transmissible spongiform encephalopathies (TSEs) is still missing. In this context, genomic approaches are unbiased and powerful tools to investigate the molecular basis of these complex diseases. Indeed, large-scale gene expression profiling of diseased vs healthy groups to identify differentially expressed genes (DEGs) may help in identifying novel genes and pathways that are switched on or off at various time points during the pathogenesis of prion diseases. Different functional genomics technologies have been used in last decades to detected genes variations including cDNA libraries, mRNA differential display, suppression subtractive hybridization, genome-wide association studies (GWAS), microarrays, and more recently next-generation sequencing (NGS). Moreover, a number of different CNS and peripheral tissues have been exploited to gain insight into the pathways that are altered upon prion infection, analyzing diverse animal models—from rodents to nonhuman primates—and human diseases, from different points of view such as transcriptional and epigenetic alterations. The majority of these studies have been performed with the aim of identifying pathways underlying the mechanism of prion-induced neurotoxicity, in order to gain insight within the pathogenesis of these diseases. A second important aim, however, has been to identify signature molecules that might act as surrogate biomarkers either for the early detection of preclinical stages or for the identification of potential drug targets, since no effective therapy is available yet.

2. GENOME-WIDE ASSOCIATION STUDIES

In last decade, after the human genome became available, GWAS have been increasingly exploited in order to discover genetic variations, such as mutations or polymorphisms, that may represent risk factors for the development of a given disease, when not being the genetic cause. Besides this, GWAS and Quantitative Trait Loci (QTL) studies in cattle, sheep, and mice have also been performed to identify genomic regions and positional candidate genes, other than the prion gene, involved in TSE pathogenesis. In particular, after the "mad cow disease" epidemic in 1980–90s and the subsequent outbreak of variant form of CJD (vCJD) in the beginning of this century, several efforts have been employed toward cattle and BSE strain studies, in order to highlight the genetic polymorphisms that could dictate the outcome of the infection. The main advantage of these methods is offered by the scanning of the whole genome without any a priori hypothesis, allowing an unbiased approach aimed at identifying only those genes that influence a given phenotype. Regardless the organism, all the strategies used for discover prion disease susceptibility genes have reinforced the key role of *PRNP* and have also led to identify some new genetic modifiers.[2]

2.1 GWAS in Animal Prion Models

Unlike other neurodegenerative disorders, mice are naturally susceptible to prion diseases, therefore representing an ideal model organism for susceptibility genes identification and hypothesis testing. Indeed, a number of GWAS have been carried in mice which, reminiscent of humans, even when exposed to the very same prion agent display very different incubation times despite harboring the same *Prnp* susceptibility alleles. This point suggests an influence of other genetic or environmental factors as disease modifiers. A couple of studies in 2000 and 2001 investigated the presence of QTLs affecting Chandler/RML susceptibility in multiple strain backcrossed mice, identifying some influencing loci in chromosomes 9 and 11[3] and also in chromosomes 2, 6, 7, 11, and 12.[4] Some of these QTL have been identified in 2002 also with mouse-adapted BSE, using the same mice model; Lloyd and colleagues identified genetic loci for BSE susceptibility in chromosomes 2 and 11, indicating that these regions may influence the incubation time regardless the prion strain.[5] In 2003, a similar work from Moreno and colleagues revealed QTL in chromosomes 4, 6, 7, 8, 5, and 17 affecting scrapie, and two of them also BSE, susceptibility in both C57BL6 and RIII mice.[6] Another work, aimed at investigating susceptibility factors for BSE prion strain in C57BL6 and RIII mice, highlighted QTLs in four chromosomal regions[2,4,8,15] and a cytoplasm interaction in the host.[7] Later, using a semirandom mating of Northport HS mice with eight different parental strains and inoculated with Chandler/RML strain, it has been possible to identify a copine family member gene, *Cpne8,* as a quantitative trait gene within the previously identified locus on chromosome 15.[8] In recent years, with a similar approach, few other genes have been identified such as the retinoic acid receptor beta, *Rarb*, and *Sod1*.[9,10] More recently, a cell-based assay using mouse neuroblastoma N2a cell line has been used for the in vitro screening of Chandler/RML infection susceptibility genes previously identified through GWAS and QTL in mice and humans. They found seven genes (*Zbtb38*, *Sorcs1*, *Stmn2*, *Hspa13*, *Fkbp9*, *Actr10*, and *Plg*) that might play key roles in the prion propagation or clearance.[11] Quite a few genome-wide studies have been performed on cattle, since the occurrence of BSE became evident. One such study, dating back to 2002, Hernandez-Sanchez and colleagues performed a genome-wide search on half-sib Holstein cattle, naturally affected by BSE, using transmission–disequilibrium tests; significant segregation distortion was found for BSE within three marker loci on chromosomes 5, 10, and 20.[12] Two years later, the same study has been performed with a lager sample size of 360 cows, and six QTLs associated with BSE were found in chromosomes 1, 6, 13, 17, 19, and X/Y.[13] In 2010, 814 naturally infected BSE Holstein cattle—one group with known family relationships and another one made of paired cases-controls—have been investigated with a genome-wide scan using more than 3000 single-nucleotide polymorphisms (SNPs). The resulting prion disease associated with loci mapped on chromosomes 2, 14, 16, 20, 21, and 28.[14] In the same year, haplotype-tagging SNPs approach has been used to test association between *PRNP* haplotypes and classical BSE disease in Holstein cattle, and haplotype defined by the alleles of four polymorphisms in the region of high LD within the *PRNP* gene was found to be associated with reduced incidence classical BSE.[15] The same group of Murdoch and colleagues performed in 2011 a GWAS study on more than 300 Holstein cattle and identified a SNP on chromosome 1 associated with BSE occurrence and a locus on chromosome 14.[16] Concerning sheep, they were the first specie in which prion disease was discovered in 1700s, it is known that three polymorphisms in the *PRNP* gene are strongly linked to the occurrence of scrapie: A136V, R154H, and Q171R/H.[17] This genetic susceptibility association is at the basis of breeding programs for scrapie resistance in different countries. In 2006, Alvarez and colleagues analyzed 5230 Spanish sheep and discovered a new haplotype carrying leucine at codon 154 (ALQ). This genetic variant lies in a region supposed to have a key role in the conversion of prions, therefore suggesting that it might influence the susceptibility to scrapie.[18] A couple of years later, Moreno and colleagues analyzed 137 sheep of two half-sib families with a susceptible *PRNP* genotype (ARQ/VRQ) and identified QTL on chromosomes 6 and 18, reinforcing the evidence that genes outside *PRNP* affect scrapie resistance in sheep.[19] In 2015 a comparative study was performed using New Zealand sheep from different breeds, harboring a wide range of susceptible and resistant *PRNP* genotypes, challenged either with BSE or with scrapie. Two polymorphisms M112T and

L141F seemed to exert a certain protective role against BSE infection; however, contradictory results have been obtained with scrapie isolate SSBP/1 in ARQ/ARQ animals depending on their origin, indicating that links between *PRNP* genotype and clinical prion disease in sheep may suffer of some variability.[20]

2.2 GWAS in Human Prion Diseases

One seminal work in this context was a genome-wide study conducted in 2009 in patients with various forms of prion diseases (variant, sporadic, iatrogenic CJD, and kuru patients), which highlighted that the risk of developing prion diseases is strongly associated with the polymorphic codon 129 of the *PRNP* gene. Previously, another polymorphism in the *PRNP* gene at codon 219 was reported to be protective against sporadic CJD.[21] More recently, the presence of SNPs in the intron of *PRNP*, retinoic acid receptor protein (*RARB*), and *SCG10*/stathmin-like 2 was found to increase the risk of prion diseases.[22] Interestingly, this kind of approach has also led to the identification of a novel *PRNP* coding change, G127V, which is located in a highly conserved and structured region of PrP. Heterozygosity at codon 127 has shown to provide strong, and possibly complete, resistance to kuru.[23] A slightly different approach has been followed by Lloyd and colleagues, conducting a hypothesis-driven SNP association study in which they tested whether genetic variation at *HECTD2* gene, an E3 ubiquitin ligase, was associated with a phenotype of variant and sporadic CJD, after having found relevant results with QTL approach in mice. In this work they highlighted an association between *HECTD2* haplotypes and susceptibility for vCJD and kuru, reinforcing the involvement of the ubiquitine–proteasome system in prion disorders.[24] Some years later, a similar study has been performed with the additional inclusion of cases presenting inherited prion diseases and resistance to kuru despite attendance at mortuary feasts; SNPs at the ZBTB38–RASA2 locus were associated with all CJD types in the UK population only, while SNP in the *CHN2* gene was associated with vCJD but not with other forms of CJD.[25] In the same year, a genome-wide association analysis with most of the vCJD samples available in the world (93 British vCJD patients and replicated in an independent population of 22 British and 20 French vCJD cases) reported that an intronic variant in the *MTMR7* gene and an intergenic region upstream of the *NPAS2* gene are significantly associated with and increased risk of developing vCJD.[26] A recent reanalysis of those GWAS data identified a gene locus on chromosome 5 corresponding to *PLCXD3* as the most significant region associated with vCJD; resequencing of three intronic SNPs in that region in 109 sCJD and 120 vCJD UK patients revealed that these were highly associated with these prion diseases.[27] More recently, another GWAS study from the same group revealed an association of the glutamate receptor pathway with sCJD. In particular, a polymorphism in an intronic region of *GRM8*, coding for a protein involved in the transduction of cellular signals triggered by the prion protein, has been identified as a novel, non-*PRNP*, genome-wide significant marker associated with heightened disease risk.[28] However, when Balendra and colleagues tried to replicate the result about *PLCXD3* in a larger cohort of 2500 sCJD and 129 vCJD patients, with more than 10,000 controls, they found only weak association for one the SNPs and none for the *PLCXD3* gene, indicating that these approaches might be useful for identifying regions involved in the susceptibility to prion disease but have to be taken with caution when attempting of identifying specific SPNs as disease modifiers.[29]

3. MICROARRAY-BASED TRANCRIPTOMIC STUDIES

For early antemortem diagnosis of prion diseases, neither specific biomarkers nor effective therapeutic targets have been yet identified. As shown from GWAS and QTLs studies, increasing evidence suggests that other genes in addition to *PRNP* also contribute to the genetic susceptibility of acquired TSEs.[30] While being powerful techniques to reveal genetic loci associated with prion susceptibility, the identified regions are often too large to allow any precise identification of differentially expressed genes (DEGs). In this context, microarray-based gene expression analysis—generally followed by RT-qPCR validation of some identified DEGs—offers unique tools to approach neurodegenerative disorders. In particular, transcriptome profiling can be used to find altered transcripts in response to pathogens and select potential targets for novel therapeutic approaches.[31] In the prion biology field, several studies have been performed in order to identify DEGs in healthy and diseased systems, such as cell cultures and rodent models in addition to sheep, cattle, and monkeys.[30]

These analyses identified molecular pathways that are altered upon prion infection, including cholesterol synthesis, apoptosis, lysosomal pathway, immune and inflammation response.[31] The commonalities among prion diseases and PD and AD disorders highlight either similar neurodegenerative molecular mechanisms or similar pathological secondary events.

Comparative analyses of altered pathways in prion diseases, PD and AD could expose characteristic gene targets that could be used as potential diagnostic biomarkers for risk determination and also as general indicators of disease progression.[30,31]

3.1 Transcriptomic Studies in Animal Models

Many investigators have performed genomic analyses of brain tissues from different models of animal TSE; some of the studies involved the mRNA profiling of cattle BSE, cervids chronic wasting disease (CWD), monkey-adapted BSE, or ovine scrapie, but obviously the majority employed rodent-adapted models of prion disease.

3.1.1 Rodents Studies

In a number of prion-infected mouse models, genomic expression profiling showed increased oxidative and endoplasmic reticulum (ER) stress, activated ER, and mitochondrial apoptosis pathways together with induction of cholesterol biosynthesis in the CNS of preclinical mice.[32] For example, Booth and colleagues in 2004 performed a cDNA microarray analysis to examine gene expression profiles at different time points in the CNS of C57BL/6 mice inoculated either with ME7 or 79a mouse-adapted scrapie prion strain. They identified 138 upregulated and 20 downregulated genes at 100 dpi, of which 5 (*B2m, Clu, apoD, Egr1, Ly86*) were further validated by RT-PCR. Functional genes groups identified in this work included secreted extracellular proteins, lysosomal proteases, and host-defense and immune response-related proteins, signal transduction and cell growth-related genes. Interestingly, some of the downregulated genes were related to hematopoietic system, suggesting its involvement in the disease progression since early stages.[33] Similarly, some years later Skinner and colleagues made cDNA microarray platforms to identify DEGs prior to and after the onset of clinical symptoms in ME7, 22L, or RML-Chandler inoculated C57BL/10 mice. In symptomatic mice, over 400 DEGs were found, while only 22 at the preclinical stage and some were also differentially altered among strains. Overall, they reported an involvement of endosome/lysosome function, immunity, synapse function, metal ion binding, calcium regulation, and cytoskeletal function.[34] In order to further expand that study, in 2008 Kim and colleagues used Affymetrix microarray platforms to identify DEGs prior the onset of clinical symptoms in the same mouse model. They identified 67 upregulated genes in the brains at preclinical stage, many of which were related to the same previously identified pathways. Interestingly, they identified also 14 DEGs in the spleen prior to clinical onset; four of them, *Atp1b1, Gh, Anp32a*, and *Grn*, were upregulated since 46 dpi, suggesting their possible use as surrogate early biomarkers.[35] Riemer *et al.* analyzed different brain regions of C57Bl/6 mice intraperitoneally infected with scrapie strain 139A, at different time points after infection. This work identified 114 genes with altered expression, mainly related to inflammatory reaction and stress response, in addition to acute phase proteins induction and cholesterol metabolism.[36] In a similar work, Xiang and colleagues used Affymetrix microarray to investigate the global gene expression profiles in brains of mice intracerebrally inoculated with scrapie strains ME7 and RML. This lead to the finding of 121 genes upregulated in both mouse models, while 16 and 6 genes showed a remarkable increase in expression only in ME7- or RML-infected mouse brains, respectively, suggesting a specific association of the expression changes of these genes with the two scrapie strains studied.[37] The same group also reported a strong evidence for cholesterol metabolism and immune system involvement in the pathology of ME7-affected mice.[38] Sorensen *et al.* investigated two different mice strains: C57BL/6 mice intracerebrally challenged with ME7, 79a, and 22A strains of scrapie and VM mice inoculated with the 22A strain. Three hundred and forty-nine prion-related genes consistently altered in mouse models were identified; Gene Ontology analysis assigned many of the upregulated genes to functional groups associated with astrocytosis and gliosis, protein synthesis, inflammation, cell proliferation, and lipid metabolism.[39] In another work, Brown and colleagues applied cDNA expression arrays and RT-qPCR to study gene expression in the hippocampus of ME7/CV inoculated mouse model of scrapie. Seven genes—cathepsin D, clusterin, alpha and beta polypeptides of complement component 1q, GFAP, vimentin, and cystatin C—were confirmed differentially expressed at the terminal stages of disease.[40] Later on, they repeated the analysis applying Affymetrix high-density oligonucleotide probe arrays and identified 78 genes that are differentially expressed within the preclinical ME7-infected mouse hippocampus. The results indicated oxidative and ER stress, activated ER and mitochondrial apoptosis pathways, and activated cholesterol biosynthesis within the scrapie-infected hippocampus.[32] Laser-dissected CA1 hippocampal neurons from RML-infected CD1 mice were subjected to microarray analysis in order to determine their preclinical transcriptional response during infection. Genes involved in neuronal projection, dendrite development, and NMDA receptor signaling were found deregulated throughout the disease, as well as miR-132-3p, miR-124a-3p, miR-16-5p, miR-26a-5p, miR-29a-3p, and miR-140-5p.[41] Transcriptomic approach has been used also to analyze expression profiles from the brains of mice preclinically and clinically infected with RML, in comparison to cuprizone-treated mice. After comparative analysis, 17 transcripts unaffected by cuprizone treatment but increasing in expression from preclinical to clinical prion infection were identified, nine of which (*Hsbp1, Socs3, Ifi44, D12Ertd647e, Casp4, Agrp, Plce1, Ptpbl, Ddx58*) were upregulated preclinically and may therefore provide insights into the progressive impairment of the CNS during prion infection.[42] More recently, transgenic mice and murine cell line models have been investigated to gain insight about the influence of *Prnp* and other genes in the expression profiling landscape. For example, Tortosa *et al.* used transgenic mice expressing the bovine *Prnp* gene (BoTg110 line with B6CBAflx129/Ola background) to assess the influence of intracranial BSE inoculation on the brain gene expression profile at different time points. They identified an 87 genes that were differentially expressed at least at one time point, mainly related to neural cell metabolism and defense mechanisms.[43] More recently, Carroll *et al.* compared C57BL/10 or C57BL/6 mice infected with either 22L or RML prion strains, identifying 15 previously unreported DEGs related to inflammation or activation of the STAT signal transduction pathway. In this work, they also investigated IL-1Ra-deficient or overexpressing transgenic mice inoculated with the same scrapie strains; however, any observable effect on expression of the inflammatory genes analyzed was present.[44] Moreover, a couple of transcriptional studies have been done in order to investigate the role of prion protein in the brain development. In 2010, Chadi *et al.* compared the gene expression profile in the brain of FVB/N wild-type (WT) mice with the one of FVB/N *Prnp*$^{-/-}$ mice, either at the zygotic or at the adult stages. At the adult stage, neuronal *Prnp* disruption induced a response to an oxidative stress and a remodeling of the nervous system; however, only subtle transcriptomic differences could be evidenced.[45] Benvegnu and colleagues instead performed a microarray analysis with RNA isolated from the hippocampus of FVB mice at two different developmental stages: newborn (4.5-day-old) and adult (3-month-old) mice, both from WT and *Prnp*$^{-/-}$ animals. Also in this case, only a moderate alteration of the gene expression profile was reported, with the main affected biological pathways being cell signaling, cell–cell communication and transduction processes, calcium homeostasis, nervous system development, synaptic transmission, and cell adhesion.[46] A comprehensive comparative approach has been performed by Hwang and colleagues who tracked global gene expression in the brains of eight distinct mouse strain–prion strain combinations (C57BL/6J, C57BL/6.I-1, FVB/NCr, FVB-Tg(PrP-A)4053 (FVB × FVB.129-*Prnp*$_{\text{tm1Zrch}}$)F1, and FVB.129-*Prnp*$_{\text{tm1Zrch}}$ mice with RML or 301V prion strain) at 8–10 time points throughout the progression of the disease to capture the effects of prion strain, host genetics, and PrP concentration on disease incubation time. A core of 333 shared DEGs—belonging to different pathway such as complement activation, lysosomal proteases, cholesterol synthesis/efflux, glycosaminoglycan metabolism, iron homeostasis and heme metabolism, androgen metabolism, and sphingolipid synthesis/degradation—appeared to be central to prion disease; however, great differences among diverse genotypes were observed.[47]

While a large body of data is available for prion-infected murine mouse models, as discussed earlier, little is known about other rodents. One study from Herbst and colleagues in 2015 investigated the gene expression profile of mouse-adapted RML-infected rats with the aim of identifying common prion disease transcriptional responses. Interestingly, they identified 712 DEGs in rat prion disease, but only 33% of them were found to have similar responses in the corresponding mouse orthologs, questioning the generalization of previous mouse gene expression profiles.[48]

Also murine cell lines have been exploited in order to identity gene expression changes that are directly to prion replication. In 2005 Greenwood and colleagues used de novo RML-infected N2a and GT1 cells to compare the response of different cell lines to the same prion strain. They found that only minimal overlap in differential gene regulation between the two cell types, although there is functional overlap of the DEGs. However, when curing ScN2a cells with pentosan polysulfate, only some of the DEGs reverted to the normal profile.[49] A similar study has been performed in 2008 by Fasano *et al.*, taking advantage of a single clone of GT1 cells for de novo RML infection in order to compare the transcriptional profile of ScGT1 cells before and after treatment with quinacrine. Curiously, only a limited number of genes with moderate dysregulation were observed, with laminin being the only validated DEG.[50] Interestingly, a concomitant trancriptomic analysis conducted by Julius and colleagues on de novo RML-infected N2a, CAD, and GT1 cells did not reveal any major alteration in the gene expression profile, suggesting that there are no universal transcriptional changes induced by prion infection of neural cells in vitro.[51] Recently, Marbiah and colleagues reported a transcriptome analysis of prion-resistant revertants, isolated from highly susceptible cells, which revealed a gene expression signature associated with susceptibility and modulated by differentiation. Several of these genes encode proteins with a role in extracellular matrix remodeling, and silencing nine of these genes (fibronectin 1, integrin α8, chromogranin A, IQ motif-containing GTPase-activating protein 2, interleukin 11 receptor, alpha chain 1, Micalc C-terminal like, regulator of G-protein signaling 4, 3′-phosphoadenosine 5′-phosphosulfate synthase 2, and galactosyltransferase) significantly increased susceptibility.[52] A different experiment was performed in 2016, when Kim *et al.* administered either ferrous chloride ($FeCl_2$) [Fe(II)] or ferric ammonium citrate (FAC) [Fe(III)] to hippocampal mouse HpL3-4 cells devoided of functional *Prnp*, subsequently exposed to bovine recombinant PrP. Following Fe(III) treatment, 97 genes were differentially expressed, while Fe(II) treatment produced moderate alterations; cell growth, cell maintenance, and intra- and extracellular transport were the most affected pathways.[53]

3.1.2 Ovine Studies

Since scrapie has been the first reported prion disorder back in mid-1700, and given that it is one of the few naturally occurring TSEs—besides CWD in cervids and BSE in cattle—prion-affected ovine models has been extensively studied in order to understand the mechanism underlying idiopathic prion diseases. Moreover, despite transmission of scrapie to humans has never been reported, it cannot be ruled out that upon strain evolution it might become able to cross the species barrier. Filali and colleagues have been performed several studies on sheep; in 2011 they investigated transcriptome changes in medulla oblongata of naturally scrapie-affected symptomatic Rasa Aragonesa sheep. A custom microarray platform analysis revealed 148 DEGs, many of which related to the immune response, ion transport, cell adhesion, and transcription.[54] A similar analysis was performed also during the presymptomatic phase, where 32 DEGs mainly related to immune response, apoptosis, and phosphorylation of proteins were identified.[55] In particular, the mitochondrial pathway of apoptosis was activated with higher intensity in presymptomatic sheep than in symptomatic ones, with the implication of genes such as BAX or AIF.[56] In a following work, they presented six potential genetic biomarkers (CAPN6, COL1A2, COL3A1, GALA1, MT2A, and MTNR1B) for both the early and terminal stages of scrapie in five different brain regions. Gene transcription changes were confirmed in the medulla oblongata, and the expression profiles were generally similar in other central nervous system regions. Four of these genes (CAPN6, GALA1, MTNR1B, and MT2A) were confirmed also at the protein level, suggesting their role as potential targets for further prion disease research.[57] The same group also assessed gene expression profile of mesenteric lymph nodes at different disease stages in scrapie sheep. In the clinical stage of the disease 43 genes were upregulated and 62 downregulated, while 50 DEGs were dysregulated in the preclinical stage of the disease; some of the DEGs could be grouped into specific signaling pathways such as focal adhesion, PPAR signaling, and ECM–receptor interaction.[58] In 2014, Gossner and colleagues used Affymetrix gene arrays to quantify changes in the transcriptome of the thalamus and obex (medulla) of New Zealand Cheviot sheep—homozygous for the *PRNP* genotype VRQ—subcutaneously infected with SSBP/1 sheep brain homogenate. They identified 337 DEGs prior to the immunohistochemical detection of PrP^{Sc} in the thalamus and 243 DEGs at the clinical stage. However, no genes were common in the two comparisons, suggesting that no relationship and little progression of quantitative gene changes are present from before to after the detection of PrP^{Sc}.[59] In a similar study involving the peripheral lymphoid tissue they identified 75 and 80 DEGs at preclinical and clinical stage, respectively, 70% of which being persistently downregulated even if with only modest fold change values.[60] More recently, Sanz Rubio and colleagues investigated the expression of blood plasma circulating microRNAs in naturally scrapie-infected sheep, identifying significant scrapie-associated increase for miR-342-3p and miR-21-5p.[61]

3.1.3 Bovine and Cervid Studies

Upon the BSE epidemic in cattle in the 1980–90s, and especially after the related outbreak of a variant form of CJD in humans, studies on bovines and in particular on BSE transmission have been performed in order to identify potential diagnostic and therapeutic targets. Given that the most likely route of entry of infection following oral exposure to BSE is via the immunologically active Peyer's patches, in 2009 Khaniya and colleagues performed a microarray analysis on this tissue after orally challenging cattle with BSE. Ninety genes and 16 Expressed Sequence Tag were identified as significantly differentially expressed after 12 months since infection, mainly related to immune function, cellular, and metabolic processes.[62] Tang *et al.* instead investigated brain tissue of naturally BSE-infected cattle using Affymetrix GeneChip Bovine Genome Arrays. They reported 230 DEGs and many of these involved in immune response, apoptosis, cell adhesion, stress response, and transcription. In particular, 17 genes are associated with the

ER, and 10 of these are involved in stress-related responses including ER chaperones, Grp94 and Grp170.[63] In 2011, Almeida *et al.* analyzed caudal medulla tissues of animals infected with different doses of BSE at 12- and 45-month postinfection and identified a number of altered pathways including extracellular matrix, cell adhesion, neuroactive ligand–receptor interaction, complement and coagulation cascades, MAPK signaling, neurodegenerative disorder, SNARE interactions in vesicular transport, and the transforming growth factor (TGF) beta signaling.[64,65] In recent years it has been discovered an atypical form of BSE, named bovine amyloidotic spongiform encephalopathy (BASE) that affects cows and can be transmitted to mice and primates. BASE is biochemically different from BSE and shares some molecular and histopathological features with the MV2 subtype of human sCJD. Panelli and colleagues analyzed the gene expression of white blood cells from 5 cattle at 1 year after oral BASE challenge, identifying 39 upregulated and 17 downregulated DEGs in BASE-inoculated animals. The majority of these genes were related to immune function, particularly genes linked to T- and B-cell development and activation, and to inflammatory responses.[66,67] More recently, Xerxa *et al.* performed a microarray study on whole blood of BASE-infected cattle, prior to and after the onset of clinical signs. While 22 DEGs were found specifically in the preclinical stage samples, 32 genes were dysregulated both the preclinical and clinical stage, mainly related to immune response and general metabolism.[68]

Besides BSE and cattle, some concern has arisen about cervids CWD, and potential infectivity for humans has been recently highlighted. CWD is an invariably fatal neurologic disease that naturally infects mule deer, white tailed deer, and elk. Despite in a recent study was shown that cynomolgus macaques, when exposed either orally or intracerebrally to CWD, did not develop any symptoms after 6 years of observation, the direct relevance to humans is not definitive and transmission studies are ongoing. In a work from Basu and colleagues, microarray analysis was performed to determine changes in the gene expression profiles in six different tissues including brain, midbrain, thalamus, spleen, retropharyngeal lymph node (RPLN), and tonsil of CWD-infected elk in comparison to noninfected healthy elk. They identified 249 DEGs in the spleen (168 up- and 81 downregulated), 30 DEGs in the RPLN (18 up- and 12 downregulated), and 55 DEGs in the tonsil (21 up- and 34 downregulated), with the main altered pathways being neuronal signaling and synapse function, calcium signaling, apoptosis and cell death, and immune cell trafficking and inflammatory response.[69]

3.2 Transcriptomic Studies in Primates and Humans

As regards nonhuman primates, the *Macaca* genus is closely related to humans, thanks to a common ancestor dated back to ≈25 million years ago. Importantly, rhesus and cynomolgus macaques are evolutionary very close to humans, have a high degree of amino acid sequence identity in the PrP sequence, and bear also the polymorphism at codon 129.[70] In 2009, Montag and colleagues have investigated differential expression of miRNAs in the brains of BSE-infected cynomolgus macaques. They discovered upregulation of hsa-miR-342-3p and hsa-miR-494 which might be related with late stage prion disease since it seems that hsa-miR-342-3p was found upregulated also in sCJD human brains.[71] The same animals were also subjected to ERV expression profiling of the *basis pontis* and *vermis cerebelli* using a retrovirus-specific microarray and quantitative real-time PCR. Class I gamma retroviruses HERV-E4-1, ERV-9, and MacERV-4 increased expression was detected in BSE-infected macaques, while both the macaque ERV-K-(HML-2) Gag protein and RNA were found downregulated in the frontal/parietal cortex.[72] More recently, a large-scale transcriptional profiling of frontal cortex of these BSE-infected macaques identified 86 DEGs most of which involved in cellular development, cell death and survival, lipid homeostasis, and acute phase response signaling. In particular, HBB and HBA2 were downregulated in infected macaques, whereas TTR, APOC1, and SERPINA3 were upregulated.[73] As regards humans, in 2005 gene expression profiling of frontal cortex from 15 sCJD patients was performed using Affymetrix HGU133A microarrays. The major alterations identified in sCJD brains included upregulation of the genes encoding immune and stress-response factors and elements involved in cell death and cell cycle, as well as prominent downregulation of genes encoding synaptic proteins. In particular, they reported 79 upregulated (e.g., metallothionein-1 and -2) and 275 downregulated genes (e.g., synaptosomal-associated protein 25 (SNAP25) modulating synaptic function and plasticity) compared to healthy controls.[74] A similar study was performed in 2013 by Tian and colleagues, analyzing the gene expression profile of cortex and thalamus of FFI patients. A total of 1314 DEGs in the thalamus and 332 ones in the parietal lobe were reported, with 255 DEGs showing the same altering tendencies in both tested regions and related to transcription and DNA-dependent regulation of transcription, RNA splicing, and mitochondrial electron transport.[75] Recently, the same group also analyzed the similarity and difference of global gene expression profiles in cortex regions of sCJD, FFI, and Alzheimer's disease patients. The most commonly changes of biological processes shared among all neurodegenerative diseases were signal transduction, synaptic transmission, and neuropeptide signaling pathway, while the most commonly changed pathways were MAPK signaling pathway, Parkinson's disease-related pathways, and oxidative phosphorylation.[76]

4. SEQUENCING APPROACH IN PRION DISORDERS

Besides whole-genome/exome sequencing studies, which allow the large-scale identification of rare genetic variants, such as SNP and mutations discovery, one major application of NGS is transcriptional profiling through RNA-sequencing (RNA-seq). Indeed, it can be used to identify different splice variants and determine their relative frequencies, but also to look at coding and noncoding RNAs. Basically, the relative expression of each mRNA (once converted into cDNA) can be calculated by counting the number of sequenced tags mapping to it. Normalized gene signals can then be used to calculate the differential expression between samples.[77] Several studies have shown that RNA-seq is considerably more sensitive and robust than traditional microarrays, has a much greater dynamic range, and can detect splice variants and noncoding RNAs that would otherwise be missed. Concerning the discovery of genetic variations, in 2008 a genetic screen of the open reading frame of the gene coding for Shadoo (*SPRN*) which has protein homology and possible functional links with the prion protein. Five hundred and twenty-two patients with prion disease, including 107 with vCJD, and 861 healthy controls were sequenced and a common coding variant of SPRN, two further SNPs and three rare insertion/deletion variants were found.[78] More recently, 16,025 prion disease cases, together with 60,706 population control exomes and 531,575 individuals were sequenced in order to assess the impact of variants in *PRNP* on the risk of prion disease. Missense variants in *PRNP* previously reported to be pathogenic were found at least 30 times more common in the population than expected on the basis of genetic prion disease prevalence, assessing the relevance of these population studies.[79] One of the first studies that used parallel sequencing to investigate the gene expression changes occurring in cattle in response to BSE infection dates back to 2011, when Basu and colleagues used digital gene expression tag profiling in medulla tissue 45 months after infection. With this approach, they revealed 73 upregulated and 117 downregulated genes, some of which involved in 38 different pathways described in the Kyoto Encyclopedia of Genes and Genomes pathways.[80] More recently, RNA-seq was used to detect potential determinants of prion propagation in cultured ovine microglial cells, in order to elucidate permissiveness and resistance factors that regulate scrapie infection. Munoz-Gutierrez *et al.* inoculated two ovine microglial clones with different prion susceptibility with either scrapie-positive or scrapie-negative sheep brainstem homogenates and comparative transcriptional analyses identified 22 DEGs. In particular, in highly permissive microglia, TGF β-induced, retinoic acid receptor response 1, and phosphoserine aminotranspherase 1 gene transcripts were upregulated.[81] Another interesting work, based on RNA-seq technology, was performed by Khalife and colleagues comparing transcriptomic profile of FVB/N and FVB/N *Prnp* knockout mice at early embryonic stages. They identified 73 and 263 DEGs at E6.5 and E7.5, respectively, with the main represented functional groups being proteolysis, protease inhibition, biological adhesion, nervous system development, apoptosis, cell proliferation, and inflammatory and innate immune response.[82] Very recently, Gao *et al.* used deep sequencing technology to obtain miRNA expression profile of the brains of scrapie-infected C57BL mice. 57, 94, and 135 differentially expressed miRNAs were identified in 139A-, ME7-, and S15-infected mice, respectively, among which 22 were commonly increased and 14 were commonly decreased in the brains of all three infected models. Of these, mmu-miR-146a-5p, mmu-miR-341-3p, mmu-miR-879-5p, mmu-miR-3470a, mmu-miR-3473a, mmu-miR-3473b, mmu-miR-96-5p, mmu-miR-141-3p, mmu-miR-182-5p, mmu-miR-200a-3p, mmu-miR-200b-3p, mmu-miR-200b-5p, mmu-miR-2, and mmu-miR-20 were further validated by RT-qPCR.[83]

5. EPIGENOMIC STUDIES

Genomic tools are contributing to the emerging understanding of the increasingly relevant field of epigenetics, which addresses how a gene function can be altered without modifying the DNA sequence.[84] Indeed, neurological conditions are not only reflected on genomic mutations and transcriptomic dysregulations but also on the change of epigenome. Among the various types of epigenomic modifications, DNA methylation and histone modifications are the most widely studied. Moreover, the expression levels of microRNAs further influence the epigenomic landscape.[1] A couple of studies involving dysregulated miRNAs in the blood[61] or brain[71] of prion disease-affected animals have been mentioned previously. Very recently, a study from Dalai and colleagues investigated CpG site DNA methylation patterns of C57BL/6CrSlc and BALB/cCrSlc mice through bisulfite sequencing PCR. They observed that the CpG island (CGI) including the *Prnp* core promoter region was completely unmethylated in a wide panel of tissues analyzed, while CpG methylation in the CGI shore region occurred in various tissue- and site-specific proportions. Moreover, they showed that one CpG site methylation at −576 was negatively correlated with the PrP mRNA level, revealing that different methylation frequencies of the CGI shore region are likely to affect *Prnp* expression in a tissue-specific manner.[85]

6. PROTEOMIC STUDIES

Proteomic technologies provide powerful tools to systematically profile the peptide or protein constituents in diverse tissues, with an unbiased approach. Proteomics approach is complementary to the transcriptomic one; however, one most important difference is that while it is now possible to interrogate essentially the entire transcriptome, there is still no comparable comprehensive approach to the proteins encoded therein.[86] On the other end, it should also be considered that the overlapping between protein and mRNA expression lacks consistency, mainly due to protein stability, which is relatively higher than mRNA.

In 2008, Crecelius and colleagues compared the brain proteome of the PrP knockout Zürich I with WT mice using both two-dimensional difference gel electrophoresis (DIGE) and liquid chromatography-mass spectrometry (LC-MS)/mass spectrometry (MS) in combination with isotope-coded protein labeling isotope labeling. Both of the two analyses demonstrated a highly conserved quantitative protein profile in brain homogenates of $Prnp^{-/-}$ and $Prnp^{+/+}$ mice.[87] The same year, also Ramljak *et al.* reported a proteomic study aiming to reveal changes caused by the presence/absence or different expression levels of the endogenous cellular form of the prion protein (PrP^C). They took advantage of transient overexpression of human PrP in HEK 293 cells and transient introduction of the human *PRNP* gene into mouse neuronal $Prnp^{0/0}$ cells and performed DIGE to compare their proteomes. Transient PrP^C overexpression revealed a major alteration of proteins participating in energy metabolism (TALDO, ATPA) and cellular homeostasis (GRP78), whereas transient introduction of PrP^C into PrP^C-deficient cells mainly affected proteins involved in protection against oxidative stress (SODC) and apoptosis (Hsp27).[88] In 2016, Tahir and colleagues studied the global proteomic alterations in the cerebellum of brain from MM1 and VV2 sCJD patients. Two-dimensional gel electrophoresis coupled with mass spectrometric identification identified 40 proteins in MM1 and 43 proteins in VV2 subtype to be differentially expressed, 12 of which were in common between two subtypes. Differentially expressed proteins mainly belonged to cell cycle, gene expression and cell death, cellular stress response/oxidative stress and signal transduction, and synaptic functions. In particular, they identified DJ1 protein to be upregulated during the clinical phase in the CSF of sCJD patients and also at presymptomatic and symptomatic stages the disease in cerebellum of humanized PrP transgenic mice inoculated with sCJD MM1 and VV2 brain, suggesting the involvement of oxidative stress in the pathogenesis of sCJD.[89]

7. CONCLUDING REMARKS

This chapter illustrates various approaches that have been employed in last two decades for understanding the complexity of prion diseases. Functional genomics has been fundamental to identify several cellular- and tissue-specific pathways involved in the neurodegeneration undergone by the nervous system upon prion infection. Pathways, such as cholesterol biosynthesis, apoptosis, lysosomal function, immune and inflammation response, oxidative and ER stress, mitochondrial apoptosis, synapse function, metal ion binding, and calcium regulation among the others, have been recurrently identified using a plethora of animal models of prion disorders and/or natural occurring pathogenic changes in animal and human prion diseases.

The results of these studies illustrate the intricacy of the disease. Indeed, what clearly emerges from this wide variety of *omics* studies is that a number of different factors such as the prion strain, the route of infection, the etiology, and the host *PRNP* sequence can all strongly alter the gene and protein expression profile of an affected individual. Nevertheless, these investigations have opened new venues of research, while the complexity of the results still needs to be fully elucidated. If focusing on common altered pathways shared among different prion diseases and models with the aim of explaining why, how, and when such changes occur in the pathology, these studies could be exploited for further advancing our knowledge on the mechanisms of prion conversion, replication, and accumulation.

REFERENCES

1. Han G, Sun J, Wang J, Bai Z, Song F, Lei H. Genomics in neurological disorders. *Genomics Proteomics Bioinformatics*. 2014;12(4):156–163.
2. Lloyd SE, Mead S, Collinge J. Genetics of prion diseases. *Curr Opin Genet Dev*. 2013;23(3):345–351.
3. Stephenson DA, Chiotti K, Ebeling C, et al. Quantitative trait loci affecting prion incubation time in mice. *Genomics*. 2000;69(1):47–53.
4. Lloyd SE, Onwuazor ON, Beck JA, et al. Identification of multiple quantitative trait loci linked to prion disease incubation period in mice. *Proc Natl Acad Sci USA*. 2001;98(11):6279–6283.
5. Lloyd SE, Uphill JB, Targonski PV, Fisher EM, Collinge J. Identification of genetic loci affecting mouse-adapted bovine spongiform encephalopathy incubation time in mice. *Neurogenetics*. 2002;4(2):77–81.
6. Moreno CR, Lantier F, Lantier I, Sarradin P, Elsen JM. Detection of new quantitative trait loci for susceptibility to transmissible spongiform encephalopathies in mice. *Genetics*. 2003;165(4):2085–2091.
7. Manolakou K, Beaton J, McConnell I, et al. Genetic and environmental factors modify bovine spongiform encephalopathy incubation period in mice. *Proc Natl Acad Sci USA*. 2001;98(13):7402–7407.
8. Lloyd SE, Maytham EG, Grizenkova J, Hummerich H, Collinge J. A Copine family member, Cpne8, is a candidate quantitative trait gene for prion disease incubation time in mouse. *Neurogenetics*. 2010;11(2):185–191.
9. Akhtar S, Grizenkova J, Wenborn A, et al. Sod1 deficiency reduces incubation time in mouse models of prion disease. *PLoS One*. 2013;8(1):e54454.
10. Grizenkova J, Akhtar S, Collinge J, Lloyd SE. The retinoic acid receptor beta (Rarb) region of Mmu14 is associated with prion disease incubation time in mouse. *PLoS One*. 2010;5(12):e15019.
11. Brown CA, Schmidt C, Poulter M, et al. In vitro screen of prion disease susceptibility genes using the scrapie cell assay. *Hum Mol Genet*. 2014;23(19):5102–5108.
12. Hernandez-Sanchez J, Waddington D, Wiener P, Haley CS, Williams JL. Genome-wide search for markers associated with bovine spongiform encephalopathy. *Mamm Genome*. 2002;13(3):164–168.
13. Zhang C, De Koning DJ, Hernandez-Sanchez J, Haley CS, Williams JL, Wiener P. Mapping of multiple quantitative trait loci affecting bovine spongiform encephalopathy. *Genetics*. 2004;167(4):1863–1872.
14. Murdoch BM, Clawson ML, Laegreid WW, et al. A 2cM genome-wide scan of European Holstein cattle affected by classical BSE. *BMC Genet*. 2010;11:20.
15. Murdoch BM, Clawson ML, Yue S, et al. PRNP haplotype associated with classical BSE incidence in European Holstein cattle. *PLoS One*. 2010;5(9):e12786.
16. Murdoch BM, Murdoch GK, Settles M, McKay S, Williams JL, Moore SS. Genome-wide scan identifies loci associated with classical BSE occurrence. *PLoS One*. 2011;6(11):e26819.
17. Baylis M, Goldmann W. The genetics of scrapie in sheep and goats. *Curr Mol Med*. 2004;4(4):385–396.
18. Alvarez L, Arranz JJ, San PF. Identification of a new leucine haplotype (ALQ) at codon 154 in the ovine prion protein gene in Spanish sheep. *J Anim Sci*. 2006;84(2):259–265.
19. Moreno CR, Cosseddu GM, Schibler L, et al. Identification of new quantitative trait Loci (other than the PRNP gene) modulating the scrapie incubation period in sheep. *Genetics*. 2008;179(1):723–726.
20. Houston F, Goldmann W, Foster J, Gonzalez L, Jeffrey M, Hunter N. Comparative susceptibility of sheep of different origins, breeds and PRNP genotypes to challenge with bovine spongiform encephalopathy and scrapie. *PLoS One*. 2015;10(11):e0143251.

21. Shibuya S, Higuchi J, Shin RW, Tateishi J, Kitamoto T. Protective prion protein polymorphisms against sporadic Creutzfeldt-Jakob disease. *Lancet*. 1998;351(9100):419.
22. Mead S, Poulter M, Uphill J, et al. Genetic risk factors for variant Creutzfeldt-Jakob disease: a genome-wide association study. *Lancet Neurol*. 2009;8(1):57–66.
23. Mead S, Whitfield J, Poulter M, et al. A novel protective prion protein variant that colocalizes with kuru exposure. *N Engl J Med*. 2009;361(21):2056–2065.
24. Lloyd SE, Maytham EG, Pota H, et al. HECTD2 is associated with susceptibility to mouse and human prion disease. *PLoS Genet*. 2009;5(2):e1000383.
25. Mead S, Uphill J, Beck J, et al. Genome-wide association study in multiple human prion diseases suggests genetic risk factors additional to PRNP. *Hum Mol Genet*. 2012;21(8):1897–1906.
26. Sanchez-Juan P, Bishop MT, Aulchenko YS, et al. Genome-wide study links MTMR7 gene to variant Creutzfeldt-Jakob risk. *Neurobiol Aging*. 2012;33(7):1487:e1421–1488.
27. Bishop MT, Sanchez-Juan P, Knight RS. Splice site SNPs of phospholipase PLCXD3 are significantly associated with variant and sporadic Creutzfeldt-Jakob disease. *BMC Med Genet*. 2013;14:91.
28. Sanchez-Juan P, Bishop MT, Kovacs GG, et al. A genome wide association study links glutamate receptor pathway to sporadic Creutzfeldt-Jakob disease risk. *PLoS One*. 2014;10(4):e0123654.
29. Balendra R, Uphill J, Collinson C, et al. Variants of PLCXD3 are not associated with variant or sporadic Creutzfeldt-Jakob disease in a large international study. *BMC Med Genet*. 2016;17:28.
30. Basu U, Guan le L, Moore SS. Functional genomics approach for identification of molecular processes underlying neurodegenerative disorders in prion diseases. *Curr Genomics*. 2012;13(5):369–378.
31. Benetti F, Gasperini L, Zampieri M, Legname G. Gene expression profiling to identify druggable targets in prion diseases. *Expert Opin Drug Discov*. 2010;5(2):177–202.
32. Brown AR, Rebus S, McKimmie CS, Robertson K, Williams A, Fazakerley JK. Gene expression profiling of the preclinical scrapie-infected hippocampus. *Biochem Biophys Res Commun*. 2005;334(1):86–95.
33. Booth S, Bowman C, Baumgartner R, et al. Identification of central nervous system genes involved in the host response to the scrapie agent during preclinical and clinical infection. *J Gen Virol*. 2004;85(pt 11):3459–3471.
34. Skinner PJ, Abbassi H, Chesebro B, Race RE, Reilly C, Haase AT. Gene expression alterations in brains of mice infected with three strains of scrapie. *BMC Genomics*. 2006;7:114.
35. Kim HO, Snyder GP, Blazey TM, Race RE, Chesebro B, Skinner PJ. Prion disease induced alterations in gene expression in spleen and brain prior to clinical symptoms. *Adv Appl Bioinform Chem*. 2008;1:29–50.
36. Riemer C, Neidhold S, Burwinkel M, et al. Gene expression profiling of scrapie-infected brain tissue. *Biochem Biophys Res Commun*. 2004;323(2):556–564.
37. Xiang W, Windl O, Wunsch G, et al. Identification of differentially expressed genes in scrapie-infected mouse brains by using global gene expression technology. *J Virol*. 2004;78(20):11051–11060.
38. Xiang W, Hummel M, Mitteregger G, et al. Transcriptome analysis reveals altered cholesterol metabolism during the neurodegeneration in mouse scrapie model. *J Neurochem*. 2007;102(3):834–847.
39. Sorensen G, Medina S, Parchaliuk D, Phillipson C, Robertson C, Booth SA. Comprehensive transcriptional profiling of prion infection in mouse models reveals networks of responsive genes. *BMC Genomics*. 2008;9:114.
40. Brown AR, Webb J, Rebus S, Williams A, Fazakerley JK. Identification of up-regulated genes by array analysis in scrapie-infected mouse brains. *Neuropathol Appl Neurobiol*. 2004;30(5):555–567.

41. Majer A, Medina SJ, Niu Y, et al. Early mechanisms of pathobiology are revealed by transcriptional temporal dynamics in hippocampal CA1 neurons of prion infected mice. *PLoS Pathog*. 2012;8(11):e1003002.
42. Moody LR, Herbst AJ, Yoo HS, Vanderloo JP, Aiken JM. Comparative prion disease gene expression profiling using the prion disease mimetic, cuprizone. *Prion*. 2009;3(2):99–109.
43. Tortosa R, Castells X, Vidal E, et al. Central nervous system gene expression changes in a transgenic mouse model for bovine spongiform encephalopathy. *Vet Res*. 2011;42:109.
44. Carroll JA, Striebel JF, Race B, Phillips K, Chesebro B. Prion infection of mouse brain reveals multiple new upregulated genes involved in neuroinflammation or signal transduction. *J Virol*. 2015;89(4):2388–2404.
45. Chadi S, Young R, Le Guillou S, et al. Brain transcriptional stability upon prion protein-encoding gene invalidation in zygotic or adult mouse. *BMC Genomics*. 2010;11:448.
46. Benvegnu S, Roncaglia P, Agostini F, et al. Developmental influence of the cellular prion protein on the gene expression profile in mouse hippocampus. *Physiol Genomics*. 2011;43(12):711–725.
47. Hwang D, Lee IY, Yoo H, et al. A systems approach to prion disease. *Mol Syst Biol*. 2009;5:252.
48. Herbst A, Ness A, Johnson CJ, McKenzie D, Aiken JM. Transcriptomic responses to prion disease in rats. *BMC Genomics*. 2015;16:682.
49. Greenwood AD, Horsch M, Stengel A, et al. Cell line dependent RNA expression profiles of prion-infected mouse neuronal cells. *J Mol Biol*. 2005;349(3):487–500.
50. Fasano C, Campana V, Griffiths B, Kelly G, Schiavo G, Zurzolo C. Gene expression profile of quinacrine-cured prion-infected mouse neuronal cells. *J Neurochem*. 2008;105(1):239–250.
51. Julius C, Hutter G, Wagner U, et al. Transcriptional stability of cultured cells upon prion infection. *J Mol Biol*. 2008;375(5):1222–1233.
52. Marbiah MM, Harvey A, West BT, et al. Identification of a gene regulatory network associated with prion replication. *EMBO J*. 2014;33(14):1527–1547.
53. Kim M, Kim EH, Choi BR, Woo HJ. Differentially expressed genes in iron-induced prion protein conversion. *Biochem Biophys Res Commun*. 2016;480:734–740.
54. Filali H, Martin-Burriel I, Harders F, et al. Gene expression profiling and association with prion-related lesions in the medulla oblongata of symptomatic natural scrapie animals. *PLoS One*. 2011;6(5):e19909.
55. Filali H, Martin-Burriel I, Harders F, et al. Medulla oblongata transcriptome changes during presymptomatic natural scrapie and their association with prion-related lesions. *BMC Genomics*. 2012;13:399.
56. Hedman C, Lyahyai J, Filali H, et al. Differential gene expression and apoptosis markers in presymptomatic scrapie affected sheep. *Vet Microbiol*. 2012;159(1–2):23–32.
57. Filali H, Vidal E, Bolea R, et al. Gene and protein patterns of potential prion-related markers in the central nervous system of clinical and preclinical infected sheep. *Vet Res*. 2013;44(1):14.
58. Filali H, Martin-Burriel I, Harders F, et al. Gene expression profiling of mesenteric lymph nodes from sheep with natural scrapie. *BMC Genomics*. 2014;15:59.
59. Gossner AG, Hopkins J. Transcriptome analysis of CNS immediately before and after the detection of PrP(Sc) in SSBP/1 sheep scrapie. *Vet Microbiol*. 2014;173(3–4):201–207.
60. Gossner AG, Hopkins J. The effect of PrP(Sc) accumulation on inflammatory gene expression within sheep peripheral lymphoid tissue. *Vet Microbiol*. 2015;181(3–4): 204–211.
61. Sanz Rubio D, Lopez-Perez O, de Andres PA, et al. Increased circulating microRNAs miR-342-3p and miR-21-5p in natural sheep prion disease. *J Gen Virol*. 2017;98(2):305–310.

62. Khaniya B, Almeida L, Basu U, et al. Microarray analysis of differentially expressed genes from Peyer's patches of cattle orally challenged with bovine spongiform encephalopathy. *J Toxicol Environ Health A*. 2009;72(17–18):1008–1013.
63. Tang Y, Xiang W, Terry L, Kretzschmar HA, Windl O. Transcriptional analysis implicates endoplasmic reticulum stress in bovine spongiform encephalopathy. *PLoS One*. 2010;5(12):e14207.
64. Almeida LM, Basu U, Khaniya B, et al. Gene expression in the medulla following oral infection of cattle with bovine spongiform encephalopathy. *J Toxicol Environ Health A*. 2011;74(2–4):110–126.
65. Almeida LM, Basu U, Williams JL, Moore SS, Guan LL. Microarray analysis in caudal medulla of cattle orally challenged with bovine spongiform encephalopathy. *Genet Mol Res*. 2011;10(4):3948–3962.
66. Panelli S, Strozzi F, Capoferri R, et al. Analysis of gene expression in white blood cells of cattle orally challenged with bovine amyloidotic spongiform encephalopathy. *J Toxicol Environ Health A*. 2011;74(2–4):96–102.
67. Trovato A, Panelli S, Strozzi F, et al. Expression of genes involved in the T cell signalling pathway in circulating immune cells of cattle 24 months following oral challenge with bovine amyloidotic spongiform encephalopathy (BASE). *BMC Vet Res*. 2015;11:105.
68. Xerxa E, Barbisin M, Chieppa MN, et al. Whole blood Gene expression profiling in preclinical and clinical cattle infected with atypical bovine spongiform encephalopathy. *PLoS One*. 2016;11(4):e0153425.
69. Basu U, Almeida LM, Dudas S, et al. Gene expression alterations in Rocky Mountain elk infected with chronic wasting disease. *Prion*. 2012;6(3):282–301.
70. Glatzel M, Pekarik V, Luhrs T, Dittami J, Aguzzi A. Analysis of the prion protein in primates reveals a new polymorphism in codon 226 (Y226F). *Biol Chem*. 2002;383(6):1021–1025.
71. Montag J, Hitt R, Opitz L, Schulz-Schaeffer WJ, Hunsmann G, Motzkus D. Upregulation of miRNA hsa-miR-342-3p in experimental and idiopathic prion disease. *Mol Neurodegener*. 2009;4:36.
72. Greenwood AD, Vincendeau M, Schmadicke AC, Montag J, Seifarth W, Motzkus D. Bovine spongiform encephalopathy infection alters endogenous retrovirus expression in distinct brain regions of cynomolgus macaques (Macaca fascicularis). *Mol Neurodegener*. 2011;6(1):44.
73. Barbisin M, Vanni S, Schmadicke AC, et al. Gene expression profiling of brains from bovine spongiform encephalopathy (BSE)-infected cynomolgus macaques. *BMC Genomics*. 2014;15:434.
74. Xiang W, Windl O, Westner IM, et al. Cerebral gene expression profiles in sporadic Creutzfeldt-Jakob disease. *Ann Neurol*. 2005;58(2):242–257.
75. Tian C, Liu D, Sun QL, et al. Comparative analysis of gene expression profiles between cortex and thalamus in Chinese fatal familial insomnia patients. *Mol Neurobiol*. 2013;48(1):36–48.
76. Tian C, Liu D, Xiang W, et al. Analyses of the similarity and difference of global gene expression profiles in cortex regions of three neurodegenerative diseases: sporadic Creutzfeldt-Jakob disease (sCJD), fatal familial insomnia (FFI), and Alzheimer's disease (AD). *Mol Neurobiol*. 2014;50(2):473–481.
77. de Magalhaes JP, Finch CE, Janssens G. Next-generation sequencing in aging research: emerging applications, problems, pitfalls and possible solutions. *Ageing Res Rev*. 2010;9(3):315–323.
78. Beck JA, Campbell TA, Adamson G, et al. Association of a null allele of SPRN with variant Creutzfeldt-Jakob disease. *J Med Genet*. 2008;45(12):813–817.
79. Minikel EV, Vallabh SM, Lek M, et al. Quantifying prion disease penetrance using large population control cohorts. *Sci Transl Med*. 2016;8(322):322ra329.

80. Basu U, Almeida L, Olson NE, et al. Transcriptome analysis of the medulla tissue from cattle in response to bovine spongiform encephalopathy using digital gene expression tag profiling. *J Toxicol Environ Health A*. 2011;74(2–4):127–137.
81. Munoz-Gutierrez JF, Pierle SA, Schneider DA, Baszler TV, Stanton JB. Transcriptomic determinants of scrapie prion propagation in cultured ovine microglia. *PLoS One*. 2016;11(1):e0147727.
82. Khalife M, Young R, Passet B, et al. Transcriptomic analysis brings new insight into the biological role of the prion protein during mouse embryogenesis. *PLoS One*. 2011;6(8): e23253.
83. Gao C, Wei J, Zhang BY, et al. MiRNA expression profiles in the brains of mice infected with scrapie agents 139A, ME7 and S15. *Emerg Microbes Infect*. 2016;5(11):e115.
84. Johnson WE, Koepfli K. The role of genomics in conservation and reproductive sciences. *Adv Exp Med Biol*. 2014;753:71–96.
85. Dalai W, Matsuo E, Takeyama N, Kawano J, Saeki K. CpG site DNA methylation patterns reveal a novel regulatory element in the mouse prion protein gene. *J Vet Med Sci*. 2017;79(1):100–107.
86. Zhang J, Keene CD, Pan C, Montine KS, Montine TJ. Proteomics of human neurodegenerative diseases. *J Neuropathol Exp Neurol*. 2008;67(10):923–932.
87. Crecelius AC, Helmstetter D, Strangmann J, et al. The brain proteome profile is highly conserved between Prnp −/− and Prnp +/+ mice. *Neuroreport*. 2008;19(10):1027–1031.
88. Ramljak S, Asif AR, Armstrong VW, et al. Physiological role of the cellular prion protein (PrPc): protein profiling study in two cell culture systems. *J Proteome Res*. 2008;7(7):2681–2695.
89. Tahir W, Zafar S, Llorens F, et al. Molecular alterations in the cerebellum of sporadic Creutzfeldt-Jakob disease subtypes with DJ-1 as a key regulator of oxidative stress. *Mol Neurobiol*. 2016;1–21.

CHAPTER TWENTY

Therapeutic Approaches to Prion Diseases

Annachiara Gandini*,†, Maria Laura Bolognesi*,[1]
*Alma Mater Studiorum-University of Bologna, Bologna, Italy
†Scuola Internazionale Superiore di Studi Avanzati (SISSA), Trieste, Italy
[1]Corresponding author: e-mail address: marialaura.bolognesi@unibo.it

Contents

Abstract

Therapy of prion diseases represents an extremely challenging effort for scientists working in the field. These challenges are epitomized by 20 years of failures in clinical trials and preclinical investigations. However, the discovery that misfolded proteins involved in other neurodegenerative diseases show a prion-like mechanism of spreading, is positively impacting the prion drug discovery field. Herein, we describe those efforts that have contributed to strengthen the drug discovery process in prion diseases. Accordingly, we analyze the historical course of clinical trials that have assessed the efficacy of several chemically unrelated repositioned drugs. Unfortunately, none of them resulted successful. Thus, alternative approaches aiming at identifying innovative drugs with a completely new mechanism of action, have been recently pursued. In this respect, the multifactorial nature of prion diseases provides a strong foundation to the development of small molecules directed to two or multiple pathological targets, critically involved in the intricate disease pathogenesis (i.e., multitarget compounds). Second, the fact that misfolded proteins can be considered not only as therapeutic target, but also as neuropathological hallmark, lends support to the development of theranostics, i.e., single molecules with concomitant therapeutic and diagnostic properties. Although nobody knows whether these innovative tools will be brought to clinical trials, and the process is certainly time-consuming and demanding, the rewards are well worth the effort.

Progress in Molecular Biology and Translational Science, Volume 150
ISSN 1877-1173
http://dx.doi.org/10.1016/bs.pmbts.2017.06.013

1. INTRODUCTION

At its very worst, the title "Therapy of prion diseases" is an oxymoron. At its best, it reflects an incredibly challenging endeavor for those tirelessly working in the field.

Indeed, treatments capable of arresting or at least effectively modifying the course of human prion diseases (PrDs) do not yet exist. This is the case for Creutzfeldt–Jakob disease (CJD), fatal familial insomnia (FFI), Gerstmann–Straussler–Scheinker syndrome (GSS), kuru and variably protease-sensitive prionopathy (VPSPr), all caused by different strains of the prion protein (PrP) and grouped under the common definition of prion diseases.

PrD patients have an extremely poor prognosis; on average they die within a year of the onset of symptoms. Efforts to treat these diseases since their discovery have been largely stories of failures and frustration. The challenges of the field are bluntly illustrated by 2 decades of negative drug trials and preclinical investigations.[1]

The problem is complex, and complex problems do not usually respond to simple solutions. It has been widely discussed, including in comprehensive recent review articles.[2–6] It is clear that a solution requires coordinated efforts and sophisticated and significant investment platforms. Fortunately, the drug landscape might get brighter as interest in developing potential therapeutics for rare PrDs has greatly increased in the last decade.[7] Prior to the 2010s, this interest was limited and scattered across different medical and scientific specialisms. But a step change occurred with the publication of several studies proposing prion-like mechanisms in more common, non-infectious neurodegenerative diseases.[8–10] Remarkably, these studies not only highlighted the protein misfolding and aggregation as a trigger of the neurodegenerative process, but they also suggested that propagation of protein misfolding in these diseases may occur via a prion-like mechanism of spreading.

As expected, this highly innovative thinking is strongly impacting the field, having the potential to unify experimental and translational approaches to a vast array of increasingly common neurodegenerative disorders. In this renewed scenario, this chapter will summarize efforts that have contributed to reinvigorate the drug development pipeline for PrDs. In a general sense, these efforts can be grouped under two key approaches: screening- and knowledge-based approaches. In the first case, the availability of several cell-based high-throughput in vitro assays has allowed to screen many compounds able to reduce the amount of prion protein resistant to proteases in these cells, independently of their mechanism of action.[11] In many cases, these screens have started from libraries of approved drugs, providing new candidates ready for clinical trials quickly. It is well accepted that repositioning existing drugs for new indications could deliver productivity increases, since critical information, such as drug-likeness and pharmacokinetic features, dosing, safety, tolerability, formulation, and manufacturing are available for approved agents.[12] Applying this strategy to the neurodegeneration therapeutic area can speed cycle times and greatly reduce the risk of early clinical flops.[13] In the second case, extensive basic research and an everincreasing understanding of the mechanism underlying pathogenesis, have led to the development of several new preclinical compounds, which may be potentially turned into true disease-modifying treatment approaches. If the first approach has already resulted in investigational drugs that have been tested in clinical trials, clearly the second one has led to hit, lead compounds, or drug candidates that have been only developed up to in vivo proof of concepts. Both will be the topic of this chapter.

2. INVESTIGATIONAL DRUGS

It is difficult to discuss PrDs therapy, present and future, without reviewing the history of the treatment.

Starting from 1971, 14 drugs have been attempted for therapeutic use in humans with PrDs. Investigational disease-modifying treatments have varied over the years and have largely been influenced by the everchanging state of knowledge within the field. Interventions encompass antivirals, antimalarials, antifungals, biological response modifiers, antidepressant, anticonvulsants, and antioxidants (see Table 1).[46]

Not surprisingly, the initial designation of PrDs as "slow viral illnesses" led to the early evaluation of several antiviral treatments. Thus, amantadine, interferon, vidarabine, and acyclovir were assayed between 1971 and 1984. Two comparative studies of the antiviral and antiparkinsonian drug amantadine (Fig. 1) for the treatment of CJD were published.[14,15] However, in both cases, there was no statistically significant prolongation of survival time in treated patients with respect to the controls. These results were confirmed by a further case report,[16] while two other reports of three patients described some improvement in clinical status after amantadine treatment.[17,18] However, in these three patients, the diagnosis of CJD was not based on the

Table 1 Drug Assessed in Clinical Trials From 1971 to Date

Drug	Mechanism of Action	References
Amantadine	Antiviral/antiparkinsonian	14–18
Acyclovir	Antiviral	19,20
Interferon	Antiviral	21
Vidarabine	Antiviral	22
Amphotericin B	Antibiotic/antimycotic	23
Clomipramine and venlafaxine	Tricyclic antidepressant and serotonin and norepinephrine reuptake inhibitor	24
Quinacrine	Antimalarial	25–28
Quinacrine and chlorpromazine	Antimalarial and antipsychotic	29,30
Phenytoin and topiramate	Anticonvulsants	31
Phenytoin, quinacrine, and chlorpromazine	Anticonvulsant, antimalaria, and antipsychotic	32
Flupirtine	Nonopioid analgesic	33
Pentosan polysulfate	Heparin mimetic	34–40
Doxycycline	Antibiotic	41–45

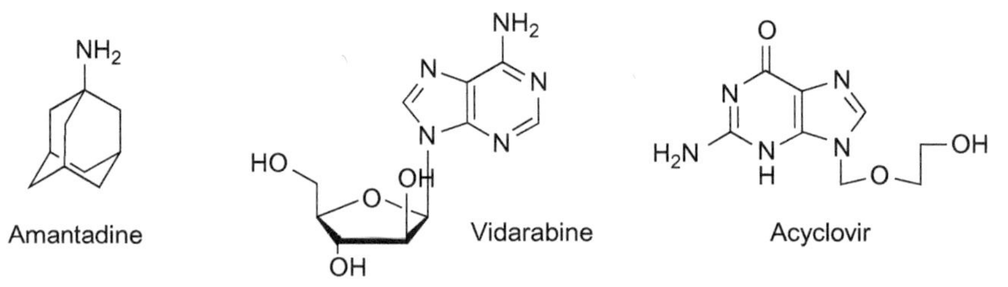

Fig. 1 Chemical structures of the antivirals assessed in clinical trials.

presence of the abnormal prion protein, thus raising questions as to diagnostic accuracy. In 1984, two case reports on the use of acyclovir (Fig. 1), a nucleoside analog, in CJD patients were published.[19,20] However, acyclovir resulted as an unsuccessful treatment option. Similarly negative results derived from two cases of CJD treated with interferon, signaling proteins known to induce a nonspecific resistance to viral infection by several mechanisms.[21] Conversely, treatment with vidarabine (Fig. 1), a nucleoside analog acting by interfering with the synthesis of viral DNA, was reported to provide transient resolution of symptoms in two patients, but no efficacy in a third.[22]

In 1992, Masullo and colleagues reported the treatment of two CJD patients with the drug amphotericin B (Fig. 2), a polyene antibiotic/antimycotic.[23] Amphotericin B was shown to successfully delay the incubation period of experimental scrapie in hamsters intracerebrally inoculated with the 263 K strain of scrapie,[47] and to prolong the clinical phase of experimental CJD in a monkey model. However, treatment of the two patients was proven ineffective. After 20 days of treatment (0.25–1 mg/g), the first patient's clinical situation worsened and she died. Similarly, the conditions of the second patient aggravated, and she died 8 months after the onset of the disease.[23] Thus, it was clear that amphotericin B had no positive effects and did not prevent the progression of the disease.

In 2001, Dervaux *et al.* reported the second French case of variant CJD (vCJD).[24] The patient was treated with clomipramine (125 mg/day), a tricyclic antidepressant (TCA), in combination with venlafaxine (200 mg/day), a serotonin and norepinephrine reuptake inhibitor, for 7 weeks (Fig. 2). No clinical improvements were observed, and the patient died 14 months after the onset of the first symptoms.

In the same year, rationally searching for blood–brain barrier (BBB) permeant molecules, Korth *et al.* decided to screen a variety of drugs known

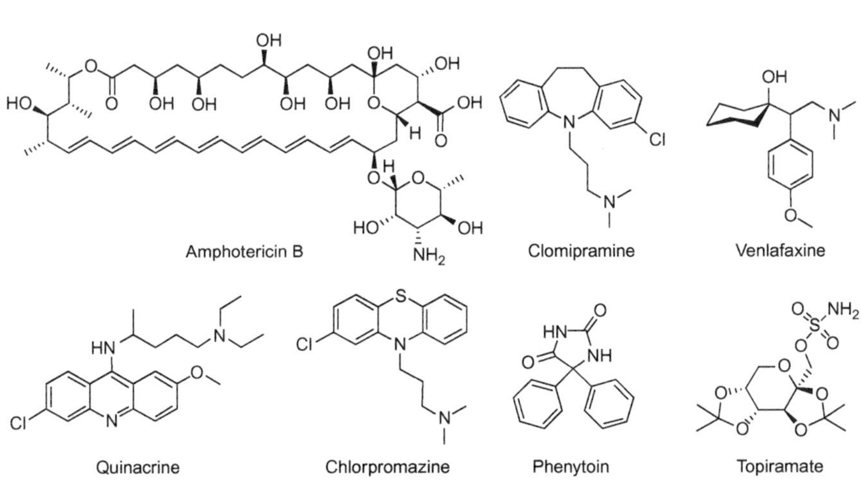

Fig. 2 Chemical structures of the antibiotics, antipsychotics, antimalarials, anticonvulsants, and antidepressants assessed in clinical trials.

to cross the BBB for their ability to inhibit PrP^{Sc} formation. Particularly, they focused on tricyclic acridine and phenothiazine derivatives. Phenothiazines are compounds with a tricyclic scaffold widely used to treat psychoses. Chlorpromazine inhibited PrP^{Sc} formation in scrapie-infected neuroblastoma cells (ScN2a) at a micromolar range, whereas quinacrine, a structural antecedent of the phenothiazines with an antimalarial profile, was 10 times more potent.[48] On the basis of these promising results, the authors suggested it as an immediate candidate for the treatment of PrDs. Moreover, in a paper published in 2003, quinacrine enantiomers demonstrated stereoselectivity on prion elimination. Indeed, (*S*)-quinacrine exhibited higher antiprion activity compared to (*R*)-quinacrine ($EC_{50(S)} = 0.25\ \mu M$, $EC_{50(R)} = 44\ \mu M$),[49] raising the possibility that treatment with one enantiomer of quinacrine may be more effective than that with the racemic mixture.

Four clinical trials have examined the effect of quinacrine (Fig. 2) in patients with PrDs. In 2002, Furukawa *et al.* reported improvements in the response to visual and auditory stimulations in some patients treated with quinacrine.[25] Particularly, a 37-year-old patient showed a partial recovery after administration of 300 mg of quinacrine. However, after 2 weeks of treatment the patient deteriorated. The authors also found that lemon-yellow discoloration of skin and liver dysfunction were common side effects of quinacrine. Two years later, an open-label trial conducted in France,[26] consisting of 32 patients treated with a loading dose of 1000 mg quinacrine over 1 day, followed by 300 mg daily, failed to demonstrate a statistically significant difference in survival time, compared with the 125 untreated controls. Once more, adverse effects such as skin discoloration, gastrointestinal symptoms, liver dysfunction, and leukopenia were described.[26] In 2009, Collinge and colleagues reported an open-label, patient-preference trial called PRION-1, conducted in the United Kingdom, and recruiting 107 patients.[27] Although adjusted mortality was lower in those patients who chose quinacrine with respect to those who did not, this was due to confounding with disease severity. Thus, there were no differences in mortality between the two groups. Hence, the authors concluded that a dose of 300 mg of quinacrine daily was reasonably tolerated, but did not significantly affect the clinical course.[27] In spite of these negative results and the limitations of the overall trial design (essentially being an observational study), PRION-1 was a well planned, detailed prospective study, thus allowing to learn important lessons that have affected the next trials in PrDs.[50,51]

Finally, a double-blind, placebo-controlled, stratified randomization treatment trial was conducted at the University of San Francisco from

2005 to 2009.[28] A total of 425 patients were randomized to quinacrine (300 mg per day) or placebo, with inpatient evaluations at baseline, and planned for months 2, 6, and 12. Once again, the results of this interventional study showed no significant differences in survival between the two groups. Surprisingly, the patients treated with quinacrine showed less decline in two out of three functional scales, than the placebo group, but this was observed only during the first 2 months of treatment. Indeed, in 2009 was founded that quinacrine, both in vitro and in vivo, might at first reduce PrP^{Sc}, but continuous treatment results in strain selection of PrP, that is resistant to the effects of quinacrine.[52] The failure of quinacrine has been attributed to several factors, including its use as a monotherapy.[41] In light of this, in 2004 Benito-Leon treated two FFI patients with quinacrine in combination with the antipsychotic drug chlorpromazine (Fig. 2).[29] Although two patients are inadequate to draw a meaningful conclusion, quinacrine, and chlorpromazine seemed not to be effective. One year later, Martínez-Lage *et al.* described a case report of a patient diagnosed with iatrogenic CJD (iCJD) in 2002,[30] who was similarly treated with quinacrine and chlorpromazine (300 mg/d). Once more, no clinical improvements were found.

The list of the 14 drugs that have been attempted for the therapeutic use in humans with PrDs, include also anticonvulsants. Surprisingly, although anticonvulsants are frequently used to control myoclonus and seizures in human PrDs, there are few published data on their disease-modifying effects. In 2003, a case report described a probable sporadic CJD (sCJD) patient, treated with phenytoin (at first 750 mg intravenously, then 300 mg a day orally) (Fig. 2).[31] Within a few days he deteriorated, with increasing poverty of speech, startle myoclonus, and generalized muscle rigidity. Thus, 50–100 mg of topiramate (Fig. 2) was added twice daily. Electroencephalogram (EEG) patterns transitioned from irregular epileptiform discharges to generalized dysrhythmia during treatment. Another case of probable sCJD was reported the next year.[32] The patient was treated with phenytoin, quinacrine, and chlorpromazine with no evident clinical improvements. Thus, none of these case reports demonstrated any significant effects of any anticonvulsant on disease course or on the patient's survival time, either in mono- or in polytherapy.

The most encouraging result in PrDs clinical trials came from flupirtine (Fig. 3). Flupirtine is a well tolerated, centrally acting nonopioid analgesic, which has been shown to display cytoprotective activity, and to decrease neurotoxicity associated with PrDs in vitro.[53] As flupirtine is a clinically

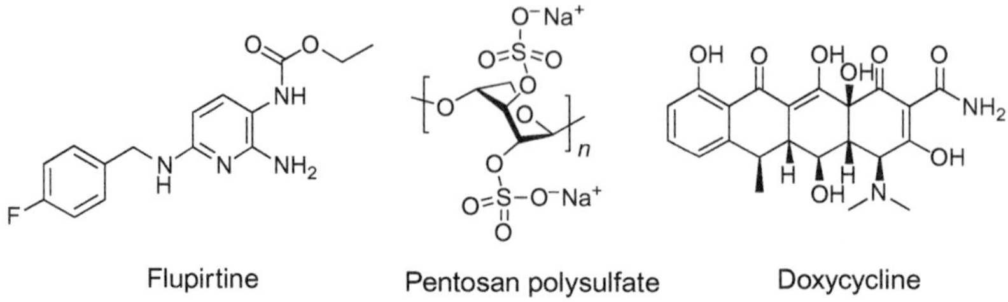

Fig. 3 Chemical structures of flupirtine, pentosan polysulfate, and doxycycline assessed in clinical trials.

well-established drug, it was investigated in a double-blind, placebo-controlled trial, without any preliminary in vivo data.[33] Twenty-eight patients with probable CJD were randomized to flupirtine (100–400 mg/d) or placebo. Cognitive tests were performed at intervals of 2, 4, 8, 12, 16, and 20 weeks. The results showed that there was less cognitive decline in the treated patients. Conversely, there were no significantly differences in mean survival time between the groups.

Another interesting investigational drug is pentosan polysulfate (PPS, in Fig. 3).

PPS is a large polyglycoside molecule, showing weak heparin-like activity. It was hypothesized that PPS acts in competition with endogenous heparin sulfate proteoglycans as coreceptor for PrP on the cell surface.[54]

The first evidence on the ability of PPS to inhibit scrapie infections in mice came from a paper published in 1984,[55] and was also confirmed 20 years later.[56]

However, as PPS is known to be unable to cross the BBB, it requires intraventricular (i.v.) administration, which is invasive and not free of complications. In light of this, in 2003, the Britain's Committee on Safety of Medicines and the CJD Therapy Advisory Group rejected PPS as a possible treatment for CJD.[50] Nevertheless, several case reports and case series were conducted. In 2005, a case report described a prolonged survival in an adolescent with vCJD who was treated i.v. with 11 mg/kg/d for 18 months.[34] Despite relative clinical stabilization, repeated brain computer tomography scans revealed continued cerebral atrophy. Two subsequent reports of two vCJD patients treated with i.v. PPS (1.1–110 μg/kg/d and 32 μg/kg/d)[35,36] demonstrated no clinical benefit in the former, but a longer survival time in the latter. In either case, no adverse events were reported. PPS has been assessed also in other types of human PrDs, as described in a case series of 23 patients, 18 CJD (8 sporadic, 7 iatrogenic, 2 genetics, and 1 variant), and 5 GSS.[37] This case series reported various responses ranging from improvement with relapse, stabilization, and slowed disease progression. In 2008 and 2009, two observational studies conducted in the United Kingdom and Japan were published.[38,39] Both studies examined the use of PPS. The first one was a 6-month study on seven patients, while in the second one PPS was administered to 11 patients. In both cases, mean survival time of all patients was longer than reported values of the controls; however, surgical complications were common. The most recent case report on the use of PPS referred to a woman with sCJD who received continuous i.v. PPS infusion (1–120 μg/kg/d) for 17 months, starting 10 months after the onset of the first symptoms.[40] The patient displayed a survival time longer than historical controls, but still within normal variations. Moreover, postmortem analysis of her brain showed a decreased cerebral PrP^{Sc} deposition, compared to controls.

Taken together, these results suggested a possible disease-modifying effect of PPS in CJD patients, despite an undoubted invasiveness.

In recent years, high expectations are placed on tetracycline doxycycline (Fig. 3), as several in vitro and in vivo studies demonstrated its potential.[57] Particularly, doxycycline reduces PrP^{Sc} proteinase-K resistance, inhibits PrP^{Sc} formation, and destabilizes amyloid aggregates probably thanks to its planar chemical structure. Moreover, PrD animal models treated with doxycycline showed decreased infectivity, and prolonged survival time.[57]

Preliminary results about the treatment of doxycycline in humans derived from two observational studies.[41,42] In the first study, conducted in Italy, 21 patients with probable CJD were authorized to receive compassionate treatment with 100 mg/d doxycycline, from the time of diagnosis to death.[41] The same treatment was adopted in the second study, conducted in Germany, on 52 patients (44 patients with sCJD and 7 with genetic CJD).[42] Both studies reported significantly prolonged survival times compared with historical controls.

These results encouraged a randomized, double-blind, placebo-controlled trial conducted in Italy and in France between 2007 and 2012.[43] Patients were treated with a single oral dose of 100 mg doxycycline daily, or with placebo. Unfortunately, the results did not confirm the previous findings. Doxycycline was well tolerated, but efficacy analyses did not show any significant difference between doxycycline- and placebo-treated patients. Thus, the study was stopped for futility.[43] In 2015, Assar *et al.* published a case report on the use of 100 mg doxycycline for 4 years, in a 54-year-old woman.[44] The long survival of the patient, associated with minor histopathological alterations, suggested doxycycline efficacy.

On the basis of these findings, Forloni *et al.* started a preventive clinical trial with doxycycline treatment in FFI patients.[45] The trial was designed with the help of individuals at genetic risk of developing the disease and asymptomatic carriers. At the time of this chapter was published, the authors were proceeding with the first follow up of the subjects, after the first 2 years of treatment. Hopefully, the study will be closed within 2023.

Sadly to say, all the drugs investigated so far have been unsuccessful. Thus, new drugs with disease-modifying potential are urgently needed.

3. PRECLINICAL DRUG CANDIDATES

The search for new drugs, in general, is a very challenging process, but it is particularly challenging for antiprion drug discovery for several reasons: (1) The lack of a certain diagnostic method. The only way to unequivocally diagnose CJD is by neuropathological examination either by brain biopsy or at autopsy. Thus, it is generally agreed that the failure of many prion drug candidates is associated to their late administration, when the pathology is too advanced. (2) Antiprion agents, as all drugs for neurodegenerative diseases need to be BBB permeable; an in-depth understanding of the structure–brain exposure relationships (built on reliable and meaningful pharmacokinetic data) might overcome the current challenging aspect of crossing the BBB by new antiprion agents.[58] (3) Furthermore, in a typical structure-based drug-design approach, the molecular details of the protein–ligand complexes or the interactive surfaces of two proteins involved in a protein–protein interaction (PPI) are needed.[59] For PrDs, the first structure of a complex between PrP^{c} and a phenothiazine ligand was disclosed only in 2013[60]; regarding prion PPI, molecular information are not available, because the precise structure of PrP^{Sc} is unknown. To overcome this limitation, recently Aguzzi and colleagues based their structure-based studies on the prion-forming domain of the fungal HET-s prion, the only self-propagating amyloid whose atomic structure has been solved.[61] (4) Antiprion drug discovery occurs, mostly, at research centers that are not ideally structured for drug discovery (i.e., academic and nonprofit research institutions).

A fundamental question arises from these premises: how can we facilitate the development of new effective drugs against PrDs? The next section will focus on two rationale approaches that have led to identify early hits with completely new mechanisms of action. Although nobody knows whether these approaches will turn into clinically useful candidates, in our opinion, they could effectively address some of the challenges posed by the current shortage of effective drugs. The first one is a multitarget drug discovery approach, which is being successfully pursued in the therapy of AD and related neurodegenerative diseases for the last 10 years.[62] It relies on the so-called multitarget drugs, i.e., single chemical entities that can interact simultaneously with multiple drug targets. The main strength of this approach is that it is inherently more adequate at addressing the multifactorial and progressive nature of neurodegenerative diseases.[63] The second one is the exploitation of theranostics, i.e., single agents with concomitant therapeutic and diagnostic properties.[62] In this case, the main strength is that it allows to optimize the efficacy and safety of therapy, as well as to streamline the long and costly drug discovery process.

3.1 Multitarget Antiprion Small Molecules

In addition to a common mechanism of misfolded protein propagation, another fundamental pathologic overlap between PrDs and neurodegenerative diseases has been increasingly evident: i.e., their multifactorial nature.[7] A set of interrelated pathological events such as excitotoxicity, mitochondrial dysfunction, oxidative stress, inflammation, apoptosis, autophagy, and transcriptional dysregulation are responsible of the underlying neurodegeneration in all these diseases.[7] On these basis, multitarget drugs can better confront such complex pathogenesis. Indeed, a pioneeristic multitarget approach in PrDs was pursued by Klingenstein and coworkers, who discovered quinpramine, a truly multitarget antiprion compound (Fig. 4). Quinpramine was rationally design by combining the molecular features of two drugs showing therapeutic potential in PrDs, i.e., quinacrine

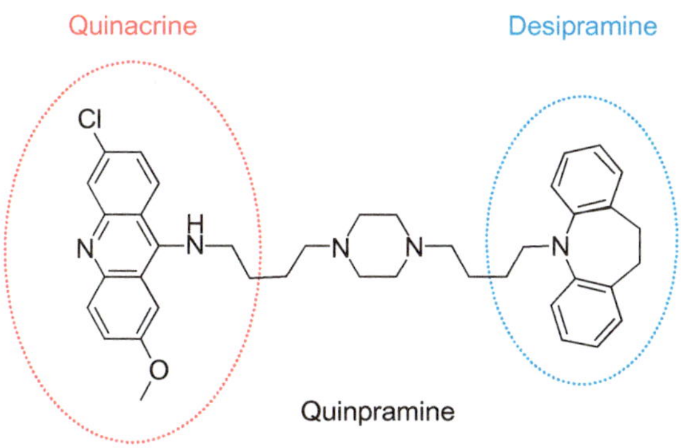

Fig. 4 Chemical structure of quinpramine.

and TCA desipramine. It could be expected that quinpramine, by simultaneously exerting the effect of the starting drugs, would show a synergistic profile. Notably, when tested against ScN2a displayed an antiprion activity that was 5-fold higher than quinacrine and 10-fold than desipramine.[64]

Motivated by this strong foundation and building on our previous experience in the field of neurodegeneration, we have recently embarked in drug discovery projects, aimed to develop PrDs drugs directed to two or multiple pathological targets, critically involved in the intricate pathogenesis.[63,65–67] These include: (1) PrP misfolding and aggregation[68]; (2) reactive oxygen species (ROS) production and lipid and protein oxidation[69]; (3) metal ions dishomeostasis[70]; and (4) brain inflammation with activation of astrocytes and microglia.[71]

Generally, drug discovery efforts in PrDs are directed to preventing or reducing the rate of PrP^{Sc} conversion. As quinacrine acts by directly binding PrP^{C} and/or blocking its conversion to PrP^{Sc},[72] it can be viewed as an effective starting prion recognition motif (PRM) in the design of new multitarget drugs.[72] On this basis, we reasoned that the linking of a quinacrine-derived PRM with a moiety endowed with antioxidant properties, might lead to the discovery of compounds capable to block both protein misfolding and oxidative stress in PrDs. In particular, a new series of antiprion compounds have been generated by linking the antioxidant nucleus of 2,5-diamino-1,4-benzoquinone to diverse PRMs (e.g., 4-amino-7-chloroquinoline). The resulting multitarget compounds were able to effectively contrast both prion fibril formation and oxidative stress in a PrDs cell model.[73] In particular, compound **1** (Fig. 5) was active against prion replication in the

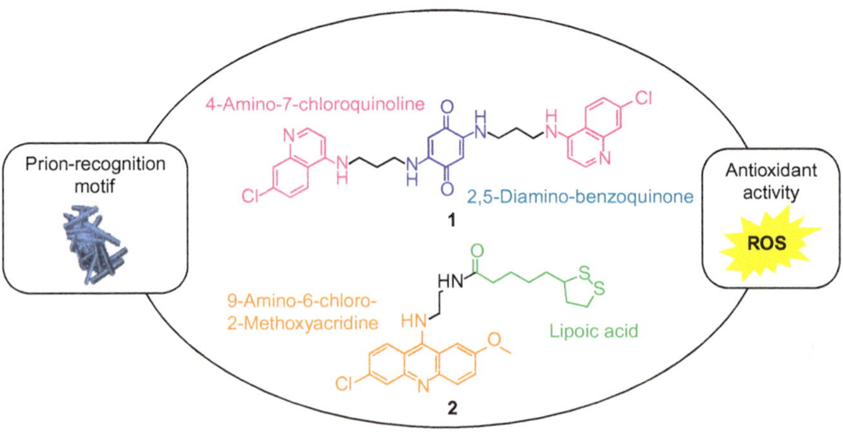

Fig. 5 Design strategy and activity profile of compounds **1** and **2**.

submicromolar range, while showing an encouraging low cytotoxicity. Notably, the antiprion activity of **1** was even lower than that of quinacrine ($EC_{50(\text{quinacrine})} = 320$ nM and $EC_{50(\mathbf{1})} = 170$ nM). Furthermore, **1** showed the largest correlation between the cellular antiprion activity and the capability of inhibiting PrP fibril formation and antioxidant activities.

Following a similar rationale, a second class of multitarget compounds has been designed by linking a PRM (e.g., 9-amino-6-chloro-2-methoxyacridine) to the antioxidant lipoic acid.[74] Importantly, lipoic acid was previously administered together with other antioxidants to a patient affected by PrDs, showing moderate therapeutic effects.[75] Compound **2** (Fig. 5), thanks to the presence of lipoic acid and the acridine PRM, was able to simultaneously interact with at least two of the multiple targets involved in prion pathology: it blocks PrP^{Sc} accumulation in ScGT1 cells ($EC_{50} = 150$ nM), delays fibril formation in amyloid seeding assay, and decreases oxidative stress.[76]

The imbalance of metal ions, Cu^{2+} in particular, has been shown to play a functional role in the PrP^C to PrP^{Sc} conformational conversion, both in vitro and in vivo.[77] In fact, Cu^{2+} may stimulate PrP^C misfolding and aggregation, inducing a β-sheet like conformational change in the unstructured *N*-terminal region of PrP^C.[78] The role of PrP^C in Cu^{2+} uptake suggests that loss of this function, as a result of aggregation to the PrP^{Sc} form, may contribute to brain metal dyshomeostasis.[70] Disturbance of the Cu^{2+} homeostasis at the neuronal synapse and in the brain of scrapie-infected mice results in neurodegeneration.[70] Moreover, PrP^C exerts Cu^{2+}-dependent neuroprotection by reducing the neurotoxic effect caused by the over activation of *N*-methyl-D-aspartate receptor.[79] Furthermore, the complex of metal ions with PrP^{Sc} is associated with a state of increased oxidative stress and neurotoxicity.[80] Ultimately, the combination of transition metals imbalance and increased oxidative stress could boost the neurotoxic effect of PrP^{Sc}.[80] Based on these findings, we purposely designed the multitarget molecule **3** (Fig. 6), by combining a heteroaromatic PRM (i.e., 7-chloroquinoline) with an 8-hydroxyquinoline (8HQ) as a metal chelator.[81]

Importantly, 8HQ derivatives have been reported to cross the BBB and to chelate metals from the brain.[82] Encouragingly, **3** showed the intended multitarget profile in vitro. In particular, **3** blocked PrP^{Sc} formation with a relevant EC_{50} of 40 nM, which makes it one of the most potent small molecule in ScGT1 prion cell model. In addition, **3** contrasted metal ion dyshomeostasis, inhibited metal-induced prion aggregation, and ROS generation.

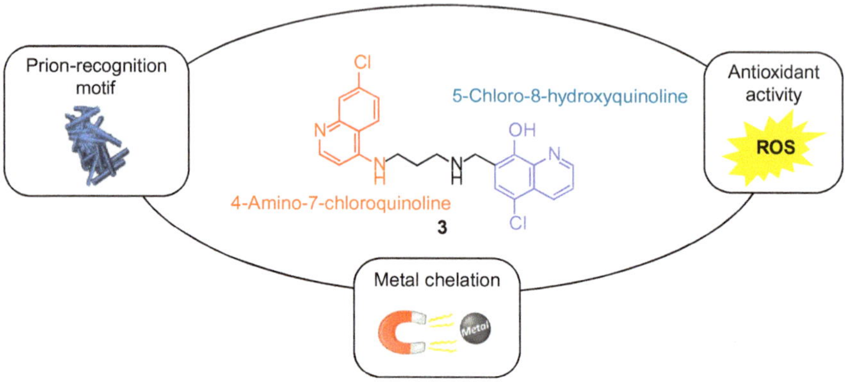

Fig. 6 Design strategy and activity profile of compound **3**.

This profile, together with acceptable PK properties and low neurotoxicity, pointed **3** as a suitable lead candidate against PrDs, and as a good starting point for a further optimization process.

3.2 Theranostic Antiprion Small Molecules

In neurodegenerative drug discovery, the misfolded proteins underlying pathogenesis have been regarded from two perspectives: as therapeutic targets and as neuropathological hallmarks. In light of this, the idea of designing single molecules with concomitant therapeutic and diagnostic properties, has gained momentum. Indeed, starting from 1998, after the term "theranostic" was coined for the first time,[83] a variety of theranostic small molecules were developed in cancer research. In the neurodegenerative field, however, as the pathological mechanisms are poorly understood in respect of cancer, the theranostic approach is still at its beginnings.[62]

Nevertheless, several fluorescent diagnostic tools have been demonstrated to prevent abnormal PrP formation in cell and animal models of PrDs, i.e., acting as theranostics.

This is the case of a very pioneering study published by Ishikawa *et al.*[84] The thioflavin T derivative 2-[4'-(methylamino)phenyl]benzothiazole (BTA-1, in Fig. 7), and the Congo Red derivative (*trans,trans*),-1-bromo2, 5-bis-(3-hydroxycarbonyl-4-hydroxy)styrylbenzene (BSB, in Fig. 7), are known to detect amyloid or amyloid-like plaques in postmortem human brains and in living mouse brains.[85,86] Thus, the authors used both compounds to label PrP deposits in PrDs brains. Both compounds were able to fluorescently labeled most of the compact PrP plaques in the cerebellar

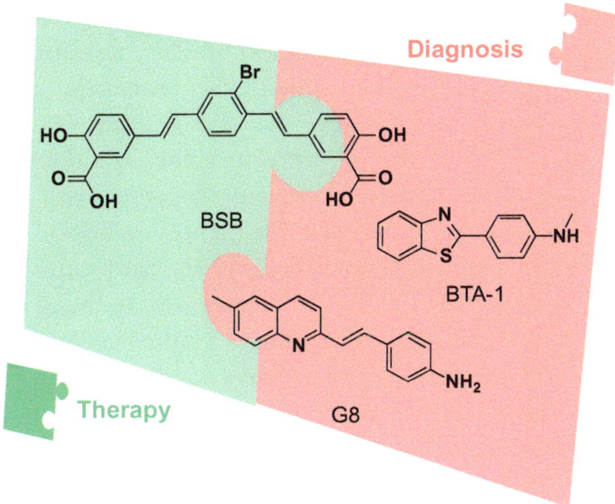

Fig. 7 Theranostic profile of BTA-1, BSB, and G8.

cortices of GSS patients. Similar results were obtained from postmortem brains of Tg7 mice, infected with the 263 K strain. As BSB and BTA-1 were known to cross the BBB, they performed in vivo experiments in mice, and demonstrated their ability to label PrP deposits also in vivo. In parallel, the authors verified the antiprion activity of both compounds in ScN2a cells. Finally, since BSB was the most potent compound with no toxicity up to 100 mM, the authors examined whether BSB could be effective in two PrDs mouse models. Intravenous injection of BSB prolonged significantly, but slightly, the incubation periods of prion-infected Tga20 mice. However, no significant survival prolongation was observed. Similarly, no substantial effects were observed in Tg7 mice infected with the 263 K strain, after the same treatment. Therefore, both BTA-1 and BSB might be theranostic candidates against PrDs. However, several limitations, such as low specificity and poor sensitivity, limit their translation into clinics.[87]

The very first small molecule intently proposed as a theranostic agent, was the styrylquinoline G8 (Fig. 7).[87] In 2013, we developed a small molecule potentially able to diagnose, deliver therapy, and monitor response to therapy in AD and PrDs. The starting point was the finding that several styryl derivatives, developed to improve the CNS properties of Congo Red, were employed as AD imaging agents in vivo. Moreover, in 2012, styryl-1H-indole and styrylquinoline derivatives were reported as SPECT imaging probes to detect Aβ plaques.[88] Thus, styryl moiety seemed a crucial framework for the labeling of pathological fibrillary aggregates, while the quinoline moiety, thanks to its good PK properties, is considered a privileged scaffold. Accordingly, we synthesized (*E*)-6-methyl-4′-amino-2-styrylquinoline (G8), and tested its ability as a theranostic molecule. G8 showed a very low neurotoxicity, together with a remarkable activity in two cell-based models of PrDs (ScN2a and ScGT1 cells), with an EC_{50} of 0.5 μM, comparable to the gold standard quinacrine. Subsequently, to confirm the labeling of PrP^{Sc} aggregates in living cells, fluorescent staining with G8 was carried out using the same ScN2a and ScGT1 cell models. Many fluorescent spots were observed in treated cells examined by fluorescent microscopy, using a ThS filter. Importantly, no spots were observed in the uninfected cells, thus confirming a specific binding. Finally, G8 also showed a good BBB permeability in an in vitro model (the parallel artificial membrane permeability assay), demonstrating its value as CNS-directed agent.

In light of these findings, G8 can be considered a truly theranostic candidate, which could contribute to the diagnosis and therapy of PrDs.

4. CONCLUSIONS

Although recent major improvements achieved in terms of basic biology, PrDs are still incurable, and prospects of having an effective drug available soon to the patients appear quite remote. Not surprisingly, this gloomy outlook applies to PrDs, such as to all the other related neurodegenerative diseases. Thus, to promote the development of effective antiprion drugs, we should try our best to improve such an unsuccessful process.

As critically discussed by Becker *et al.*,[89] we need to acknowledge that neurodegenerative drug discovery should build on a more comprehensive understanding of molecular brain mechanisms, neuronal networks, and their critical link to pathogenesis and pathology. With this understanding, drugs need to be more precisely directed at specific brain targets.

If it's in principle very productive to test in a clinical trial a repurposed drug, we should be aware of the risks we are running. The FDA and EMEA consider only safety and efficacy as essential prerequisite for a drug to be marketed, whereas they do not require any understanding of the mechanism of action. Thus, it is very tempting to move drug candidates into clinical trials without this information.

In this respect, we should learn from the failures of AD drug research. In fact, if PrDs can help to understand AD and related neurodegenerative disorders, then the opposite should also apply. In AD, the lack of knowledge of mechanism of action might have contributed to the clinical failure of a promising investigational, repurposed drug, dimebon.[90] Undoubtedly, a comprehensive understanding of a drug's mechanism of action and target engagement allows better dosing and monitoring drug's effects in clinical trials. It is also critical to stratify patients and select those who are more likely respondent to the treatment.[90] However, we have experienced that rational and mechanistic studies in the neurodegenerative disease therapeutic area are time-consuming and extremely complex, and not always greeted with enthusiasm by clinicians and patients. It seems a lot of effort, but we are convinced it is the best approach in the long run: mechanistic knowledge and rational approaches are certainly critical and valuable to discover novel molecules with antiprion activity, and guiding the difficult steps required to develop them into clinically useful drugs.

REFERENCES

1. Appleby BS. Many challenges of conducting clinical trials in CJD. *J Neurol, Neurosurg Psychiatry*. 2017;88(2):98.
2. Teruya K, Doh-Ura K. Insights from therapeutic studies for PrP prion disease. *Cold Spring Harb Perspect Med*. 2016.
3. Vetrugno V, Puopolo M, Cardone F, Capozzoli F, Ladogana A, Pocchiari M. The future for treating Creutzfeldt–Jakob disease. *Expert Opin Orphan Drugs*. 2015;3(1):57–74.
4. Panegyres PK, Armari E. Therapies for human prion diseases. *Am J Neurodegener Dis*. 2013;2(3):176–186.
5. Tsuboi Y. Current trends in the treatment of prion disease. *Brain Nerve*. 2015;67(7):941–946.
6. Forloni G, Artuso V, Roiter I, Morbin M, Tagliavini F. Therapy in prion diseases. *Curr Top Med Chem*. 2013;13(19):2465–2476.
7. Bolognesi ML, Legname G. Approaches for discovering anti-prion compounds: lessons learned and challenges ahead. *Expert Opin Drug Discovery*. 2015;10(4):389–397.
8. Brundin P, Melki R, Kopito R. Prion-like transmission of protein aggregates in neurodegenerative diseases. *Nat Rev Mol Cell Biol*. 2010;11(4):301–307.
9. Frost B, Diamond MI. Prion-like mechanisms in neurodegenerative diseases. *Nat Rev Neurosci*. 2010;11(3):155–159.
10. Jucker M, Walker LC. Pathogenic protein seeding in Alzheimer disease and other neurodegenerative disorders. *Ann Neurol*. 2011;70(4):532–540.
11. Kocisko DA, Caughey B. Searching for anti-prion compounds: cell-based high-throughput in vitro assays and animal testing strategies. *Methods Enzymol*. 2006;412:223–234.
12. Cummings JL, Zhong K. Repackaging FDA-approved drugs for degenerative diseases: promises and challenges. *Expert Rev Clin Pharmacol*. 2014;7(2):161–165.
13. Cummings JL, Morstorf T, Zhong K. Alzheimer's disease drug-development pipeline: few candidates, frequent failures. *Alzheimer's Res Ther*. 2014;6(4):37.

14. Terzano MG, Montanari E, Calzetti S, Mancia D, Lechi A. The effect of amantadine on arousal and EEG patterns in Creutzfeldt–Jakob disease. *Arch Neurol.* 1983;40(9): 555–559.
15. Neri G, Figa-Talamanca L, Di Battista GC, Lo Russo F. Amantadine in Creutzfeldt–Jakob disease. Review of the literature and case contribution. *Riv Neurobiol.* 1984;30(1):47–56.
16. Ratcliffe J, Rittman A, Wolf S, Verity MA. Creutzfeldt–Jakob disease with focal onset unsuccessfully treated with amantadine. *Bull Los Angeles Neurol Soc.* 1975;40(1):18–20.
17. Sanders WL, Dunn TL. Creutzfeldt–Jakob disease treated with amantidine. A report of two cases. *J Neurol Neurosurg Psychiatry.* 1973;36(4):581–584.
18. Sanders WL. Creutzfeldt–Jakob disease treated with amantadine. *J Neurol Neurosurg Psychiatry.* 1979;42(10):960–961.
19. David AS, Grant R, Ballantyne JP. Unsuccessful treatment of Creutzfeldt–Jakob disease with acyclovir. *Lancet.* 1984;1(8375):512–513.
20. Newman PK. Acyclovir in Creutzfeldt–Jakob disease. *Lancet.* 1984;1(8380):793.
21. Kovanen J, Haltia M, Cantell K. Failure of interferon to modify Creutzfeldt–Jakob disease. *Br Med J.* 1980;280(6218):902.
22. Furlow Jr TW, Whitley RJ, Wilmes FJ. Repeated suppression of Creutzfeldt–Jakob disease with vidarabine. *Lancet.* 1982;2(8297):564–565.
23. Masullo C, Macchi G, Xi YG, Pocchiari M. Failure to ameliorate Creutzfeldt–Jakob disease with amphotericin B therapy. *J Infect Dis.* 1992;165(4):784–785.
24. Dervaux A, Vicart S, Lopes F, Le Borgne MH. Psychiatric manifestations of a new variant of Creutzfeldt–Jakob disease. Apropos of a case. *L'Encephale.* 2001;27(2): 194–197.
25. Furukawa H, Takahashi M, Nakajima M, Yamada T. Prospects of the therapeutic approaches to Creutzfeldt–Jakob disease: a clinical trial of antimalarial, quinacrine. *Nihon Rinsho.* 2002;60(8):1649–1657.
26. Haik S, Brandel JP, Salomon D, et al. Compassionate use of quinacrine in Creutzfeldt–Jakob disease fails to show significant effects. *Neurology.* 2004;63(12): 2413–2415.
27. Collinge J, Gorham M, Hudson F, et al. Safety and efficacy of quinacrine in human prion disease (PRION-1 study): a patient-preference trial. *Lancet Neurol.* 2009;8(4):334–344.
28. Geschwind MD, Kuo AL, Wong KS, et al. Quinacrine treatment trial for sporadic Creutzfeldt–Jakob disease. *Neurology.* 2013;81(23):2015–2023.
29. Benito-Leon J. Combined quinacrine and chlorpromazine therapy in fatal familial insomnia. *Clin Neuropharmacol.* 2004;27(4):201–203.
30. Martínez-Lage JF, Rábano A, Bermejo J, et al. Creutzfeldt–Jakob disease acquired via a dural graft: failure of therapy with quinacrine and chlorpromazine. *Surg Neurol.* 2005;64(6):542–545.
31. Flöel A, Reilmann R, Frese A, Lüdemann P. Anticonvulsants for Creutzfeldt–Jakob disease? *Lancet.* 2003;361(9353):224.
32. Arruda WO, Bordignon KC, Milano JB, Ramina R. Creutzfeldt–Jakob disease, Heidenhain variant: case report with MRI (DWI) findings. *Arq Neuropsiquiatr.* 2004;62(2a):347–352.
33. Otto M, Cepek L, Ratzka P, et al. Efficacy of flupirtine on cognitive function in patients with CJD: a double-blind study. *Neurology.* 2004;62(5):714–718.
34. Todd NV, Morrow J, Doh-ura K, et al. Cerebroventricular infusion of pentosan polysulphate in human variant Creutzfeldt–Jakob disease. *J Infect.* 2005;50(5):394–396.
35. Whittle IR, Knight RS, Will RG. Unsuccessful intraventricular pentosan polysulphate treatment of variant Creutzfeldt–Jakob disease. *Acta Neurochir.* 2006;148(6):677–679. discussion 679.
36. Parry A, Baker I, Stacey R, Wimalaratna S. Long term survival in a patient with variant Creutzfeldt–Jakob disease treated with intraventricular pentosan polysulphate. *J Neurol Neurosurg Psychiatry.* 2007;78(7):733–734.

37. Rainov NG, Tsuboi Y, Krolak-Salmon P, Vighetto A, Doh-Ura K. Experimental treatments for human transmissible spongiform encephalopathies: is there a role for pentosan polysulfate? *Expert Opin Biol Ther*. 2007;7(5):713–726.
38. Bone I, Belton L, Walker AS, Darbyshire J. Intraventricular pentosan polysulphate in human prion diseases: an observational study in the UK. *Eur J Neurol*. 2008;15(5):458–464.
39. Tsuboi Y, Doh-Ura K, Yamada T. Continuous intraventricular infusion of pentosan polysulfate: clinical trial against prion diseases. *Neuropathology*. 2009;29(5):632–636.
40. Terada T, Tsuboi Y, Obi T, et al. Less protease-resistant PrP in a patient with sporadic CJD treated with intraventricular pentosan polysulphate. *Acta Neurol Scand*. 2010;121(2):127–130.
41. De Luigi A, Colombo L, Diomede L, et al. The efficacy of tetracyclines in peripheral and intracerebral prion infection. *PLoS One*. 2008;3(3): e1888.
42. Zerr I. Therapeutic trials in human transmissible spongiform encephalopathies: recent advances and problems to address. *Infect Disord Drug Targets*. 2009;9(1):92–99.
43. Haik S, Marcon G, Mallet A, et al. Doxycycline in Creutzfeldt–Jakob disease: a phase 2, randomised, double-blind, placebo-controlled trial. *Lancet Neurol*. 2014;13(2): 150–158.
44. Assar H, Topakian R, Weis S, et al. A case of variably protease-sensitive prionopathy treated with doxycyclin. *J Neurol Neurosurg Psychiatry*. 2015;86(7):816–818.
45. Forloni G, Tettamanti M, Lucca U, et al. Preventive study in subjects at risk of fatal familial insomnia: innovative approach to rare diseases. *Prion*. 2015;9(2):75–79.
46. Stewart LA, Rydzewska LH, Keogh GF, Knight RS. Systematic review of therapeutic interventions in human prion disease. *Neurology*. 2008;70(15):1272–1281.
47. Pocchiari M, Schmittinger S, Masullo C. Amphotericin B delays the incubation period of scrapie in intracerebrally inoculated hamsters. *J Gen Virol*. 1987;68(pt 1):219–223.
48. Korth C, May BC, Cohen FE, Prusiner SB. Acridine and phenothiazine derivatives as pharmacotherapeutics for prion disease. *Proc Natl Acad Sci USA*. 2001;98(17): 9836–9841.
49. Ryou C, Legname G, Peretz D, Craig JC, Baldwin MA, Prusiner SB. Differential inhibition of prion propagation by enantiomers of quinacrine. *Lab Invest*. 2003;83(6): 837–843.
50. Appleby BS, Lyketsos CG. Rapidly progressive dementias and the treatment of human prion diseases. *Expert Opin Pharmacother*. 2011;12(1):1–12.
51. Geschwind MD. Clinical trials for prion disease: difficult challenges, but hope for the future. *Lancet Neurol*. 2009;8(4):304–306.
52. Ghaemmaghami S, Ahn M, Lessard P, et al. Continuous quinacrine treatment results in the formation of drug-resistant prions. *PLoS Pathog*. 2009;5(11): e1000673.
53. Perovic S, Schroder HC, Pergande G, Ushijima H, Muller WE. Effect of flupirtine on Bcl-2 and glutathione level in neuronal cells treated in vitro with the prion protein fragment (PrP106-126). *Exp Neurol*. 1997;147(2):518–524.
54. Larramendy-Gozalo C, Barret A, Daudigeos E, et al. Comparison of CR36, a new heparan mimetic, and pentosan polysulfate in the treatment of prion diseases. *J Gen Virol* 2007;88(pt 3):1062–1067.
55. Ehlers B, Diringer H. Dextran sulphate 500 delays and prevents mouse scrapie by impairment of agent replication in spleen. *J Gen Virol*. 1984;65(pt 8):1325–1330.
56. Doh-ura K, Ishikawa K, Murakami-Kubo I, et al. Treatment of transmissible spongiform encephalopathy by intraventricular drug infusion in animal models. *J Virol*. 2004;78(10):4999–5006.
57. Forloni G, Salmona M, Marcon G, Tagliavini F. Tetracyclines and prion infectivity. *Infect Disord Drug Targets*. 2009;9(1):23–30.
58. Rankovic Z. CNS drug design: balancing physicochemical properties for optimal brain exposure. *J Med Chem*. 2015;58(6):2584–2608.

59. Satyanarayanajois SD, Hill RA. Medicinal chemistry for 2020. *Future Med Chem.* 2011;3(14):1765–1786.
60. Baral PK, Swayampakula M, Rout MK, et al. Structural basis of prion inhibition by phenothiazine compounds. *Structure.* 2014;22(2):291–303.
61. Herrmann US, Schutz AK, Shirani H, et al. Structure-based drug design identifies polythiophenes as antiprion compounds. *Sci Transl Med.* 2015;7(299):299ra123.
62. Bolognesi ML, Gandini A, Prati F, Uliassi E. From companion diagnostics to theranostics: a new avenue for Alzheimer's disease? *J Med Chem.* 2016;59(17): 7759–7770.
63. Bolognesi ML. Polypharmacology in a single drug: multitarget drugs. *Curr Med Chem.* 2013;20(13):1639–1645.
64. Klingenstein R, Lober S, Kujala P, et al. Tricyclic antidepressants, quinacrine and a novel, synthetic chimera thereof clear prions by destabilizing detergent-resistant membrane compartments. *J Neurochem.* 2006;98(3):748–759.
65. Cavalli A, Bolognesi ML, Minarini A, et al. Multi-target-directed ligands to combat neurodegenerative diseases. *J Med Chem.* 2008;51(3):347–372.
66. Bolognesi ML, Simoni E, Rosini M, Minarini A, Tumiatti V, Melchiorre C. Multitarget-directed ligands: innovative chemical probes and therapeutic tools against Alzheimer's disease. *Curr Top Med Chem.* 2011;11(22):2797–2806.
67. Prati F, Uliassi E, Bolognesi ML. Two diseases, one approach: multitarget drug discovery in Alzheimer's and neglected tropical diseases. *MedChemComm.* 2014;5(7):853–861.
68. Aguzzi A, Calella AM. Prions: protein aggregation and infectious diseases. *Physiol Rev.* 2009;89(4):1105–1152.
69. Pamplona R, Naudi A, Gavin R, et al. Increased oxidation, glycoxidation, and lipoxidation of brain proteins in prion disease. *Free Radic Biol Med.* 2008;45(8): 1159–1166.
70. Lehmann S. Metal ions and prion diseases. *Curr Opin Chem Biol.* 2002;6(2):187–192.
71. Perry VH, Cunningham C, Boche D. Atypical inflammation in the central nervous system in prion disease. *Curr Opin Neurol.* 2002;15(3):349–354.
72. Bongarzone S, Bolognesi ML. The concept of privileged structures in rational drug design: focus on acridine and quinoline scaffolds in neurodegenerative and protozoan diseases. *Expert Opin Drug Discovery.* 2011;6(3):251–268.
73. Bongarzone S, Tran HN, Cavalli A, et al. Parallel synthesis, evaluation, and preliminary structure–activity relationship of 2,5-diamino-1,4-benzoquinones as a novel class of bivalent anti-prion compound. *J Med Chem.* 2010;53(22):8197–8201.
74. Bolognesi ML, Minarini A, Tumiatti V, Melchiorre C. Lipoic acid, a lead structure for multi-target-directed drugs for neurodegeneration. *Mini Rev Med Chem.* 2006;6(11):1269–1274.
75. Drisko JA. The use of antioxidants in transmissible spongiform encephalopathies: a case report. *J Am Coll Nutr.* 2002;21(1):22–25.
76. Bongarzone S, Tran HN, Cavalli A, et al. Hybrid lipoic acid derivatives to attack prion disease on multiple fronts. *ChemMedChem.* 2011;6(4):601–605.
77. Gaggelli E, Kozlowski H, Valensin D, Valensin G. Copper homeostasis and neurodegenerative disorders (Alzheimer's, prion, and Parkinson's diseases and amyotrophic lateral sclerosis). *Chem Rev.* 2006;106(6):1995–2044.
78. Brown DR, Guantieri V, Grasso G, Impellizzeri G, Pappalardo G, Rizzarelli E. Copper (II) complexes of peptide fragments of the prion protein. Conformation changes induced by copper(II) and the binding motif in C-terminal protein region. *J Inorg Biochem.* 2004;98(1):133–143.
79. Gasperini L, Meneghetti E, Pastore B, Benetti F, Legname G. Prion protein and copper cooperatively protect neurons by modulating NMDA receptor through S-nitrosylation. *Antioxid Redox Signal.* 2015;22(9):772–784.

80. Singh N, Singh A, Das D, Mohan ML. Redox control of prion and disease pathogenesis. *Antioxid Redox Signal.* 2010;12(11):1271–1294.
81. Bolognesi ML, Bongarzone S, Aulic S, et al. Rational approach to an antiprion compound with a multiple mechanism of action. *Future Med Chem.* 2015;7(16):2113–2120.
82. Prachayasittikul V, Prachayasittikul S, Ruchirawat S, Prachayasittikul V. 8-Hydroxyquinolines: a review of their metal chelating properties and medicinal applications. *Drug Des Devel Ther.* 2013;7:1157–1178.
83. Kelkar SS, Reineke TM. Theranostics: combining imaging and therapy. *Bioconjug Chem.* 2011;22(10):1879–1903.
84. Ishikawa K, Doh-ura K, Kudo Y, et al. Amyloid imaging probes are useful for detection of prion plaques and treatment of transmissible spongiform encephalopathies. *J Gen Virol.* 2004;85(6):1785–1790.
85. Mathis CA, Bacskai BJ, Kajdasz ST, et al. A lipophilic thioflavin-T derivative for positron emission tomography (PET) imaging of amyloid in brain. *Bioorg Med Chem Lett.* 2002;12(3):295–298.
86. Skovronsky DM, Zhang B, Kung MP, Kung HF, Trojanowski JQ, Lee VM. In vivo detection of amyloid plaques in a mouse model of Alzheimer's disease. *Proc Natl Acad Sci USA.* 2000;97(13):7609–7614.
87. Staderini M, Aulic S, Bartolini M, et al. A fluorescent Styrylquinoline with combined therapeutic and diagnostic activities against Alzheimer's and prion diseases. *ACS Med Chem Lett.* 2013;4(2):225–229.
88. Yang Y, Jia HM, Liu BL. (E)-5-styryl-1H-indole and (E)-6-styrylquinoline derivatives serve as probes for beta-amyloid plaques. *Molecules.* 2012;17(4):4252–4265.
89. Becker RE, Seeman MV, Greig NH, Lahiri DK. What can triumphs and tribulations from drug research in Alzheimer's disease tell us about the development of psychotropic drugs in general? *Lancet Psychiatry.* 2015;2(8):756–764.
90. Mechanism matters. *Nat Med.* 2010;16(4):347.

CHAPTER TWENTY-ONE

Biosafety of Prions

Edoardo Bistaffa*,†, Martina Rossi†, Chiara M.G. De Luca*,‡, Fabio Moda*,1

*IRCCS Foundation Carlo Besta Neurological Institute, Milan, Italy
†Scuola Internazionale Superiore di Studi Avanzati (SISSA), Trieste, Italy
‡Università degli Studi di Pavia, Pavia, Italy
[1]Corresponding author: e-mail address: fabio.moda@istituto-besta.it

Contents

Abstract

Prions are the infectious agents that cause devastating and untreatable disorders known as Transmissible Spongiform Encephalopathies (TSEs). The pathologic events and the infectious nature of these transmissible agents are not completely understood yet. Due to the difficulties in inactivating prions, working with them requires specific recommendations and precautions. Moreover, with the advent of innovative

Progress in Molecular Biology and Translational Science, Volume 150
ISSN 1877-1173
http://dx.doi.org/10.1016/bs.pmbts.2017.06.017

technologies, such as the Protein Misfolding Cyclic Amplification (PMCA) and the Real Time Quaking-Induced Conversion (RT-QuIC), prions could be amplified in vitro and the infectious features of the amplified products need to be carefully assessed.

1. INTRODUCTION

Prion diseases, or Transmissible Spongiform Encephalopathies (TSEs), are a group of neurodegenerative diseases that afflict mammalian species.[1,2] They are caused by an uncommon infectious agent known as "prion" (or PrP^{Sc}) which derives from the conformational conversion of the physiological prion protein (PrP^{C}).[3] PrP^{C} is a glycophosphatidylinositol-anchored glycoprotein encoded by the *PRNP* gene located on chromosome 20.[4] The biological functions of PrP^{C} are not completely understood but is involved in signal transduction,[5,6] metal metabolism,[7–9] anti-apoptotic processes,[10] and cell protection from oxidative stress.[11,12] Moreover, PrP^{C} interacts with the laminin receptor (LPR),[13] the *N*-methyl-D-aspartate (NMDA) receptor,[14–16] the α7 nicotinic acetylcholine receptor,[17] and the metabotropic glutamate receptor mGluR1.[18] It is highly expressed by neurons and is concentrated in the synapses.[19]

PrP^{C} is synthesized in the rough endoplasmic reticulum (ER)[4,20] where it undergoes three main modifications: (1) the addition of a GPI anchor, (2) the formation of a disulfide bond between Cys179 and Cys214,[4,21] and (3) the N-linked glycosylation at Asn181 and Asn197.[22,23] Subsequently, the protein goes to the Golgi network and the oligosaccharide chains are further modified.[24] Mature PrP^{C} is composed of 208 amino acids and comprises three major forms differing in their glycosylation degree (glycoform ratio): a diglycosylated one (containing two glycan chains), a monoglycosylated one, and an unglycosylated form.[22,24–26] PrP^{C} is rich in α-helix structures, is soluble in detergent, and is sensitive to proteinase-K (PK) digestion. On the contrary, PrP^{Sc} is characterized by higher β-sheet structures, insolubility in nondenaturing detergent, and partial resistance to PK digestion.[27,28] Different PrP^{Sc} isoforms, also referred to as prion "strains," can be characterized by alternative (i) conformations, (ii) glycosylations, and (iii) sizes of the fragments resistant to PK digestion.[29–31]

One of the most important features of PrP^{Sc} is the ability to act as template for the conversion of newly synthesized PrP^{C} into PrP^{Sc}.[32] The initial mechanism which induces PrP^{Sc} formation is unknown but, once generated, PrP^{Sc} can spread (i) in different tissues of the same hosts, (ii) among different hosts of the same species or, sometimes, (iii) between hosts of different species (as detailed in Section 4).Spreading and accumulation of PrP^{Sc} in the central nervous system (CNS) induce neurodegeneration, through a mechanism which is still not well understood.[33]

With the advent of two innovative techniques, known as Protein Misfolding Cyclic Amplification (PMCA) and Real-Time Quaking-Induced Conversion (RT-QuIC), the process of prion conversion was reproduced in vitro. PMCA consists of cycles of incubation and sonication of samples containing small amount of PrP^{Sc} in the presence of an excess of PrP^{C}.[34] During the incubation phase, PrP^{Sc} aggregates grow through recruitment and conversion of PrP^{C} molecules. The sonication is responsible for fragmenting these polymers to create new PrP^{Sc} seeds, which can induce further conversion of PrP^{C}. Similarly, RT-QuIC alternates phases of incubation to phases of vigorous shaking where soluble recombinant prion protein (rPrP-sen) is used as substrate to amplify different strains of PrP^{Sc}.[35] The addition of PrP^{Sc} to the reaction induces rPrP-sen to aggregate and form fibrils. The fibril formation is monitored with the use of Thioflavin-T (ThT), which binds to amyloid structures. Thus, the increase in fluorescence is proportional to the formation of amyloids. Given their sensitivity, both techniques are able to detect and amplify trace amount of PrP^{Sc} and have now been extensively used in the field of basic research and diagnosis.[36,37] Using both techniques it was possible to detect minute amount of PrP^{Sc} in tissues of animal and human with TSEs that were considered not infectious using conventional assays (e.g., histology, biochemistry, cell-based assay, and animal bioassays). This is of fundamental importance in terms of biosafety, because human tissues that were considered not infectious in the past, contained, instead, trace amount of PrP^{Sc} detectable after PMCA or RT-QuIC analysis.[38–40] An updated list of PrP^{Sc} distribution in different animal and human tissues is reported in Section 3.

2. PRION CLASSIFICATION

2.1 Natural Prions

Animal prion diseases include scrapie in sheep, goats, and mufflons,[41] Chronic Wasting Disease (CWD) in mule deer and elk,[42] Transmissible Mink Encephalopathy (TME) in mink,[43] Bovine Spongiform Encephalopathy (BSE) in cow,[44] and Feline Spongiform Encephalopathy (FSE) in felines (including domestic cats).[45]

Human prion diseases can be sporadic and genetic or can be acquired by infection. Sporadic forms include Creutzfeldt-Jakob disease (sCJD),[46] fatal insomnia (sFI),[47] and variably protease-sensitive prionopathy (VPSPr), the latter being characterized by a PrP^{Sc} which is sensitive to PK digestion.[48,49] Genetic forms are associated to mutations (point mutations, insertional mutations, etc.) in the prion protein gene (*PRNP*)[47,50] and include E200K, V210I, D178N (causing Fatal Familial Insomnia),[51] and the P102L (causing the Gerstmann-Sträussler-Scheinker syndrome).[52,53] Acquired forms are caused by the transmission of PrP^{Sc} from human to human (see Section 4 for details) or from animals to humans (zoonosis), as in the case of BSE transmission to human causing the variant form of CJD (vCJD).[54]

The discovery that PrP^{Sc} is present in excretions of some cases of human or animal prion diseases (including scrapie and CWD in animals or vCJD in humans) enhances the risks of transmissibility and represents a serious problem for public health.[38,39,55–59] Indeed, prions can persist for long periods of time in the environment while maintaining their infectious properties and this phenomenon might favor their transmission between individuals.[57,60,61]

All these diseases are characterized by a long incubation period (ranging from months to years) during which PrP^{Sc} accumulates in the CNS and peripheral tissues, in the absence of clinical signs.[55,56,62–64]Once symptoms appear, a severe brain damage has already occurred. Knowing the distribution of PrP^{Sc} is extremely important for researchers who work with prion-infected samples. Especially, considering that some prion agents can be transmitted to humans (e.g., BSE, vCJD), handling prion material requires special precautions and dedicated facilities that guarantee appropriate protection of humans and environment (see Section 6 for details).

2.2 Synthetically Generated Prions

In 2004, Legname G. and colleagues were able to generate the first synthetic mammalian prions able to behave as natural prions do. In particular, bacterially expressed recombinant mouse PrP was induced to misfold in vitro, under specific biochemical conditions, to form amyloid structures.[65] These amyloid structures were then intracerebrally inoculated in transgenic animals which developed prion pathology characterized by long incubation time. Subsequently, other groups were able to misfold rPrP (with mouse and hamster amino acid sequence) that was capable of inducing prion pathology when injected in susceptible animals[66–69] (see Section 4.1.3 for details). Misfolded rPrP acquired slightly different conformations that were associated with different biochemical, structural, and infectious features when challenged in animals.[70–72] According to these observations, it seems that synthetically generated prions might possess infectious properties that are strictly associated to their abnormal conformation. Unfortunately, there are still limited information about the infectivity of synthetic material at present and special precautions should be adopted when handling such products.

3. LEVELS OF INFECTIVITY IN DIFFERENT TISSUES OF ANIMAL AND HUMAN TSEs

3.1 Tissue Distribution of PrP^{Sc} in Animal TSEs

The distribution of PrP^{Sc} in peripheral tissues of human and animal with TSEs is extremely variable. For this reason, the World Health Organization (WHO) has generated a table (last updated in 2010) where tissues of human and animals with prion diseases were classified as highly, mildly, or not infectious (http://www.who.int/bloodproducts/tablestissueinfectivity.pdf). With the generation of innovative and extremely sensitive methodologies, such as the PMCA and RT-QuIC assays, minute amounts of PrP^{Sc} were recently detected in tissues and bodily fluids that were classified as not infectious. Thus, the list reported below indicates the PrP^{Sc} distribution by taking into account the more recent results obtained with PMCA and RT-QuIC other than common observations of naturally occurring disease, or experimental infection by the oral route.

- BSE: PrP^{Sc} is mainly confined to the CNS, spinal cord, spinal ganglia, retina, bone marrow, distal ileum,[73–75] peripheral nerves,[76,77] and skeletal muscle.[78]

In animals with scrapie, CWD, TME, and FSE, PrP^{Sc} is more disseminated in nonnervous and nonlymphoid organs compared to BSE.[79]

- Scrapie: PrP^{Sc} is found in CNS, brainstem, palatine tonsils, ileum Peyer's patches, spleen, adrenal gland, pancreas, heart, skin, urinary bladder, mammary glands, salivary gland,[80] lung, liver, kidney, skeletal muscle,[81] blood,[64] milk,[82] saliva,[59] and oral cavity[83];
- CWD: PrP^{Sc} is found in CNS, brainstem, retina, peripheral nerves, spleen, lymph nodes, tonsils, esophagus, rectum, skin, pancreas,[84–86] nasal mucosa, saliva,[58,87] urine,[58] feces,[56] and blood[87];
- TME: PrP^{Sc} is found in CNS, spleen, mesenteric and retropharyngeal lymph nodes, thymus, kidney, liver, intestine, and salivary gland[88,89];

- FSE: this disease is thought to be linked to the consumption of meat from BSE infected cattle. One case of domestic cat with FSE was reported and PrP^{Sc} was found in CNS, spinal cord, spinal ganglia, nerve of the tongue, sciatic nerve, optic nerve, enteric nervous system, and adrenal gland.[90]

3.2 Tissue Distribution of PrP^{Sc} in Human TSEs

As for animal TSE, PMCA and RT-QuIC were adopted to study the distribution of PrP^{Sc} in different forms of human prion disorders. Below we report a list of PrP^{Sc} distribution in the sporadic and variant forms of CJD. The presence of PrP^{Sc} in peripheral tissues of patients with VPSPr is currently under investigation, as well as their ability to be amplified by means of PMCA or detected with RT-QuIC.[91]

- Sporadic CJD (sCJD): PrP^{Sc} was found in CNS, retina,[92] optic nerve,[93] spinal cord,[94] peripheral nerves,[95] blood,[40] urine,[96] CSF,[36,97] olfactory mucosa[98] and, in some cases, in spleen and muscles after concentration by sodium phosphotungstic acid (PTA) precipitation to increase immunoassay sensitivity.[99,100]
- Variant CJD: PrP^{Sc} is more diffuse in peripheral tissues than sCJD and can be found in spleens, tonsils, lymph nodes, retina, pituitary gland, thymus, proximal optic nerve,[101] appendix, skeletal muscle,[100] ileum, adrenal gland, pancreas, blood,[39,55] and urine.[38]

It is likely that, in the near future, PrP^{Sc} will be detected in other tissues that were considered not infectious and this classification might be further revised.

4. RISK ASSESSMENT

4.1 Transmissibility of Natural and Synthetic Prions

4.1.1 Natural Transmission of Prions

Animal TSEs are thought to propagate *naturally* after consumption of prion-infected food or through contaminated environments where prions can persist for long periods of time.[102] In this regard, BSE and TME prion transmission has been controlled by the removal of sources of contamination from animal's foodstuff. Scrapie and CWD are more difficult to control since prions are excreted in the environment (through saliva,[59,103] urine,[104] and feces[56,105]) where they can survive for a long period of time and favor prion transmission between animals. In this regard, interventions specifically aimed at blocking the propagation of the disease are necessary.

Cases of human TSEs transmission (iCJD) have been reported from corneal transplants,[106] dura mater grafts,[107] stereotactic EEG recordings,[108] use of contaminated surgical instruments,[109] and after treatment with human-derived growth hormone.[110] The first example of prion transmission between humans is represented by the kuru. This disease developed in the twentieth century in Papua New Guinea and was transmitted between humans after cannibalistic rituals.[111] The variant form of CJD (vCJD) represents another example of natural prion transmission to human. vCJD has been associated to the consumption of BSE contaminated foodstuff, thus demonstrating that this type of animal prion can efficiently infect humans (zoonosis).[112] Recently, five cases of human-to-human vCJD transmission by blood transfusion were reported,[113–116] thus demonstrating the highest transmissibility of this prion strain between humans. Currently, there are no evidences suggesting that human prion diseases can spread from person to person by close contact.

4.1.2 Experimental Transmission of Natural Prions: The "Species Barrier"

Prions can be *experimentally* transmitted by different routes, including intracerebral, intraocular, intraspinal, intraperitoneal,[117] intravenous,[118] subcutaneous, dental,[119] and oral.[117] Recent evidences suggest that prions can be also transmitted by aerosol and intranasal inoculations[120–123] in susceptible animals; thus the respiratory system might be an efficient route for prion disease transmission and should be included and updated in the biosafety recommendations. By taking advantage of their infectious properties, different types of human prions (sporadic, acquired, and genetic) were used to perform experimental infection of wild-type and transgenic animal models of the disease[46,53,111,124–126] with the aim of (i) studying their molecular, biochemical, and infectious features and (ii) planning specific therapeutic interventions. The more recent forms of prion diseases, known as VPSPr, were shown to be experimentally transmissible to transgenic mice expressing the human prion protein but with very limited efficiency.[127,128]

The efficiency of all these transmissions depends on multiple factors including, among the others, the route of infection (being the intracerebral the most efficient) and the species where the infectious agent is transmitted. Indeed, the transmission is more efficient within the same species and much less efficient between different species. This phenomenon is known as "species barrier" and is closely related to differences in prion protein sequence between donor and acceptor organisms[124,129,130] Generally, cross-species transmission of prions often results in a faithful propagation of the original inoculum.[131] In some cases, after transmission, the original strain seems to undergo a process of adaptation and acquires a conformation which is different from that of the original one.[31,132] The reasons of this phenomenon are still unclear but strongly indicate that prions are able to "adapt" when transmitted between different species.

4.1.3 Experimental Transmission of Synthetic Prions

The infectious properties of synthetic prions represent one of the most puzzling aspects of this innovative field of research. Infectivity experiments are generally performed by inoculating cell models or susceptible animals with synthetic material.[133] Unfortunately, results are often controversial.[71] This might be due to different factors, including (i) the amount of infectious material inoculated that might be too low to trigger a prion-like pathology, and (ii) the use of inappropriate cellular or animal models.[134,135] Therefore, genetically modified cells or transgenic animals expressing the same PrP sequence (truncated and/or mutated) of the rPrP used as inoculum are prevalently adopted for testing the infectivity of different synthetic materials.[65,136] Transgenic animals were much prone to develop prion pathology (characterized by peculiar clinical and neuropathological alterations) if compared to wild-type animals which hardly developed a prion-like pathology, especially at the first passage.[71,133] In 2010, Makarava and colleagues reported that after incubation of misfolded hamster rPrP with normal hamster brain homogenate for 1 min at 80°C followed by 1 min at 37°C (five cycles of annealing) the rPrP was then able to infect wild-type Golden Syrian hamsters and led to the generation of a new prion disease.[66] This indicated that the annealing was able to increase the efficiency of the synthetic material to induce prion pathology in wild-type animals.

With the advent of PMCA, synthetic rPrP was incubated with brain homogenates of wild-type animals and subjected to serial cycles of incubation and sonication. Under these conditions, some rPrP induced the conformational conversion of brain-derived PrP^C into a proteinase-K (PK) resistant form (PrP^{res}). Inoculation of this latter in wild-type animals was able to induce severe prion pathology.[71] Taken together, these data suggest that synthetic prions are able to efficiently misfold PrP^C, as natural prions do, but the conditions and the environment for the conversion might play an important role in promoting or inhibiting this process. Examples of the infectivity of synthetic prions amplified by means of PMCA are detailed in Section 5.1.

All the aforementioned observations support the "protein only hypothesis" which sustains that PrP alone is the causative agent of prion diseases and that different disease phenotypes could be associated with different structural conformations of the rPrP.[1,137] Moreover, neither the presence of glycans nor glycosylphosphatidylinositol (GPI)-anchor seem to be required to confer infectious properties to rPrP.[138,139]

Therefore, special precautions should be taken when handling synthetically generated materials.

4.2 Infectious Dose

Generally, the risk of infection is almost zero if the dose of exposure to the biological agent is low enough. For prion diseases, defining a dose under which prions are not infectious is extremely challenging. Indeed, prion infectivity relies on different factors, including the species barrier[140] (see Section 4.1.2 for details) and the strain of prion which the operator is exposed to.[141] Currently, there are no clear relationships between the levels of TSE exposure and the probability of infection. Using several mathematical models, it was possible to demonstrate the presence of a threshold dose, where only in the presence of PrP^{Sc} polymers with a certain length the conversion of PrP^{C} to PrP^{Sc} may occur.[142,143] Animal studies revealed that the probability of infection becomes smaller as the dose of infection decreases.[144] Additional studies established that the higher the size of the aggregates, the lower is the efficiency of transmission and infectivity. Indeed, the most infectious particles seem to be those with a size ranging from 17 to 27 nm (300–600 kDa) while particles of higher size (e.g., oligomers and large fibrils) are less or even not infectious. Thus, molecules with lower size are thought to be the most efficient initiators of TSE disease.[145]

5. INFECTIOUS PROPERTIES OF PMCA AND RT-QuIC REACTION PRODUCTS

In the last years, PMCA and RT-QuIC were widely adopted in many laboratories involved in prion research and diagnosis but only limited studies were performed for assessing the infectious properties of the final products of amplification. Here we report the results of few experiments performed for evaluating the infectivity of the final PMCA and RT-QuIC reaction products.

5.1 Infectivity of PMCA Reaction Products

PMCA can amplify either natural (from human or animal origins) or synthetic prions using brain homogenates of susceptible animals. A limited number of studies aimed at verifying the infectious properties of the final products of amplification were carried out and provided controversial results. In 2011, Shikiya R.A. and coworkers demonstrated that after PMCA, serially diluted hyper (HY) prion strain was able to amplify and reach an infectious titer similar to that present in the brain of terminally sick hamster.[146] Similarly, in 2004, Moudjou and colleagues[147] demonstrated that one round of PMCA was enough to completely restore the infectivity of serial dilutions of 127S scrapie prion strain to a level comparable to that contained in the brain of a terminally sick mouse. By contrast, other studies performed with 263K hamster prion strain showed that the infectivity diminished progressively at each round of PMCA.[148,149] In particular, there was an increase in PrP^{res} at each round of amplification but with lower specific infectivity (infectivity tier/amount of PrP^{res}). These findings support the hypothesis that not all the PrP^{res} proteins generated during PMCA are infectious. Of note, even with a lower concentration, the infectious proteins maintained the pathological features of the original inoculum.

Therefore, the products of PMCA amplification might have the same or lower (but apparently not higher) infectious titer than that contained in the brain homogenate of terminally sick animals. As a consequence, PMCA products need to be manipulated with the same precautions used for treating the infectious tissue from which PrP^{Sc} has been amplified.

Recently, PMCA was successfully applied for amplifying trace amounts of prion in urine and blood of patients affected by vCJD.[38,39,55] Similarly, we have recently optimized PMCA for detecting prions in the olfactory mucosa of patients with Fatal Familial Insomnia (http://tandfonline.com/doi/full/10.1080/19336896.2016.1162644). In none of these cases, the final products of amplification were inoculated in susceptible animals for assessing their infectious properties. According to several data already present in the literature, PMCA product is infectious and can induce pathology in susceptible animals.[147,149–151] Moreover, these products can be considered as infectious as the original inoculum and the pathological features of the inoculum are faithfully maintained.

Recently, PMCA was used to assess whether synthetic prions might be amplified in vitro. For example, Wang and colleagues, in 2010, were able to use a modified version of the PMCA technique to produce PK-resistant rPrP able to produce severe prion pathology when inoculated in wild-type mice.[67] Recently, our group has amplified synthetic prions by means of PMCA using the brain homogenate of wild-type animal that induced prion pathology when inoculated in mice. Notably, not all synthetic prions were able to efficiently convert brain-derived PrP^C to a proteinase-K (PK)-resistant isoform (PrP^{res}). Once inoculated in our mice, PMCA products generated two different disease phenotypes suggesting an involvement of other factors on prion replication that might have influenced the selection and adaptation of different prion isolates in vivo.[71]

Thus, also in this case, PMCA products acquired infectious properties and need to be manipulated with the same precautions used when handling natural prion strains (see Section 6 for details).

5.2 Infectivity of RT-QuIC Reaction Products

RT-QuIC can detect minute amounts of infectious prions contained in biological samples using bacterially expressed prion protein (rPrP) from different species (e.g., mouse, hamster, bank vole, human) as substrate for the reaction. For this reason, given its high sensitivity and specificity, RT-QuIC has been recently employed as supportive test for the diagnosis of different forms of prion disease.[97,152–154] After the addition of infected tissue samples collected from human or animals, rPrP starts to aggregate and form amyloid fibrils the growing of which can be monitored using a fluorescent dye (Thioflavin T) in a time-dependent manner.

Currently, there are extremely limited information about the infectivity of the final RT-QuIC products. Few indications were collected from the group of McGuire L. and coworkers which reported that the final RT-QuIC products obtained by seeding hamsters rPrP with an hamster scrapie brain homogenate was not able to induce any prion pathology when inoculated in animals.[97] On the contrary, Sano K. and colleagues demonstrated that two different mouse prion strains (22L and Chandler) were able to trigger the aggregation of rPrP with the formation of final amyloid rPrP fibrils characterized by distinct biochemical and structural features. When injected in wild-type mice these different amyloids gave rise to a mutant strain able to induce prion-like pathology.[155]

Since some laboratories use the RT-QuIC assay as supportive test for a diagnosis of prion diseases, it is of essential importance to perform additional studies with the aim of assessing the infectious features of the final reaction products, especially when rPrP with human sequences is used as substrate for the reaction.

6. GENERAL BIOSAFETY RECOMMENDATIONS

Despite intensive research, the infectious nature of prions has not been well characterized. Moreover, considering the difficulties to inactivate them, working with these agents requires special recommendations and precautions. Specific guidelines have been established to minimize the exposure risk for the operator and the environment but they differ from country to country. For this reason, researchers and scientists interested in working with prion should contact their institutional biosafety office to obtain all the information on how to comply with the regulations related to the use of these agents. Here we describe general indications mostly excerpted from:

1. *WHO/CDS/CSR/APH/2000.3*
   (WHO Infection Control Guidelines for Transmissible Spongiform Encephalopathies. Report of a WHO consultation)
   http://www.who.int/csr/resources/publications/bse/whocdscsraph2003.pdf?ua=1
2. *CDC/NIH BMBL, 5th edition*
   (Biosafety in Microbiological and Biomedical Laboratories, 5th edition)
   https://www.cdc.gov/biosafety/publications/bmbl5/BMBL.pdf
3. *WHO/EMP/QSM/2010.1*
   (WHO Tables on Tissue Infectivity Distribution in Transmissible Spongiform Encephalopathies)
   http://www.who.int/bloodproducts/tablestissueinfectivity.pdf
4. Recommended Biosafety Practices for Handling Prions and Prion-Infected Tissues.
   http://www.orcbs.msu.edu/biological/programs_guidelines/prions/working_with_prions.pdf

Specific containment of the infectious agents and prevention of accidental exposure of the operators are accomplished by the presence of three barriers:

1. *Individual Protective Devices (IPDs)*: standard caps, face masks, face shields/glasses, gowns, gloves, and shoes cover are the minimal IPD adequate to minimize the risk of working with TSE;
2. *Laboratory area*: places where the infectious agents are manipulated. They are grouped in four biosafety levels (BSLs) with the most hazardous classified as BSL-4. All TSEs agents require a BSL-2 or BSL-3 facility,

depending on the prion strain. Indeed, considering the recent evidences showing the ability of BSE and vCJD to infect humans, more stringent safety measures should be adopted when working with these strains and a BSL-3 facility should be preferred to a BSL-2 which is generally required for minimizing the risk of working with the other TSE (see Tables1 and 2 for details);

3. *Building in which the laboratory area is located*: BSL-2 and BSL-3 laboratories should be located away from public areas and lockable doors are required to restrict the access to authorized personnel only.

The equipment used to work with prions should be specifically dedicated to that task alone. Researchers must receive all the information and specific training required to work with prions. Standard Operating Procedures (SOPs) must be distributed to all employees involved in prion research and easily accessible from all the operators.

7. OPERATING STANDARD PROCEDURES FOR WORKING IN BSL-2 OR BSL-3 FACILITIES

Researchers working in BSL-2 or BSL-3 facilities must receive periodical and specific formation and training related to the risks of working with these infectious agents. Each laboratory must be equipped with specific biosafety manuals and standard operating procedures (SOPs). These manuals should be accessible to all personnel and adapted for each facility. Examples of SOPs are reported below:

7.1 Biosafety Level-2 Laboratories (BSL-2)

Recommended for scrapie, Chronic Wasting Disease (CWD), Transmissible Mink Encephalopathy (TME), and Feline Spongiform Encephalopathy (FSE), with the exception of Bovine Spongiform Encephalopathy (BSE) where a BSL-3 level is preferable (see Tables1 and 2 for details).The access must be restricted to authorized personnel. Universal biohazard symbol and specific information (such as type of agents, laboratory's biosafety level, supervisor's name, telephone number, and entry/exit procedures) must be placed at the entrance of the laboratory.

All researchers must be equipped with specific individual protection devices (IDPs), including:

- laboratory coats;
- gloves (to be frequently changed);
- goggles;

Table 1 Classification of the Most Common Mammalian Prions

Disease	Host	Pathogenetic Mechanism	Transmissibility to Humans	Transmissibility to Mice Expressing Human PrP	Biosafety Level
Scrapie	Sheep, goats, and mufflons	Infection in susceptible animals	Unknown	Experimentally *transmitted* to transgenic mice expressing human PrP[156]	2
Chronic Wasting disease (CWD)	Deer, mule, elk	Not clear: possibly by direct contact or indirectly from the environment (e.g., contaminated water, food)	Unknown	*Not transmissible* to transgenic mice overexpressing human PrP[157]	2
Transmissible Mink Encephalopathy (TME)	Mink	Infection with prion-contaminated food	Unknown	Experimentally *transmitted* to transgenic mice expressing human PrP[158]	2
Bovine Spongiform Encephalopathy (BSE)	Cattle	Infection with prion-contaminated food	*Yes* By consumption of BSE-contaminated food	Experimentally *transmitted* to wild type[159] or transgenic mice expressing human PrP[112]	3
Feline Spongiform Encephalopathy (FSE)	Felines (cats, cheetah)	Infection with BSE-contaminated food	TSEs were simultaneously found in a cat and its owner in 1998.[160] It is not known whether these prions might have been transmitted between man and cat.	Not proven	2

Sporadic CJD (sCJD) and sporadic Familial Insomnia (sFI)	Humans	Not clear: possibly by spontaneous conversion of PrP^{C} to PrP^{Sc}	Unknown	Experimentally *transmitted* to transgenic mice expressing human PrP[161,162]	2/3
Variably Protease-Sensitive Prionopathy (VPSPr)	Humans	Not clear: possibly by spontaneous conversion of PrP^{C} to PrP^{Sc}	Unknown	Experimentally *transmitted* to transgenic mice expressing human PrP[49,127]	2/3
Familial CJD	Humans	Mutations in *PRNP* gene	Unknown	Some forms were experimentally *transmitted* to transgenic mice expressing human PrP[163]	2/3
Fatal Familial Insomnia (FFI)	Humans	Mutation in *PRNP* gene in coupling phase with methionine homozygosis or methionine/valine heterozygosis at codon 129 of *PRNP*	Unknown	Experimentally *transmitted* to transgenic mice expressing human PrP[163]	2/3
Gerstmann-Sträussler-Scheinker syndrome (GSS)	Humans	Mutation in *PRNP* gene	Unknown	Experimentally *transmitted* to wild-type or transgenic mice expressing human PrP[164]	2/3

Continued

Table 1 Classification of the Most Common Mammalian Prions—cont'd

Disease	Host	Pathogenetic Mechanism	Transmissibility to Humans	Transmissibility to Mice Expressing Human PrP	Biosafety Level
Iatrogenic CJD (iCJD)	Humans	Infection from contaminated corneal and dura mater grafts, neurosurgical instruments, and growth hormone	*Yes*	Experimentally *transmitted* to transgenic mice expressing human PrP[165]	2/3
Variant CJD (vCJD)	Humans	Infection with BSE contaminated food	*Yes* By blood transfusion from donors who later developed vCJD[113–116]	Experimentally *transmitted* to wild-type or transgenic mice expressing human PrP[166]	3

Table 2 General BSL-2 and BSL-3 Facilities Requirements

	Biosafety Level	
Facility Requirements	**2**	**3**
Restricted access to authorized personnel only	Recommended	Yes
Biosafety manual with specific SOPs and medical surveillance policies	Yes	Yes
Working area separated from corridors and other work areas	No	Recommended
HEPA filtration of the laboratory effluent and exhaust air	No	Yes, on exhaust air
Biohazard warning signal	Yes	Yes
Specific ventilation requirements (negative-pressure working areas)	No	Recommended
Sealed windows	No	Yes
Self-closing double-door access	No	Yes
Ventilated anteroom with doors interlocked	No	Yes
Dedicated autoclave, equipment, and materials	Yes	Yes
Laboratory inspection window	Recommended	Yes
Dedicated protective equipment	Recommended	Yes
Work surfaces resistant to acids, bases, solvents, and disinfectants	Recommended	Yes
Water-repellent work surfaces	Yes, for the bench	Yes, for bench and floor
Specific decontamination procedures	Yes	Yes
Manipulation of infected materials (tissue homogenates, biological fluids, PMCA, and RT-QuIC reaction products) under class II biosafety cabinet	Recommended	Yes
Manipulation of infected materials which causes the formation of aerosol or splashes under class II biosafety cabinet	Yes	Yes
Manipulation of infected animals and surgical procedures and necropsy under class II biosafety cabinet.	Recommended	Yes

Continued

Table 2 General BSL-2 and BSL-3 Facilities Requirements—cont'd

	Biosafety Level	
Facility Requirements	**2**	**3**
Liquid waste treatment, decontamination and neutralization	Yes	Yes
Dry waste incineration	Yes	Yes
Water waste treatment	No	Yes
Hand washing sink near the exit of the laboratory	Yes	Yes
Decontamination area for personnel	Yes	Yes
Disinfecting footbath	No	Recommended

- masks or face shields;
- eye and face protection (recommended to avoid splashes or aerosols of infectious material).

Two pairs of gloves are recommended when handling material with high level of infectivity (e.g., nervous tissues). Protective clothing need to be removed before leaving the laboratory and disposed appropriately.

The BSL-2 facility should be equipped with:

- biosafety cabinets (preferably class II);
- sink for hand washing near the exit door;
- eyewash station;
- self-closing doors with locks;
- bench tops resistant to water, heat, organic solvents, and other chemicals;
- HEPA filters on vacuum lines.

The laboratory should be frequently cleaned and decontaminated. All laboratory wastes should be decontaminated according to specific institutional procedure (see Table 3 for WHO suggestions). Biosafety cabinets must be placed away from doors, windows, and traveled laboratory areas to avoid airflow disruptions.

Examples of BSL-2 SOPs are reported below:

- IPDs need to be worn all time. During special hazardous proceedings, eye and face protection is recommended.
- Procedures which generate aerosols or splashes and procedures that involve manipulation of infectious tissues must be conducted in class II biosafety cabinets (HEPA filtered) and following procedures that can minimize aerosol or splashes production;

Table 3 Decontamination Procedures Recommended by WHO Guidelines

	NaOH	Bleach	Notes
Liquid waste	Mix with NaOH to reach a final concentration of 1 N and keep at room temperature for 1 h	Mix with bleach to reach a final concentration of 20,000 ppm available chlorine and keep at room temperature for 1 h	Store in a chemical fume hood for the duration of the treatment
Reusable heat-resistant instruments	Immerse in 1 N NaOH for 1 h, rinse with water, and autoclave for 1 h at 134°C	Immerse in bleach solution (20,000 ppm available chlorine) for 1 h, rinse with water and autoclave for 1 h at 134°C	In case of slight contaminations, autoclave at 134°C for 18 min
Surfaces and heat-sensitive instruments	Flood with 2 N NaOH and let stand for 1 h	Flood with undiluted bleach and let stand for 1 h	After treatment, surfaces should be carefully rinsed with clear water
Dry waste	—	—	All contaminated dry waste should be incinerated
Sharp waste	—	—	Prion contaminated sharp waste must be identified as "prion contaminated sharps for incineration only" on the hazardous waste pickup request to assure incineration of these materials

- After use, needles and sharps must be discarded in special containers used for sharp disposals and properly decontaminated (as indicated in Section 9). Not-disposable sharps must be decontaminated as indicated in Section 9;
- Instruments used for histological processing of tissue samples (e.g., microtome) must be cleaned avoiding the near contact of the operator hands with sharps and knives;

- Potentially contaminated tissues (fresh, fixed, or frozen) must be placed in waterproof containers and individually labeled with the universal biohazard symbol;
- Work surfaces should be covered with disposable absorbent pads to facilitate cleaning and decontamination. After work, pads must be discarded and surfaces must be decontaminated with the appropriate method (as detailed in Section 9);
- Reusable instruments must be kept wet until decontamination (see Section 9);
- Eating, drinking, smoking, applying cosmetics and contact lenses, and storing food are strictly forbidden;
- Operators must wash their hands before and after leaving the laboratory.

7.2 Biosafety Level-3 Laboratories (BSL-3)

Recommended for human prions and required for BSE and vCJD. Indeed, according to the actual regulations, processing of many prion-infected human tissues (or infected tissues from transgenic animals expressing human prion protein) can be processed with extreme care in a BSL-2 facility utilizing BSL-3 practices.

The BSL-3 safety protocol is based on BSL-2 with some implementations for enhancing the environmental and personal protection.

In particular, BSL-3 laboratories must have sealed windows, floors, walls, and ceiling surfaces and must be designed to be easily cleaned and decontaminated, with spaces between benches accessible for cleaning. Two self-closing doors must restrict the access to the facility. The sink for hand washing must be positioned immediately outside the exit door and has to be hands-free or automatically operated. The laboratory exhaust air must circulate outside from the building through HEPA filters. Exhaust air resulting from class II biosafety cabinets HEPA filters could recirculate into the laboratory if the cabinet is correctly tested and certified. The laboratory must be equipped with a method to decontaminate all laboratory wastes, such as autoclave (specifically adapted for prion), chemical disinfection, incineration, or other validated procedures.

The following environmental improvements are recommended:

- Anteroom for clothing change and clean storage of IDPs with specific entry/exit procedures;
- HEPA filtration of the laboratory exhaust air;
- Gas-tight dampers to improve laboratory isolation;
- Laboratory effluent decontamination.

BSL-3 IDPs include: shoes cover, two pairs of gloves, two pairs of gowns, and eye/face protection devices.

Special recommendations include:

- Manipulate prion-infected materials under biosafety cabinets (preferably Class II or Class III) or in other physical containment devices;
- Minimize aerosol or splashing production;
- Centrifuges should be equipped with safety cup or sealed rotor;
- Eye, face, and respiratory protection is required;
- Laboratory equipment has to be frequently decontaminated (as detailed in Section 9).

8. BIOCHEMICAL AND HISTOLOGICAL ANALYSIS

Biochemical analysis of prion-infected samples (fresh, frozen, or homogenized tissues) must be performed under a class II biosafety cabinet, using dedicated equipment. Sampling procedures (including manipulation and brain cut) should be performed with the use of disposable blade and cut-resistant gloves. Since the homogenization and sonication of tissues produce the formation of aerosol, these procedures must be performed under a class II biosafety cabinet (whenever possible) or at least using all the precautions for avoiding and reducing the formation of aerosol and its dispersion in the environment. Pipetting of infectious materials should be performed with a dedicated set of pipettes (daily decontaminated as reported in Table 3) and filter tips to avoid pipette contamination. Biochemical analysis (enzymatic digestion, Western blot, ELISA, and immunoprecipitation) that involve manipulation of infectious material should be performed under a class II biosafety cabinet. Homogenates of infectious material should be stored in special containers with secure closing to prevent any leakage of infected material. These containers can be then stored in dedicated freezers (−80°C or −20°C) and properly labeled with the biohazard symbol.

Fixation of tissues with formalin or glutaraldehyde does not reduce their infectivity and should be considered as infectious as the fresh ones.[167] The infectivity of fixed tissues may be reduced after treatment of the samples with 98% formic acid.[168] All the steps required for tissues processing (dehydration, paraffin embedding, cut, sections hydration and staining) must be performed under biological hoods (or class II biosafety cabinets when possible) using dedicated equipment. In particular, a dedicated microtome for prion-infected tissue should be used since it is very difficult to efficiently decontaminate this type of instrument. Once tissue sections are stained (or immunostained) they should be sealed with a coverslip and the external surface should be decontaminated (as reported in Table 3), labeled as "hazardous" and stored for examination. Fixed material can be stored in containers with secure closing, labeled with hazardous signal, and placed in clean plastic bag (http://www.who.int/csr/resources/publications/biosafety/en/Biosafety7.pdf).

All the exhaust fixatives and alcohols used for histological processing of samples, paraffin residues coming from tissues cutting, and all reagents used for the staining (e.g., hematoxylin and eosin staining, buffers, and water used for immunohistochemistry) should be decontaminated as infectious waste according to specific laboratory biosafety guideline (as suggested in Table 3). All the reagents, buffers, and antibodies used for biochemical investigations must be decontaminated and discarded as hazardous liquid waste (as reported in Table 3). Disposable dry material used for experimental procedures should be marked as infectious waste and incinerated (as reported in Table 3).

9. INACTIVATION OF PRIONS AND IATA REGULATIONS

A striking feature of prions is their extraordinary resistance to conventional sterilization procedures (including standard autoclave treatment at 121°C) and their capacity to bind to surfaces of metal and plastic without losing infectivity.[117,169] Recommended practices for prion inactivation diverge from country to country.[170,171] Here, we report the WHO Infection Control Guidelines for Transmissible Spongiform Encephalopathies (WHO/CDS/CSR/APH/2000.3 WHO Infection Control Guidelines for Transmissible Spongiform Encephalopathies: http://www.who.int/csr/resources/publications/bse/whocdscsraph2003.pdf?ua=1).

Determination of the best decontamination procedure depends on the level of infectivity of the tissue to be analyzed and whether the instruments will be reused or discarded. For tissues with high infectivity, single-use devices are strongly recommended and should be immediately discarded by incineration. If disposable instruments are not available, maximum safety is attained by total destruction of reusable instruments. Where destruction is not practical, reusable instruments must be decontaminated as summarized in Table 3 and handled as reported below:

- Devices should be kept humid until decontamination;
- Devices used for manipulating tissues with no detectable infectivity should not be mixed with devices used for tissues with high infectivity;

- Devices should be reused only after deep decontamination (by methods summarized in Table 3);
- Work benches and surfaces should be covered with disposable material (e.g., adsorbent pads), which can be easily removed and incinerated; otherwise cleaning and decontamination of the surfaces should be performed using recommended decontamination procedures (as summarized in Table 3).
- Observe manufacturers' recommendations regarding care and maintenance of equipment.

All disposable instruments, materials, and wastes exposed to high infectivity tissues should be properly incinerated. In case of contaminated reusable instruments, liquid waste, and contaminated surfaces, the most effective procedure involves chemical inactivation with Sodium Hydroxide (NaOH) or Sodium Hypochlorite (NaOCl solution or bleach) as detailed in Table 3.

Regarding the operators, in case of intact skin exposure to infected material wash with 1 N NaOH or 10% bleach for 2–3 min then wash with water. For needle sticks or lacerations, encourage bleeding and wash with water. In the event of contact with the eye, rinse the eye with abundant amount of saline solution. In all these cases, the event has to be reported to the biosafety office and the exposed subject should be followed up and assisted with actions to reduce future occurrence.

9.1 IATA Guidelines for the Transport of Infectious Materials

The Department of Transportation (DOT) and the International Air Transport Association (IATA) established specific guidelines for the transport of infectious materials (http://www.un3373.com/info/regulations/). In particular, the primary containers must be decontaminated and placed in another secondary cleaned container. Absorbent material able to absorb the entire contents of all primary containers must be placed between the primary and the secondary packaging according to the UN3373 international regulations.

REFERENCES

1. Prusiner SB. Novel proteinaceous infectious particles cause scrapie. *Science*. 1982;216(4542):136–144.
2. Bolton DC, McKinley MP, Prusiner SB. Identification of a protein that purifies with the scrapie prion. *Science*. 1982;218(4579):1309–1311.
3. Cohen FE, Prusiner SB. Pathologic conformations of prion proteins. *Annu Rev Biochem*. 1998;67:793–819.

4. Stahl N, Borchelt DR, Hsiao K, Prusiner SB. Scrapie prion protein contains a phosphatidylinositol glycolipid. *Cell.* 1987;51(2):229–240.
5. Westergard L, Christensen HM, Harris DA. The cellular prion protein (PrP(C)): its physiological function and role in disease. *Biochim Biophys Acta.* 2007;1772(6): 629–644.
6. Mouillet-Richard S, Ermonval M, Chebassier C, et al. Signal transduction through prion protein. *Science.* 2000;289(5486):1925–1928.
7. Gasperini L, Meneghetti E, Legname G, Benetti F. In absence of the cellular prion protein, alterations in copper metabolism and copper-dependent oxidase activity affect iron distribution. *Front Neurosci.* 2016;10:437.
8. Millhauser GL. Copper and the prion protein: methods, structures, function, and disease. *Annu Rev Phys Chem.* 2007;58:299–320.
9. Viles JH, Klewpatinond M, Nadal RC. Copper and the structural biology of the prion protein. *Biochem Soc Trans.* 2008;36(pt 6):1288–1292.
10. Roucou X, Gains M, LeBlanc AC. Neuroprotective functions of prion protein. *J Neurosci Res.* 2004;75(2):153–161.
11. Milhavet O, Lehmann S. Oxidative stress and the prion protein in transmissible spongiform encephalopathies. *Brain Res Brain Res Rev.* 2002;38(3):328–339.
12. Brown DR, Wong BS, Hafiz F, Clive C, Haswell SJ, Jones IM. Normal prion protein has an activity like that of superoxide dismutase. *Biochem J.* 1999;344(pt 1):1–5.
13. Rieger R, Edenhofer F, Lasmezas CI, Weiss S. The human 37-kDa laminin receptor precursor interacts with the prion protein in eukaryotic cells. *Nat Med.* 1997;3(12):1383–1388.
14. Khosravani H, Zhang Y, Tsutsui S, et al. Prion protein attenuates excitotoxicity by inhibiting NMDA receptors. *J Gen Physiol.* 2008;131(6):i5.
15. Gasperini L, Meneghetti E, Pastore B, Benetti F, Legname G. Prion protein and copper cooperatively protect neurons by modulating NMDA receptor through S-nitrosylation. *Antioxid Redox Signal.* 2015;22(9):772–784.
16. Black SA, Stys PK, Zamponi GW, Tsutsui S. Cellular prion protein and NMDA receptor modulation: protecting against excitotoxicity. *Front Cell Dev Biol.* 2014;2:45.
17. Beraldo FH, Arantes CP, Santos TG, et al. Role of alpha7 nicotinic acetylcholine receptor in calcium signaling induced by prion protein interaction with stress-inducible protein 1. *J Biol Chem.* 2010;285(47):36542–36550.
18. Beraldo FH, Arantes CP, Santos TG, et al. Metabotropic glutamate receptors transduce signals for neurite outgrowth after binding of the prion protein to laminin gamma1 chain. *FASEB J.* 2011;25(1):265–279.
19. Sales N, Rodolfo K, Hassig R, Faucheux B, Di Giamberardino L, Moya KL. Cellular prion protein localization in rodent and primate brain. *Eur J Neurosci.* 1998;10(7):2464–2471.
20. Turk E, Teplow DB, Hood LE, Prusiner SB. Purification and properties of the cellular and scrapie hamster prion proteins. *Eur J Biochem.* 1988;176(1):21–30.
21. Maiti NR, Surewicz WK. The role of disulfide bridge in the folding and stability of the recombinant human prion protein. *J Biol Chem.* 2001;276(4):2427–2431.
22. Rudd PM, Endo T, Colominas C, et al. Glycosylation differences between the normal and pathogenic prion protein isoforms. *Proc Natl Acad Sci U S A.* 1999;96(23): 13044–13049.
23. Locht C, Chesebro B, Race R, Keith JM. Molecular cloning and complete sequence of prion protein cDNA from mouse brain infected with the scrapie agent. *Proc Natl Acad Sci U S A.* 1986;83(17):6372–6376.
24. Caughey B, Race RE, Ernst D, Buchmeier MJ, Chesebro B. Prion protein biosynthesis in scrapie-infected and uninfected neuroblastoma cells. *J Virol.* 1989;63(1):175–181.

25. Monari L, Chen SG, Brown P, et al. Fatal familial insomnia and familial Creutzfeldt-Jakob disease: different prion proteins determined by a DNA polymorphism. *Proc Natl Acad Sci U S A*. 1994;91(7):2839–2842.
26. Petersen RB, Parchi P, Richardson SL, Urig CB, Gambetti P. Effect of the D178N mutation and the codon 129 polymorphism on the metabolism of the prion protein. *J Biol Chem*. 1996;271(21):12661–12668.
27. Oesch B, Westaway D, Walchli M, et al. A cellular gene encodes scrapie PrP 27-30 protein. *Cell*. 1985;40(4):735–746.
28. Meyer RK, McKinley MP, Bowman KA, Braunfeld MB, Barry RA, Prusiner SB. Separation and properties of cellular and scrapie prion proteins. *Proc Natl Acad Sci U S A*. 1986;83(8):2310–2314.
29. Safar J, Prusiner SB. Molecular studies of prion diseases. *Prog Brain Res*. 1998;117:421–434.
30. Parchi P, Zou W, Wang W, et al. Genetic influence on the structural variations of the abnormal prion protein. *Proc Natl Acad Sci U S A*. 2000;97(18):10168–10172.
31. Peretz D, Williamson RA, Legname G, et al. A change in the conformation of prions accompanies the emergence of a new prion strain. *Neuron*. 2002;34(6):921–932.
32. Ma J, Lindquist S. Conversion of PrP to a self-perpetuating PrPSc-like conformation in the cytosol. *Science*. 2002;298(5599):1785–1788.
33. Soto C, Satani N. The intricate mechanisms of neurodegeneration in prion diseases. *Trends Mol Med*. 2011;17(1):14–24.
34. Saborio GP, Permanne B, Soto C. Sensitive detection of pathological prion protein by cyclic amplification of protein misfolding. *Nature*. 2001;411(6839):810–813.
35. Atarashi R, Wilham JM, Christensen L, et al. Simplified ultrasensitive prion detection by recombinant PrP conversion with shaking. *Nat Methods*. 2008;5(3):211–212.
36. Atarashi R, Satoh K, Sano K, et al. Ultrasensitive human prion detection in cerebrospinal fluid by real-time quaking-induced conversion. *Nat Med*. 2011;17(2):175–178.
37. Saa P, Castilla J, Soto C. Cyclic amplification of protein misfolding and aggregation. *Methods Mol Biol*. 2005;299:53–65.
38. Moda F, Gambetti P, Notari S, et al. Prions in the urine of patients with variant Creutzfeldt-Jakob disease. *N Engl J Med*. 2014;371(6):530–539.
39. Concha-Marambio L, Pritzkow S, Moda F, et al. Detection of prions in blood from patients with variant Creutzfeldt-Jakob disease. *Sci Transl Med*. 2016;8(370):370ra183.
40. Douet JY, Zafar S, Perret-Liaudet A, et al. Detection of infectivity in blood of persons with variant and sporadic Creutzfeldt-Jakob disease. *Emerg Infect Dis*. 2014;20(1): 114–117.
41. Wood JL, Lund LJ, Done SH. The natural occurrence of scrapie in moufflon. *Vet Rec*. 1992;130(2):25–27.
42. Williams ES, Young S. Spongiform encephalopathy of Rocky Mountain elk. *J Wildl Dis*. 1982;18(4):465–471.
43. Barlow RM. Transmissible mink encephalopathy: pathogenesis and nature of the aetiological agent. *J Clin Pathol Suppl (R Coll Pathol)*. 1972;6:102–109.
44. Wells GA, Scott AC, Johnson CT, et al. A novel progressive spongiform encephalopathy in cattle. *Vet Rec*. 1987;121(18):419–420.
45. Wyatt JM, Pearson GR, Smerdon TN, Gruffydd-Jones TJ, Wells GA, Wilesmith JW. Naturally occurring scrapie-like spongiform encephalopathy in five domestic cats. *Vet Rec*. 1991;129(11):233–236.
46. Gibbs Jr CJ, Gajdusek DC, Asher DM, et al. Creutzfeldt-Jakob disease (spongiform encephalopathy): transmission to the chimpanzee. *Science*. 1968;161(3839):388–389.
47. Gambetti P, Kong Q, Zou W, Parchi P, Chen SG. Sporadic and familial CJD: classification and characterisation. *Br Med Bull*. 2003;66:213–239.

48. Gambetti P, Dong Z, Yuan J, et al. A novel human disease with abnormal prion protein sensitive to protease. *Ann Neurol.* 2008;63(6):697–708.
49. Zou WQ, Puoti G, Xiao X, et al. Variably protease-sensitive prionopathy: a new sporadic disease of the prion protein. *Ann Neurol.* 2010;68(2):162–172.
50. Mastrianni JA. The genetics of prion diseases. *Genet Med.* 2010;12(4):187–195.
51. Medori R, Tritschler HJ, LeBlanc A, et al. Fatal familial insomnia, a prion disease with a mutation at codon 178 of the prion protein gene. *N Engl J Med.* 1992;326(7): 444–449.
52. Young K, Clark HB, Piccardo P, Dlouhy SR, Ghetti B. Gerstmann-Straussler-Scheinker disease with the PRNP P102L mutation and valine at codon 129. *Brain Res Mol Brain Res.* 1997;44(1):147–150.
53. Masters CL, Gajdusek DC, Gibbs Jr CJ. Creutzfeldt-Jakob disease virus isolations from the Gerstmann-Straussler syndrome with an analysis of the various forms of amyloid plaque deposition in the virus-induced spongiform encephalopathies. *Brain.* 1981;104(3):559–588.
54. Bruce ME, Will RG, Ironside JW, et al. Transmissions to mice indicate that 'new variant' CJD is caused by the BSE agent. *Nature.* 1997;389(6650):498–501.
55. Bougard D, Brandel JP, Belondrade M, et al. Detection of prions in the plasma of presymptomatic and symptomatic patients with variant Creutzfeldt-Jakob disease. *Sci Transl Med.* 2016;8(370):370ra182.
56. Tamguney G, Miller MW, Wolfe LL, et al. Asymptomatic deer excrete infectious prions in faeces. *Nature.* 2009;461(7263):529–532.
57. Gough KC, Maddison BC. Prion transmission: prion excretion and occurrence in the environment. *Prion.* 2010;4(4):275–282.
58. Haley NJ, Seelig DM, Zabel MD, Telling GC, Hoover EA. Detection of CWD prions in urine and saliva of deer by transgenic mouse bioassay. *PLoS One.* 2009;4(3): e4848.
59. Tamguney G, Richt JA, Hamir AN, et al. Salivary prions in sheep and deer. *Prion.* 2012;6(1):52–61.
60. Wiggins RC. Prion stability and infectivity in the environment. *Neurochem Res.* 2009;34(1):158–168.
61. Pritzkow S, Morales R, Moda F, et al. Grass plants bind, retain, uptake, and transport infectious prions. *Cell Rep.* 2015;11(8):1168–1175.
62. Gill ON, Spencer Y, Richard-Loendt A, et al. Prevalent abnormal prion protein in human appendixes after bovine spongiform encephalopathy epizootic: large scale survey. *BMJ.* 2013;347: f5675.
63. Lacroux C, Comoy E, Moudjou M, et al. Preclinical detection of variant CJD and BSE prions in blood. *PLoS Pathog.* 2014;10(6): e1004202.
64. Terry LA, Howells L, Hawthorn J, et al. Detection of PrPsc in blood from sheep infected with the scrapie and bovine spongiform encephalopathy agents. *J Virol.* 2009;83(23):12552–12558.
65. Legname G, Baskakov IV, Nguyen HO, et al. Synthetic mammalian prions. *Science.* 2004;305(5684):673–676.
66. Makarava N, Kovacs GG, Bocharova O, et al. Recombinant prion protein induces a new transmissible prion disease in wild-type animals. *Acta Neuropathol.* 2010;119(2): 177–187.
67. Wang F, Wang X, Yuan CG, Ma J. Generating a prion with bacterially expressed recombinant prion protein. *Science.* 2010;327(5969):1132–1135.
68. Colby DW, Giles K, Legname G, et al. Design and construction of diverse mammalian prion strains. *Proc Natl Acad Sci U S A.* 2009;106(48):20417–20422.
69. Piro JR, Wang F, Walsh DJ, Rees JR, Ma J, Supattapone S. Seeding specificity and ultrastructural characteristics of infectious recombinant prions. *Biochemistry.* 2011;50(33):7111–7116.

70. Legname G, Nguyen HO, Baskakov IV, Cohen FE, Dearmond SJ, Prusiner SB. Strain-specified characteristics of mouse synthetic prions. *Proc Natl Acad Sci U S A*. 2005;102(6):2168–2173.
71. Moda F, Le TN, Aulic S, et al. Synthetic prions with novel strain-specified properties. *PLoS Pathog*. 2015;11(12): e1005354.
72. Bessen RA, Marsh RF. Distinct PrP properties suggest the molecular basis of strain variation in transmissible mink encephalopathy. *J Virol*. 1994;68(12):7859–7868.
73. Wells GA, Dawson M, Hawkins SA, et al. Infectivity in the ileum of cattle challenged orally with bovine spongiform encephalopathy. *Vet Rec*. 1994;135(2):40–41.
74. Wells GA, Hawkins SA, Green RB, et al. Preliminary observations on the pathogenesis of experimental bovine spongiform encephalopathy (BSE): an update. *Vet Rec*. 1998;142(5):103–106.
75. Wells GA, Hawkins SA, Green RB, Spencer YI, Dexter I, Dawson M. Limited detection of sternal bone marrow infectivity in the clinical phase of experimental bovine spongiform encephalopathy (BSE). *Vet Rec*. 1999;144(11):292–294.
76. Terry LA, Marsh S, Ryder SJ, Hawkins SA, Wells GA, Spencer YI. Detection of disease-specific PrP in the distal ileum of cattle exposed orally to the agent of bovine spongiform encephalopathy. *Vet Rec*. 2003;152(13):387–392.
77. Buschmann A, Groschup MH. Highly bovine spongiform encephalopathy-sensitive transgenic mice confirm the essential restriction of infectivity to the nervous system in clinically diseased cattle. *J Infect Dis*. 2005;192(5):934–942.
78. Okada H, Miyazawa K, Fukuda S, et al. The presence of disease-associated prion protein in skeletal muscle of cattle infected with classical bovine spongiform encephalopathy. *J Vet Med Sci*. 2014;76(1):103–107.
79. Leunda A, Van Vaerenbergh B, Baldo A, Roels S, Herman P. Laboratory activities involving transmissible spongiform encephalopathy causing agents: risk assessment and biosafety recommendations in Belgium. *Prion*. 2013;7(5):420–433.
80. Vascellari M, Nonno R, Mutinelli F, et al. PrPSc in salivary glands of scrapie-affected sheep. *J Virol*. 2007;81(9):4872–4876.
81. Garza MC, Monzon M, Marin B, Badiola JJ, Monleon E. Distribution of peripheral PrP(Sc) in sheep with naturally acquired scrapie. *PLoS One*. 2014;9(5): e97768.
82. Maddison BC, Baker CA, Rees HC, et al. Prions are secreted in milk from clinically normal scrapie-exposed sheep. *J Virol*. 2009;83(16):8293–8296.
83. Maddison BC, Rees HC, Baker CA, et al. Prions are secreted into the oral cavity in sheep with preclinical scrapie. *J Infect Dis*. 2010;201(11):1672–1676.
84. Sigurdson CJ, Aguzzi A. Chronic wasting disease. *Biochim Biophys Acta*. 2007;1772(6):610–618.
85. Selariu A, Powers JG, Nalls A, et al. Utero transmission and tissue distribution of chronic wasting disease-associated prions in free-ranging Rocky Mountain elk. *J Gen Virol*. 2015;96(11):3444–3455.
86. Haley NJ, Hoover EA. Chronic wasting disease of cervids: current knowledge and future perspectives. *Annu Rev Anim Biosci*. 2015;3:305–325.
87. Mathiason CK, Powers JG, Dahmes SJ, et al. Infectious prions in the saliva and blood of deer with chronic wasting disease. *Science*. 2006;314(5796):133–136.
88. Schneider DA, Harrington RD, Zhuang D, et al. Disease-associated prion protein in neural and lymphoid tissues of mink (*Mustela vison*) inoculated with transmissible mink encephalopathy. *J Comp Pathol*. 2012;147(4):508–521.
89. Hadlow WJ, Race RE, Kennedy RC. Temporal distribution of transmissible mink encephalopathy virus in mink inoculated subcutaneously. *J Virol*. 1987;61(10): 3235–3240.
90. Eiden M, Hoffmann C, Balkema-Buschmann A, Muller M, Baumgartner K, Groschup MH. Biochemical and immunohistochemical characterization of feline

spongiform encephalopathy in a German captive cheetah. *J Gen Virol.* 2010;91(pt 11):2874–2883.

91. Peden AH, Sarode DP, Mulholland CR, et al. The prion protein protease sensitivity, stability and seeding activity in variably protease sensitive prionopathy brain tissue suggests molecular overlaps with sporadic Creutzfeldt-Jakob disease. *Acta Neuropathol Commun.* 2014;2:152.
92. Head MW, Northcott V, Rennison K, et al. Prion protein accumulation in eyes of patients with sporadic and variant Creutzfeldt-Jakob disease. *Invest Ophthalmol Vis Sci.* 2003;44(1):342–346.
93. Head MW, Peden AH, Yull HM, et al. Abnormal prion protein in the retina of the most commonly occurring subtype of sporadic Creutzfeldt-Jakob disease. *Br J Ophthalmol.* 2005;89(9):1131–1133.
94. Goodbrand IA, Ironside JW, Nicolson D, Bell JE. Prion protein accumulation in the spinal cords of patients with sporadic and growth hormone associated Creutzfeldt-Jakob disease. *Neurosci Lett.* 1995;183(1–2):127–130.
95. Ishida C, Okino S, Kitamoto T, Yamada M. Involvement of the peripheral nervous system in human prion diseases including dural graft associated Creutzfeldt-Jakob disease. *J Neurol Neurosurg Psychiatry.* 2005;76(3):325–329.
96. Luk C, Jones S, Thomas C, et al. Diagnosing sporadic Creutzfeldt-Jakob disease by the detection of abnormal prion protein in patient urine. *JAMA Neurol.* 2016;73(12): 1454–1460.
97. McGuire LI, Peden AH, Orru CD, et al. Real time quaking-induced conversion analysis of cerebrospinal fluid in sporadic Creutzfeldt-Jakob disease. *Ann Neurol.* 2012;72(2):278–285.
98. Bongianni M, Orru C, Groveman BR, et al. Diagnosis of human prion disease using real-time quaking-induced conversion testing of olfactory mucosa and cerebrospinal fluid samples. *JAMA Neurol.* 2017;74:155–162.
99. Glatzel M, Abela E, Maissen M, Aguzzi A. Extraneural pathologic prion protein in sporadic Creutzfeldt-Jakob disease. *N Engl J Med.* 2003;349(19):1812–1820.
100. Peden AH, Ritchie DL, Head MW, Ironside JW. Detection and localization of PrPSc in the skeletal muscle of patients with variant, iatrogenic, and sporadic forms of Creutzfeldt-Jakob disease. *Am J Pathol.* 2006;168(3):927–935.
101. Ramasamy I, Law M, Collins S, Brooke F. Organ distribution of prion proteins in variant Creutzfeldt-Jakob disease. *Lancet Infect Dis.* 2003;3(4):214–222.
102. Miller MW, Williams ES, Hobbs NT, Wolfe LL. Environmental sources of prion transmission in mule deer. *Emerg Infect Dis.* 2004;10(6):1003–1006.
103. Henderson DM, Manca M, Haley NJ, et al. Rapid antemortem detection of CWD prions in deer saliva. *PLoS One.* 2013;8(9): e74377.
104. John TR, Schatzl HM, Gilch S. Early detection of chronic wasting disease prions in urine of pre-symptomatic deer by real-time quaking-induced conversion assay. *Prion.* 2013;7(3):253–258.
105. Safar JG, Lessard P, Tamguney G, et al. Transmission and detection of prions in feces. *J Infect Dis.* 2008;198(1):81–89.
106. Duffy P, Wolf J, Collins G, DeVoe AG, Streeten B, Cowen D. Letter: possible person-to-person transmission of Creutzfeldt-Jakob disease. *N Engl J Med.* 1974;290(12): 692–693.
107. Centers for Disease C, Prevention. Creutzfeldt-Jakob disease associated with cadaveric dura mater grafts—Japan, January 1979-May 1996. *MMWR Morb Mortal Wkly Rep.* 1997;46(45):1066–1069.
108. Bernoulli C, Siegfried J, Baumgartner G, et al. Danger of accidental person-to-person transmission of Creutzfeldt-Jakob disease by surgery. *Lancet.* 1977;1(8009):478–479.

109. Williams D. New thoughts on CJD, surgical instruments and disease transmission. *Med Device Technol.* 2006;17(1):7–8.
110. Gibbs Jr CJ, Joy A, Heffner R, et al. Clinical and pathological features and laboratory confirmation of Creutzfeldt-Jakob disease in a recipient of pituitary-derived human growth hormone. *N Engl J Med.* 1985;313(12):734–738.
111. Gajdusek DC, Gibbs CJ, Alpers M. Experimental transmission of a kuru-like syndrome to chimpanzees. *Nature.* 1966;209(5025):794–796.
112. Hill AF, Desbruslais M, Joiner S, et al. The same prion strain causes vCJD and BSE. *Nature.* 1997;389(6650):448–450. 526.
113. Davidson LR, Llewelyn CA, Mackenzie JM, Hewitt PE, Will RG. Variant CJD and blood transfusion: are there additional cases? *Vox Sang.* 2014;107(3):220–225.
114. Hewitt PE, Llewelyn CA, Mackenzie J, Will RG. Creutzfeldt-Jakob disease and blood transfusion: results of the UK transfusion medicine epidemiological review study. *Vox Sang.* 2006;91(3):221–230.
115. Turner ML, Ironside JW. New-variant Creutzfeldt-Jakob disease: the risk of transmission by blood transfusion. *Blood Rev.* 1998;12(4):255–268.
116. Diack AB, Head MW, McCutcheon S, et al. Variant CJD. 18 years of research and surveillance. *Prion.* 2014;8(4):286–295.
117. Weissmann C, Enari M, Klohn PC, Rossi D, Flechsig E. Transmission of prions. *Proc Natl Acad Sci U S A.* 2002;99(Suppl 4):16378–16383.
118. Andreoletti O, Litaise C, Simmons H, et al. Highly efficient prion transmission by blood transfusion. *PLoS Pathog.* 2012;8(6): e1002782.
119. Ingrosso L, Pisani F, Pocchiari M. Transmission of the 263K scrapie strain by the dental route. *J Gen Virol.* 1999;80(pt 11):3043–3047.
120. Denkers ND, Seelig DM, Telling GC, Hoover EA. Aerosol and nasal transmission of chronic wasting disease in cervidized mice. *J Gen Virol.* 2010;91(pt 6):1651–1658.
121. Hamir AN, Kunkle RA, Richt JA, Miller JM, Greenlee JJ. Experimental transmission of US scrapie agent by nasal, peritoneal, and conjunctival routes to genetically susceptible sheep. *Vet Pathol.* 2008;45(1):7–11.
122. Sbriccoli M, Cardone F, Valanzano A, et al. Neuroinvasion of the 263K scrapie strain after intranasal administration occurs through olfactory-unrelated pathways. *Acta Neuropathol.* 2009;117(2):175–184.
123. Haybaeck J, Heikenwalder M, Klevenz B, et al. Aerosols transmit prions to immunocompetent and immunodeficient mice. *PLoS Pathog.* 2011;7(1): e1001257.
124. Prusiner SB, Scott M, Foster D, et al. Transgenetic studies implicate interactions between homologous PrP isoforms in scrapie prion replication. *Cell.* 1990;63(4):673–686.
125. Roos R, Gajdusek DC, Gibbs Jr CJ. The clinical characteristics of transmissible Creutzfeldt-Jakob disease. *Brain.* 1973;96(1):1–20.
126. Prusiner SB. Genetic and infectious prion diseases. *Arch Neurol.* 1993;50(11):1129–1153.
127. Diack AB, Ritchie DL, Peden AH, et al. Variably protease-sensitive prionopathy, a unique prion variant with inefficient transmission properties. *Emerg Infect Dis.* 2014;20(12):1969–1979.
128. Notari S, Xiao X, Espinosa JC, et al. Transmission characteristics of variably protease-sensitive prionopathy. *Emerg Infect Dis.* 2014;20(12):2006–2014.
129. Hill AF, Joiner S, Linehan J, Desbruslais M, Lantos PL, Collinge J. Species-barrier-independent prion replication in apparently resistant species. *Proc Natl Acad Sci U S A.* 2000;97(18):10248–10253.
130. Race R, Raines A, Raymond GJ, Caughey B, Chesebro B. Long-term subclinical carrier state precedes scrapie replication and adaptation in a resistant species: analogies to

bovine spongiform encephalopathy and variant Creutzfeldt-Jakob disease in humans. *J Virol.* 2001;75(21):10106–10112.
131. Nonno R, Di Bari MA, Cardone F, et al. Efficient transmission and characterization of Creutzfeldt-Jakob disease strains in bank voles. *PLoS Pathog.* 2006;2(2): e12.
132. Tanaka M, Chien P, Yonekura K, Weissman JS. Mechanism of cross-species prion transmission: an infectious conformation compatible with two highly divergent yeast prion proteins. *Cell.* 2005;121(1):49–62.
133. Schmidt C, Fizet J, Properzi F, et al. A systematic investigation of production of synthetic prions from recombinant prion protein. *Open Biol.* 2015;5(12): 150165.
134. Castilla J, Gutierrez-Adan A, Brun A, et al. Subclinical bovine spongiform encephalopathy infection in transgenic mice expressing porcine prion protein. *J Neurosci.* 2004;24(21):5063–5069.
135. Jackson WS. Selective vulnerability to neurodegenerative disease: the curious case of prion protein. *Dis Model Mech.* 2014;7(1):21–29.
136. Kaneko K, Ball HL, Wille H, et al. A synthetic peptide initiates Gerstmann-Straussler-Scheinker (GSS) disease in transgenic mice. *J Mol Biol.* 2000;295(4):997–1007.
137. Griffith JS. Self-replication and scrapie. *Nature.* 1967;215(5105):1043–1044.
138. Chesebro B, Trifilo M, Race R, et al. Anchorless prion protein results in infectious amyloid disease without clinical scrapie. *Science.* 2005;308(5727):1435–1439.
139. Taraboulos A, Rogers M, Borchelt DR, et al. Acquisition of protease resistance by prion proteins in scrapie-infected cells does not require asparagine-linked glycosylation. *Proc Natl Acad Sci U S A.* 1990;87(21):8262–8266.
140. Sharma A, Bruce KL, Chen B, et al. Contributions of the prion protein sequence, strain, and environment to the species barrier. *J Biol Chem.* 2016;291(3):1277–1288.
141. Tanaka M, Chien P, Naber N, Cooke R, Weissman JS. Conformational variations in an infectious protein determine prion strain differences. *Nature.* 2004;428(6980): 323–328.
142. Eigen M. Prionics or the kinetic basis of prion diseases. *Biophys Chem.* 1996;63(1): A1–18.
143. Masel J, Jansen VA, Nowak MA. Quantifying the kinetic parameters of prion replication. *Biophys Chem.* 1999;77(2–3):139–152.
144. Fryer HR, McLean AR. There is no safe dose of prions. *PLoS One.* 2011;6(8): e23664.
145. Silveira JR, Raymond GJ, Hughson AG, et al. The most infectious prion protein particles. *Nature.* 2005;437(7056):257–261.
146. Shikiya RA, Bartz JC. Vitro generation of high-titer prions. *J Virol.* 2011;85(24):13439–13442.
147. Moudjou M, Sibille P, Fichet G, et al. Highly infectious prions generated by a single round of microplate-based protein misfolding cyclic amplification. *MBio.* 2013;5(1): e00829-00813.
148. Klingeborn M, Race B, Meade-White KD, Chesebro B. Lower specific infectivity of protease-resistant prion protein generated in cell-free reactions. *Proc Natl Acad Sci U S A.* 2011;108(48):E1244–E1253.
149. Castilla J, Saa P, Hetz C, Soto C. In vitro generation of infectious scrapie prions. *Cell.* 2005;121(2):195–206.
150. Gao C, Han J, Zhang J, et al. Protein Misfolding cyclic amplification cross-species products of mouse-adapted scrapie strain 139A and hamster-adapted scrapie strain 263K with brain and muscle tissues of opposite animals generate infectious prions. *Mol Neurobiol.* 2017;54:3771–3782.
151. Chianini F, Fernandez-Borges N, Vidal E, et al. Rabbits are not resistant to prion infection. *Proc Natl Acad Sci U S A.* 2012;109(13):5080–5085.
152. Orru CD, Bongianni M, Tonoli G, et al. A test for Creutzfeldt-Jakob disease using nasal brushings. *N Engl J Med.* 2014;371(6):519–529.

153. Schmitz M, Cramm M, Llorens F, et al. The real-time quaking-induced conversion assay for detection of human prion disease and study of other protein misfolding diseases. *Nat Protoc.* 2016;11(11):2233–2242.
154. Orru CD, Groveman BR, Raymond LD, et al. Bank vole prion protein as an apparently universal substrate for RT-QuIC-based detection and discrimination of prion strains. *PLoS Pathog.* 2015;11(6): e1004983.
155. Sano K, Atarashi R, Ishibashi D, Nakagaki T, Satoh K, Nishida N. Conformational properties of prion strains can be transmitted to recombinant prion protein fibrils in real-time quaking-induced conversion. *J Virol.* 2014;88(20):11791–11801.
156. Cassard H, Torres JM, Lacroux C, et al. Evidence for zoonotic potential of ovine scrapie prions. *Nat Commun.* 2014;5:5821.
157. Sandberg MK, Al-Doujaily H, Sigurdson CJ, et al. Chronic wasting disease prions are not transmissible to transgenic mice overexpressing human prion protein. *J Gen Virol.* 2010;91(pt 10):2651–2657.
158. Comoy EE, Mikol J, Ruchoux MM, et al. Evaluation of the zoonotic potential of transmissible mink encephalopathy. *Pathogens.* 2013;2(3):520–532.
159. Capobianco R, Casalone C, Suardi S, et al. Conversion of the BASE prion strain into the BSE strain: the origin of BSE? *PLoS Pathog.* 2007;3(3): e31.
160. Zanusso G, Nardelli E, Rosati A, et al. Simultaneous occurrence of spongiform encephalopathy in a man and his cat in Italy. *Lancet.* 1998;352(9134):1116–1117.
161. Bishop MT, Will RG, Manson JC. Defining sporadic Creutzfeldt-Jakob disease strains and their transmission properties. *Proc Natl Acad Sci U S A.* 2010;107(26):12005–12010.
162. Mastrianni JA, Nixon R, Layzer R, et al. Prion protein conformation in a patient with sporadic fatal insomnia. *N Engl J Med.* 1999;340(21):1630–1638.
163. Telling GC, Parchi P, DeArmond SJ, et al. Evidence for the conformation of the pathologic isoform of the prion protein enciphering and propagating prion diversity. *Science.* 1996;274(5295):2079–2082.
164. Asante EA, Grimshaw A, Smidak M, et al. Transmission properties of human PrP 102L prions challenge the relevance of mouse models of GSS. *PLoS Pathog.* 2015;11(7): e1004953.
165. Telling GC, Scott M, Hsiao KK, et al. Transmission of Creutzfeldt-Jakob disease from humans to transgenic mice expressing chimeric human-mouse prion protein. *Proc Natl Acad Sci U S A.* 1994;91(21):9936–9940.
166. Bishop MT, Hart P, Aitchison L, et al. Predicting susceptibility and incubation time of human-to-human transmission of vCJD. *Lancet Neurol.* 2006;5(5):393–398.
167. Brown P, Liberski PP, Wolff A, Gajdusek DC. Resistance of scrapie infectivity to steam autoclaving after formaldehyde fixation and limited survival after ashing at 360 degrees C: practical and theoretical implications. *J Infect Dis.* 1990;161(3):467–472.
168. Taylor DM, Brown JM, Fernie K, McConnell I. The effect of formic acid on BSE and scrapie infectivity in fixed and unfixed brain-tissue. *Vet Microbiol.* 1997;58(2–4): 167–174.
169. Taylor DM. Inactivation of transmissible degenerative encephalopathy agents: a review. *Vet J.* 2000;159(1):10–17.
170. Rutala WA, Weber DJ. Creutzfeldt-Jakob disease: recommendations for disinfection and sterilization. *Clin Infect Dis.* 2001;32(9):1348–1356.
171. Hughson AG, Race B, Kraus A, et al. Inactivation of prions and amyloid seeds with hypochlorous acid. *PLoS Pathog.* 2016;12(9): e1005914.

INDEX

Note: Page numbers followed by "*t*" indicate tables.

D

I

K

L

M

N

Q

R

S

T

U

V

W

Z